METHODS IN MOLECULAR BIOLOGY

Monoclonal Antibodies

Methods and Protocols

Second Edition

Edited by

Vincent Ossipow

Département de Biochimie, University of Geneva, Geneva, Switzerland

Nicolas Fischer

NovImmune SA, Plan-les-Ouates, Switzerland

Editors
Vincent Ossipow
Département de Biochimie
University of Geneva
Geneva, Switzerland

Nicolas Fischer
NovImmune SA
Plan-les-Ouates
Switzerland

ISSN 1064-3745 ISSN 1940-6029 (electronic)
ISBN 978-1-62703-991-8 ISBN 978-1-62703-992-5 (eBook)
DOI 10.1007/978-1-62703-992-5
Springer New York Heidelberg Dordrecht London

Library of Congress Control Number: 2013958232

Printed on acid-free paper

Humana Press is a brand of Springer
Springer is part of Springer Science+Business Media (www.springer.com)

Preface

Over the past decades, monoclonal antibodies (monoclonals) have become invaluable for basic research, diagnostics, clinicians, and thousands of patients suffering from severe afflictions. Monoclonals are widely used laboratory reagents, and it is fair to say that the availability of a good monoclonal has led to increased understanding of the target biology as many experimental approaches, ranging from classical western blots to chromatin immnunoprecipitation assays, are enabled. Beyond basic research, therapeutic monoclonals are increasingly used as drugs and will account for over 50 billion USD in sales in 2013, and this figure is forecast to grow at double-digit rate, higher than any other therapeutic class. Since OKT3 (Muromonab-CD3, Johnson & Jonson/Ortho Biotech), the first monoclonal approved for human use, 34 other monoclonals have been approved. This is the tip of the iceberg, as it is estimated that over 400 monoclonals are in clinical development worldwide.

This brisk success is explained by several factors. First and most importantly, their specificity and low off-target toxicity often provide monoclonals with an exceptional "therapeutic window," i.e., the ability to dose them effectively with acceptable side effects. Monoclonals are also embraced because of their prolonged half-life and, more generally, as their pharmacokinetic properties are more predictable than for other classes of drugs. Consequently, the rate of success for the clinical development of monoclonals is significantly higher than for small molecule drugs or for other biologicals.

With the progress of genetic and protein engineering, academic labs and the biopharmaceutical industry have advanced many novel antibody formats and antibody-based approaches. These comprise multi-specific antibodies, fragments, antibody–drug conjugates, antibodies with enhanced effector function, and so on. However, despite these evolutions, being able to generate high-quality monoclonals against carefully selected epitopes remains the absolute foundation for subsequent improvements. Equally important is the meticulous characterization of candidate monoclonals. We actually contend that because of the greatly expanded toolkit for improving monoclonals, epitope selection and biochemical characterization has now become even more important for generating well-differentiated monoclonals. One can directly witness this notion at the bench with antibodies raised against the same target that perform very differently in various assays but also in the clinic, where monoclonals against different epitopes can exert very different responses in patients (e.g., the three anti-CD20 monoclonals Rituximab, Ofatumumb, and GA-101 by Roche, Genmab/GSK, and Roche respectively).

The purpose of this new edition of "Monoclonal Antibodies, Methods in Molecular Biology" is to offer modern approaches to indeed generate high-quality monoclonals against carefully selected epitopes, and meticulously characterize them. With a few exceptions, we deliberately concentrated on the basic IgG format. All the key steps from antigen generation to some final applications are described in these 36 chapters and should provide the reader with multiple useful methods to generate an appropriate monoclonal. We divided the book into four parts corresponding to four distinct objectives. Part I covers monoclonal antibody generation, Part II deals with monoclonal antibody expression and purification, Part III presents methods for monoclonal antibody characterization and modification, and Part IV describes some applications of monoclonal antibodies. For each Part we strived to balance "must-have" protocols and recent innovative approaches, all "debugged" in the author's laboratories. In Part I, we included, for instance, protocols for the generation of monoclonals using natural sources such as mouse, rabbit, or immortalized human B-cells, but also in vitro selection methodologies such as phage and yeast display as well as antibody repertoire mining by deep sequencing. In Part II, several approaches are proposed for downstream purification of IgG as well as some alternative formats that should satisfy different requirements and downstream uses. In Part III, epitope mapping with various astute technologies such as phage- or bacterial display or extensive mutagenesis are well covered, as well as strategies for examining the primary sequence and structural and physicochemical properties of monoclonals. The latter are often overlooked but are important, as experiments performed with aggregated or unstable monoclonals can lead to erroneous conclusions. In Part IV, we provide some examples of use of monoclonals including immunofluorescence, crystallization chaperoning and the generation of solid-state arrays.

By no means is our selection of protocols exhaustive, a task impossible within the context of such a book. On the other hand, some topics are covered in more than one chapter, providing alternatives for the readers to select the most appropriate method for her/his use. We hope that our protocol choice will fulfill its intended goal of covering the crucial initial steps of monoclonal antibody generation and characterization with state-of-the art protocols.

Geneva, Switzerland *Vincent Ossipow*
Plan-les-Ouates, Switzerland *Nicolas Fischer*

Contents

Preface . . . . *v*
Contributors . . . . *xi*

PART I ANTIBODY GENERATION

1 Antigen Production for Monoclonal Antibody Generation 3
Giovanni Magistrelli and Pauline Malinge

2 Method for the Generation of Antibodies Specific for Site and Posttranslational Modifications 21
Hidemasa Goto and Masaki Inagaki

3 Immunization, Hybridoma Generation, and Selection for Monoclonal Antibody Production 33
Hyung-Yong Kim, Alexander Stojadinovic, and Mina J. Izadjoo

4 Hybridoma Technology for the Generation of Rodent mAbs via Classical Fusion 47
Efthalia Chronopoulou, Alejandro Uribe-Benninghoff, Cindi R. Corbett, and Jody D. Berry

5 Generation of Rabbit Monoclonal Antibodies 71
Pi-Chen Yam and Katherine L. Knight

6 Screening Hybridomas for Cell Surface Antigens by High-Throughput Homogeneous Assay and Flow Cytometry 81
Alejandro Uribe-Benninghoff, Teresa Cabral, Efthalia Chronopoulou, Jody D. Berry, and Cindi R. Corbett

7 Screening and Subcloning of High Producer Transfectomas Using Semisolid Media and Automated Colony Picker 105
Suba Dharshanan and Cheah Swee Hung

8 Design and Generation of Synthetic Antibody Libraries for Phage Display 113
Gang Chen and Sachdev S. Sidhu

9 Selection and Screening Using Antibody Phage Display Libraries 133
Patrick Koenig and Germaine Fuh

10 Yeast Surface Display for Antibody Isolation: Library Construction, Library Screening, and Affinity Maturation 151
James A. Van Deventer and Karl Dane Wittrup

11 Human B Cell Immortalization for Monoclonal Antibody Production 183
Joyce Hui-Yuen, Siva Koganti, and Sumita Bhaduri-McIntosh

12 Using Next-Generation Sequencing for Discovery of High-Frequency Monoclonal Antibodies in the Variable Gene Repertoires from Immunized Mice 191
Ulrike Haessler and Sai T. Reddy

PART II ANTIBODY EXPRESSION AND PURIFICATION

13 Cloning, Reformatting, and Small-Scale Expression of Monoclonal Antibody Isolated from Mouse, Rat, or Hamster Hybridoma 207
Jeremy Loyau and François Rousseau

14 Cloning of Recombinant Monoclonal Antibodies from Hybridomas in a Single Mammalian Expression Plasmid 229
Nicole Müller-Sienerth, Cécile Crosnier, Gavin J. Wright, and Nicole Staudt

15 Monoclonal Antibody Purification by Ceramic Hydroxyapatite Chromatography 241
Larry J. Cummings, Russell G. Frost, and Mark A. Snyder

16 Rapid Purification of Monoclonal Antibodies Using Magnetic Microspheres. 253
Pauline Malinge and Giovanni Magistrelli

17 Generation of Cell Lines for Monoclonal Antibody Production 263
Krista Alvin and Jianxin Ye

18 Expression and Purification of Recombinant Antibody Formats and Antibody Fusion Proteins 273
Martin Siegemund, Fabian Richter, Oliver Seifert, Felix Unverdorben, and Roland E. Kontermann

19 Purification of Antibodies and Antibody Fragments Using CaptureSelect™ Affinity Resins 297
Pim Hermans, Hendrik Adams, and Frank Detmers

20 Reformatting of scFv Antibodies into the scFv-Fc Format and Their Downstream Purification. 315
Emil Bujak, Mattia Matasci, Dario Neri, and Sarah Wulhfard

PART III ANITBODY CHARACTERIZATION AND MODIFICATION

21 Antibody V and C Domain Sequence, Structure, and Interaction Analysis with Special Reference to IMGT® 337
Eltaf Alamyar, Véronique Giudicelli, Patrice Duroux, and Marie-Paule Lefranc

22 Measuring Antibody Affinities as Well as the Active Concentration of Antigens Present on a Cell Surface 383
Palaniswami Rathanaswami

23 Determination of Antibody Structures. 395
Robyn L. Stanfield

24 Affinity Maturation of Monoclonal Antibodies
by Multi-site-Directed Mutagenesis. 407
Hyung-Yong Kim, Alexander Stojadinovic, and Mina J. Izadjoo

25 Epitope Mapping with Membrane-Bound Synthetic Overlapping Peptides . . . 421
Terumi Midoro-Horiuti and Randall M. Goldblum

26 Epitope Mapping by Epitope Excision, Hydrogen/Deuterium Exchange,
and Peptide-Panning Techniques Combined with In Silico Analysis. 427
Nicola Clementi, Nicasio Mancini, Elena Criscuolo,
Francesca Cappelletti, Massimo Clementi, and Roberto Burioni

27 Fine Epitope Mapping Based on Phage Display and Extensive Mutagenesis
of the Target Antigen 447
Gertrudis Rojas

28 Epitope Mapping with Random Phage Display Library 477
Terumi Midoro-Horiuti and Randall M. Goldblum

29 Epitope Mapping of Monoclonal and Polyclonal Antibodies
Using Bacterial Cell Surface Display 485
Anna-Luisa Volk, Francis Jingxin Hu, and Johan Rockberg

30 Ion Exchange-High-Performance Liquid Chromatography (IEX-HPLC) 501
Marie Corbier, Delphine Schrag, and Sylvain Raimondi

31 Size Exclusion-High-Performance Liquid Chromatography (SEC-HPLC). . . . 507
Delphine Schrag, Marie Corbier, and Sylvain Raimondi

32 N-Glycosylation Characterization by Liquid Chromatography
with Mass Spectrometry 513
Song Klapoetke

33 Fc Engineering of Antibodies and Antibody Derivatives
by Primary Sequence Alteration and Their Functional Characterization 525
Stefanie Derer, Christian Kellner, Thies Rösner, Katja Klausz,
Pia Glorius, Thomas Valerius, and Matthias Peipp

Part IV Applications of Monoclonal Antibodies

34 Labeling and Use of Monoclonal Antibodies in Immunofluorescence:
Protocols for Cytoskeletal and Nuclear Antigens 543
Christoph R. Bauer

35 Generation and Use of Antibody Fragments for Structural Studies
of Proteins Refractory to Crystallization 549
Stephen J. Stahl, Norman R. Watts, and Paul T. Wingfield

36 Antibody Array Generation and Use 563
Carl A.K. Borrebaeck and Christer Wingren

Index *573*

Contributors

HENDRIK ADAMS • *BAC BV, Life Technologies, Leiden, The Netherlands*
ELTAF ALAMYAR • *The International ImMunoGenetics information system, Laboratoire d'ImmunoGénétique Moléculaire, Institut de Génétique Humaine IGH, Université Montpellier 2, Montpellier, France*
KRISTA ALVIN • *BioProcess Development, Merck & Co., Inc., Kenilworth, NJ, USA*
CHRISTOPH R. BAUER • *Bioimaging Platform, University of Geneva, Geneve, Switzerland*
JODY D. BERRY • *Antibody Discovery, BD Biosciences, La Jolla, CA, USA; Department of Medical Microbiology or Immunology, University of Manitoba, Winnipeg, MB, Canada*
SUMITA BHADURI-MCINTOSH • *Pediatric Infectious Diseases, Department of Pediatrics, Stony Brook University School of Medicine, Stony Brook, NY, USA; Department of Molecular Genetics and Microbiology, Stony Brook University School of Medicine, Stony Brook, NY, USA*
CARL A.K. BORREBAECK • *Department of Immunotechnology and CREATE Health, Lund University, Lund, Sweden*
EMIL BUJAK • *Philochem AG, Zurich, Switzerland*
ROBERTO BURIONI • *Microbiology and Virology Unit, "Vita-Salute" San Raffaele University, Milan, Italy*
TERESA CABRAL • *Bioforensics Assay Development and Diagnostics Section, National Microbiology Laboratory, Public Health Agency of Canada, Winnipeg, MB, Canada*
FRANCESCA CAPPELLETTI • *Microbiology and Virology Unit, "Vita-Salute" San Raffaele University, Milan, Italy*
GANG CHEN • *University of Toronto, Toronto, ON, Canada*
EFTHALIA CHRONOPOULOU • *Antibody Discovery, BD Biosciences, La Jolla, CA, USA*
NICOLA CLEMENTI • *Microbiology and Virology Unit, "Vita-Salute" San Raffaele University, Milan, Italy*
MASSIMO CLEMENTI • *Microbiology and Virology Unit, "Vita-Salute" San Raffaele University, Milan, Italy*
CINDI R. CORBETT • *Bioforensics Assay Development and Diagnostics Section, National Microbiology Laboratory, Public Health Agency of Canada, Winnipeg, MB, Canada*
MARIE CORBIER • *NovImmune S.A., Plan-Les-Ouates, Switzerland*
ELENA CRISCUOLO • *Microbiology and Virology Unit, "Vita-Salute" San Raffaele University, Milan, Italy*
CÉCILE CROSNIER • *Cell Surface Signalling Laboratory, Wellcome Trust Sanger Institute, Hinxton, Cambridge, UK*
LARRY J. CUMMINGS • *Bio-Rad Laboratories, Inc., Hercules, CA, USA*
STEFANIE DERER • *Division for Stem Cell Transplantation and Immunotherapy, Christian-Albrechts-University Kiel and University Hospital Schleswig-Holstein, Kiel, Germany*
FRANK DETMERS • *BAC BV, Life Technologies, Leiden, The Netherlands*

SUBA DHARSHANAN • *Protein Science Department, Inno Biologics, Nilai, Negeri Sembilan, Malaysia*
PATRICE DUROUX • *The International ImMunoGenetics information system, Laboratoire d'ImmunoGénétique Moléculaire, Institut de Génétique Humaine IGH, Université Montpellier 2, Montpellier, France*
RUSSELL G. FROST • *Bio-Rad Laboratories, Inc., Hercules, CA, USA*
GERMAINE FUH • *Antibody Engineering Department, Genentech Inc., South San Francisco, CA, USA*
VÉRONIQUE GIUDICELLI • *The International ImMunoGenetics information system, Laboratoire d'ImmunoGénétique Moléculaire, Institut de Génétique Humaine IGH, Université Montpellier 2, Montpellier, France*
PIA GLORIUS • *Division for Stem Cell Transplantation and Immunotherapy, Christian-Albrechts-University Kiel and University Hospital Schleswig-Holstein, Kiel, Germany*
RANDALL M. GOLDBLUM • *Department of Pediatrics, University of Texas Medical Branch, Galveston, TX, USA*
HIDEMASA GOTO • *Aichi Cancer Center Research Institute, Nagoya, Aichi, Japan*
ULRIKE HAESSLER • *Department of Biosystems Science and Engineering, ETH Zürich, Basel, Switzerland*
PIM HERMANS • *BAC BV, Life Technologies, Leiden, The Netherlands*
FRANCIS JINGXIN HU • *School of Biotechnology, AlbaNova University Center, KTH – Royal Institute of Technology, Stockholm, Sweden*
JOYCE HUI-YUEN • *Pediatric Rheumatology, Morgan Stanley Children's Hospital of New York Presbyterian, Columbia University, New York, NY, USA*
CHEAH SWEE HUNG • *Faculty of Medicine, Department of Physiology, University of Malaya, Kuala Lumpur, Malaysia*
MASAKI INAGAKI • *Nagoya University Graduate School of Medicine, Nagoya, Aichi, Japan*
MINA J. IZADJOO • *Diagnostics and Translational Research Center, Henry M. Jackson Foundation for the Advancement of Military Medicine, Gaithersburg, MD, USA*
CHRISTIAN KELLNER • *Division for Stem Cell Transplantation and Immunotherapy, Christian-Albrechts-University Kiel and University Hospital Schleswig-Holstein, Kiel, Germany*
HYUNG-YONG KIM • *Diagnostics and Translational Research Center, Henry M. Jackson Foundation for the Advancement of Military Medicine, Gaithersburg, MD, USA*
SONG KLAPOETKE • *KBI Biopharma, Durham, NC, USA*
KATJA KLAUSZ • *Division for Stem Cell Transplantation and Immunotherapy, Christian-Albrechts-University Kiel and University Hospital Schleswig-Holstein, Kiel, Germany*
KATHERINE L. KNIGHT • *Stritch School of Medicine, Loyola University Chicago, Maywood, IL, USA*
PATRICK KOENIG • *Antibody Engineering Department, Genentech Inc., South San Francisco, CA, USA*
SIVA KOGANTI • *Pediatric Infectious Diseases, Department of Pediatrics, Stony Brook University School of Medicine, Stony Brook, NY, USA*
ROLAND E. KONTERMANN • *Institut für Zellbiologie und Immunologie, Universität Stuttgart, Stuttgart, Germany*

MARIE-PAULE LEFRANC • *The International ImMunoGenetics information system, Laboratoire d'ImmunoGénétique Moléculaire, Institut de Génétique Humaine IGH, Université Montpellier 2, Montpellier, France*
JEREMY LOYAU • *NovImmune S.A., Plan-Les-Ouates, Switzerland*
GIOVANNI MAGISTRELLI • *NovImmune S.A., Plan-Les-Ouates, Switzerland*
PAULINE MALINGE • *NovImmune S.A., Plan-Les-Ouates, Switzerland*
NICASIO MANCINI • *Microbiology and Virology Unit, "Vita-Salute" San Raffaele University, Milan, Italy*
MATTIA MATASCI • *Philochem AG, Zurich, Switzerland*
TERUMI MIDORO-HORIUTI • *Department of Pediatrics, University of Texas Medical Branch, Galveston, TX, USA*
NICOLE MÜLLER-SIENERTH • *Cell Surface Signalling Laboratory, Wellcome Trust Sanger Institute, Hinxton, Cambridge, UK*
DARIO NERI • *Department of Chemistry and Applied Biosciences, Institute of Pharmaceutical Sciences; ETH Zurich, Zurich, Switzerland*
MATTHIAS PEIPP • *Division for Stem Cell Transplantation and Immunotherapy, Christian-Albrechts-University Kiel and University Hospital Schleswig-Holstein, Kiel, Germany*
SYLVAIN RAIMONDI • *NovImmune S.A., Plan-Les-Ouates, Switzerland*
PALANISWAMI RATHANASWAMI • *Amgen British Columbia, Burnaby, Canada*
SAI T. REDDY • *Department of Biosystems Science and Engineering, ETH Zürich, Basel, Switzerland*
FABIAN RICHTER • *Institut für Zellbiologie und Immunologie, Universität Stuttgart, Stuttgart, Germany*
JOHAN ROCKBERG • *School of Biotechnology, AlbaNova University Center, KTH – Royal Institute of Technology, Stockholm, Sweden*
GERTRUDIS ROJAS • *Systems Biology Department, Center of Molecular Immunology, La Habana, Cuba*
THIES RÖSNER • *Division for Stem Cell Transplantation and Immunotherapy, Christian-Albrechts-University Kiel and University Hospital Schleswig-Holstein, Kiel, Germany*
FRANÇOIS ROUSSEAU • *NovImmune S.A., Plan-Les-Ouates, Switzerland*
DELPHINE SCHRAG • *NovImmune S.A., Plan-Les-Ouates, Switzerland*
OLIVER SEIFERT • *Institut für Zellbiologie und Immunologie, Universität Stuttgart, Stuttgart, Germany*
SACHDEV S. SIDHU • *University of Toronto, Toronto, ON, Canada*
MARTIN SIEGEMUND • *Institut für Zellbiologie und Immunologie, Universität Stuttgart, Stuttgart, Germany*
MARK A. SNYDER • *Bio-Rad Laboratories, Inc., Hercules, CA, USA*
STEPHEN J. STAHL • *Protein Expression Laboratory, National Institute of Arthritis and Musculoskeletal and Skin Diseases, National Institutes of Health, Bethesda, MD, USA*
ROBYN L. STANFIELD • *Department of Molecular Biology, The Scripps Research Institute, La Jolla, CA, USA*
NICOLE STAUDT • *Cell Surface Signalling Laboratory, Wellcome Trust Sanger Institute, Hinxton, Cambridge, UK*
ALEXANDER STOJADINOVIC • *Department of Surgery, Walter Reed National Military Medical Center, Bethesda, MD, USA*

FELIX UNVERDORBEN • *Institut für Zellbiologie und Immunologie, Universität Stuttgart, Stuttgart, Germany*
ALEJANDRO URIBE-BENNINGHOFF • *Antibody Discovery, BD Biosciences, La Jolla, CA, USA*
THOMAS VALERIUS • *Division for Stem Cell Transplantation and Immunotherapy, Christian-Albrechts-University Kiel and University Hospital Schleswig-Holstein, Kiel, Germany*
JAMES A. VAN DEVENTER • *Department of Chemical Engineering, Koch Institute for Integrative Cancer Research, Massachusetts Institute of Technology, Cambridge, MA, USA*
ANNA-LUISA VOLK • *School of Biotechnology, AlbaNova University Center, KTH – Royal Institute of Technology, Stockholm, Sweden*
NORMAN R. WATTS • *Protein Expression Laboratory, National Institute of Arthritis and Musculoskeletal and Skin Diseases, National Institutes of Health, Bethesda, MD, USA*
PAUL T. WINGFIELD • *Protein Expression Laboratory, National Institute of Arthritis and Musculoskeletal and Skin Diseases, National Institutes of Health, Bethesda, MD, USA*
CHRISTER WINGREN • *Department of Immunotechnology and CREATE Health, Lund University, Lund, Sweden*
KARL DANE WITTRUP • *Department of Chemical and Biological Engineering, Koch Institute for Integrative Cancer Research, Massachusetts Institute of Technology, Cambridge, MA, USA*
GAVIN J. WRIGHT • *Cell Surface Signalling Laboratory, Wellcome Trust Sanger Institute, Hinxton, Cambridge, UK*
SARAH WULHFARD • *Philochem AG, Zurich, Switzerland*
PI-CHEN YAM • *Stritch School of Medicine, Loyola University Chicago, Maywood, IL, USA*
JIANXIN YE • *BioProcess Development, Merck & Co., Inc., Kenilworth, NJ, USA*

Part I

Antibody Generation

Chapter 1

Antigen Production for Monoclonal Antibody Generation

Giovanni Magistrelli and Pauline Malinge

Abstract

The quality of the target antigen is very important in order to generate a good antibody, in particular when binding to a conformational epitope is desired. The use of mammalian cells for recombinant protein expression provides an efficient machinery for the correct folding and posttranslational modification of proteins. In this chapter, we describe a process to rapidly generate semi-stable human cell lines secreting a recombinant protein of interest into the culture medium. Simple disposable bioreactors that can be used in any standard cell culture laboratory enable the production of recombinant protein in the multi-milligram range. The protein can be readily purified from the culture supernatant by immobilized metal affinity chromatography. In addition, by inserting a tag recognized by a co-expressed biotin ligase, the protein can be biotinylated during the secretion process. This greatly facilitates the immobilization of the protein for assay development or for antibody isolation using in vitro selection technologies.

Key words CELLine bioreactor, Protein purification, IMAC, Single site biotinylation, Episomal vector, Semi-stable cell line

1 Introduction

Having access to pure antigen in sufficient quantities is a first and important step in the development of a monoclonal antibody (mAb). As access to the native protein is usually limited, recombinant protein expression and purification is a method of choice to generate the antigen, although alternative approaches, such as the use of cells or DNA for immunization, can also be envisaged [1–3]. When the objective is to generate antibodies that are biologically active or recognize the antigen in its native state, it is particularly important to generate a properly folded protein that closely mimics the structure of the native protein [4].

Three types of cells are commonly used for recombinant protein production: bacteria (*Escherichia coli*), insect cells and mammalian cells. The choice of an expression system can be guided by the structure and origin of the protein to be expressed, the presence of disulfide bonds or if posttranslational modifications are important for its functionality. The recombinant protein can be expressed

Vincent Ossipow and Nicolas Fischer (eds.), *Monoclonal Antibodies: Methods and Protocols*, Methods in Molecular Biology, vol. 1131, DOI 10.1007/978-1-62703-992-5_1, © Springer Science+Business Media New York 2014

either in the cytoplasm or can be secreted in the culture medium by the addition of an appropriate signal peptide. In addition, the antigen can be expressed as a fusion with a peptide or a protein that can increase its solubility [5–10]. Protein tags, such as the Fc of an antibody or a hexahistidine sequence, are also routinely included at one extremity of the protein to facilitate purification using affinity chromatography. All these available options enable the design and testing of a large number of expression constructs.

In this chapter we describe the use of an episomal expression system for the rapid generation of semi-stable human cell lines secreting a recombinant protein of interest into the medium [11]. We have found this approach very versatile and it often allowed for the successful expression of soluble recombinant protein in the milligram range, even for challenging eukaryotic proteins. This system combines several elements that facilitate and accelerate the overall process.

First, the gene encoding the protein of interest is cloned into an expression vector capable of autonomous episomal replication that can be maintained within cells without integration into the genome. The vector also contains a puromycin resistance gene for efficient selection of transfected cells and for the establishment of semi-stable pools of cells harboring the expression vector. The multi-cistronic design of the expression construct enables the simultaneous expression of multiple proteins. The protein of interest is secreted into the medium and carries a hexahistidine tag for single step purification and an AviTag™ for single site biotinylation. The internal ribosome entry site (IRES) enables the co-expression and secretion of biotin ligase (LsBirA) driving site-specific biotinylation and enhanced green fluorescent protein (EGFP) expression for easy monitoring of the transfection efficiency and the expression within a cell pool [12]. Single site-biotinylation on the AviTag™ facilitates protein immobilization and detection using streptavidin based reagents [13]. Second, the use of two-compartment bioreactors (CELLine) allow for a straightforward scale-up of production using equipment found in any standard cell culture laboratory [14]. The benefit of this system is that the two compartments are separated by a membrane that allows nutrients and waste products to exchange. In this way serum-containing medium can be used in the main compartment to support optimal cell growth and viability, without contamination of the cell compartment with bovine serum albumin (BSA) and bovine IgG, facilitating the purification process. In addition, the protein can be recovered from the culture in a small volume avoiding the need for concentration steps before purification. Finally, an optimized protocol for affinity purification via immobilized metal ion affinity chromatography (IMAC) enables efficient purification of the recombinant protein in a single step from the mammalian cell culture supernatant.

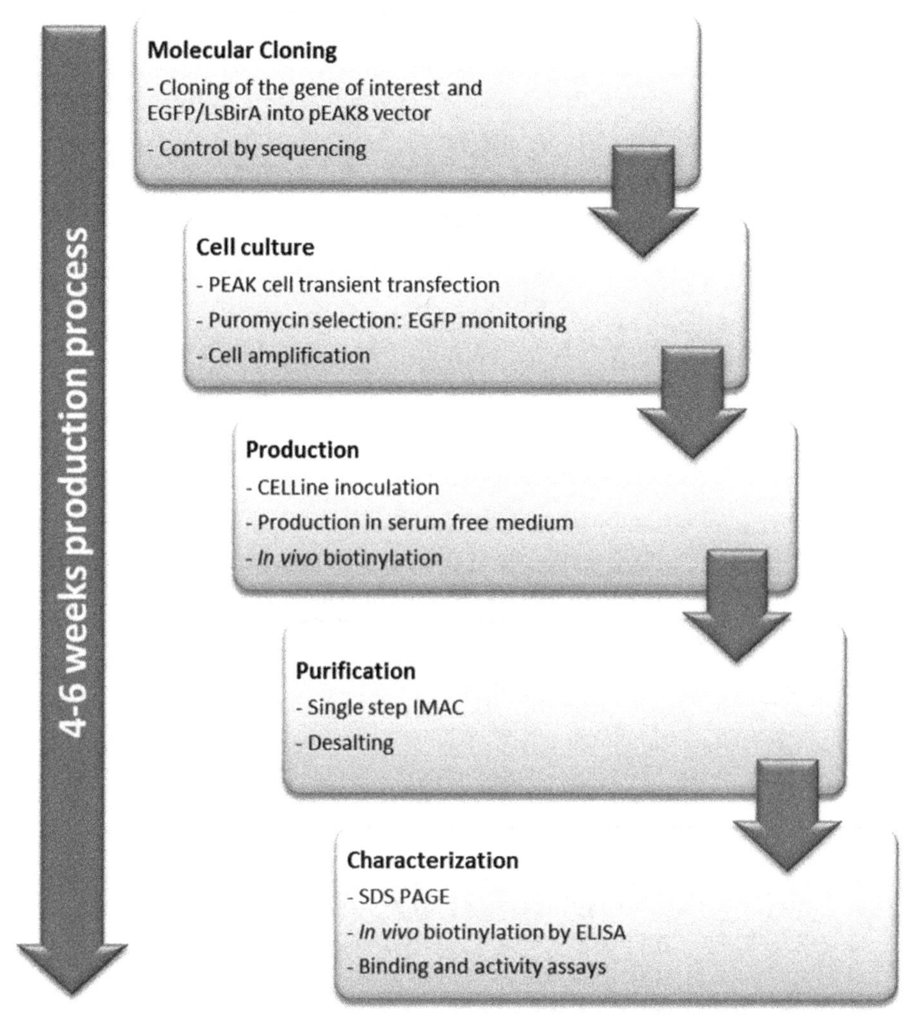

Fig. 1 Schematic representation of the overall process

Over 20 human and mouse recombinant proteins have been expressed and purified using this approach, greatly facilitating the production and characterization of monoclonal antibodies. The overall process is schematically represented in Fig. 1.

2 Materials

All reagents and kits must be used and stored as described by the manufacturer instructions. Experiments using cells are performed under a sterile hood and chemical reagents are manipulated under a fume hood when necessary.

2.1 Molecular Cloning

1. pEAK8 plasmid (Edge Biosystems).
2. pVector IRES EGFP and pVector IRES LsBirA IRES EGFP plasmids (available upon request from the authors).
3. *Hin*dIII, *Eco*RI, *Xba*I, *Mfe*I, and *Nhe*I restriction enzymes and appropriate buffers (New England Biolabs) (*see* **Note 1**).
4. Rapid DNA Dephos & Ligation Kit (Roche).
5. E-Gel® Agarose gels and power-pack (Invitrogen).
6. E-Gel® 1 Kb Plus DNA ladder (Invitrogen).
7. MinElute® PCR purification Kit (Qiagen).
8. MinElute® Gel Extraction Kit (Qiagen).
9. QIAprep Spin Miniprep Kit (Qiagen).
10. XL1-Blue competent *E. coli* cells.
11. Ampicillin (AppliChem).
12. Terrific Broth (TB) medium (Invitrogen): Dissolve 47 g of TB powder in 1 L of demineralized water; add 4 mL of glycerol; autoclave for 15 min at 121 °C.
13. LB-Amp agar plates: Dissolve 10 g of Bacto™ Peptone (Becton Dickinson), 10 g of NaCl, 5 g of Bacto™ Yeast (Becton Dickinson), and 15 g of Bacto™ Agar (Becton Dickinson) in 950 mL of deionized H_2O; adjust the pH to 7.5 by adding 10 mM NaOH; adjust the volume of the solution to 1 L. Sterilize by autoclaving for 20 min at 121 °C and let cool in a water bath at 55 °C for 30–60 min; add ampicillin at a final concentration of 100 μg/mL; distribute 20 mL of medium per petri dish and let solidify at room temperature (RT). Store the plates at 4 °C.
14. Shaking Incubator.
15. UV imaging system.
16. Temperature controlled water bath.
17. Scalpel.
18. Thermomixer.
19. NanoDrop® ND-1000 Spectrophotometer (Witec AG) or equivalent spectrophotometer.
20. Sequencing primers: Forward primer 5′ TGCGATGGAG TTTCCCCACACTG 3′ and Reverse primer 5′ CACCCGGG CAGACCTGAGGAAGAGATG 3′.

2.2 Cell Culture, Transfection, and Selection

1. Transformed human embryo kidney monolayer epithelial cells (PEAK cells; Edge Biosystems) (*see* **Note 2**).
2. Dulbecco's Modified Eagle Medium (DMEM; Sigma-Aldrich).
3. Fetal Bovine Serum (FBS; Sigma-Aldrich).
4. L-glutamine solution 200 mM (Sigma-Aldrich).

5. Gentamicin solution 50 mg/mL (Sigma-Aldrich).
6. Complete DMEM: DMEM supplemented with 10 % of FBS; 2 mM L-glutamine; 50 μg/mL of gentamicin.
7. 6-Well cell culture plates (Multidishes Nunclon™ Δ Surface, Nunc).
8. ThermoForma Steri-Cycle CO_2 incubator (Thermo Scientific).
9. TransIT®-LT1 transfection reagent (Mirus).
10. Puromycin dihydrochloride solution 10 mg/mL (Sigma-Aldrich).
11. Fluorescence microscope (Axiovert 40CFL; Zeiss).
12. FACS (FACSCalibur; BD Biosciences).
13. Phosphate Buffered Saline (PBS; Sigma-Aldrich).
14. Bovine Serum Albumin (BSA; Sigma-Aldrich).
15. Vi-CELL XR counter (Beckman Coulter) or equivalent cell counter.
16. Dimethyl sulfoxide (DMSO; Sigma-Aldrich).
17. Freezing medium: FBS; 10 % DMSO.
18. Biobanking and cell culture cryogenic tubes (Nunc).
19. Cool-box: CoolCell® Cryopreservation cell freezing container (VWR).

2.3 Protein Production

1. Serum-free medium: 293 SFM II liquid medium (Gibco).
2. CELLine 1000 bioreactor (CL1000; Integra) (*see* **Note 3**).
3. Biotin solution 5 mM (Avidity).
4. TC Flask 175 CM2 SI Filter cap (Thermo Scientific).
5. TC Flask 80 CM2 SI Filter cap N (Thermo Scientific).

2.4 Protein Purification and Desalting

1. Stericup filter unit with a 0.22 μm filter (Millipore).
2. Imidazole powder (Sigma-Aldrich).
3. Binding/Washing buffer: PBS; 20 mM imidazole; pH 7.4; filtered on a Stericup filter unit.
4. Elution buffer: PBS; 400 mM imidazole pH 7.4; filtered on a Stericup filter unit.
5. Ni-NTA Superflow Cartridge, 5 mL (Qiagen).
6. ÄKTA Prime chromatography system (GE Healthcare) or equivalent.
7. PrimeView 5.0 software (GE Healthcare).
8. 5 mL Polystyrene round-bottom tubes (BD Falcon™).
9. Amicon filter device: Ultra-4 PLTK Ultracel-PL membrane (Millipore) (*see* **Note 4**).
10. Siliconized tubes (Sigma-Aldrich) (*see* **Note 5**).

2.5 Protein Characterization

2.5.1 Proteins Analysis by Gel Electrophoresis

1. Electrophoresis Novex mini-cell (Invitrogen).
2. NuPAGE® 4–12 % Bis-Tris Mini Gel (Invitrogen).
3. NuPAGE® MES SDS Running Buffer (Invitrogen).
4. SeeBlue® Plus2 Pre-Stained Standard (Invitrogen).
5. NuPAGE® LDS Sample Buffer (Invitrogen).
6. β-Mercaptoethanol (Sigma Aldrich) (*see* **Note 6**).
7. Power supply (Fischer Scientific).
8. SimplyBlue™ SafeStain (Invitrogen).

2.5.2 Capture ELISA for Biotinylation Analysis

1. Tween 20 (Sigma-Aldrich).
2. PBST: PBS; 0.05 % Tween 20.
3. PBST–BSA: PBST; 1 % BSA.
4. StreptaWell transparent 96-well microplate (Roche).
5. Antigen specific mouse monoclonal antibody (*see* **Note 7**).
6. Horseradish Peroxidase (HRP)-coupled goat anti-mouse IgG antibody (Jackson ImmunoResearch).
7. Penta-His HRP Conjugate (Qiagen).
8. 3,3′,5,5′-Tetramethylbenzidine (TMB; Sigma-Aldrich).
9. Sulfuric acid H_2SO_4 2 N (Sigma-Aldrich).
10. Microplate reader.

3 Methods

3.1 Molecular Cloning

The molecular cloning approach relies on the simultaneous ligation of two DNA fragments into the episomal pEAK8 vector: (a) a DNA fragment encoding the recombinant protein to be expressed; (b) a DNA fragment containing either one or two IRES for the co-expression of EGFP alone or EGFP and LsBirA. The IRES containing expression cassettes are easily retrieved by restriction enzyme digestion from existing plasmids. The cloning strategy to generate the final expression constructs is schematically described in Fig. 2.

3.1.1 Preparation of the DNA Fragment Encoding the Protein of Interest

1. Generate a DNA fragment encoding the protein of interest, either by gene synthesis or by PCR amplification. Include a leader sequence for secretion into the medium as well as a hexahistidine tag at the N or C-terminus. An AviTag™ sequence can also be included if single site biotinylation is desired. Also include *Hin*dIII and *Nhe*I restriction sites at the 5′ and 3′ ends of the DNA fragment, respectively (*see* **Notes 8** and **9**).
2. Digest the DNA fragment with *Hin*dIII and *Nhe*I: Mix 5 μL of NEBuffer 2 10×; 0.5 μL of BSA 100×; 1 μg of DNA fragment;

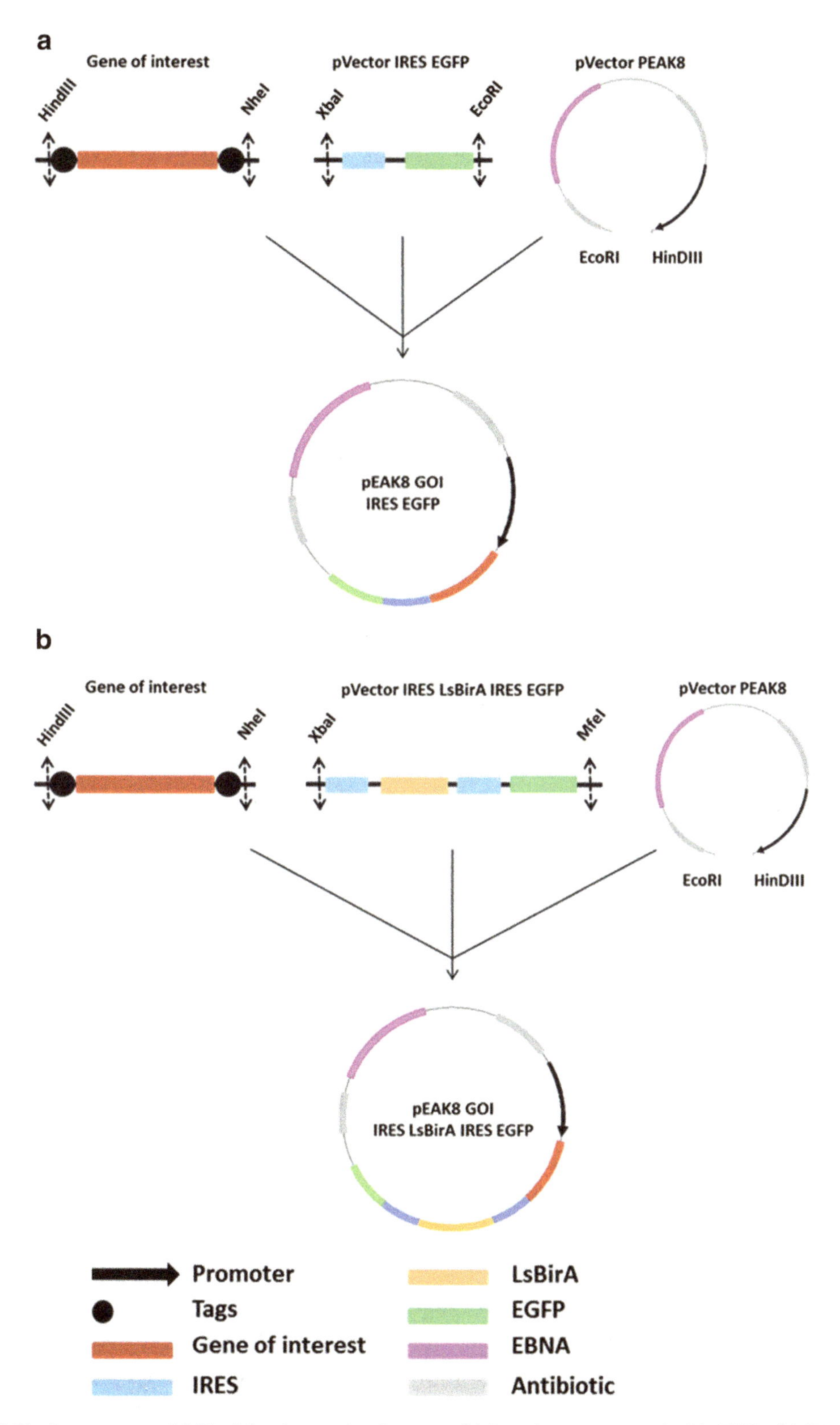

Fig. 2 Cloning process. (**a**) Bi-cistronic construct: gene of interest co-expressed with EGFP; (**b**) Tri-cistronic construct: gene of interest co-expressed with LsBirA and EGFP. GOI: Gene of Interest

10 U of each restriction enzyme in a final volume of 50 μL adjusted with sterile water.

3. Incubate the reaction for 2 h at 37 °C.
4. Purify DNA using the MinElute® PCR Purification Kit as follows.
5. Add 5 volumes of PB buffer to 1 volume (50 μL) of the digestion reaction.
6. Apply the sample to MinElute column and centrifuge for 1 min at full speed in a micro centrifuge.
7. Discard flow-through and wash the column with 750 μL of reconstituted PE buffer.
8. Centrifuge for 1 min at full speed and discard the flow-through. Centrifuge the column at full speed for two additional minutes to dry the filter membrane.
9. Transfer the column to a 1.5 mL tube, add 10 μL of EB buffer to the filter membrane, wait for 1 min, and centrifuge the column for 1 min at full speed to elute DNA.
10. Measure DNA concentration using a Nanodrop (R).

3.1.2 Preparation of the DNA Fragments Encoding EGFP and LsBirA

1. Digest either the plasmid pVector IRES EGFP (for co-expression of EGFP) or pVector IRES LsBirA IRES EGFP (for co-expression of EGFP and LsBirA) with *Xba*I/*Eco*RI and *Xba*I/*Mfe*I respectively: Mix 5 μL of NEBuffer 4 10×; 0.5 μL of BSA 100×; 1–5 μg of plasmid; 10 U of each restriction enzyme in a final volume of 50 μL adjusted with sterile water.
2. Incubate the reaction for 2 h at 37 °C.
3. Load the digested vector product and 10 μL of E-Gel® 1 Kb Plus DNA ladder on an E-Gel® 1.2 % (*see* **Note 10**).
4. After migration, extract the band corresponding to the IRES-EGFP or IRES-LsBisA-IRES-EGFP expression cassette (the expected size are approximately 1,500 bp and 3,000 bp, respectively).
5. Purify the band containing vector DNA using the MinElute® Gel Extraction Kit as follows.
6. Using a scalpel cut the bands from the gel and add 3 volumes of QG buffer for 1 volume of gel.
7. Incubate at 50 °C during 10 min under agitation with a thermomixer to dissolve gel slice.
8. Apply the sample to a MinElute column.
9. Centrifuge for 1 min at full speed. Discard the flow-through.
10. Wash the column with 750 μL of reconstituted PE buffer. Discard the flow-through and centrifuge the column at full speed for 1 additional minute to dry the filter membrane.

11. Transfer the column to a 1.5 mL tube, add 20 μL of EB buffer to the filter membrane, wait for 1 min, and centrifuge the column for 1 min at full speed to elute DNA.
12. Measure DNA concentration using a Nanodrop (R).

3.1.3 Preparation of Digested pEAK8 Vector

1. Digest the pEAK8 vector with *Hind*III and *Eco*RI: Mix 5 μL of NEBuffer 2 10×, 0.5 μL of purified BSA 100×, 1 μg of pEAK8 vector DNA, 10 U of each enzymes in a final volume of 50 μL adjusted with sterile water.
2. Incubate the reaction for 2 h at 37 °C.
3. Add 1 μL of Alkaline Phosphatase provided with the Rapid DNA Dephos & Ligation Kit and incubate for 15 additional minutes at 37 °C.
4. Load the digested vector product and 10 μL of E-Gel® 1 Kb Plus DNA ladder on an E-Gel® 1.2 % (*see* **Note 11**).
5. After migration, extract the band containing vector DNA using the MinElute® Gel Extraction Kit as described in Subheading 3.1.2.
6. Measure DNA concentration using a Nanodrop (R).

3.1.4 DNA Ligation

1. The two digested DNA fragments are co-ligated into the digested and dephosphorylated pEAK8 vector (*see* **Note 12**): Mix 2 μL of DNA dilution Buffer 5×, 50 ng of vector and 20–50 ng of each insert in a final volume of 10 μL adjusted with sterile water (*see* **Note 13**).
2. Add 10 μL of T4 DNA Ligation Buffer 2× and 1 μL of T4 DNA Ligase.
3. Incubate the reaction mix for 15 min at RT.
4. Transform the ligation products and negative control into chemically competent XL1-blue *E. coli* by adding directly 50 μL of cells to the ligation mix (*see* **Note 14**).
5. Mix gently and incubate for 30 min on ice. Heat the cells in a 42 °C water bath for 1 min and then replace them for 2 min on ice.
6. Add 500 μL of TB medium and incubate for 1 h at 37 °C under agitation at 1,250 rpm.
7. Spread 1/10 of the transformed bacteria on LB-Amp agar plates.
8. Incubate the plates overnight at 37 °C in a static incubator.
9. Prepare plasmid DNA from individual colonies using any standard kit such as QIAprep Spin Miniprep Kit.
10. Control the sequence of the inserts by DNA sequencing using the forward and reverse primers located upstream of the gene of interest and downstream of the EGFP sequence, respectively.

3.2 Cell Transfection and Selection

PEAK cells are cultured in complete DMEM in a static incubator at 37 °C, 5 % CO_2, and humidified atmosphere (*see* **Note 15**).

1. Plate 2×10^5 cells per well in 6-well plates, in 2 mL of complete DMEM medium per well.
2. After 24 h, transfect cells using the TransIT®-LT1 transfection reagent. For each well mix: 97 μL of DMEM; 3 μL of transfection reagent; 2 μg of vector DNA. Mix the reagents as follows.
3. Add the transfection reagent to DMEM and incubate at RT for 5 min.
4. Add this mix to the DNA and incubate at RT for 30 min.
5. Gently add, drop by drop, 100 μL of the mix to each well of the 6-well plate.
6. Mix gently the medium by carefully rocking the plate.
7. One day after transfection, evaluate transfection efficiency by monitoring EGFP expression by fluorescence microscopy or by FACS (*see* **Note 16**).
8. 1–2 days after transfection, add 0.5 μg/mL puromycin to the medium for selection and semi-stable clone generation.
9. Culture and propagate until an adherent layer of cells resistant to puromycin is obtained (*see* **Note 17**). Propagate the selected semi-stable cells in puromycin containing medium until enough cells for cryopreservation and seeding of culture flasks for amplification and production are obtained (*see* Subheading 3.3).
10. Cryopreserve the semi-stable cell lines: prepare aliquots of 1×10^7 cells in 1 mL of freezing medium.
11. Place cryotubes in the cool-box and incubate at −80 °C for 1–2 h (*see* **Note 18**).
12. Transfer cryotubes in liquid nitrogen for long-term storage.

3.3 Cell Amplification and Protein Production in CELLine Bioreactor (Fig. 3)

1. Amplify the semi-stable cells in DMEM containing 0.5 μg/mL puromycin in T75 or T175 flasks depending on the number of bioreactor that will be seeded (5×10^7 cells for each CELLine) (*see* **Note 19**).
2. Detach the cells from the flasks by vigorous manual shaking of the flask.
3. Count the cells using a Vi-Cell counter or equivalent.
4. For each CELLine to be seeded, centrifuge 5×10^7 cells at $500 \times g$ for 10 min.
5. Discard the supernatant and resuspend the cells in 10 mL of serum-free medium.
6. For bioreactor conditioning, add 200 mL of complete DMEM to the upper compartment of a CELLine and 10 mL of serum-free medium in the lower compartment (*see* **Note 20**).

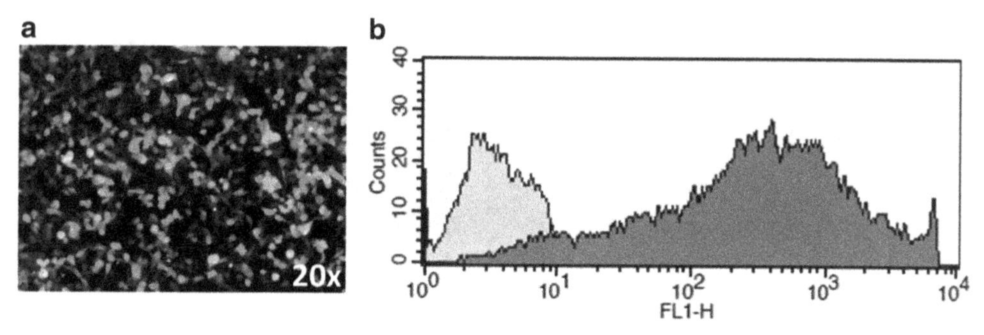

Fig. 3 (**a**) EGFP expression detected by fluorescence microscopy in PEAK cells; (**b**) Analysis of PEAK cell transfection by monitoring EGFP expression by FACS: control of untransfected PEAK cells (*light grey*); PEAK cells 48 h after transfection (*dark grey*)

7. Add the 10 mL cell suspension in the lower compartment and fill the upper compartment up to 700 mL with complete DMEM.
8. Add 5–50 μM of biotin in the upper compartment for in vivo biotinylation (*see* **Note 21**).
9. After 3 to 4 days of culture, add 200–300 mL of complete medium in the upper compartment to provide fresh nutrients to cells (Fig. 4).

3.4 Purification

1. Harvest cell culture supernatant after 7 to 10 days of production: recover with a pipette the medium from the lower compartment. Add fresh serum-free medium to wash the cell compartment and combine this wash fraction with the first harvest.
2. Clarify the harvest by centrifugation at 1,000 × *g* for 10 min.
3. Filter the clarified supernatant on a 0.22 μm Stericup filter unit to remove remaining cells and cell debris.
4. Add imidazole to the harvest to a final concentration of 100 mM.
5. Equilibrate the Ni-NTA chromatography column at RT with 5 column volumes of PBS and then 5 column volumes of Binding/Washing buffer at a flow rate of 2 mL/min using a peristaltic pump (*see* **Note 22**).
6. Load the harvest on the column at a flow rate of 2 mL/min (*see* **Note 23**).
7. Wash the column with Binding/Washing buffer (*see* **Note 24**).
8. At the same time, wash the ÄKTA Prime system and tubing with water and then corresponding inlet tubing with Binding/ Washing buffer and Elution buffer. Insert 45 collection tubes into the ÄKTA fraction collector.

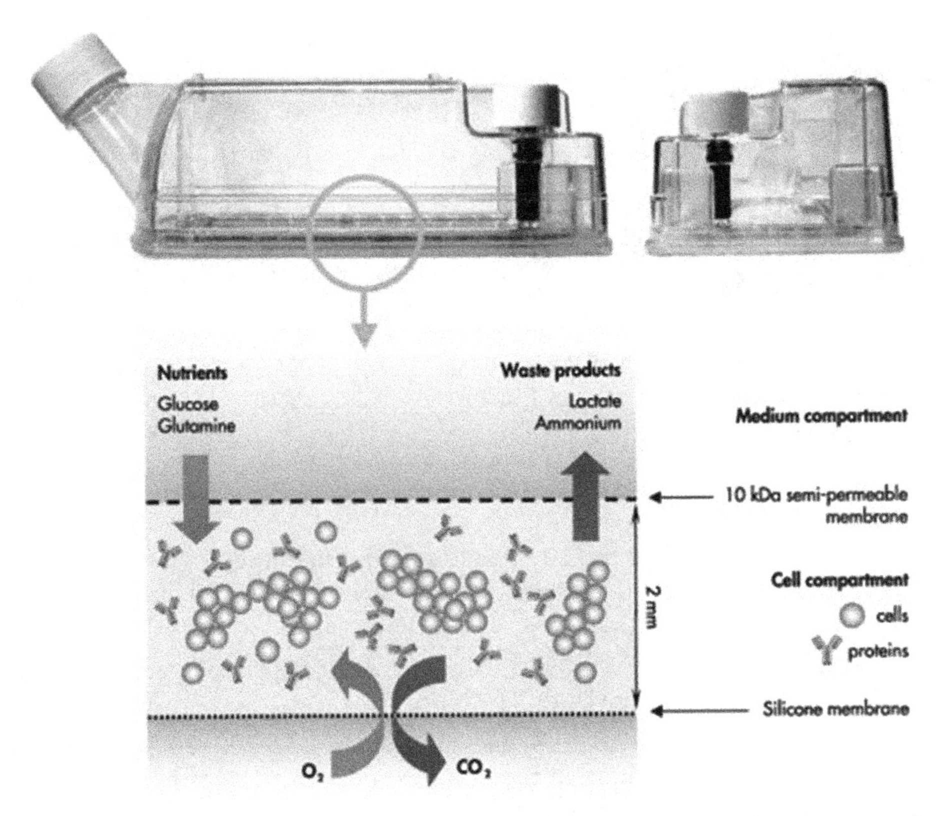

Fig. 4 CELLine—Two-compartment Technology (adapted from http://www.integra-biosciences.com)

9. Connect the column to the ÄKTA (*see* **Note 25**) and elute proteins applying the following parameters: flow rate of 2 mL/min, 30 mL imidazole gradient (from 20 to 400 mM of imidazole, i.e., from 100 % Binding/Washing buffer to 100 % Elution buffer), collect 1 mL fractions (45 fractions in total). The chromatogram displayed using the Prime View software shows the absorption at 280 nm, the buffer gradient as well as the collected fractions.
10. Pool fractions corresponding to the elution peak (*see* **Note 26**) (Fig. 5).

3.5 Desalting by Diafiltration (See Note 27*)*

1. Equilibrate the Amicon filter device with PBS (or an appropriate formulation buffer) by centrifugation at 2,000 × *g* for 4–5 min.
2. Apply the sample to the filter device and fill with PBS.
3. Centrifuge at 2,000 × *g* for 5–7 min, discard the flow-through, and fill the device with PBS to dilute the sample. Repeat this step 2–3 times to completely exchange elution buffer with PBS.
4. Recover the sample from the device by careful pipetting using a side-to-side sweeping motion for complete recovery.

a

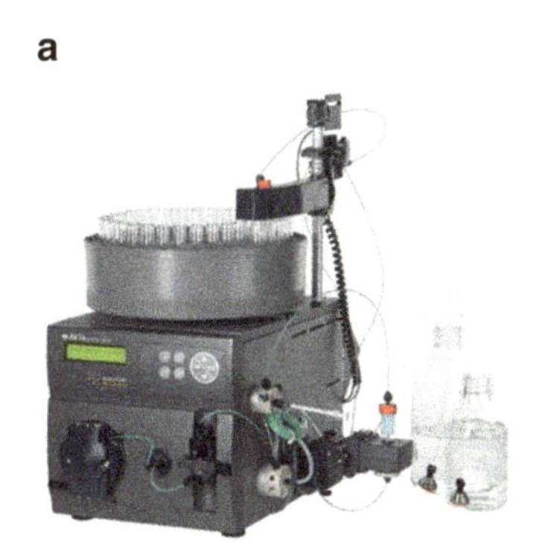

b

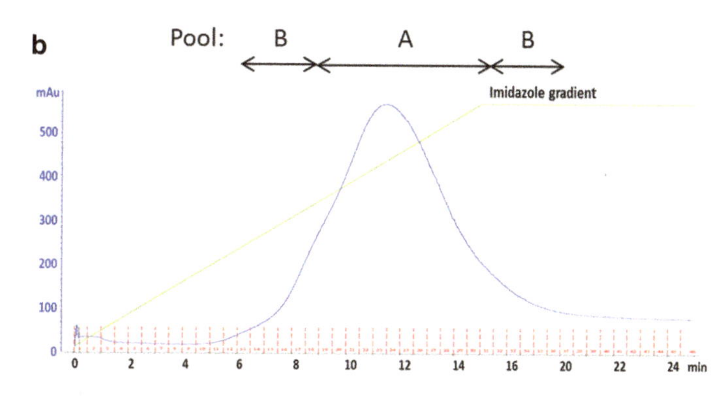

Fig. 5 (**a**) ÄKTA Prime system used for elution and fraction recovery; (**b**) Chromatogram of human IL-6R purification on a nickel chromatography resin, with fractions pooled to obtain a high (**a**) and a low (**b**) concentration batch

5. Aliquot the final product in siliconized tubes.
6. Quantify the purified protein by measuring the absorption at 280 nm using the Nanodrop (R) spectrophotometer.

3.6 Protein Characterization

3.6.1 Proteins Analysis by Gel Electrophoresis

1. Dilute 1 μg of purified protein sample into 20 μL of PBS (*see* **Note 28**).
2. Add 5 μL of LDS sample buffer, containing or not β-mercaptoethanol for loading in reducing or non-reducing conditions, respectively. A 1 mL reducing solution contains 960 μL of LDS sample buffer and 40 μL of β-mercaptoethanol.
3. Heat samples at 95 °C for 5 min for reducing conditions or 2 min for non-reducing conditions.
4. Load samples on the gel (*see* **Note 29**).
5. Start migration in MES running buffer at 180 V for 35 min.
6. After migration, recover the gel and wash with distilled water.
7. Add SimplyBlue™ SafeStain and stain the gel overnight (*see* **Note 30**).
8. Discard the SimplyBlue™, wash with distilled water and destain gel in distilled water under gentle agitation until bands appear (Fig. 6).

3.6.2 ELISA for Biotinylation Analysis (See Note 31)

1. Dilute the purified proteins in PBS to a concentration between 1 and 5 μg/mL.
2. Add 100 μL of sample per well of a StreptaWell plate and incubate for 1 h at RT (*see* **Note 32**).
3. Wash each well three times with 200 μL of PBST.
4. Add 100 μL per well of antigen specific mouse monoclonal antibody diluted at 1 μg/mL or Penta-His HRP conjugate antibody diluted 1:2,000, in PBST–BSA (*see* **Note 33**).
5. Incubate for 1 h at RT.

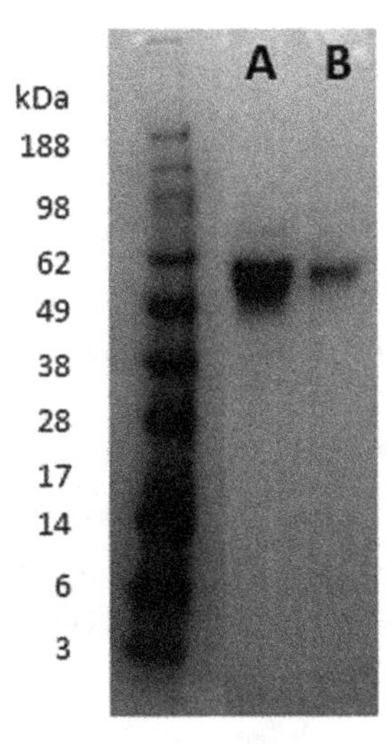

Fig. 6 SDS PAGE gel of purified human IL-6R in denaturing and reducing conditions on a 4–12 % Bis-Tris acrylamide gel. The two pools described in Fig. 5 were analyzed (Pool A and Pool B)

6. Wash each well three times with 200 μL of PBST.
7. For antigen specific antibody, add 100 μL per well of HRP-coupled goat anti-mouse IgG antibody diluted at 1:2,000 in PBST–BSA.
8. Incubate for 1 h at RT.
9. Wash each well three times with 200 μL of PBST.
10. Add 100 μL of TMB chromogenic substrate for revelation.
11. After 1 to 10 min of revelation time, add 100 μL of H_2SO_4 2 N to stop the reaction.
12. Measure the absorption at 450 nm on a microplate reader.

4 Notes

1. Specific reagents necessary for digestion reactions, such as BSA, buffers, are provided with enzymes.
2. PEAK cells stably express the EBNA-1 gene allowing for episomal replication. These cells are semi-adherent and can be easily detached from the flask by a vigorous shaking.
3. A CELLine is composed of two compartments separated by a 10 kDa semi-permeable membrane. Nutrients from the medium can migrate through the membrane to the cell compartment and waste products (lactate, ammonium) from cell culture in the lower compartment can diffuse to the upper compartment. A silicone membrane at the bottom of the bioreactor allows for gas exchange, as illustrated in Fig. 4.

4. The cutoff of the filter device must be chosen according to the size of the protein to desalt. This cutoff should be at least three times higher than the protein molecular weight to avoid product loss during this centrifugation process. The device exists also in different sizes according to the volume of sample to be centrifuged. During the successive centrifugations, store the flow-through in the event of protein loss during desalting.
5. Siliconized tubes must be used to avoid loss of protein on the inner wall of tubes. The mentioned tubes' caps are not locked. For freezing, keep tubes in a vertical position to avoid opening of the tube and product loss. Other locked tubes can be used such as the LoBind tubes from Eppendorf.
6. The β-mercaptoethanol must be manipulated under a fume hood.
7. In this chapter, a mouse primary antibody is used and so the secondary antibody is an anti-mouse antibody. If the primary antibody comes from another species, it is necessary to adapt the secondary antibody to the corresponding species.
8. The AviTag™ (GLNDIFEAQKIEWHE) from Avidity allows for a targeted enzymatic conjugation of a single biotin molecule on the tag using the biotin ligase. Here we describe the co-expression of the biotin ligase and in vivo biotinylation by addition of biotin in the production medium. The co-expression of several proteins can have a negative impact on the expression of the protein of interest. In this case, an in vitro biotinylation can be performed on the purified product using the biotin protein ligase kit from Avidity. A desalting step is necessary after in vitro biotinylation.
9. We highly recommend generating constructions, with tags at the N or at the C-terminus of the protein. In some cases, the position of the tags can influence protein expression, folding, functionality, but also tag accessibility for purification or biotinylation. By generating both constructs simultaneously, the impact of the tags is quickly identified, and protein is efficiently produced at the same time with the appropriate construction [15].
10. Maximum 1 μg of DNA can be loaded in one well to obtain a nice band. If different DNA samples are loaded on the same gel for purification, each sample must be separated by one empty well to avoid DNA contamination.
11. It is recommended to load in parallel 1 μg of non-digested vector to ensure that the vector was correctly digested.
12. *Nhe*I and *Xba*I are compatible sites; *Mfe*I and *Eco*RI are compatible sites.
13. 20 ng of insert is sufficient for the shorter insert (IRES-EGFP) and 50 ng are necessary for the longer insert (IRES-LsBirA-IRES-EGFP). Appropriate controls must be prepared during

the ligation such as the ligation mix with vector but without insert to obtain the background number of colonies corresponding to empty vector.

14. Untransformed competent cells must be spread on an LB agar dish to check the absence of contamination in the bacteria preparation.

15. PEAK cells are semi-adherent, avoiding the need for agitation and gassing. Therefore, no complicated fermentation system was necessary to produce the antigen. A CO_2 humidified incubator was sufficient for all cell culture.

16. Usually, between 70 and 90 % of the cells are transfected and are fluorescent under a fluorescence microscope or by FACS.

17. In some cases, the cells need a longer time for adaptation under puromycin and for growth to confluence. In this case, the medium can be exchanged before and usually, 5 to 10 days are necessary for selection.

18. The use of a cool-box allows for a slow freezing of cells, avoiding for the crystallization which could destruct cells. Different systems are available, containing either alcohol or a solid core. They provide a consistent and controlled decrease of the temperature at a rate of –1 °C per minute.

19. Usually, one flask T175 with cells at confluence is used to inoculate one bioreactor CL1000.

20. For a new CELLine bioreactor, it is necessary to humidify the membrane on both sides (cell and medium compartments) with serum-free medium and serum-containing medium in the lower and the upper compartment, respectively, before adding the cell suspension. No air should remain in the lower compartment after cell suspension inoculation; air bubbles can be eliminated by pipetting and tilting the CELLine. Additional information about the use of the CELLine bioreactor can be found at http://www.integra-biosciences.com/sites/celline_lit_e.html.

21. Due to its small size (244 Da), the biotin is added in the upper medium compartment and can diffuse across the CELLine membrane (cutoff 10 kDa).

22. We use a peristaltic pump for loading and washing steps as it allows for increased flexibility (for instance, several columns can be loaded in parallel). However, all purification steps (loading, washing, and elution) can also be performed on the ÄKTA system.

23. The flow-through must be stored until the end of the process in case of problems or if the capacity of the column is exceeded.

24. Usually, 3 to 5 column volumes are necessary to wash the resin after sample loading. The column must get back to its original blue color after washing.

25. To avoid insertion of air bubbles in the column and tubing during connection of the column to the ÄKTA, add binding/washing buffer on the top of the column before connecting the tubing. Artifacts in the elution can be caused by the presence of air bubbles in the system.
26. Fractions can be pooled in two different samples, one containing fractions from the high absorption region of the peak and one corresponding to the fractions of lower concentration (lower OD), as shown in Fig. 5b.
27. The desalting step can also be performed by gravity using gel filtration columns such as PD-10, NAP-10 or NAP-5 from GE Healthcare.
28. 1 ug of purified protein loaded per well of a NuPAGE gel is sufficient to obtain a clear band on the gel. To perform semi-quantification on gel, different amounts (5 μg to 0.1 μg) of a reference protein can be loaded in parallel. The concentration of the sample can be estimated by comparing the band intensities of the standards and the samples.
29. NuPAGE pre-casted gels are used for protein electrophoresis. After installation of the gel in the electrophoresis chamber, fill the buffer compartment with MES buffer, remove the comb, and wash wells by pipetting MES buffer into wells.
30. The staining can be shorter, for some proteins 2–3 h are sufficient to observe the bands on the gel.
31. The biotinylation of the protein can be verified by an ELISA. However, the result will confirm that biotinylated protein is present in the final sample but doesn't give information about the amount of biotinylated protein. The final product can contain biotinylated and non-biotinylated molecules. To evaluate the percentage of biotinylation, a pull-down assay using streptavidin-coated magnetic particles must be performed.
32. StreptaWell plates are pre-blocked, and no additional blocking step is required.
33. The concentration of the primary antibody can be adjusted according to the antibody used.

References

1. Takatsuka S, Sekiguchi A, Tokunaga M, Fujimoto A, Chiba J (2011) Generation of a panel of monoclonal antibodies against atypical chemokine receptor CCX-CKR by DNA immunization. J Pharmacol Toxicol Methods 63:250–257
2. Zhao Q, Zhu J, Zhu W, Li X, Tao Y, Lv X, Wang X, Yin J, He C, Ren X (2013) A monoclonal antibody against transmissible gastroenteritis virus generated via immunization of a DNA plasmid bearing TGEV S1 gene. Monoclon Antib Immunodiagn Immunother 32:50–54
3. Roivainen M, Alakulppi N, Ylipaasto P, Eskelinen M, Paananen A, Airaksinen A, Hovi T (2005) A whole cell immunization-derived monoclonal antibody that protects cells from coxsackievirus A9 infection binds to both cell surface and virions. J Virol Methods 130: 108–116

4. Brooks SA (2006) Protein glycosylation in diverse cell systems: implications for modification and analysis of recombinant proteins. Expert Rev Proteomics 3:345–359
5. Esposito D, Chatterjee DK (2006) Enhancement of soluble protein expression through the use of fusion tags. Curr Opin Biotechnol 17:353–358
6. Li W, Gao M, Liu W, Kong Y, Tian H, Yao W, Gao X (2012) Optimized soluble expression and purification of an aggregation-prone protein by fusion tag systems and on-column cleavage in Escherichia coli. Protein Pept, Lett
7. Peroutka Iii RJ, Orcutt SJ, Strickler JE, Butt TR (2011) SUMO fusion technology for enhanced protein expression and purification in prokaryotes and eukaryotes. Methods Mol Biol 705:15–30
8. Qi Y, Zou Z, Zou H, Fan Y, Zhang C (2011) [Intrinsic prokaryotic promoter activity of SUMO gene and its applications in the protein expression system of Escherichia coli. Sheng Wu Gong Cheng Xue Bao 27:952–962
9. Sun P, Tropea JE, Waugh DS (2011) Enhancing the solubility of recombinant proteins in Escherichia coli by using hexahistidine-tagged maltose-binding protein as a fusion partner. Methods Mol Biol 705:259–274
10. Tirat A, Freuler F, Stettler T, Mayr LM, Leder L (2006) Evaluation of two novel tag-based labelling technologies for site-specific modification of proteins. Int J Biol Macromol 39:66–76
11. Magistrelli G, Malinge P, Lissilaa R, Fagete S, Guilhot F, Moine V, Buatois V, Delneste Y, Kellenberger S, Gueneau F, Ravn U, Kosco-Vilbois M, Fischer N (2010) Rapid, simple and high yield production of recombinant proteins in mammalian cells using a versatile episomal system. Protein Expr Purif 72:209–216
12. Sung K, Maloney MT, Yang J, Wu C (2011) A novel method for producing mono-biotinylated, biologically active neurotrophic factors: an essential reagent for single molecule study of axonal transport. J Neurosci Methods 200:121–128
13. Magistrelli G, Malinge P, Anceriz N, Desmurs M, Venet S, Calloud S, Daubeuf B, Kosco-Vilbois M, Fischer N (2012) Robust recombinant FcRn production in mammalian cells enabling oriented immobilization for IgG binding studies. J Immunol Methods 375:20–29
14. Bruce MP, Boyd V, Duch C, White JR (2002) Dialysis-based bioreactor systems for the production of monoclonal antibodies–alternatives to ascites production in mice. J Immunol Methods 264:59–68
15. Kimura A, Adachi N, Horikoshi M (2003) Series of vectors to evaluate the position and the order of two different affinity tags for purification of protein complexes. Anal Biochem 314:253–259

Chapter 2

Method for the Generation of Antibodies Specific for Site and Posttranslational Modifications

Hidemasa Goto and Masaki Inagaki

Abstract

Protein phosphorylation plays critical roles in multiple aspects of cellular events. Site- and phosphorylation state-specific antibodies are indispensable to analyze spatially and temporally distribution of protein phosphorylation in cells. Such information provides some clues of its biological function. Here, we describe a strategy to design a phosphopeptide as an antigen for a site- and phosphorylation state-specific antibody. Importantly, this strategy is also applicable to the production of other types of antibodies, which specifically recognize the site-specific modification, such as acetylation, methylation, and proteolysis. This protocol also focuses on the screening for monoclonal version of a site- and phosphorylation state-specific antibody.

Key words Phosphopeptide, Site- and phosphorylation state-specific antibody, ELISA, Immunoblotting, Immunocytochemistry

1 Introduction

Protein phosphorylation is known to change the affinity of a protein towards its interacting partner, its enzymatic activity, or its subcellular localization. Such functional changes of proteins by phosphorylation are implicated in multiple aspects of cellular events such as signal transduction [1, 2], cell cycle progression/checkpoint [3, 4], cytoskeletal rearrangements [5], etc.

In order to investigate the biological role(s) of protein phosphorylation in cells, it is of great importance to analyze the cellular protein phosphorylation. In the past, labeling of cells with radioactive phosphate was a widely used strategy to monitor in vivo phosphorylation of proteins. However, there are several problems related to radiation exposure in this method. Recently, mass spectrometry (MS) analysis is a convenient alternative to the above method because it does not require radioisotopes. However, these strategies have two major difficulties. First, it is impossible to obtain clear images of the spatial and temporal distribution of

Vincent Ossipow and Nicolas Fischer (eds.), *Monoclonal Antibodies: Methods and Protocols*, Methods in Molecular Biology, vol. 1131, DOI 10.1007/978-1-62703-992-5_2,

protein phosphorylation in cells, since these analyses require cell lysis. Second, in order to analyze the site-specific protein phosphorylation, these methods require many steps, such as the purification of a protein of interest from cells, the fragmentation of the protein by a protease [6], etc.

In 1983, Sternberger's group reported that a subset of their neuron-specific monoclonal antibodies recognized specifically phosphorylated forms of proteins but not non-phosphorylated forms [7]. This study also demonstrated that use of such an antibody in immunocytochemistry could lead to visualization of the intracellular distribution of protein phosphorylation [7]. However, for generating an antibody that can recognize a protein phosphorylated specifically at targeted residue(s), immunizing with a phosphorylated whole protein has little chance of being successful.

To overcome this difficulty, we immunize rabbits or mice with a phosphorylated peptide corresponding to a target phosphorylated residue and its surrounding sequence of amino acids in 1990. This method, which we first established [8–10], has both a greater chance of obtaining a phosphoepitope-specific antibody and the advantage that one can predesign a targeted phosphorylation site(s) [11–14]. By 1994, several companies were able to synthesize a phosphopeptide chemically. This chemical production enables us to use this technology more easily. Our method with a modified peptide has also been applied to the production of antibodies that can specifically recognize the other types of site-specific protein modification, such as acetylation [15] and methylation [16]. In this protocol, we describe not only the strategy to design a phosphopeptide as an antigen but also the screening for monoclonal version of a site- and phosphorylation state-specific antibody.

2 Materials

2.1 Antigen Production

1. Synthetic peptides: we usually design phosphorylated and non-phosphorylated versions of peptides to contain targeted residue(s) [phosphorylation site(s)] and the flanking 5 amino acids at both sides, because 5 or 6 amino acid residues are considered to constitute an antigen epitope recognized by an antibody molecule. In order to conjugate it to a carrier protein such as keyhole limpet hemocyanin (KLH), we usually introduce a Cys (C) residue at the amino-terminal side of the synthetic peptide. As an example, we show synthetic peptides for the production of an antibody against phosphoSer296 or phosphoSer345 on Chk1 (Fig. 1). Some variations are allowable in amino acid length from a phosphorylation site [13, 14]. However, we do not recommend that you use peptides containing over 10 amino acids at either side of a phosphorylation

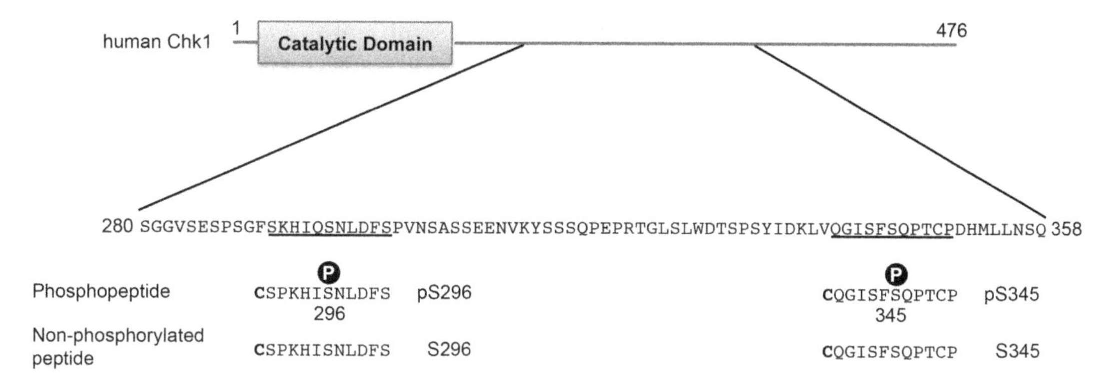

Fig. 1 Design of synthetic peptides for the production of each anti-phospho-specific antibody. We show a set of synthetic peptides in the case of production of an antibody against phosphoSer296 or phosphoSer345 on Chk1. Each peptide sequence corresponding to human Chk1 is underlined. Additional Cys residue in the amino-terminal site is indicated as a *bold letter*. Each phosphate group in phosphoSer is also indicated as P within a *circle*

site, because such peptides may elevate the possibility of the production of antibodies against non-phosphorylated epitope. The above method in the peptide design can be also applicable to the production of an antibody that can specifically recognize the other types of site-specific protein modification, such as acetylation [15] and methylation [16]. Nowadays, many companies can perform not only the synthesis of non-modified peptides but also that of peptides including phosphorylated, acetylated, and/or methylated amino acid(s). We usually order 15 mg non-modified peptide and 25 mg phosphopeptide from Peptide Institute Inc. (Osaka, Japan; *see* **Note 1**).

2. Phosphopeptide-conjugated carrier protein: Many companies also perform the conjugation of phosphopeptide to carrier protein such as KLH. On ordering peptides, we usually request Peptide Institute to conjugate 5 mg of phosphopeptide to KLH. We also request to leave the peptide-conjugated KLH aqueous, because lyophilized KLH is difficult to be dissolved in the aqueous buffer such as phosphate-buffered saline (PBS). Store in aliquots at −80 °C before use.

2.2 ELISA

1. Peptide dilution buffer: 0.1 M $Na_2HPO_4 \cdot NaH_2PO_4$ (pH 7.4).
2. ELISA blocking buffer: 10 mM $Na_2HPO_4 \cdot NaH_2PO_4$ (pH 8.0), 5 % (w/v) BSA, 5 % (w/v) sucrose, 0.1 % NaN_3.
3. ELISA second antibody solution: HRP-conjugated anti-mouse or -rat IgG (Life Technologies, Gaithersburg, MD) diluted at 1: 1,000 in 10 mM $Na_2HPO_4 \cdot NaH_2PO_4$ (pH 8.0), 100 mM NaCl, 1 % (w/v) BSA, 0.1 % (w/v) p-hydroxy phenylacetic acid, 0.025 % (w/v) thimerosal.

4. ELISA Reaction buffer: 0.4 mg/ml o-phenylenediamine, 5 % (v/v) methanol, 0.01 % (v/v) H_2O_2 (*see* **Note 2**).
5. 2 N H_2SO_4.
6. ELISA 96-well plates (Nunc-Immuno Plates Maxisorb, Thermo Fisher Scientific, Inc., Roskilde, Denmark).

2.3 Immunoblotting

1. IB transfer buffer: 25 mM Tris, 192 mM Glycine, 20 % (v/v) methanol.
2. Tris buffered saline containing Tween-20 (TBS-T): 20 mM Tris–HCl (pH 7.6), 150 mM NaCl, 0.1 % (v/v) Tween-20.
3. IB blocking solution: 5 % (w/v) skim milk in TBS-T.
4. IB second antibody solution: HRP-conjugated appropriate secondary antibody (Life Technologies) diluted at 1: 20,000 in TBS-T.
5. PVDF membrane (Immobilon P, Millipore, Bedford, MA).
6. Extra Thick Blot Paper (Bio-Rad, Hercules, CA).
7. Trans-blot® SD semi-dry transfer cell (Bio-Rad).
8. Can Get Signal™ (Toyobo, Osaka, Japan).
9. Electrochemiluminescence detection liquid (ECL; Life Technologies).
10. FUNA-UV-LINKER (Funakoshi, Tokyo, Japan).

2.4 Immunocytochemistry

1. PHEM: 60 mM Pipes, 25 mM HEPES, 10 mM EGTA, 4 mM $MgSO_4$ (adjust pH 7.0 with KOH) [17].
2. IF blocking buffer: 5 % heat-inactivated donkey serum in PHEM (*see* **Note 3**).
3. Mouse monoclonal anti-Chkl antibody (Clone G4, Santa Cruz Biotechnology, Santa Cruz, CA).
4. IF second antibody solution: Alexa Fluor 488-conjugated donkey anti-mouse or -rat IgG (Invitrogen, Carlsbad, CA) diluted at 1: 1,000 and 0.5 μg/ml of 4′, 6-diamidine-2′-phenylindole-dihydrochloride (DAPI; Dojindo Laboratories, Kumamoto, Japan) in IF blocking buffer.

3 Methods

3.1 Immunization and Hybridoma Production

For each immunization, we use 100 μg of the conjugated protein (containing approximately 20 μg of synthetic peptide) per a mouse or a rat. Animal immunization, cell fusion, HAT selection, and limiting dilution for the establishment of monoclones are performed according to ordinary protocols for monoclonal antibody production (*see* other Chapters).

3.2 First Screening for Monoclonal Version of a Site and Phosphorylation State-Specific Antibody by ELISA

1. Dilute phosphorylated or non-phosphorylated version of peptide to 0.3 μg/ml with Peptide dilution buffer (*see* **Note 4**).
2. Add 50 μl of the peptide solution into each well of ELISA 96-well plates.
3. Incubate for 2 h at room temperature (RT) or overnight at 4 °C.
4. Remove the peptide solution from each well and wash with 100 μl of PBS/well three times.
5. Add 300 μl of ELISA blocking buffer into each well.
6. Incubate for 4 h at 37 °C or overnight at 4 °C.
7. Remove ELISA blocking buffer (*see* **Note 5**).
8. 5–9 days after cell fusion (*see* **Note 6**), collect each culture supernatant from each well in the 96-well culture dish. After the collection, add fresh growing medium into each well.
9. Use 2 types of ELISA plates per each supernatant; one is coated with phosphopeptide and the other with non-phosphorylated version of peptide as described above. Add 50 μl of each culture supernatant per well and then incubate for 1 h at 37 °C.
10. Remove the supernatant and then wash with 100 μl of PBS 5 times.
11. Add 100 μl of ELISA second antibody solution into each well and then incubate for 1 h at 37 °C.
12. Remove the supernatant and then wash with 200 μl of PBS 5 times.
13. Add 100 μl of ELISA reaction buffer (*see* **Note 2**) into each well and incubate for 1 h at RT.
14. Stop the reaction by the adding 100 μl of 2 N H_2SO_4 and measure the absorbance at 492 nm with an ELISA plate reader.
15. Pick up and propagate each desirable monoclone (*see* **Note 7**), and then collect additional culture supernatant (0.5–1 ml) for the second and third screening.

3.3 Second Screening by Immunoblotting

1. Prepare positive and negative control samples for SDS-PAGE. In the case of a monoclonal antibody against Chk1 phosphorylated at Ser296 [18], we prepared lysates of HeLa cells irradiated with (positive control) or without (negative control) 254-nm ultraviolet (UV) light (*see* **Note 8**).
2. Load a set of positive and negative control samples per each supernatant (*see* **Notes 9** and **10**). For the cut of transferred membrane, load pre-stained protein marker between sets of samples.
3. After SDS-PAGE, soak the gel in IB transfer buffer for 10 min at RT with rocking.

4. Wet PVDF membrane with methanol and transfer it to IB transfer buffer.
5. Soak an Extra Thick Blot Paper with IB transfer buffer. Place it in the lower electrode (anode) of the Bio-Rad Trans-blot® SD semi-dry transfer cell.
6. Place the PVDF membrane and then the gel on the top.
7. Soak another an Extra Thick Blot Paper with IB transfer buffer, put it on top, and mount the cathode.
8. Transfer for 1 h at RT. Set voltage to 25 V (actual starting voltage is around 6–7 V) and limit current to 1.5–2 mA/cm^2 gel area (*see* **Note 11**).
9. Soak the transferred membrane in IB blocking solution for 1 h or over night at RT with rocking.
10. Wash the membranes 3 times in TBS-T at RT with rocking, briefly. Then, cut the membrane on lanes of the pre-stained marker (*see* **Note 12**).
11. Soak a piece of membrane in each culture supernatant diluted at 1: 100 in TBS-T (*see* **Note 13**) at RT with rocking.
12. Wash the membrane twice (for 10 min each) in TBS-T at RT with rocking.
13. Soak the membranes in IB second antibody solution (*see* **Note 13**) for 30 min at RT with rocking.
14. Wash the membranes twice (for 10 min each) in TBS-T at RT with rocking.
15. For chemiluminescent detection, incubate the membranes in electrochemiluminescence detection liquid (equal volume of detection reagents 1 and 2; *see* **Note 14**) for 1 min.
16. Place the membranes in a film cartridge between two overhead transparency sheets and expose an X-ray film to the membrane (*see* **Note 15**).
17. Select desirable culture supernatant(s) which immunoreact specifically with band(s) corresponding to the position of a protein of interest in the positive control sample but not in the negative control sample (Fig. 2).
18. Pick up and propagate each desirable monoclone (*see* **Note 16**).

3.4 Third Screening by Immunocytochemistry

1. Place the sterilized coverslip in a suitable tissue culture dish. Plate the cell suspension into the above dish (or chamber slide) and culture for at least 24 h (*see* **Note 17**).
2. To fix cells, transfer coverslips into wells of 3.7 % ice-cold formaldehyde in PBS (*see* **Note 18**). Incubate cells for 10 min.
3. Wash in ice-cold PBS, twice for 10 min each.
4. Permeabilize cell membrane by removing PBS and applying ice-cold 0.1 % Triton X-100 in PBS (*see* **Note 19**) for 10 min.

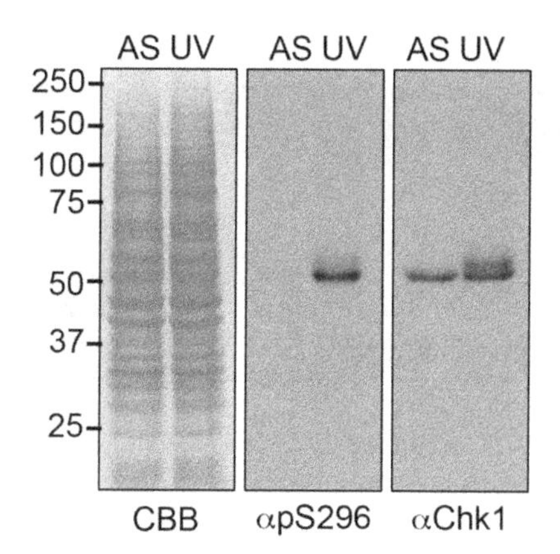

AS: non-treated, asynchronous HeLa cells (Negative Control)
UV: HeLa cells irradiated with UV light (Positive Control)

Fig. 2 The second screening by immunoblotting in the case of an antibody against phosphoSer296 on Chk1 (αpS296). As a positive control, we treated HeLa cells as follows. The culture medium was removed and the cells were irradiated in uncover tissue culture dishes with 254-nm ultraviolet (UV) light at a dose of 10 J/m^2. Fresh culture medium was added back and the cells were incubated for an additional 2 h. We used non-treated (asynchronous; AS) HeLa cells as a negative control

5. Wash in ice-cold PBS, twice for 10 min each.
6. Transfer each coverslip into humidified chamber. Block non-specific binding sites by applying 50 μl of IF blocking buffer per one 13-mm coverslip (*see* **Note 20**). Incubate for 30–60 min at RT.
7. Transfer coverslips into wells of PBS and then into humidified chamber. Apply 50 μl of each culture supernatant per one 13-mm coverslip (*see* **Note 20**). Incubate for 1 h at RT.
8. Transfer coverslips into wells of PBS and then wash in PBS, twice for 10 min each.
9. Transfer each coverslip into humidified chamber and immediately apply 50 μl of IF second antibody solution per one 13-mm coverslip (*see* **Note 20**). Incubate for 15 min at RT in the dark.
10. Transfer coverslips into wells of PBS and then wash in the dark, twice for 10 min each.
11. Gently dip the coverslip into deionized water to rinse the PBS salt from the sample.
12. Mount the coverslip and then analyze using fluorescent microscopy.
13. Select culture supernatant(s) showing desired specificity in all the screening. After the selection, we recommend co-staining with an antibody against a protein of interest if possible.

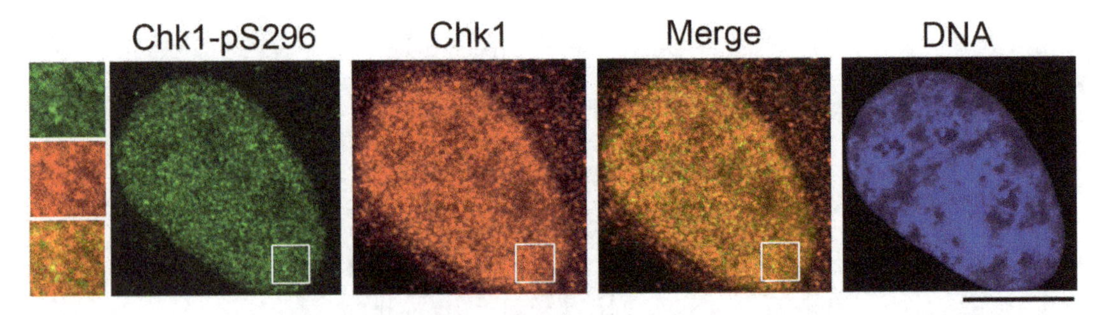

Fig. 3 Immunostaining of UV-irradiated HeLa cells with a rat monoclonal antibody αpS296 (a site and phosphorylation state-specific antibody for Ser296 on Chk1) and a mouse monoclonal anti-Chk1 antibody. HeLa cells were irradiated with UV light as described in the legend of Fig. 2. The immunoreactivity with αpS296 (*green*) and anti-Chk1 (*red*) was visualized through the incubation with Alexa Fluor 488-conjugated donkey anti-rat IgG and Alexa Fluor 555-conjugated donkey anti-mouse IgG, respectively. The nucleus was also stained with DAPI. Magnifications of insets are shown in the *left*. We observed no αpS296-immunoreactivity in non-treated cells (not depicted)

The specificity of a phospho-specific antibody is strongly supported by the colocalization of signals of both antibodies (Fig. 3).

14. Store each desirable monoclone in liquid nitrogen. Collect each culture supernatant enough to carry out any additional assays and then freeze it in aliquots.

4 Notes

1. The usage of highly pure peptide (over 90 % purity) is the key to the production of excellent antibodies.
2. The solution should be freshly prepared.
3. Use the serum from the same organism from which you buy the secondary antibody.
4. The pH of the peptide solution should be 7.0–7.5. If not, use 50 mM $Na_2CO_3 \cdot NaHCO_3$ (pH 9.0) for dilution.
5. If the plates are not immediately used, leave the plates to air-dry for 5–10 min and then store them at 4 °C. In most cases, they can be stored at 4 °C for years.
6. ELISA assay should be performed before hybridomas in any well start to die.
7. Positive clones are defined as their supernatants showing at least 0.3 higher absorbance unit in phosphopeptide-coated plate than in its non-phosphorylated version.

8. Purified protein of interest (e.g., recombinant protein, immunoprecipitated protein) can be alternatively used for the screening. The protein for positive control should be phosphorylated at a target site in vitro or in vivo. Use a non-phosphorylated (dephosphorylated) protein or a protein mutated at a target phosphorylation site to Ala/Val/Phe as a negative control.
9. A protein of interest in the positive and negative control samples should be loaded in the same amounts. The loading amounts should be optimized for each screening. If the specific signal is too weak, increase the loading amounts. However, loading excess amounts may lead to non-specific reaction against other proteins or non-phosphorylated protein of interest.
10. We recommend loading one more set for an antibody against a protein of interest to detect its migrating position on the transferred membrane.
11. The transfer condition should be optimized for each protein. We recommend using an antibody against a protein of interest for this optimization.
12. This procedure should be quick so that the membrane stays moist.
13. Alternatively, dilute the primary and secondary antibodies with the immunoreaction enhancer solution such as Can Get Signal™, instead of TBS-T.
14. The mixture should be enough to cover the membranes completely.
15. We recommend exposing several X-ray films for different periods to determine the optimal exposure time (0.5–5 min exposure is used for most of our assays).
16. If it is difficult to determine desirable clones in the second screening, we recommend performing the third screening (using all ELISA-positive samples) preferentially.
17. Prepare both positive and negative control cells. In the case of a monoclonal antibody against Chk1 phosphorylated at Ser296 [18], we prepared UV-irradiated HeLa cells as a positive control and the non-treated cells as a negative control.
18. Before the third screening, we recommend determining which fixative and permeabilization reagent (see below) are most appropriate for a protein of interest using an antibody against it. Formaldehyde fixation masks some epitopes from recognition of some antibodies. Ice-cold methanol or 10 % trichloroacetic acid (TCA) is an alternative to formaldehyde and tends to preserve epitopes recognized by some antibodies [19]. In the case of methanol fixation, the treatment with permeabilization reagent is not required (see the chapter by Bauer in this book)

19. Various permeabilization reagents (e.g., Triton X-100, Saponin, methanol, acetone, etc.) perforate the cell membrane by different mechanisms.
20. This procedure should be quick so that the coverslips are not dried up. The coverslip should be completely covered with each solution.

Acknowledgements

This work was supported in part by Grants-in-Aid for Scientific Research on Innovative Areas from the Ministry of Education, Culture, Sports, Science and Technology, Japan; by Grants-in-Aid for Scientific Research from Japan Society for the Promotion of Science, Japan; by a Grant-in-aid for the Third Term Comprehensive 10-Year Strategy for Cancer Control from the Ministry of Health, Labour and Welfare, Japan; by the Uehara Memorial Foundation; by the Astellas Foundation for Research on Metabolic Disorders; by the Naito Foundation; by the Daiichi-Sankyo Foundation of Life Science; and by the Takeda Science Foundation.

References

1. Graves JD, Krebs EG (1999) Protein phosphorylation and signal transduction. Pharmacol Ther 82(2–3):111–121
2. Nishizuka Y (1992) Intracellular signaling by hydrolysis of phospholipids and activation of protein kinase C. Science 258(5082):607–614
3. Nurse P (2000) A long twentieth century of the cell cycle and beyond. Cell 100(1):71–78
4. Zhou BB, Elledge SJ (2000) The DNA damage response: putting checkpoints in perspective. Nature 408(6811):433–439
5. Bishop AL, Hall A (2000) Rho GTPases and their effector proteins. Biochem J 348(Pt 2):241–255
6. Boyle WJ, van der Geer P, Hunter T (1991) Phosphopeptide mapping and phosphoamino acid analysis by two-dimensional separation on thin-layer cellulose plates. Methods Enzymol 201:110–149
7. Sternberger LA, Sternberger NH (1983) Monoclonal antibodies distinguish phosphorylated and nonphosphorylated forms of neurofilaments in situ. Proc Natl Acad Sci U S A 80(19):6126–6130
8. Nishizawa K, Yano T, Shibata M, Ando S, Saga S, Takahashi T, Inagaki M (1991) Specific localization of phosphointermediate filament protein in the constricted area of dividing cells. J Biol Chem 266(5):3074–3079
9. Yano T, Taura C, Shibata M, Hirono Y, Ando S, Kusubata M, Takahashi T, Inagaki M (1991) A monoclonal antibody to the phosphorylated form of glial fibrillary acidic protein: application to a non-radioactive method for measuring protein kinase activities. Biochem Biophys Res Commun 175(3):1144–1151
10. Matsuoka Y, Nishizawa K, Yano T, Shibata M, Ando S, Takahashi T, Inagaki M (1992) Two different protein kinases act on a different time schedule as glial filament kinases during mitosis. EMBO J 11(8):2895–2902
11. Nagata K, Izawa I, Inagaki M (2001) A decade of site- and phosphorylation state-specific antibodies: recent advances in studies of spatiotemporal protein phosphorylation. Genes Cells 6(8):653–664
12. Izawa I, Inagaki M (2006) Regulatory mechanisms and functions of intermediate filaments: a study using site- and phosphorylation state-specific antibodies. Cancer Sci 97(3):167–174. doi:10.1111/j.1349-7006.2006.00161.x
13. Goto H, Inagaki M (2007) Production of a site- and phosphorylation state-specific antibody. Nat Protoc 2(10):2574–2581. doi:10.1038/nprot.2007.374
14. Czernik AJ, Girault JA, Nairn AC, Chen J, Snyder G, Kebabian J, Greengard P (1991) Production of phosphorylation state-specific antibodies. Methods Enzymol 201:264–283
15. White DA, Belyaev ND, Turner BM (1999) Preparation of site-specific antibodies to

acetylated histones. Methods San Diego, Calif 19(3):417–424

16. Perez-Burgos L, Peters AH, Opravil S, Kauer M, Mechtler K, Jenuwein T (2004) Generation and characterization of methyl-lysine histone antibodies. Methods Enzymol 376:234–254
17. Howell BJ, McEwen BF, Canman JC, Hoffman DB, Farrar EM, Rieder CL, Salmon ED (2001) Cytoplasmic dynein/dynactin drives kinetochore protein transport to the spindle poles and has a role in mitotic spindle checkpoint inactivation. J Cell Biol 155(7):1159–1172. doi:10.1083/jcb.200105093
18. Kasahara K, Goto H, Enomoto M, Tomono Y, Kiyono T, Inagaki M (2010) 14-3-3gamma mediates Cdc25A proteolysis to block premature mitotic entry after DNA damage. EMBO J 29(16):2802–2812. doi: 10.1038/emboj. 2010.157, emboj 2010157 [pii]
19. Hayashi K, Yonemura S, Matsui T, Tsukita S (1999) Immunofluorescence detection of ezrin/radixin/moesin (ERM) proteins with their carboxyl-terminal threonine phosphorylated in cultured cells and tissues. J Cell Sci 112(Pt 8):1149–1158

Chapter 3

Immunization, Hybridoma Generation, and Selection for Monoclonal Antibody Production

Hyung-Yong Kim, Alexander Stojadinovic, and Mina J. Izadjoo

Abstract

Monoclonal antibodies (MAbs) produced by a single clone of cells with homogeneous binding specificity for an antigenic determinant have been used in diagnostics and therapeutics. Many new methods have been devised by scientists for making hybridomas and MAbs. The three major steps for producing MAbs are immunization, immortalization, and isolation. Here, we describe technical details of the three important steps for generating mouse hybridomas and MAbs.

Key words Hybridomas, Monoclonal antibodies, Antigenic determinant, Binding specificity, Immunization, Immortalization, Isolation

1 Introduction

Monoclonal antibodies (MAbs) have become increasingly important in biomedical research, diagnosis, and biotherapy in the last few decades. Technological progress in generating MAbs has been made since Georges J. F. Köhler (German scientist) and César Milstein (Argentine-British immunologist) first developed a method in 1974 for producing MAb with a single antigen binding specificity [1]. Ten years later, they shared the Nobel Prize in Physiology or Medicine with Niels K. Jerne for their significant contribution to science. Drs. Shulman, Wilde, and Köhler developed a mouse myeloma Sp2/0-Ag14 cell line as a fusion partner for obtaining hybridomas, which produced MAbs of a desired specificity [2]. By fusing lymphocytes of an immunized mouse with mouse myeloma cells, researchers could produce unlimited quantities of MAbs. In this chapter, we present the three critical steps needed for successful generation of mouse hybridomas and MAbs.

The first step of MAb production is stimulation of the animal immune system. This is done using a standard immunization protocol with antigen of choice in BALB/c mice.

Vincent Ossipow and Nicolas Fischer (eds.), *Monoclonal Antibodies: Methods and Protocols*, Methods in Molecular Biology, vol. 1131, DOI 10.1007/978-1-62703-992-5_3, © Springer Science+Business Media New York 2014

Many immunization regimens have been used for the fusion of myeloma cells and immune spleen cells [3]. The antigen is usually mixed with an adjuvant (e.g., Freund's adjuvant) to enhance immune response through slow release of the antigen and longer exposure time of the mouse immune system to the antigens. This injected antigen mixture is slowly released and travels up to the mouse perithymic lymph node, leading to activation of antibody secreting plasma cells.

The second step in MAb production is immortalization. This is achieved in one of two ways: (a) constraints of lymphoid cells and (b) cell transformation. Today, the most common immortalization method utilizes genetic information obtained from transformed or immortalized myeloma cell lines (Sp2/0-Ag14 or P3 × 63-Ag8.653). The transformed cell is fused with antigen-specific lymphoid cells. These myeloma cell lines are of mouse B-lymphoid origin with all the cellular machinery needed for antibody production and secretion [4].

The third and final step of MAb production is isolation of single antibody-producing cells. This is done by elimination of the unfused parent cell types. The parent myeloma cells are eliminated using hypoxanthine, aminopterin, and thymidine (HAT) selection medium [5, 6]. The screening process uses enzyme-linked immunosorbent assay (ELISA) and dot blot analysis [7, 8]. After determining which culture supernatants are of interest, these cells are harvested, diluted, and then replated. Since these clones may contain more than one MAb-secreting hybridoma clone in a well at this point, the cultures of interest must be subcloned by limiting dilution technique, resulting in isolating a MAb clone of interest. About 2–3 months of dilution, plating, and growth will complete the cloning process. Desirable and stable clones can be grown in large quantity for production of sufficient amounts of MAbs. MAb-producing clones are preserved in 90 % fetal bovine serum (FBS) containing 10 % dimethylsulfoxide (DMSO) by freezing in liquid nitrogen for future use. MAbs produced by a single clone of cells, having a homogeneous binding specificity for an antigenic determinant, can be further characterized and used as a specific reagent for diagnostics or therapeutics.

2 Materials

Please note all disposables, materials, and reagents used for MAb production must be sterile. All tissue culture work is conducted under a biosafety cabinet. Extreme care is taken to prevent contamination (e.g., use of disinfectants in water baths, routine disinfection of incubators, centrifuges).

2.1 Immunization

1. Stock pure antigens (soluble or cellular antigens) are dissolved or resuspended in PBS (2.7 mM KCl, 1.8 mM KH_2PO_4, 137 mM NaCl, and 10 mM Na_2HPO_4, pH 7.4) to achieve

desired concentrations. Total protein concentration is measured by bicinchoninic acid (BCA) protein assay (*see* **Note 1**).

2. Freund's complete adjuvant (FCA) and Freund's incomplete adjuvant (FIA) are from Sigma (St. Louis, MO, USA) to boost mouse immune responses.
3. BALB/c mice.

2.2 Mice Anesthetic

Saline-diluted mixture of ketamine (100 mg/ml) at 0.15 mg/kg of body weight, acepromazine (5 mg/ml) at 30 mg/kg, and xylazine (20 mg/ml) at 3.5 mg/kg. The anesthetic mixture is administered at a dose of 0.1 ml/kg of mouse body weight.

2.3 Myeloma Cells, Culture Media, and Others

1. For successful fusion process (*see* **Note 2**), myeloma cells [Sp2/0-Ag14, American Type Culture Collection (ATCC), Manassas, VA] are maintained in complete Dulbecco's modified Eagle's medium (cDMEM) (Invitrogen, Grand Island, NY) supplemented with 10 % FBS and 2 mM L-glutamine. The culture is incubated at 37 °C in a humidified 5 % CO_2–95 % air atmosphere.
2. Sterile tissue grinders (Thermo Fisher Scientific, Inc., Barrington, IL).
3. Pristane (Sigma).
4. Isotyping IsoQuick Kit (Sigma-Aldrich, St. Louis, MO).

3 Methods

3.1 Preparation of Antigens

1. Mix well 100–200 μg antigens in either FCA (first immunization via intraperitoneal route) or FIA (second boost immunization via subcutaneous route) (1:1; v:v) or cellular antigens (approximately 2×10^7 cells per mouse) depending on the immunization schedule (Fig. 1) to enhance mouse immune responses. The antigen is sucked into a syringe and mixed with FCA or FIA in a test tube by pushing the mixture back and forth.
2. The antigen is completely mixed until a thick creamy solution is obtained (*see* **Note 3**).

3.2 Immunization

1. The first step for MAb production is to immunize mice with the antigen. Depending on the antigen type, different immunization routes and schedules are used. Five-week-old BALB/c mice are purchased. The mice are kept in cage at animal isolator system (22 ± 2 °C, humidity 55 ± 10 %) and are fed food and water ad libitum.
2. To restrain and immunize, the mouse can be placed on the wire mesh of the cage lid. This allows the operator to grasp the mouse gently and firmly in one hand, using thumb and forefinger to grasp the tail, and neck. After swabbing injection site with 70 %

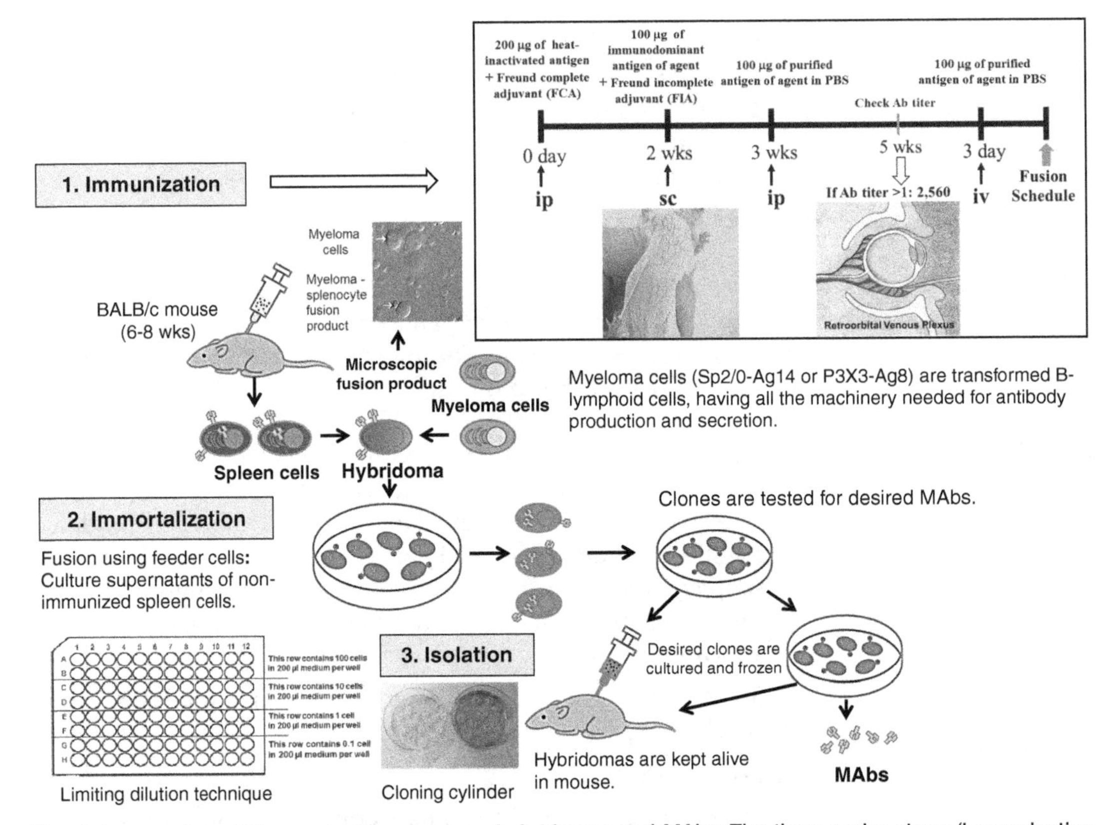

Fig. 1 An overview of the production of mouse hybridomas and MAbs. The three major steps (immunization, immortalization, and isolation) are key challenges for a successful generation of hybridomas and MAbs

alcohol, the antigen (0.2 ml antigen mixture in 200 µl volume in FCA) is injected intraperitoneally.

3. Two weeks after the first immunization, each mouse is boosted subcutaneously with 0.2-ml mixture of equal volumes of 100 µg of the purified antigen in PBS and FIA (*see* **Note 4**).
4. The ear of the mouse can be spotted with a permanent ink pen to identify the mouse that has been immunized.
5. The mice are intraperitoneally injected with a purified antigen in PBS (100 µg/mouse) without adjuvant on the 7th day after the second immunization. Antigen administered intraperitoneally is trapped by the spleen, which then attracts retains all antigen-specific lymphocytes that circulate through it. Thus, at the time of fusion, the majority of antigen-specific lymphocytes are located within spleen.
6. To check antibody titer, 5 weeks after the first inoculation the mice are first anesthetized (*see* **Note 5**), and then a small blood sample is collected from the retro-orbital venous plexus with care and without any harm to the mouse. If repeated blood samples may be needed, restraining the mouse as shown in

Fig. 1 causes the eye to protrude slightly, allowing for insertion of a Pasteur pipette tip behind the eyeball. Gently turning the pipette tip breaks retro-orbital blood vessels, and releases blood flows up and into the pipette. The pipette tip is then drawn. The eyeball returns to the normal position undamaged (*see* **Note 6**).

7. The collected blood is allowed to clot, and serum is removed from top of tube by centrifugation for 10 min at 400 × *g*. The serum is tested for the presence of specific antibodies by ELISA (*see* **Note 7**).
8. Three to four days prior to fusion, the mice are intravenously boosted with purified antigen (100 μg/mouse in 0.2 ml of PBS) without adjuvant. For this injection, the animal is first warmed with hot water for several min to dilate blood vessel.
9. Immersion of the tail into warm water helps soften the hard skin of the tail and also helps further dilate the tail vein. The mouse is restrained in the mouse restraint tube device, which allows the tail to protrude and then injection can be given into the lateral tail vein.
10. The antigen material is slowly injected in to the tail vein using a 26–30 Gauge needle. The dark venous blood is temporarily displaced in the tail vein, indicating that antigen injection has been done correctly.

3.3 Immortalization

1. Five weeks after antigen inoculation, if antibody titers of the mice are adequate (>1:2,560) (*see* **Note 8**), then the mice are euthanized by cervical dislocation prior to cell fusion.
2. The mouse is thoroughly disinfected with ethanol and then placed in the hood on a sterile gauze pad. The spleen, which may be two to three times the normal size, is dissected out using aseptic techniques. The excess fat and connective tissue are trimmed away from the spleen in a sterile dish (60 × 10 mm). The spleen is then teased apart using sterile curved forceps (*see* **Note 9**).
3. After addition of 10 ml of RPMI medium, tissue materials are passed through the sieve. This is rinsed with additional RPMI and added to the suspension, and the remaining tissues are meshed through the sieve with handle of a sterile 5 ml syringe, and the sieve is rinsed with 10 ml of RPMI media.
4. The entire 20 ml of cell suspensions is transferred to a 50 ml conical tube. The spleen cell suspension has red blood cells. Red blood cells do not interfere with fusion process, and therefore, they do not need to be eliminated (*see* **Note 10**).
5. The appropriate volume of each cell type is calculated and cells are pooled in one tube.

6. To fuse spleen cells with myeloma cells (Sp2/0-Ag14), myeloma cells (3.5×10^7) are added to 10^8 of spleen cells in a sterile 15 ml conical centrifuge tube (*see* **Note 11**).
7. The tubes are allowed to sit for 1 min, centrifuged at room temperature at $250 \times g$ for 10 min, and the supernatants are then aspirated. The tube is held in one hand and, the pellets are completely resuspended by flickering the bottom of tubes with index finger of the opposite hand (*see* **Note 12**).
8. One milliliter of 40 % PEG 4,000 containing 7 % DMSO and PBS is gently added to the tube.
9. The tubes are swirled for 2 min. The PEG is slowly diluted by adding 5 ml of warm serum-free DMEM along the side of the tube to stop the fusion process and reduce the concentration of DMSO which can be destructive (*see* **Note 13**).
10. An additional 13 ml of warmed DMEM are added to the tube. The tube is kept mixing while adding medium to bring up the volume to 20 ml, and then let it to sit at room temperature for 30 min.
11. The fusion suspension is gently centrifuged for 5 min. at $250 \times g$, the supernatant is discarded and the pellet is gently dispersed. The fusion suspension is added to 500 ml of cDMEM containing 20 % FBS, 1 % sodium pyruvate, 1 % nonessential amino acid, 2 mM L-glutamine, and freshly reconstituted HAT supplement, diluted 1:100 from stock of 10 mM sodium hypoxanthine, 40 μM aminopterin, and 1.6 mM thymidine.
12. Two milliliters of fusion suspension is aliquoted into each well of ten 24-well culture plates ($2–4 \times 10^5$ spleen cells in each well) containing feeder cells (*see* **Note 14** and Subheading 3.4).
13. Once cultures have grown at 37 °C for 14 days without additional feeding, transfer them to 6-well culture plates with cDMEM containing HAT supplement.
14. Subcultures are gradually weaned off the HAT medium to HT medium and then onto regular tissue culture medium. Therefore, from the third week after fusion, cDMEM containing 20 % FBS and HT supplement is used (*see* **Note 15**).
15. The plates are placed in the incubator at 37 °C in an atmosphere of 5 % CO_2. They are monitored daily for signs of contamination, death of myeloma cells, and growth of hybridomas. This is done using an inverted microscope as well as by checking for the characteristics such as color change in the medium. Microscopic examination of successful fusion at this point shows many large clumps of cells (*see* **Note 16**).

3.4 Preparation of Feeder Cells

1. To condition cDMEM and to stabilize hybridomas, feeder cells (2-day culture supernatants of normal spleen cells from two normal BALB/c mice) at every fusion process are aseptically

prepared in the laminar flow hood 4 days before fusion schedule.

2. The spleens are placed in 15 ml sterile tissue grinders and aseptically homogenized in 2 ml of sterile serum-free DMEM.
3. The cell suspensions are passed through a cell-dispersing screen (100 mesh with 0.14 mm pores) and transferred into a sterile 15 ml centrifuge tube.
4. Serum-free DMEM is continuously poured through the mesh until the final volume is 15 ml.
5. The tube is centrifuged at $250 \times g$ for 5 min at 4 °C, and the supernatants are aspirated aseptically.
6. The pellets are resuspended in 1 ml of serum-free DMEM.
7. Five milliliters of sterile 0.83 % NH_4Cl is added to each pellet to lyse red blood cells.
8. The suspension is kept on ice for 5 min, and 10 ml of cold DMEM is added to the suspension.
9. The tube is centrifuged at $250 \times g$ for 10 min and supernatants are aspirated.
10. Pellets are resuspended with 2 ml of serum-free DMEM.
11. The spleen cells are counted using a hemocytometer, and cell viability is determined by staining using a 0.4 % (w/v) trypan blue.
12. Normal splenocytes (10^5/ml/well) are plated into ten culture plates (24-well) per each fusion.
13. All plates are incubated for 2 days at 37 °C in a humidified incubator at 5 % CO_2.
14. The feeder cells or culture supernatants are used in culturing the initial fusion mixture with diluted (1:2) volumes of the culture supernatants.

3.5 Isolation of MAb-Producing Cells

1. The third and final step of MAb production is to isolate single MAb-producing cells. This is done by first eliminating the unfused parent cells. The parent myeloma cells are eliminated using the HAT selection medium.
2. The HAT selection process is essential to the overall success of the procedure (*see* **Note 17**). Since only 1 % of the original cells are fused, only $1–10 \times 10^5$ cells form viable hybrids. These hybrids are relatively fragile and easily overgrown by parental cells that have not been killed.
3. Single antibody producing hybrid cells are then isolated by limiting dilutions, followed by growth and cell culture of hybrid cells producing a single type of antibody clone. Fusion of cells obtained from one spleen can result in many different clones, antibody-producing cells, which then must be screened

to identify the antibody of interest. Established clones grow very well in HAT medium.

4. After a subculture, they are given HT medium in the place of HAT medium. Actively growing cells produce acid, which changes the pH of the medium causing a dye color change from red to yellow. This is an indication of successful hybridoma production. At this time, hybridoma cells are given normal culture medium and are expected to grow tremendously.
5. The supernatants of growth positive wells are aspirated and screened for the presence of antibodies. This is possibly the most laborious part of the entire procedure. The screening process uses ELISA (*see* **Note 18**).
6. After determining which supernatants are of interest, the corresponding wells are harvested, diluted, and then replated on new plates. It is possible that these wells may contain more than one antibody-secreting hybridoma clone in a well at this point.
7. If cultures are of interest, then these wells must be subcloned by limiting dilution, resulting in individual clones producing MAbs. About 60–80 % of the initial total wells can be expected to yield growing hybridomas. Subcloning may be repeated after counting MAb mixtures. About 2 or 3 months of dilution, plating, and growth will complete the cloning process. Desirable and stable clones can be grown at large quantity for production of sufficient quantities of MAbs.
8. For screening of candidate hybridoma clones, ELISA is performed. Normally, wells of interest (>1.5 at OD450 nm) are used in subcloning of hybridoma via the limiting dilution technique (Fig. 1).
9. Candidate clones are preserved in 90 % FBS containing 10 % DMSO by freezing in liquid nitrogen for future research and analysis.

3.6 Production of Ascites

1. Optionally, ascites can be prepared for larger scale MAb production (note that in several jurisdictions this procedure is banned). For rapid production of ascites containing MAbs, 3–5 days prior to hybridoma injection, 0.5 ml of pristane (Sigma) as an inflammatory agent is injected intraperitoneally into 10 male BALB/c mice (6–8 week) per clone, using 22-G needles.
2. Thereafter, actively growing hybridomas are intraperitoneally inoculated into the mice ($2–3 \times 10^6$ cells/mouse in 0.2 ml sterile PBS).
3. All mice are sacrificed by cervical dislocation 12 days after injection.
4. Ascitic fluid containing MAbs are harvested from the intraperitoneal cavity using a 21-G needle and centrifuged at $10{,}000 \times g$ at 4 °C for 10 min.

5. Ascites are pooled and inactivated at 56 °C for 30 min, and their antibody titers are determined by ELISA (*see* **Note 18**).
6. Ascites containing the MAbs are diluted at 1:10 and filter-sterilized through 0.45 μm filters.
7. To avoid repeated freezing and thawing, 1 ml each of ascites is aliquoted to microfuge tubes and frozen at −20 °C or lower temperature.

3.7 Isotyping of MAbs

Using either ascitic fluids or culture supernatants, isotypes of MAbs are determined by the IsoQuick Kit (Sigma-Aldrich, St. Louis, MO) which is based on a lateral flow immunoassay (LFI) technique.

4 Notes

1. To measure total protein concentration of the pure antigens in PBS, (1) make serial dilution of sample and standard albumin solution (2 mg/ml) in a 96-well plate as follows: (a) dispense 50 μl PBS in well no. 1–6 of the first row of 96-well plate, and dispense 50 μl PBS in well no. 2–4 of the second row, (b) take 50 μl of stock albumin, add first well in the first row of 96-well plate, and dilute standard solution serially to be 1:2-folds, (c) take 100 μl of purified antigen, place in the first well of the second row of this plate, transfer 50 μl to the second well, make a serial dilution of unknown protein suspension as same as standard solution. (2) Make working BCA reagent as recommended by the manufacturer (Normally, use the ratio 1:50, BCA reagent B solution to BCA reagent A solution. (3) Take 5 ml of reagent A to 15 ml centrifuge tube, add 0.1 ml of reagent B, and then mix it well). (4) Place 200 μl of standard working solution into the third and fourth well required for the BCA reaction, and transfer 10 μl of each dilution from the wells, which are corresponding to the first and second row dilution from low concentration to the high concentration. (5) Mix the plate well and incubate for 30 min at 37 °C. (6) Read 96-well plate at OD595 nm wavelength to analyze unknown antigen concentration of purified antigens using curve fitting software. [Alternatively, to quickly measure total protein concentration of bacteria, you can measure OD400 nm (Normally, OD400 nm = 1 means 100 μg/ml by the Beer's Law). Although it is different depending on bacterial morphology, you can estimate the protein concentration very roughly]. Both methods can be used for the measurement of protein concentration. The BCA protein assay is much more accurate than using OD400 nm value.
2. Important criteria for generation of hybridomas are summarized in Table 1. Myeloma cells for successful fusion are chosen due to three main characteristics. First, myeloma cells are

Table 1
Important criteria for successful generation of hybridomas

Immunization: Antigen should be pure; purified antigen in PBS is used
Use of feeder cells: Supernatants of splenocytes from normal mouse
Antibody titer 3 days before fusion schedule: >1:2,560[a]
Ratio of splenocytes and myeloma cells: 3–4:1
PEG concentration in fusion mixtures: 40 % PEG plus 7 % DMSO
Culture plates: 24, 48, and 96 wells
Serum percentage in HAT medium: >15 %

[a]Refer to ref. 9

transformed cells, which have originated from B-lymphoid cells, and thus have all the machinery needed for antibody production and secretion. The reason for this is that if the fused myeloma cell has full complements of chromosomes from both parents, then it is inheritantly unstable "heterokaryon," so that it loses chromosomes randomly, thus duplicated genes of antibody production are of increased likelihood, and at least one complete copy will remain in the hybrid cell. Second, myeloma cells are mutated, so that they do not secrete antibodies of their own. This greatly simplifies the later step of screening and cloning as the only cell secreting antibody will be hybridoma cells. Third, the cell line is also mutated to allow the selective killing of unfused parent myeloma cell, remaining in the fusion mixture. These cells will ordinary grow indefinitely, and possibly overgrow more delicate hybridomas in the early stage of culture. This selection step is essential in recovering a high proportion of hybridized cells [4].

3. If the solution is properly mixed, it should remain as one phase after a 1 h-incubation at 4 °C. If two phases are observed, the solution was not properly homogenized.
4. This route around the rib cage may prevent leakage of antigen materials from the subcutaneous site. This is also safer for the operator. No formation of blood is made as antigen material is injected (Fig. 1).
5. Mice are anesthetized by intramuscular injection.
6. Approximately, 0.5 ml of blood at a time is collected in this way without harm.
7. The convenient assay method is chosen to detect as many different kinds of specific antibody to the antigen as possible. For example, MAbs do not precipitate, or agglutinate antigen, unless multiple identical antigen binding sites are present. If antibody titers are adequate (>1:2,560) by ELISA, the fusion is scheduled (Fig. 1).

8. The antibody titer of mice is tested by ELISA [9] on the 35th day after the first antigen inoculation. The microfuge tubes of blood collected from the retro-orbital venous plexus of two immunized mice with a Pasteur pipette are allowed to sit for 30 min in a cold room, and the tube is centrifuged at $400 \times g$ at 4 °C for 10 min.
9. It is important to prevent excessive cell damage, although more individual splenocytes can be obtained as the spleen is bladed on the dispersing screen mesh.
10. One spleen usually yields about 1×10^8 lymphocytes. Using more spleen cells, resulting in hybridomas producing more than one type of antibodies, may lead to increase fusion yields, while using fewer spleen cells may reduce the number of successful fusion products. The resulting pellet of splenocytes is red due to red blood cells. The RBCs can be removed using a sterile 5 ml of 0.83 % NH_4Cl.
11. Three days before fusion schedule, myeloma cells are thawed and started to culture in bacteriological grade petri dish rather than tissue culture grade to prevent adherence of cells to the plastic. The freshly grown myeloma cells are plated at cell density 5×10^5 in 10 ml of cDMEM containing 20 % FBS, and then are grown in CO_2 atmosphere at 37 °C. They are harvested just before fusion by swirling the plate. The resuspended cells are then pipetted into the suspension into the 50 ml conical tube. The cells are inspected under a microscope for viability and approximating the numbers. Dead cells may have rough membranes that allow penetration of the "trypan blue" dye. The minimum number of viable cells needed for fusion is about 2×10^7. The safe starting point cell count is 3×10^7. The appropriate ratio of myeloma cells and splenocytes is about 1:3 in one tube.
12. The suspension mixture of myeloma cells and splenocytes is centrifuged to bring the two cell types into close proximity. The initial speed is $200 \times g$, which is gradually increased to $300 \times g$ due to the greater fragility of splenocytes.
13. The pellets are gently dispersed over the bottom of the tube to better contact the cell pellet with fusion solution. One milliliter pipette is placed with tip of bottom conical polypropylene tube, and the fusing agents are slowly added by rotating the tube. The fusing agent contains 40 % PEG 4,000 in PBS plus 7 % DMSO, which can increase membrane permeability.
14. To condition the media and stabilize the growth of hybridoma cells, non-immunized spleen cells are incubated for 2 days to produce "feeder layers," which conditions the media and stabilizes the growth of hybridomas. About half of feeder cells are aspirated out of each well, and then 1 ml of fusion suspension

is pipetted into each well. Numbers of plates are handled in the same order each time, making it easier to identify any source of problems, such as contamination. For successful fusion, if more spleen cells are used, hybridomas producing more than one type of antibodies can be cloned; this may lead to increased fusion yields. On the contrary, using fewer spleen cells may reduce the number of successful fusion products. The appropriate ratio of myeloma cell to splenocytes is about 1:3.

15. The lymphocyte bud is incorporated into the larger myeloma cells. Fusion of nuclei will begin producing unstable heterokaryon. These cells will gradually stabilize with a lot of chromosomes, until stable hybrid cells are produced (Fig. 1). HT medium helps dilute out aminopterin (A), since normal cells have some toxicity to it, and cultures are fed again at 14 days post fusion after aspiration of about 1.5 ml of medium in each well of 24-well culture plates.
16. Myeloma–splenocyte fusion products can be seen with apparent morphological similarity to budding yeast in a very short time.
17. The myeloma cell selection process is based on the fact that there are two pathways of nucleic acid synthesis: De novo and *Salvage* pathway. The chemical "aminopterin" called A blocks the de novo pathway by interfering with folate metabolism. Normal cells will grow in the presence of A by the salvage pathway as long as purine and pyrimidine substrates, hypoxanthine called "H" and thymidine called "T", are supplied. Cells lacking essential enzymes for the salvage pathway will die. The myeloma cell line used here lacks necessary salvage pathway enzymes, hypoxanthine guanine phosphoribosyl transferase, called HGPRT, thus they will die in the presence of A. If the myeloma cell is successfully fused with the normal cell containing the HGPRT gene, then the fusion product will grow in the presence of A as long as H and T are supplied. These substances H, A, and T are supplied in the selective growth medium, called HAT used after fusion. Growing cells produce acid, which changes the pH of medium, thereby causing a change in the medium color from red to yellow [6].
18. For ELISA, 96-well plates are coated with purified antigen at 200 ng of protein per well in 50 mM Na-bicarbonate buffer (pH 9.6) for 2 h. The wells are washed three times with PBS containing 5 % non-fat dry milk and blocked with 5 % non-fat dry milk in the same buffer for 1 h. One hundred microliters of undiluted hybridoma supernatant is added to each well, including 100 μl each of both the positive and negative (normal BALB/c mouse serum) controls at 1:10 dilutions in PBS containing 5 % non-fat dry milk. All the plates are incubated at 37 °C for 1 h and rinsed three times with PBS containing 0.05 % Tween 20. Peroxidase-conjugated goat anti-mouse

immunoglobulins (IgG + IgA + IgM) at a 1:1,000 dilution in PBS containing 5 % non-fat dry milk are added and incubated at 37 °C for 1 h. The wells are washed three times with PBS containing 5 % non-fat dry milk and substrate, ABTS (KPL) is added and incubated at room temperature for 5 min. Finally, all hybridomas identified as positive clones by ELISA are subcultured into 6-well culture plates. When cells are confluent, these clones are frozen in 90 % FBS and 10 % DMSO at −80 °C. The wells identified as positive clones by primary screening are subcloned three times via a limiting dilution technique [9].

Acknowledgment

We thank Tommy Kim for editorial assistance and Nakita Devenish for her drawing.

Disclaimer: The views expressed in this manuscript are those of the authors and do not reflect the official policy of the Department of the Army, the Department of Defense, the United States Government or the Henry M Jackson Foundation for the Advancement of Military Medicine.

References

1. Köhler G, Milstein C (1975) Continuous cultures of fused cells secreting antibody of predefined specificity. Nature 256:495–497
2. Shulman M, Wilde CD, Köhler G (1978) A better cell line for making hybridomas secreting specific antibodies. Nature 276:269–270
3. Gustafsson B (1990) Immunizing schedules for hybridoma production. Methods Mol Biol 5: 597–599
4. Wood JN (1984) Immunization and fusion protocols for hybridoma production. Methods Mol Biol 1:261–270
5. Davis JM, Pennington JE, Kubler AM, Conscience JF (1982) A simple, single-step technique for selecting and cloning hybridomas for the production of monoclonal antibodies. J Immunol Methods 50:161–171
6. Goding JW (1996) Monoclonal antibodies: principles and practice. Academic, San Diego, CA, pp 116–191
7. Noteboom WD, Knurr KE, Kim HS, Richmond WG, Martin AP, Vorbeck ML (1984) An ELISA for screening hybridoma cultures for monoclonal antibodies against a detergent solubilized integral membrane protein. J Immunol Methods 75:141–148
8. Stya M, Wahl R, Beierwaltes WH (1984) Dot-based ELISA and RIA: two rapid assays that screen hybridoma supernatants against whole live cells. J Immunol Methods 73:75–81
9. Kim H-Y, Rikihisa Y (1998) Characterization of monoclonal antibodies to the 44-kilodalton major outer membrane protein of the human granulocytic ehrlichiosis agent. J Clin Microbiol 36:3278–3284

Chapter 4

Hybridoma Technology for the Generation of Rodent mAbs via Classical Fusion

Efthalia Chronopoulou, Alejandro Uribe-Benninghoff, Cindi R. Corbett, and Jody D. Berry

Abstract

Monoclonal antibodies (mAbs) have proven to be instrumental in the advancement of research, diagnostic, industrial vaccine, and therapeutic applications. The use of mAbs in laboratory protocols has been growing in an exponential fashion for the last four decades. Described herein are methods for the development of highly specific mAbs through traditional hybridoma fusion. For ultimate success, a series of simultaneously initiated protocols are to be undertaken with careful attention to cell health of both the myeloma fusion partner and immune splenocytes. Coordination and attention to detail will enable a researcher with basic tissue culture skills to generate mAbs from immunized rodents to a variety of antigens (including proteins, carbohydrates, DNA, and haptens) (*see* **Note 1**). Furthermore, in vivo and in vitro methods used for antigen sensitization of splenocytes prior to somatic fusion are described herein.

Key words Monoclonal antibody, Hybridoma, Fusion, Myeloma, Immunogen, Antigen, Adjuvants

1 Introduction

This chapter describes procedures used for the somatic cell fusion of immune rodent (mouse, rat, hamster) antigen-sensitized splenocytes to myeloma cells (B cell tumor cells). The fusion of these two cell types result in the generation of antibody secreting hybridoma cells. Such hybridomas can proliferate indefinitely in tissue culture and upon cell cloning and expansion produce large quantities of antibodies of the same specificity (monoclonal antibodies, mAbs). Somatic cell fusion can be accomplished using a variety of techniques such as chemical (polyethylene glycol), viral (Sendai virus), or electrofusion [1, 2], of the cell membranes of splenocyte and myeloma fusion partner.

A variety of methods have been used for animal inoculation aiming to generate in vivo antigen-primed splenocytes for cell fusion. The methods may differ in the type and amount of antigen used, adjuvant, route of antigen administration, and immunization schedule [3–6].

Vincent Ossipow and Nicolas Fischer (eds.), *Monoclonal Antibodies: Methods and Protocols*, Methods in Molecular Biology, vol. 1131, DOI 10.1007/978-1-62703-992-5_4,

In cases where the antigen availability is limited or is highly conserved across species (therefore posing challenges for in vivo inoculation) alternative approaches are possible. For example, in vitro boosting of primed splenocytes via short term stimulation with a minute amount of specific antigen [7] or genetic knockouts can be made to help overcome tolerance for conserved antigens. The immunization schedules outlined herein should be used as the method of choice for splenocyte sensitization prior to fusion for the majority of target immunogens (*see* **Note 2**).

1.1 Immunogens

Immunogens are substances that when introduced into the body generate an immune response. Those used for animal inoculation in antibody development include proteins (native or recombinant), peptides (map peptides or linear peptides conjugated to a protein carrier), carbohydrates, complex lipids, DNA, haptens, whole cells and cell extracts. The choice of immunogen preparation used for animal inoculation depends on the downstream applications of the target antibodies (i.e., reactivity to intracellular or surface markers, proteins, or lipids, etc.). Other factors affecting the choice of preparation for animal inoculation are the size and complexity of the target molecule which in turn affect its immunogenicity (ability of an antigen to induce an immune response). For small immunogens (<10 kDa) it is possible to increase their immunogenicity by conjugating them to large immunogenic carrier proteins such as keyhole limpet hemocyanin (KLH), bovine serum albumin (BSA), or albumin [5] (*see* **Notes 3** and **4**). One factor that should be considered when choosing immunogens for inoculation is the application the researcher would like to develop the antigen specific antibody for, as the immunogen preparation could affect the antibody's performance in various applications. For example, if the researcher would like to produce a neutralizing mAb, the antigen used for animal inoculation should be in its native conformation. Alternatively, if the researcher were interested in producing an antibody to detect a molecule in Western Immunoblot or immunohistochemistry applications, the immunogen should be used in its denatured form [8].

1.2 Antigens

Antigens are the corresponding molecule used for the screening and identification of mAbs generated against the immunogen. In some cases the immunogen and the antigen are identical. In other cases the immunogen is the antigen conjugated to a carrier protein or helper epitopes via chemical conjugation or molecular techniques. The antigen can also be a sub fragment of the immunogen. For example, when whole virus is used as the immunogen and recombinant viral glycoprotein is used as the antigen in screening. It is not uncommon to use multiple antigens during different types of screening.

1.3 Adjuvants

Adjuvants are agents known for their ability to promote or augment an immune response and are regularly used in conjunction with immunogens for in vivo inoculation. Adjuvants commonly used are Complete Freund's Adjuvant (CFA) (*Mycobacterium bovis* or attenuated *M. tuberculosis* H37Ra), TiterMax®, Ribi (MPL® + TDM), alum, and Gerbu adjuvants. Selection of the adjuvant depends on the immunogen used, the route of inoculation, and the level and type of inflammatory response the researcher would like to induce [5, 9].

1.4 Hybridomas

Hybridomas are hybrid cells produced by fusing the cell membranes of sensitized B cell splenocytes with a myeloma cell partner. Hybridoma cell lines are generated by academic, government and commercial laboratories as a means of creating mAbs to a variety of targets used for research, diagnostic or therapeutic purposes. Hybridomas are only one method for generating mAbs. Other methods that could be considered are yeast, phage display libraries and single B cell sorting and cloning. However, the robust protocols surrounding hybridoma fusion technology (which date back to its original discovery by Kohler and Milsten) [1] have been time tested and proven to be successful for decades suitable to most mainstream mAb development. The display methods require a different understanding of molecular techniques as well as access to display vectors and *E. coli* or yeast culture. These tools may not always be available to all laboratories and researchers are encouraged to read the following review to help determine what avenue to take for creating mAbs successfully [10].

1.5 Screening

Newly produced hybridoma cells must be screened in an antibody binding assay to identify the hybridoma cells producing antibodies of the desired specificity. The majority of the hybridomas produced will not secrete antibody of the desired specificity, some may not secrete antibody at all. Typically, if the fusion is plated in 96-well plates, hybridomas that secrete antigen specific antibody range from 1 to 5 % of wells and can be identified and subcloned via screening.

The most common, efficient and high throughput method used to screen hybridomas post fusion is ELISA. A simple ELISA screening protocol is outlined below. Other methods such as indirect immunofluorescence, or capture antibody based microarray assays have also been described in the literature [11]. One single method or a variety of methods can be used for screening taking into account the end use application the antibody needs to perform. The earlier the antibodies are screened on the application of choice, the higher the probability to isolate a well performing antibody for that application [10]. The screening method should be rapid, reliable, sensitive, and easy to perform with hundreds or thousands of clones and amenable to robotic adaptation. The assay results should be available within 1–2 days following screening to ensure early subcloning of hybridomas secreting antibody of the desired specificity. If subcloning becomes protracted other co-plated cells, potentially with a growth advantage, could overtake the culture and result in loss of the desired hybridomas from a fusion.

2 Materials

2.1 Immunization

1. Phosphate buffer saline (PBS) (*see* **Note 5**).
2. Antigen.
3. CFA (Complete Freund's Adjuvant; or H37Ra) available from DIFCO suppliers.
4. IFA (Incomplete Freund's Adjuvant) are available from DIFCO suppliers.
5. Gerbu Adjuvant MM, Accurate Chemical & Scientific (rapid intrasplenic immunization protocol).
6. Ketamine-HCL (1.2 mg) and xylazine (0.39 mg) solution for animal anesthesia (rapid intrasplenic immunization protocol).
7. 2 % isoflurane solution (RIMMS immunization protocol).
8. Precision vaporizer (RIMMS immunization protocol).
9. Animal restrainer.
10. Animals for immunization (BALB/c, C57/B6, or other strain at 6–8 weeks old, Lou rats, Armenian or Syrian hamsters).
11. 3 ml syringes with locking hubs (Luer-lock syringes, Beckton Dickinson).
12. Eppendorf tubes.
13. Double-ended locking hub connector.
14. 22 g needles.
15. Amalgamator.
16. Insulin syringes.
17. 2-mercaptoethanol (2-ME) for in vitro splenocyte priming.
18. Thymocyte conditioned media (TCM) for in vitro splenocyte priming (*see* **Note 6**).

2.2 Spleen Harvesting

1. 50 ml conical tubes (Falcon or Corning sterile).
2. Sterilized forceps and scissors.
3. Dissection stage (can be styrofoam shipping box lid wrapped in aluminum foil).
4. 70 % ethanol for dissection tool sterilization.
5. Small plastic or glass beaker.

6. Basal Media; one can use IMDM, RPMI 1640, or DMEM without serum.
7. CO_2 chamber for animal euthanasia.

2.3 Preparation of Splenocytes, Lymph Node Cell Suspension and Myeloma Cells

1. Tabletop centrifuge.
2. 50 ml conical tubes (Falcon or Corning sterile).
3. 3 ml sterile disposable syringe, no needle attached (splenocyte preparation via the syringe plunger mashing method).
4. Sterile cell strainers (70–100 μm; BD Falcon).
5. Frosted-end glass slides (frosted glass slide splenocyte preparation method).
6. Scalpel (scalpel splenocyte preparation method).
7. 21 g syringe needle (splenocyte preparation via spleen perfusion).
8. Sterile disposable petri dishes (150 × 22 mm, 60 × 15 mm, and 100 × 15 mm).
9. Basal Media; one can use IMDM, RPMI 1640, or DMEM without serum.
10. Sterile serological pipettes.
11. Pipet-Aid.
12. Trypan blue (0.4 %) for cell counting and viability determination.
13. Cell counting device such as a hemocytometer and inverted microscope, Cellometer™ Auto 2000, or other comparable device.
14. Indelible waterproof marker.

2.4 Fusion

1. 50 ml conical tubes (Falcon or Corning sterile).
2. 96-well sterile tissue culture plates.
3. Small petri dishes (60 × 15 mm) and large petri dishes (150 × 22 mm).
4. Pipet-Aid.
5. Sterile serological pipettes.
6. Multichannel pipettor and appropriate sterile tips.
7. Reagent reservoir.
8. Sterile waste bottle (500 ml).
9. Kim wipes.
10. 37 ± 2 °C large water bath to accommodate several bottles of media.
11. Humidified 37 ± 2 °C, 5 % CO_2 dedicated tissue culture incubator.

12. Class IIb Biosafety Cabinet (BSC) or laminar flow hood (not openly exposed work surface area).
13. Benchtop Centrifuge, i.e., Beckman table top or Hettich model Rotofix 32A with 6 position swinging-bucket rotor or equivalent.
14. Laboratory timer or equivalent.
15. Indelible waterproof marker.
16. Basal Media; one can use IMDM, RPMI 1640, BD Quantum Yield, or DMEM.
17. Distilled Water, autoclaved prior to the fusion date.
18. Polyethylene glycol (PEG).
19. Semisolid Agarose for Hybridoma Plating—Add 5 ml of BioVeris Hybridoma Cloning to each bottle of ClonaCell™-HY Medium D (Stem cell technologies, Vancouver) used in the fusion and then place at 37 °C in an CO_2 incubator with cap slightly loosened to bring down the pH until needed for fusion.
20. HT media: IMDM (alternate media RPMI 1640, BD Cell Quantum Yield, or DMEM), 10 % FBS, 1 % penicillin/streptomycin/gentamycin, 10 % HCF, plus 10 ml HT (50×, Sigma).
21. Hybridoma Growth Medium: HT (NO AMINOPTERIN), IMDM basic media (RPMI 1640, BD Cell Media, or DMEM), 10 % FBS (characterized and pre-screened for its ability to support hybridoma growth), 4 % BioVeris Hybridoma Cloning 4 mM Factor containing 4 mM L-glutamine and antibiotics to be used for semisolid agarose method and fusion maintenance (*see* **Notes 7–9**).

2.5 ELISA Screening

1. ELISA plates—Due to the diverse nature of antigens, make sure to select plates that are intended for quantitative and qualitative solid phase immunoassays and/or binding assays.
2. ABTS—2,2′-azino-bis(3-ethylbenzothiazoline-6-sulphonic acid) (*see* **Note 10**).
3. Blocking solution—to prevent antibodies from passively adsorbing binding to the ELISA plate. Commonly used blocking agents are 1 % BSA, serum, nonfat dry milk, casein, gelatin in PBS (*see* **Note 11**).
4. 1 % BSA (*see* **Note 12**).
5. Hydrogen Peroxide at 30 %.
6. PBS.
7. Secondary Antibody—Depending on the animals immunized, use Anti-mouse/anti-rat IgG Subclass specific coupled to HRP with minimum cross-reactivity to other species (i.e., human, bovine, rabbit, etc.).

8. Wash solution—PBS or Tris-buffered saline (pH 7.4) with detergent (Tween20) at 0.05 % (v/v).
9. Plate spectrophotometer such as Molecular Devices Spectramax 384 which is an Absorbance (optical density) microplate reader.
10. Multichannel pipettor and appropriate sterile tips.
11. Reagent reservoir (disposable plastic multichannel reservoir or regular sized petri dish).

2.6 Hybridoma Maintenance

1. 96- and 24-well sterile tissue culture plates.
2. T25 flasks.
3. Multichannel pipettor and appropriate sterile tips.
4. Reagent reservoir.
5. Humidified 37 ± 2 °C, 5 % CO_2 dedicated tissue culture incubator.
6. Hybridoma Growth Medium: HT (NO AMINOPTERIN), IMDM basic media (RPMI 1640, BD Cell Media, or DMEM), 10 % FBS (characterized and pre-screened for its ability to support hybridoma growth), 4 % BioVeris Hybridoma Cloning Factor containing 4 mM L-glutamine and antibiotics to be used for semisolid agarose method and fusion maintenance (*see* **Notes 6–8**).

2.7 Subcloning

1. 96- and 24-well sterile tissue culture plates.
2. Pipet-Aid.
3. Sterile serological pipettes.
4. Multichannel pipettor and appropriate sterile tips.
5. Reagent reservoir.
6. 37 ± 2 °C large water bath to accommodate several bottles of media.
7. Humidified 37 ± 2 °C, 5 % CO_2 dedicated tissue culture incubator.
8. Class IIb BSC or laminar flow hood (not openly exposed work surface area).
9. Subcloning media: IMDM (RPMI 1640, BD Quantum Yield, or DMEM), 15–20 % FBS, 10 % HCF, penicillin/streptomycin/gentamycin, 4 mM L-glutamine.
10. Inverted Microscope for tissue culture.

2.8 Cryopreservation

1. Sterile cryovials.
2. Pipet-Aid.
3. Sterile serological pipettes.
4. Indelible waterproof marker for cryovial labelling.

5. Eppendorf tubes.
6. Freezing container, Nalgene® Mr. Frosty H × diam. 86 mm × 117 mm
7. Benchtop Centrifuge, i.e., Beckman table top.
8. Cell counting device such as a hemocytometer and inverted microscope, Cellometer™ Auto 2000, or other comparable device.
9. Class IIb BSC or laminar flow hood (not openly exposed work surface area).
10. Ice bucket.
11. −80 freezer.
12. Liquid nitrogen tank for long term hybridoma storage.
13. Trypan blue (0.4 %) for cell counting and viability determination.
14. Freezing media: 10 % Basic Media, 80 % Heat-inactivated Non-irradiated FBS, and 10 % DMSO.

3 Methods

3.1 Immunization of Rodents

Several methods can be used for in vivo splenocyte priming for subsequent fusion.

3.1.1 Conventional Immunization Method

1. Set aside animals to be immunized, usually 3–5 mice per target since individual animals respond differently to a given immunogen.
2. Prepare an antigen and Complete Freund's Adjuvant (CFA) emulsion (200 μg by linking two locking syringes with a double ended lock connector) (*see* **Note 13**). Equal volumes of PBS containing 1 to 50 μg (protein or peptide conjugate) (Peptides are conjugated to a carrier protein such as ovalbumin or BSA and CFA are mixed) (*see* **Notes 14–16**).
3. Press syringes back and forth for about 5–10 min until a stable emulsion is prepared. Alternatively combine antigen and CFA in an eppendorf tube and use an amalgamator for 30 s on high setting to prepare the antigen/CFA emulsion.
4. Inject animals with the immunogen-adjuvant emulsion intraperitoneally (i.p.) or subcutaneously (s.c.) at 0.2 ml per mouse, 0.5–1 ml per rat and 0.2–0.4 ml per hamster using a 22-G needle (*see* **Notes 17** and **18**).
5. After a period of 10 to 14 days, boost the animals with 50 % of the amount of the original priming dose dissolved in Incomplete Freund's Adjuvant (IFA).

6. Seven days after the boost injection, test the antibody titers by ELISA, WB, or flow cytometry to determine reactivity to the specific antigen.
7. Continue boosting the primed animals an additional two to three times with the IFA emulsified-immunogen at 2 week intervals, until the antibody antigen specific titer has been achieved (determined by ELISA).
8. When the immunization is performed with whole cells, use 2×10^6 to 5×10^7 cells for animal inoculation. Wash cells in serum-free medium 3× prior to immunization [12].
9. Allow animals to rest 2–4 weeks following the last boost. Spleens derived from animals with the highest antibody titers are selected for fusion (*see* **Note 18**).
10. Boost the selected animals 3–4 days prior to fusion with immunogen in PBS. Injection can be delivered i.p. or intravenously (i.v.).

3.1.2 Rapid Intrasplenic Immunization (See ***Note 19****)*

1. Anesthetise animals with a mixture of Ketamine-HCL (1.2 mg) and xylazine (0.39 mg) solution injected i.p prior to immunization.
2. Shave the left side of the animal with clippers to visualize the spleen.
3. Thirty minutes following anesthesia inject the animal directly into the spleen with a mixture of 50–100 μg of immunogen plus Gerbu (*see* **Note 20**).
4. Inject the immunogen-adjuvant mixture into the spleen using an insulin syringe on days 0, 4 and 11.
5. Harvest the sensitized spleen 3–5 days following the last boost [13, 14].

3.1.3 Repetitive Immunization at Multiple Sites (RIMMS) (See ***Note 19****)*

1. Anesthetise animals prior to immunization with 2 % isoflurane solution. The isoflurane solution is delivered in oxygen using a precision vaporizer.
2. Inject immunogen emulsified in CFA at 50 μl/site at sites that drain to the popliteal, inguinal, axillary, and brachial lymph nodes on days 0, 4, and 11.
3. On day 13 following the first immunization, collect the bilateral and lumbar lymph nodes for fusion [15].

3.1.4 In Vitro Splenocyte Priming (See ***Note 19****)*

1. Inject mice i.p. with 20 μg of antigen emulsified in CFA as described in **step 2** of Subheading 3.1.1.
2. Boost animals i.p. with 20 μg antigen in PBS.
3. Harvest spleens from immunized mice and prepare splenocyte cell suspensions as described in Subheadings 3.2 and 3.3.

4. Activate pre-primed immune splenocytes in vitro at 10^7 cells/ml in serum-free media supplemented with an equal volume of thymocyte conditioned medium, 50 μM 2-ME, and sterile filtered antigen at the desired concentration.
5. After 3 days harvest the cells following sedimentation via centrifugation and use them for somatic cell hybridization [16].

3.2 Spleen Harvesting

1. Exsanguinate the animal as per your approved institutional animal care protocol. Avoid using anesthetics when sacrificing the animal to avoid introduction of foreign chemicals into the bloodstream that can travel to the spleen prior to harvesting. Spray the left side of the animal with 70 % alcohol.
2. In a laminar flow hood and using sterile forceps and scissors (instruments can be sterilized by keeping them in 70 % alcohol) cut away the fur along the left side of the animal about halfway between the front and the back legs.
3. With a second set of sterile instruments cut open the body cavity of the animal and remove the spleen (dark red kidney shaped organ). Remove excess fat and place it in a tube in cold basic serum-free media (IMDM, RPMI 1940, or DMEM).
4. The tissue should be collected asceptically and kept on ice until further processing.

3.3 Preparation of Splenocytes

A variety of methods can be used to prepare a cell suspension from the isolated spleen prior to fusion.

3.3.1 Frosted Glass Slide Method

1. Place isolated spleen in a petri dish in serum-free basic media.
2. Sterilize the frosted end of glass slides by spraying or dipping in alcohol.
3. Homogenize the spleen between the frosted ends of the slides by making sure to touch only the non-frosted (clear) ends.
4. Pass the spleen cell suspension through a cell strainer mounted on a 50 ml conical tube to remove cell aggregates, cell debris and connective tissue (*see* **Note 21**).

3.3.2 Spleen Perfusion Method

1. Place spleen in a petri dish containing 10 ml of serum-free basic media.
   Holding the spleen by sterile forceps perform spleen perfusion with serum-free media through 21 g syringe needle.
2. Repeat the injections about 50–100 times until the spleen is essentially an empty membrane.
3. Pass splenocyte suspension through sterile cell strainer mounted on a 50 ml conical tube.

3.3.3 Plunger Mashing Method

1. Cut spleen in small pieces with sterile scissors and place the pieces in a sterile strainer.
2. Use the plunger end of a syringe to mash gently the dissected spleen through the strainer mounted on a 50 ml conical tube.
3. Wash splenocytes through the strainer with sterile serum-free media.

3.3.4 Scalpel Method

1. With the spleen in a sterile petri dish containing sterile media (no FBS) use a scalpel and cut a very small portion of the bottom of the spleen off.
2. Then use the scalpel to cut the membrane along one side of the spleen using sterile forceps to aid in holding the spleen. Essentially the spleen will resemble an open book, such that you can use the scalpel to gently scrape the spleen cells out of the bottom of the spleen moving from top to bottom. Eventually you will be left with essentially the open membrane of the spleen.
3. Use a 10 ml serological pipette to place the splenocytes into a 50 ml conical tube, to maximize cell recovery, ensure to wash the bottom of the petri dish with media.
4. Wash isolated splenocytes with any of the four methods indicated above, twice in basic media at 210 × *g* for 8 min and perform a cell count. For successful fusion the splenocyte viability should be over 80 %. One or more spleens can be combined if factored into cell counts accordingly.

3.4 Preparation of Lymph Node Suspension

1. Collect the bilateral and lumbar lymph nodes asceptically in a 15 ml conical tube containing serum-free media.
2. Place the lymph nodes in a sterile strainer mounted to a 50 ml conical tube containing 3 ml of serum-free media.
3. Use the plunger end of a syringe to mash gently the dissected spleen through the strainer.
4. Wash lymph node cells through the strainer with sterile serum-free media.
5. Wash isolated lymph node cells twice in basic media at 210 × *g* for 8 min and perform a cell count. For successful fusion the lymph node viability should be over 80 %.

3.5 Preparation of Myeloma Cells

Mouse myeloma cells commonly used for both mouse and for rat fusions include NSO [17], P3-X63/Ag8.653 [18], or Sp2/0 [6, 19]. SP2/0 has also been used successfully for hamster fusions [6]. Yb2/0 is a myeloma cell line derived from the LOU rat strain that is commonly used for rat fusions. The above myeloma cell lines have been selected for not secreting endogenous antibodies, and they lack full length myeloma kappa light chains [20] (*see* **Notes 22** and **23**).

1. Grow the various myeloma cell lines of the range 5×10^4 cells/ml to 1×10^6 cells/ml in RPMI 1640 (alternate media IMDM, DMEM) media plus 10 % FBS.
2. Split myeloma cells the day prior to fusion to ensure that they are growing vigorously. To ensure a successful fusion, avoid growing the myeloma cells in culture for a lengthy period of time.
3. Wash mid log phase myeloma cells three times in sterile media (IMDM, RPMI 1640, or DMEM) with no FBS, count and determine their viability prior to fusion.
4. Expand the myeloma cells as needed to have adequate numbers of cells available for fusion. Some investigators have offered the following ratios which we have tabulated below with the point being the splenocytes should be slightly in excess of myeloma cells [21]. Thus culture enough myeloma cells to meet the desired ratio, an estimate of $0.5–1 \times 10^8$ cells/murine spleen is a good starting point to determine amount of myeloma cells to culture. Even 10:1 has shown success (*see* **Note 23**).

Species	Splenocytes–myeloma
Mouse–mouse	5:1
Hamster–mouse	4:1
Rat–mouse	5:1

3.5.1 Procedure for Hybridoma Fusion

1. Remove the PEG (*see* **Note 25**) from fridge and bring to RT during cell preparation. Pre-warm the Hypoxanthine Thymidine (HT) or Hypoxanthine Aminopterin Thymidine (HAT) in the 37 °C water bath. Wipe off the outside of the PEG vial with 70 % ethanol and place in BSC. Remove media from incubator, also wipe dry and spray with 70 % ethanol, and place in BSC. Place a sterile waste bottle capable of holding at least 500 ml into the hood (*see* **Notes 26–28**).
2. Obtain the two tubes containing both the washed, counted and pelleted immune spleen and myeloma cells as processed above and bring these into the same BSC hood. Note, at this point the amount of myeloma cells should be at your chosen splenocyte to myeloma ratio. If for some reason there is an unexpected delay the spleen cells should be stored on ice to keep viability high. However, if well timed and the wash steps are executed in unison they can remain at room temperature and should be at room temperature for the fusion step (*see* **Notes 29** and **30**).
3. Wipe off gloves with copious amounts of 70 % Ethanol. Dry hands off in the air-stream at the front of the BSC.

4. Carefully pour off the supernatant from the cell pellets into a waste bottle in the BSC.
5. Gently resuspend the myeloma cells with 10 ml of IMDM (RPMI 1640 or DMEM) basic media and transfer the myeloma cells into the 50 ml centrifuge tube containing the washed splenocytes.
6. A small amount (~100 μl) of myeloma cells should be kept in the tube to be used for the myeloma control plate for confirming the HAT selection kills any non-fused myelomas (*see* **Note 30**).
7. Mix the spleen/myeloma cell suspension thoroughly with gentle pipetting up and down with an additional 30 ml of basic cell media. (At least 10 times.)
8. Centrifuge the spleen-myeloma cells at 210×*g* for 8 min at room temperature.
9. Gently pour off the supernatant into the waste bottle. It is crucial that the cells be as free of serum proteins prior to fusion as possible due to the propensity of PEG to cause precipitation. After the last wash pour off the supernatant into the waste container and leave the tube inverted on a Kim-wipe pre-soaked with 70 % Ethanol within the BSC for at least 3 min (to remove residual supernatant prior to addition of PEG).
10. Have a multi-timing stopwatch/timer pre-set for several times or recruit another lab staff for assistance. Set the timers for 90 s; another for 150 s; and the last for 4 min and 30 s. Have a sterile 1 ml syringe with a 21 gauge needle filled with 1 ml of PEG ready (leave needle and syringe in PEG bottle until ready to begin). Pre load a sterile 10 ml disposable pipette onto a Multi-Speed Pipet-Aid (set on low) within easy access before starting the fusion procedure.
11. Spread out the pelleted myeloma and splenocytes in the bottom of the tube via gentle and repeated firm tapping of the tube on the stainless steel surface of the BSC hood in an angular vertical motion. It is important to try to ensure that all of the cells have access to the PEG.
12. Start all 3 channels on the timer and add 1 ml of PEG 1500 (prewarmed to RT) dropwise over 1 min using the 1 ml syringe and needle, while gently rotating and tapping the tube to resuspend the cells.
13. Continuously tap the tube on the BSC surface for another 30 s.
14. Immediately fill up a 10 ml serological pipette with 9 ml of pre warmed basal media (RPMI or BD Cell mAb or IMDM) (must have NO FBS) and gently add 1–2 ml (dropwise) to the PEG-cell mixture over ~1 min. Continuously tap the tube while adding the media (*see* **Note 31**).

15. Add the remaining 7–8 ml of basal media slowly over 2 min to slowly dilute the PEG solution. Place the tube containing the cell pellet in a 37 °C water bath for 10 min.
16. Centrifuge the cells at 200×*g* for 10 min at room temperature.
17. Pour off the supernatant and add 5 ml of HAT containing complete media (with FBS etc). Do NOT resuspend the pellet by pipetting up and down as they are extremely fragile at this stage. Just dispensing the media into the tube is enough.
18. Place the tube containing the cell pellet in a 37 °C water bath for at least 15–25 min. The cells will slowly disperse and become loose during this time. If they do not disperse, let the tube incubate for an additional 15 min until they do (Note: If not dispersed after 35 min continue on with next steps) (*see* **Note 32**).
19. Dilute and plate out the cells in semisolid methyl cellulose with HAT or in complete HAT media to 96-well plates as per below (Subheadings 3.5.2 or 3.5.3, respectively).

3.5.2 Procedure for Plating Fusion Using ClonaCell™-HY Medium D Method

1. Remove the cells from the water bath and gently resuspend with media already in the tube. Transfer into 95 ml of ClonaCell™-HY Medium D (HAT media) containing 5 ml of HCF and plate out 10 ml per small petri dish (60×15 mm). Be sure to mix the cell suspension two to four times between plates.
2. Ensure to set aside 2 ml of ClonaCell™-HY Medium D selection media before adding the cells. This will be used for the myeloma control.
3. Using approximately 2 ml of ClonaCell™-HY Medium D, aliquot 1 ml into 2 wells in a 24-well plate. Pipette approximately 1,000 myeloma cells into each well (usually 50–100 μl of concentrated myeloma cells is sufficient). This is the myeloma/HAT control.
4. For incubating, place two small petri plates (60×15 mm) inside of a large petri plate (150×22 mm), with an additional small petri plate full of autoclaved distilled water without a lid for humidity. Leave for 10–18 days in a CO_2 incubator at 37 °C (*see* **Note 33**).
5. Observe only one of the plates under the inverted microscope daily for cell growth (do NOT disturb the other plates). Profound cell death should be observed by days 2 and 3. Only those cells that can survive in the HAT media will expand.
6. Pick clones from semisolid agarose into 96-well flat bottom tissue culture plates containing 200 μl of Hybridoma Growth Media (with HT supplement) and screen all wells 3–6 days later (*see* **Note 34**).

3.5.3 Procedure for Plating Fusion by Limiting Dilution

1. For plating a fusion in 96-well culture plates, first dilute the dispersed cell pellet to 40 ml with Hybridoma Growth Media—HAT. Count the cells with a hemocytometer and determine the number of live cells per ml in the tube. Total the number of cells per ml volume and calculate the total number of cells.
2. Dilute the cell suspension to a final cell count of ~5–8 × 10^4/ml with Hybridoma Growth Media—HAT pre-warmed to RT. This is the seeding concentration for the fusion with is aiming at obtaining only 1 fusion per well (1/10,000 B cells randomly fuse).
3. Plate out ~150 μl per well to 96-well cell culture plates using a multichannel pipette and sterile 200 μl tips. Use sterile reservoirs for dispensing. Mix the cell suspension each time before starting a new plate.
4. P3-X63-Ag8 myelomas prefer to grow well at higher cell densities so they should be plated into 15–20 plates in general (~8 × 10^4 cells/ml).
5. SP2/0 myelomas prefer to grow well at lower cell densities so may need to be plated to more plates (25–35 for example or ~5 × 10^4 cells/ml). If the spleen(s) are very large, the researcher may need to plate into an even larger number of plates.
6. Plate out non-fused myelomas in the same Hybridoma Growth Media—HAT at the same or higher concentration as the hybridomas were plated. This is a control to ensure that the aminopterin is killing the non-fused myeloma cells.
7. Label each 96-well plate with the fusion number, plate number, antigen and date and initial. Place the cells in a designated CO_2 incubator at 37 °C (*see* **Note 35**).
8. Observe several wells under the inverted microscope daily for cell growth from one of the plates only (do NOT disturb the other plates). Profound cell death should be observed by days 2 and 3. Only those cells that can survive in the HAT media will survive and expand. If media in culture plates is low by day 6 add 100 μl/well of HT media using a multichannel pipette.
9. For 96-well plates, screen the hybridoma supernatants via ELISA between days 7 and 9 for mouse fusions, day 11 for rat–mouse fusions, and day 14 for hamster–mouse fusions. Many of the wells will not have any growth and thus are expectedly negative in the screen (good fusions may have growth in 30–80 % of the wells depending upon the myeloma and the fusion). Recovery is also dependent upon the growth rate of the cells which is a factor of plating density and survival. If many wells are turning yellow due to lactic acid production the fusion may need earlier screening (*see* **Note 36**).

3.6 ELISA Screening

3.6.1 Procedure for ELISA Plate Preparation

1. Dilute the antigen in PBS to a final concentration of 0.5 μg/ml for recombinant proteins or 2.0 μg/ml for peptides. Make sure that you prepare enough antigen solution to coat the appropriate number of plates that will be used for the whole fusion screen process. This step may require some level of optimization due to the nature of the antigen.
2. Coat plates using 100 μl/well of antigen solution. Stack the plates and cover the first plate with an adhesive plastic, so as to protect them from contamination.
3. Incubate the plates for at least 1 h at room temperature, or overnight at 4 °C.
4. Remove the coating solution and wash the plate five times by filling the wells with 200 μl of wash solution. Removal of the coating solution and wash buffer can be done by flicking the plate over a sink and then patting the plates over a paper towel, or by using an automated plate washer.
5. Block the remaining antigen-binding sites in the coated plates by adding 200 μl/well of blocking solution for at least 1 h at room temperature. Blocking can also be done overnight at 4 °C.
6. Remove the blocking solution by flicking the plates over a sink and tapping them dry over a paper towel. Store at 20 °C freezer until use.

3.6.2 Procedure for ELISA Screening (Fusion and Subcloning)

1. Seven to ten days post fusion for fusions plated into 96-well plates or 3–4 days post transfer into a 96-well plate from semi-solid agarose, remove 100 μl of tissue culture supernatant from each well and transfer to the ELISA plates pre-coated with immobilized antigen using a multichannel pipette. Use sterile tips for the supernatant transfers and make sure to use one tip per well as cross-contamination could lead to false positive results during screening. Also, add fresh media to the wells of the tissue culture plate and return it to the incubator to enable cell harvesting once positive clones are identified.
2. Incubate the supernatant for at least 1 h at room temperature.
3. Remove the supernatant and wash the plate five times by filling the wells with 200 μl of wash solution. This can be done manually removing the supernatant and wash buffer by flicking the plate over a sink and then patting the plates over a paper towel, or by using an automated plate washer.
4. Prepare the secondary antibody solution (Anti-mouse/anti-rat IgG Subclass specific HRP) per vendor specifications. The researcher must make sure that there is enough of the solution for all the plates in the screening run that day.
5. Dispense 100μl/well of secondary antibody solution and incubate for 1 h at room temperature.

6. Wash the plates five times with wash solution.
7. Prepare the substrate solution by adding 0.1 % of Hydrogen Peroxide (30 % solution) to ABTS. Calculate enough volume to have enough for all plates of the experiment run, usually 10 ml/plate.
8. Dispense 100 μl/well of the substrate and incubate for 30 min at room temperature.
9. Read the absorbance (optical density at 405 nm) of each well in a plate reader. Wells that have an OD higher than three times the background should be considered positive and those wells require moving up into the next stage of growth.

3.7 Fusion Growth and Maintenance

3.7.1 Procedure for Fusion Maintenance

1. Once hybridomas secreting antibody of the desired specificity have been identified they will need to be expanded from the 96-well to the 24-well plate followed by subcloning via limiting dilution to establish hybridoma lines. Note a T-25 (small flask) should be seeded at this stage to allow for the token freeze of the positive cells to be created for backup against loss.
2. Growing hybridomas should be 25–50 % confluent at the 96-well plate stage prior to expansion.
3. Expand ELISA positive wells by resuspending the entire content of each 96 well and transferring 100 μl to the well of a 24-well plate. Feed hybridoma cells in the 24-well plate with 1 to 1.5 ml of Hybridoma growth-HT media making sure to use a different tip for each well transferred. Ensure that the 96-well plate is also fed with 100 μl of Hybridoma growth-HT media as it serves as a backup to the expanded cells. Incubate plates for 2–3 days in humidified 37 °C, 5 % CO2 incubator until cells are 25–50 % confluent. RNA isolation backups can be taken at this stage for labs with molecular V gene cloning capability.
4. Screen expanded hybridoma supernatant to confirm the original screening test or run additional applications that would allow further antibody characterization. Subclone selected clones via limiting dilution as soon as possible after they were tested positive for specific antibody in a given assay and freeze original remaining cells in the well.
5. In cases where a large number of wells produce positive hits to the target being developed (signifying a specific IgG antibody response), a culling step is recommend. During the subcloning stages, dilutions of the supernatants that were positive by ELISA can be compared to each other and a cutoff chosen (i.e., those with a higher optical density are selected). Not only will this reduce the workload, but it will also ensure that the clones with the best possibilities of success go through.

3.8 Subcloning

3.8.1 Procedure for Fusion Subcloning

1. Hybridoma subcloning is achieved by limiting dilution aiming to dilute the hybridoma culture so that there is one cell present per well. Subsequent division of the single cell will yield a cell colony that produces antibody of a single specificity (monoclonal).
2. For high-efficiency cloning the hybridoma culture should be in mid-log phase. Count cells hybridoma cells to be subcloned and determine their cell concentration and viability using a cell counting device such as a hemocytometer, the Cellometer™ Auto 2000, or equivalent.
3. Prepare 10 ml cell suspensions of 50 viable cells/ml (giving 10 cells/well) and 5 viable cells/ml (giving 1 cell/well) in hybridoma subcloning medium. It requires two dilutions to take a cell suspension to 50 cells/ml. At first dilute the hybridoma suspension to 50,000 cells/ml and then make a second dilution 1:999 dilution to achieve a cell suspension of 50 cells/ml.
4. Plate 200 μl/well of each cell suspension into 96-well plate labelled with the fusion name, target, subcloning stage, number of subcloned cells/well. Incubate cultures in a humidified 37 °C, 5 % CO_2 incubator for 7–10 days.
5. On about day 4 observe individual wells for cell growth and presence of tight clusters of cells under the inverted microscope. The microtiter plates plated with 10 cells/well should grow faster than those plated at 1 cell/well since they will have more than one colony of cells growing per well.
6. Screen wells for the desired antibody at approximately 50 % confluency, about 7–10 days following subcloning. If there are no colonies growing in the microtiter plates seeded with 1 cell/well examine the plates seeded with 10 cells/well for presence of cell colonies and screen them for presence of the desired antibody.
7. In deciding which monoclonals to select for further expansion consider the following factors: size of cell colony, rate of cell growth, antibody yield, and antibody performance on end-use application.
8. To guarantee that one hybridoma cell population is clonal, it should be subcloned twice via limiting dilution (*see* **Note 37**).

3.9 Cryopreservation

3.9.1 Procedure for Cryopreservation

1. Label cryovials with indelible waterproof marker including is the clone name, specificity, freezing date, investigator's name, and subcloning stage.
2. After taking cells for cloning via limiting dilution transfer the remaining cells in the well in a 15 ml conical tube and wash them with basic medium. For optimal results cells should be in mid log phase for freezing.

3. Centrifuge the cells at 200 × g for 10 min and resuspend them in freezing cold cryopreservation media (10 % Basic Media, 80 % Heat-inactivated Non-irradiated FBS, and 10 % DMSO) media at 4 × 106/ml to 2 × 107/ml.
4. Distribute cell suspension in labelled cryovials (1 ml/vial) which are maintained on ice for 30 min prior to transfer to the −80 °C freezer. It is important to slowly freeze the cells. An alternative to placing on ice is to place in Styrofoam tube racks and transfer to the −80 °C freezer for a slower freeze.
5. For long-term storage cryovials should be transferred from the −80 °C freezer to the liquid nitrogen freezer within 24 h.

4 Notes

1. Antigen—The molecule used for screening antibody reactivity.
2. Immunogen—The substance used to generate an antibody response in vivo.
3. KLH—Keyhole limpet hemocyanin is a large glycosylated protein (MW 350–390 kDa) derived from the hemolymph of *Megathura crenulata* that is used as a protein carrier.
4. BSA—used as a protein carrier for antigen.
5. PBS—Phosphate buffered saline solution. A water based solution containing sodium chloride, sodium phosphate, potassium phosphate, and potassium chloride used as an isotonic buffer (pH 7.4) in cell applications. The easiest way to prepare a PBS solution is to use PBS buffer tablets. They are formulated to give a ready to use PBS solution upon dissolution in a specified quantity of distilled water.
6. Thymocyte conditioned media is prepared by culturing thymocytes at 4×10^6 cells/ml in DMEM, 1 mM pyruvate in 18 mM HEPES buffer with 12 % type 100 rabbit serum for 48 h at 37 °C. The thymocyte-conditioned medium (TCM) is then harvested by centrifugation (500 × *g* for 5 min to pellet the thymus cells). The supernatant (TCM) is subsequently stored at −70 °C.
7. Hypoxanthine-guanine phosphoribosyl transferase (HGPRT)—The HGPRT enzyme is responsible for recovering purine nucleotides through the purine salvage pathway of nucleotide metabolism.
8. HAT—Hypoxanthine aminopterin, thymidine: Hypoxanthine and thymidine are metabolic intermediates of nucleotide metabolism provided as growth factors. Aminopterin is a drug which inhibits the de Novo pathway of nucleotide metabolism.

The de novo pathway, is necessary for survival/cell growth of myeloma lines defective in HGPRT. Aminopterin kills HGPRT mutant myelomas usually within a couple of days.

9. HT—Hypoxanthine and thymidine are metabolic intermediates provided as growth factors without the selection drug aminopterin to prevent limiting amounts of these precursors from slowing growth during hybridoma recovery post fusion.
10. ABTS—2, 2′-azino-bis(3-ethylbenzothiazoline-6-sulphonic acid). A chemical compound used in the enzyme-linked immunosorbent assay (ELISA) to detect, in this specific case, binding of antibodies to a specific antigen.
11. Antibody dilution buffer—Secondary antibody should be diluted in 1× blocking solution to reduce nonspecific binding.
12. BSA—A serum albumin protein derived from cows.
13. The preparation of a stable emulsion is critical for the adjuvant to be effective. Before the immunogen-adjuvant emulsion is prepared, make sure to mix thoroughly the CFA to ensure that the mycobacteria contained in the adjuvant are uniformly distributed. The emulsion is prepared by mixing an equal volume of aqueous solution with the equal volume of oil-based adjuvant using two syringes with a double hub needle. The goal is to introduce the aqueous phase into the oil without introduction of air. The stability of the emulsion can be tested by releasing a drop on cold water. If the drop remains cohesive the emulsion is ready for inoculation.
14. Use extreme care while handling CFA. Personal protective equipment should be used at all times as accidental injection or exposure to the eyes could lead to a granulomatous reaction and even loss of eyesight.
15. Instead of CFA, other oil-based adjuvant such as Ribi and TiterMax can be used for immunization.
16. Protein carriers are proteins used to conjugate haptens, peptides and small proteins to increase their immunogenicity. Examples of protein carriers are KLH-Keyhole limpet hemocyanin is a large glycosylated protein (MW 350–390 kDa) derived from the hemolymph of *Megathura crenulata* and BSA.
17. Mice and hamsters are generally easier to handle with one hand during immunization while rats are easier to manipulate if they are put under general anesthesia.
18. Frequent immunization with high amounts of immunogen should be avoided as it can in some cases lead to immune tolerance or clonal exhaustion rather than the desired immune response.

19. The rapid intrasplenic immunization, RIMMS, and in vitro splenocyte priming methods have been used successfully to generate immune splenocytes after exposure to low antigen amounts or antigens highly conserved across species.
20. Gerbu is a water-soluble adjuvant. Once it is added to the immunogen it can be mixed thoroughly by vortexing. CFA is not recommended for intrasplenic inoculations since it is prone to induce granulomas.
21. If you do not have disposable cell strainers, than try to remove the large chunks of cellular debris as follows. Allow the cells in the newly transferred 50 ml conical tube to settle for 5 min and then transfer the supernatant (cells and media) to a new sterile 50 ml conical tube. This leaves any large aggregates or membrane pieces in the original tube.
22. Myeloma cells should be treated periodically with 6-Thio-Guanine and then 8-Aza-Guanine to ensure that the cells are HGPRT negative and Thymidine Kinase negative. In addition myeloma cells should be treated periodically with plasmocin to maintain them mycoplasma free.
23. The number of myeloma cells prepared for cell fusion should be planned carefully as it has to be adequate to the number of spleens used for cell fusion. Factors to be taken into account for planning purposes are:
    - Spleen cell–myeloma cell ratio for the fusion procedure should be within the range from 5:1 to 10:1.
    - The estimated splenocyte number per spleen is $0.5–1 \times 10^8$ cells.
24. Asceptic technique and "clean" (contamination free) tissue culture rooms are critical for successful hybridoma development. All initial manipulations of animal tissue should be carried out in a Class IIb BSC in the designated animal room or an isolated hood so the cells are effectively quarantined until such time as they are found to be "clean". A different hood should be used for fusions and downstream sub cloning and scale up.
25. PEG—polyethylene glycol: A polymer of ethylene oxide chemical used to fuse cell membranes.
26. HAT and HT media should be prepared fresh or no longer than 1 week prior to the fusion. It is good practice for investigators to make their own tissue culture media and not to share media among laboratory associates to avoid contamination.
27. No one should enter the designated animal rooms or disturb the technical staff during these procedures.

28. Since the exact B cell phenotype which survives fusion to become a hybridoma is unknown, it is recommended to use splenocytes from at least one spleen for cell fusion. Spleens are isolated from animals that demonstrate high antibody titres/responses to the antigen of interest by ELISA or other desired application (flow cytometry, western immunoblotting). It is recommended to use the same assay for screening the immune response at the serum level that will be used for further fusion screening to ensure that the antibodies generated following fusion would work in specific assay systems and conditions.
29. Spleen cells from two animals with high antibody titer levels can be fused simultaneously, if the main concern is successfully obtaining a high affinity IgG mAb. Commercial labs might refrain from this practice for cost savings on animals.
30. It is imperative to treat the myeloma cell line with 8-Aza Guanidine (*see* **Note 22**). If this step is not performed, it will cause a huge number of myeloma cells to grow post fusion. Regular mycoplasma testing is also recommended for both the myeloma partner and the hybridomas post subcloning. Aminopterin-sensitive myelomas will die within days.
31. Be cautious with the cells in the steps proceeding and during the fusion. These cells are precious and have taken many months of effort to get to this stage. Other activities should be avoided so that uninterrupted attention is given to the fusion. Do not leave the PEG solution to sit too long on the cells or the viability will be adversely affected. Be sure to add back media with FBS to the tight pellet post fusion and removal of PEG with basal media and allow it to rehydrate pellet slowly by fusion at the post fusion step to let it gently come apart (do NOT pipette up and down nor vortex mix). Be cautious not to throw away the immune spleen cells or myeloma during the lengthy washing steps.
32. For antigens obtained from a bovine source, medium supplemented with 10 % Horse Serum must be used for the dispensation of the spleen cell–myeloma cell suspension into petri dishes containing ClonaCell™-HY Medium D.
33. No one except authorized personnel should be allowed to use/open or move these incubators to prevent contamination or disturbing the sensitive cells post fusion and during the clonal expansion.
34. If picking clone foci from semisolid agarose, pre-wet each pipette tip in media from the corresponding well of the 96-well plate. This will allow better dissociation of the ball of cells from the tip.
35. It is also recommended to use media containing antibiotics and antimycotics for fusion maintenance, subcloning and fur-

ther clone scale up. To avoid cell line cross-contamination, all materials and reagents used for this procedure must be sterile and disinfected prior to use. Wipe down the BSC with 70 % ethanol and allow at least 10 min hood clearance time. In instances when more than one cell line is being worked on, make sure that the blower in the BSC has been turned on in between the work performed on each cell line.

36. When subcloning larger numbers of individual positive clones (>10 per fusion) a simpler serial dilution strategy can be used to save time without having to count all of the clones. Essentially, dilute an aliquot of cells (20–30 μl) from the well or T-25 across the top (twofold) of a new 96-well plate filled with complete hybridoma media (from left to right) without changing the tip in this case. Next dilute the entire plate down the columns twofold and top off the wells to 200 μl per well. Screening of the right hand side of the plate usually is all that is required. Grow the plates at 37 °C for 7–10 days under 5 % CO_2 and rescreen as desired.

37. Do not lose your confidence if the first fusion does not work; It may take a few tries to get the timing correct and to avoid errors. Not all fusions work.

References

1. Kohler G, Milstein C (1975) Continuous cultures of fused cells secreting antibody of predefined specificity. Nature 256:495–497
2. Lane RD (1985) A short duration polyethylene glycol fusion technique for increasing production of monoclonal antibody-secreting hybridomas. J Immunol Meth 81:223–228
3. Jennings MV (1995) Review of selected adjuvants used in antibody production. ILAR J V37(3):1–9
4. Harlow E, Lane D (1988) Antibodies: a laboratory manual. Cold Spring Harbour Laboratory Press, New York
5. Schunk MK, Macallum GE (2005) Applications and optimization of immunization procedures. ILAR J 46(3):241–257
6. Sanchez-Madrid F, Szklut P, Springer TA (1983) Stable hamster-mouse hybridomas producing IgG and IgM hamster monoclonal antibodies of defined specificity. J Immunol 130:309
7. Borrebaeck CA (1983) In vitro immunization for the production of antigen-specific lymphocyte hybridomas. Scand J Immunol 18(1):9–12
8. Brown MC, Joaquim TR, Chambers R, Onisk DV, Yin F et al (2011) Impact of immunization technology and assay application on antibody performance—a systematic comparative evaluation. PLoS ONE 6(12):e28718. doi:10.1371/journal.pone.0028718
9. Ferber PC, Ossent P, Homberger FR, Fischer RW (1999) The generation of monoclonal antibodies in mice: influence of adjuvants on the immune response, fusion efficiency and distress. Lab Anim 33(4):334–350
10. Berry JD (2005) Rational monoclonal antibody (mAb) development to agents of foreign animal disease, bio-threat agents (BTA), and emerging infectious threats: The antigen scale. The Veterinary Journal Vol 170(2):193–211
11. Rieger M, Cervino C, Sauceda JC, Niessner R, Knopp D (2009) Efficient hybridoma screening technique using capture antibody based microarrays. Anal Chem 81:2373–2377
12. Yokoyama WM, Christensen M, Santos GD, Miller D. Production of Monoclonal Antibodies 2006. Curr Protoc Immunol 2:25
13. Nilsson BO, Larsson A (1990) Intrasplenic immunization with minute amounts of antigen. Trends Immunol 11(1):13–20
14. Simkins SG, Knapp SL, Brough GH, Lenz KL, Barley-Maloney L, Baker JP, Dekking L, Wai H, Dixon EP (2011) Generation of monoclonal antibodies to the AML1-ETO fusion protein: strategies for overcoming high homology. Hybridoma (Larchmt) 30(5):433–443

15. Kilpatrick KE, Wring SA, Walker DH, Macklin MD, Payne JA, Su J-L, Champion BR, Caterson B, Mclntyre GD (1997) Rapid development of affinity matured monoclonal antibodies using RIMMS. Hybridoma 16: 381–389, http://www.plosone.org/article/info%3Adoi%2F10.1371%2Fjournal.pone.0028718
16. De Boer M, Ten Voorde GH, Ossendorp FA, Van Duijn G, Tager JM (1988) Requirements for the generation of memory B cells in vivo and their subsequent activation in vitro for the production of antigen-specific hybridomas. J Immunol Methods 113(1):143–149
17. Galfre G, Milstein C (1981) Immunochemical techniques Part B. Methods Enzymol 73:3
18. Kearney JF, Radbruch A, Liesegang B, Rajewsky K (1979) A new mouse myeloma cell line that has lost immunoglobulin expression but permits the construction of antibody-secreting hybrid cell lines. J Immunol 124(4):1548–1550
19. Leo O, Foo M, Sachs DH, Samelson LE, Bluestone JA (1987) Identification of a monoclonal antibody specific for murine T3 polypeptide T-cell antigen receptor/mouse/T-cell activation). Proc Natl Acad Sci U S A 84: 1374–1378
20. Yuan X, Gubbins MJ, Berry JD (2004) A simple and rapid protocol for the sequence determination of functional kappa light chain cDNAs from aberrant-chain-positive murine hybridomas. J Immunol Methods 294((1-2)): 199–207
21. http://pathology.wustl.edu/research/hybridoma.php?page=fusion

Chapter 5

Generation of Rabbit Monoclonal Antibodies

Pi-Chen Yam and Katherine L. Knight

Abstract

Rabbit hybridomas are gaining wide acceptance as serologic reagents to identify a variety of antigens, including proteins, small peptides, phosphorylated proteins, and polysaccharides. Rabbits make high-affinity IgG antibodies, all of which bind with high affinity to Protein A from *Staphylococcus aureus* and Protein G from Group G *Streptococcus*. Consequently, rabbit monoclonal antibodies of desired specificity can be rapidly detected using Protein A/G as secondary reagents. Here we describe the method for generating rabbit monoclonal antibodies using the rabbit hybridoma fusion partner 240E-1. The method begins with the immunization of rabbits and ends with the cloning the antigen-specific hybridomas.

Key words Hybridoma, Rabbit, Monoclonal antibody, Adjuvant

1 Introduction

Rabbits are a preferred animal species for generation of polyclonal antibodies because they are small, and yet large amounts of serum can be obtained at regular intervals. Most importantly, rabbits make high-affinity antibody against a wide variety of antigens including proteins, phosphorylated proteins, small peptides, and polysaccharides. Several attempts to make stable mouse–rabbit heterohybridomas for the purpose of producing high-affinity rabbit monoclonal antibodies were made [1–3], but a rabbit hybridoma fusion partner was not successfully developed until 1995. Spieker-Polet et al. [4] generated a cell line, 240E-1 from a plasmacytoma-like tumor that arose in a c-*myc*/*v-abl* transgenic rabbit. From this 240E-1 cell line, the authors obtained an HGPRT mutant, allowing the use of HAT medium for selection of hybridomas. This 240E-1 fusion partner is the original source of all rabbit hybridomas made in both commercial and noncommercial settings.

The advantages of rabbit monoclonal antibodies are that they have increased sensitivity and specificity compared to murine monoclonal antibodies. Rossi et al. [5] compared rabbit and murine monoclonal antibodies in immunohistochemistry experiments to

Vincent Ossipow and Nicolas Fischer (eds.), *Monoclonal Antibodies: Methods and Protocols*, Methods in Molecular Biology, vol. 1131, DOI 10.1007/978-1-62703-992-5_5,

identify tumor cells in paraffin-embedded tissues, and they directly demonstrated the increased efficacy of rabbit monoclonal antibodies. Similarly, Nunes et al. [6] showed that a rabbit monoclonal antibody was more sensitive than mouse monoclonal antibodies for detection of Her2 in breast cancer; and Cheuk et al. [7] showed that rabbit monoclonal antibody against cyclin D1 can be used readily to detect mantle cell lymphoma. The high sensitivity of rabbit monoclonal antibodies is undoubtedly due to their high affinity as reported by Yu et al. [8] and Zhu and Yu [9]. Rabbit IgG exists as a single subclass [10] which binds Protein A/G with high affinity, making protein A/G an excellent secondary reagent for immunofluorescence, ELISA, or western blot analyses of rabbit monoclonal antibodies. Not only these antibodies provide an alternative to mouse monoclonal antibodies, but they can also be used to make antibodies against murine antigens. Hundreds of rabbit monoclonal antibodies against a wide range of antigens are now commercially available.

In general, the process of generating rabbit monoclonal antibodies using the 240E-1 hybridoma fusion partner is similar to that for generating mouse monoclonal antibodies. Several laboratories have experienced difficulty in developing hybridomas with the 240E-1 fusion partner, and we highly recommend that the following protocol be followed carefully.

The procedure we describe for the production of monoclonal antibodies (Fig. 1) includes the following steps: (1) Immunization of rabbits; (2) Preparation of the parental fusion partner, 240E-1; (3) Fusion of primary lymphocytes with the fusion partner; (4) Selection and expansion of stable hybridomas; (5) Freezing of select hybridomas; and (6) Cloning of select hybridomas.

2 Materials

2.1 Immunization Reagents

1. Freund's adjuvant, Complete (Sigma-Aldrich).
2. TiterMax® Gold Adjuvant (Sigma-Aldrich).
3. Freund's adjuvant, Incomplete (Sigma-Aldrich).
4. 6-month to 2-year-old rabbits.

2.2 Reagents for Preparation of the Hybridoma Fusion Partner

1. 240E-1 rabbit hybridoma fusion partner (available from the authors).
2. 8-azaguanine (Sigma-Aldrich).
3. Enriched RPMI 1640 powder medium (Life Technologies) dissolved in 900 mL ultrapure water; add the following to 1 L total (*see* **Note 1**):

   L-glutamine 200 mM (100×), (Life Technologies), 10 mL

   Hepes 1 M, (Life Technologies), 10 mL

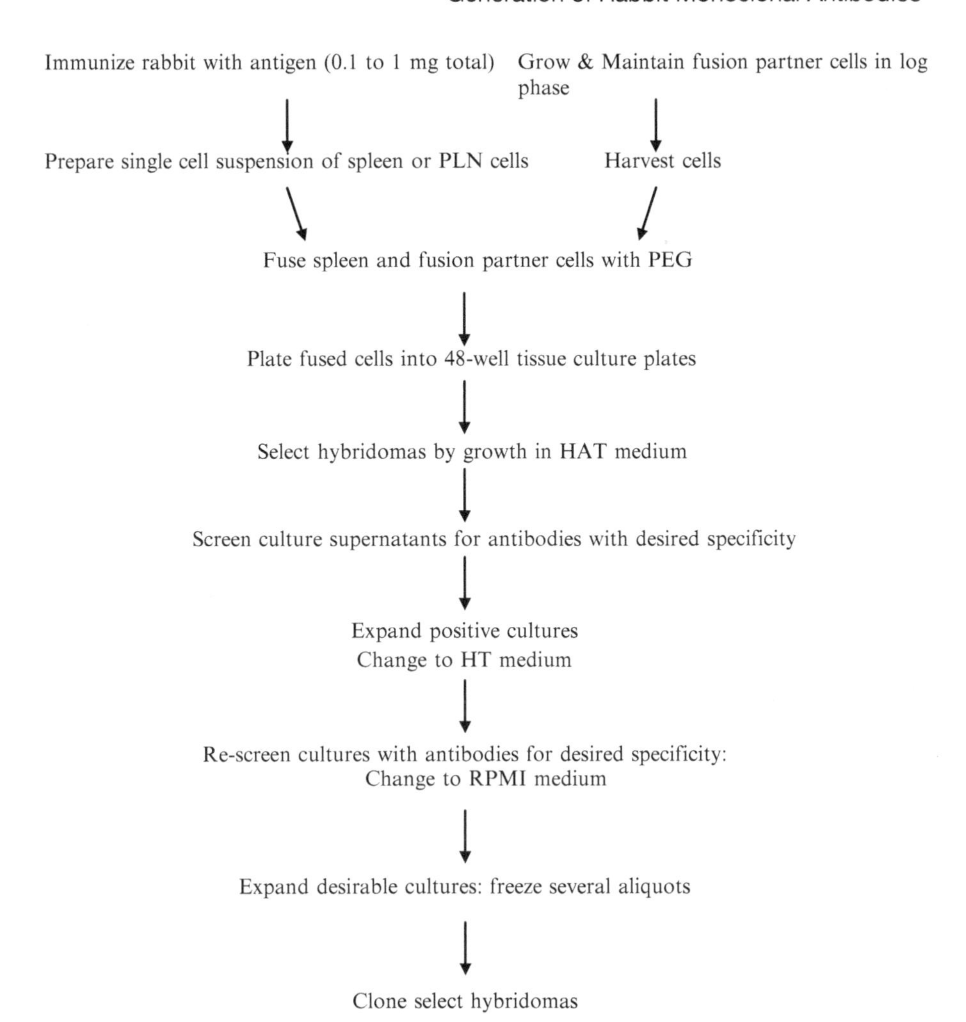

Fig. 1 Flow diagram for generation of rabbit monoclonal antibodies

MEM Amino Acids, 50× (Life Technologies) 10 mL

MEM Nonessential Amino Acids100×, (Life Technologies), 10 mL

Sodium Pyruvate, (Life Technologies) 10 mL

MEM vitamins 100×, (Life Technologies) 10 mL

β-mercaptoethanol (pre-dilution of 50 mM) 1 mL

Sodium bicarbonate, 7.5 %, (Life Technologies) 25–30 mL (pH ~7.2; judge by color—pinkish-red)

Gentamicin (Life Technologies 15750) 10 mg/mL 3 mL

Penicillin (10,000 U/mL) and Streptomycin (10,000 μg/mL)

(Life Technologies 15140) 5 mL

Fungizone (Life Technologies 15290) 250 μg/mL 2 mL

Fetal calf serum (FCS) 15 % unless otherwise stated (*see* **Note 2**).

2.3 Preparation of Spleen Cell Suspension

1. Forceps.
2. Scissors.
3. Razor blades.
4. 100 mm petri dishes.
5. 100 μm BD Falcon Cell Strainers.

2.4 Fusion

1. PEG 4000 (EM Science, Cherry Hill, N.J.): make 50 % PEG (w/v in HBSS or serum-free medium) (*see* **Note 3**).
2. Serum-free medium at room temperature (RT) for initial washing of cells.
3. 50 mL of pre-warmed (37 °C) medium for washing PEG after the fusion.

2.5 Selection and Expansion of Hybridoma

1. HAT supplement (50×) (Life Technologies).
2. HT supplement (50×) (Life Technologies).

2.6 Freezing Select Hybridomas

1. Freezing medium: Enriched RPMI 1640 medium with 20 % FCS and 10 % DMSO.

3 Methods

3.1 Immunization

1. Inject rabbits with 0.1–1.0 mg of protein (total) (or 2×10^7 antigen-expressing cells) with Freund's Complete Adjuvant or TiterMax® Gold Adjuvant (Sigma-Aldrich) subcutaneously, intramuscularly, and if using TiterMax® Gold Adjuvant, intraperitoneally.
2. Give one or two booster immunizations at 2–3 week intervals in the same manner using incomplete Freund's adjuvant.
3. Give final boost (0.2 mg) in saline, intraperitoneally and intravenously 4 days prior to the fusion.

3.2 Preparation of the Hybridoma Fusion Partner

1. Culture 240E-1 (non-adherent) cells in enriched RPMI 1640 containing 15 % FCS in T75 flasks at a concentration of $0.2–0.6 \times 10^6$/mL (*see* **Note 4**).
2. 4–6 days before a fusion, feed the cells daily, either by adding fresh medium or by dilution. It is important that cells are in log phase growth at the time of fusion. Cells that have been frozen in liquid nitrogen are grown in the presence of 8-azaguanine (20 μg/mL) for several days to insure that all cells are defective for HGPRT and will die in selective HAT medium.
3. 3 days before fusion, pellet the cells and resuspend in medium without 8-azaguanine. Do not use cells that have been in culture longer than 4 weeks.

3.3 Preparation of Spleen Cell Suspension

1. Wash the entire spleen several times in sterile medium (*see* **Note 5**).
2. After the final wash, place the spleen in a sterile petri dish containing 10 mL medium with serum.
3. Cut the spleen transversely into 3–4 mm strips using forceps and either scissors or a razor. Place 2–3 strips into a cell strainer over a 50 mL tube.
4. Gently tease the cells from the tissue using the rubber end of a 3 mL syringe plunger. Use a circular motion pressing the spleen chunks between the syringe plunger and the strainer mesh.
5. Add 2–3 mL of the medium to obtain the cells that were released into the tube below the strainer.
6. Continue releasing cells from the tissue until all the chunks have been processed.
7. After the entire spleen has been processed, spin the cells down at $300 \times g$ for 8 min and place suspension of cells in flasks overnight to remove adherent cells (*see* **Note 6**).
8. Wash non-adherent cells twice in serum-free medium prior to fusion (*see* **Note 7**).

3.4 Fusion

1. Fuse approximately $1.5–3 \times 10^8$ lymphocytes with 240E cells at a ratio of 2:1 with 50 % PEG 4000 at 37 °C in serum-free medium [11].
2. Wash splenocytes and 240E-1 fusion partner cells twice in separate tubes by pelleting at $300 \times g$ for 8 min at RT and suspending the cells in serum-free medium. Use 25 mL of serum-free medium for each wash.
3. Pool splenic and fusion partner cells together in ratio of 2:1, and wash once more.
4. Discard as much of the supernatant as possible.
5. Loosen the dry cell pellet and place tube in 37 °C water bath.
6. Add 50 % PEG (1–2 mL) with a syringe and 19 gauge needle in the following manner (keep tube immersed in a 37 °C water bath the entire 5 min):

   First minute—add all PEG, drop by drop, while mixing

   Second minute—continue mixing the cell pellet in the PEG by flicking the tube

   Third minute—add 1–2 mL serum-free medium, while mixing

   Fourth minute—add 5 mL serum-free medium; continue mixing

   Fifth minute—add the remaining 30–40 mL of serum-free medium.
7. Centrifuge the fused cells at $300 \times g$ for 8 min and wash once.
8. Plate the cells in 48-well microtiter plates ($\leq 2 \times 10^5$ lymphocytes/0.5 mL/well).

3.5 Selection and Expansion of Hybridomas

1. Plate the cells in 48-well microtiter plates at $\leq 2 \times 10^5$ lymphocytes per well in 0.5 mL medium with 15 % FCS (*see* **Note 8**).
2. After 3–5 days, 0.5 mL of 2× HAT diluted in the medium is added to the medium to kill unfused 240E-1 cells.
3. Change the medium every 4–5 days by aspirating ~50 % of the medium and replacing with fresh medium containing 1× HAT.
4. Observe clones between 10 and 14 days after fusion.
5. Test supernatants for antibody in 3–5 weeks (*see* **Notes 9** and **10**).
6. After positive hybridomas are identified, expand into 2 mL cultures and transfer to 12-well plates. At this point the culture can be switched from HAT medium into HT medium for the purpose of removing aminopterin, the folic acid antagonist in HAT that inhibits de novo nucleoside biosynthesis (*see* **Note 11**).
7. Further expand the cultures into 2–3 wells for each positive hybridoma and maintain them in HT medium. Do several changes with HT/RPMI medium to insure removal of HAT; trace amounts of aminopterin will kill cells if HT is not provided.

3.6 Freezing Select Hybridomas

1. Let hybridomas grow in HT medium with several changes of medium until ~5×10^6 cells have accumulated.
2. Centrifuge the cells at $300 \times g$ and resuspend in 1.2–1.5 mL freeze medium/vial.
3. Freeze at least 1×10^6 cells/vial and store in liquid nitrogen for future use and cloning (*see* **Note 12**).

3.7 Cloning Select Hybridomas

1. Clone by limiting cell dilution in 48-well microtiter plates (*see* **Note 13**).
2. Add 5×10^4 fusion partner 240E-1 cells per well as feeder cells (these cells will die in 5–6 days in HAT medium.
3. Plate 0.5 mL hybridoma cells (1–2 cells/mL in medium) in each well.
4. 5 days later, add 0.5 mL of medium with 2× HAT (the goal is to obtain colonies in less than 30 % of the wells to ensure the likelihood that they will be clonal.
5. Feed cultures every 3–4 days by removing one-half of the medium and adding equal amount of fresh medium with 1× HAT.
6. When the wells becomes confluent, and the supernatant fluid is at least 5–7 days. Old, test supernatant for secretion of specific antibody. Positive clones can again be expanded, weaned from HAT, and frozen as described previously.

4 Notes

1. We use RPMI 1640 medium powder which contains L-glutamine but no bicarbonate. Filter-sterilized medium can be stored frozen, but it must be mixed after thawing. FCS (15 %) is added and the medium is filter-sterilized again, before use. If necessary, adjust pH to 7.2–7.4 with 1 N HCl or 1 N NaOH. (It is better to use medium that is slightly more acidic than basic.)
2. If the 240E-1 cells grow slower than expected (i.e., less than the doubling time of 2–3 days), it may help to try different batches of FCS. We find that different lots of FCS vary in the capacity to support the growth of 240E-1.
3. Make 50 % PEG immediately before use. Depending on the size of the pellet of fusion partner and spleen cells you will need a total of 1–2 mL of PEG. Make twice as much as you will need to accommodate the loss of volume after filtration:
   (a) Melt PEG at 68 °C
   (b) Add serum-free medium (pre-warmed to 68 °C)
   (c) Vortex to mix well
   (d) Adjust pH to 7.2–7.4, if necessary (pinkish-red)
   (e) Filter (0.2 μm or 0.45 μm) medium to sterilize while still warm
   (f) Keep tube of PEG in 37 °C water bath until ready to use
4. The 240E-1 cell line grows slowly with a doubling time of 2–3 days; the cells may increase less than 50 % in number after 24 h in culture. These cells do, however, metabolize heavily and the medium may become very yellow. One may be surprised by how few cells are present when the medium is yellow. We find it easier to grow 240E-1 cells in standing flasks rather than in flasks lying side-ways because the medium can be changed every 1–2 days without having to spin down the cells. This is especially important when large numbers of log-phase cells are needed for a fusion.
5. Peripheral lymph node (PLN) cells from immunized rabbits can also be used to generate hybridomas. The advantage of PLNs is that there are many fewer adherent cells to interfere with growth of the hybridomas. Rabbits are immunized with 0.1–1 mg (total) by subcutaneous injection in several places of the leg below the PLN. The first injection is given with Freund's Complete Adjuvant or TiterMax® Gold Adjuvant (Sigma-Aldrich); three or four subsequent injections are given at 5 day intervals with the antigen dissolved in saline; the fusion is performed 4 days after the last boost.

6. The purpose of the overnight incubation is to minimize the excessive growth of adherent cells after the fusion. If adherent cells grow and interfere with the growth of the hybridomas, it will be necessary to transfer the hybridomas into another well as soon as possible. We also find that the fusion efficiency is increased if the isolated lymphocytes are incubated for 24–48 h on plastic dishes or in flasks with the antigen (about 50 μg/mL) for in vitro boosting. The efficiency of fusion can also be increased by the addition of CD40L-transfected cells (X-irradiated) in a ratio of 1:10 (CD40L-transfected cells to spleen cells) during this step of pre-incubation.
7. If the fusion cannot be performed on the day the rabbit is sacrificed, the intact spleen can be maintained for 1 day on ice (in a tube without medium) and used successfully for fusion 24 h later. We have little experience using frozen spleen cells from immunized rabbits for the fusion; however, if the cells remain viable, the fusion should be successful.
8. The aim is to obtain no more than 25 % positive wells, thereby insuring that the emerging hybridomas will likely be clonal.
9. The 48-well format may require longer incubation time before the supernatant fluids can be tested, but the advantage is that more supernatant will be available for multiple tests. Since not all hybridomas will be ready for testing simultaneously, this format also allows time between retesting positive clones, e.g., for re-screening and cloning. In a typical fusion, we obtain 100–300 positive wells from 1,000 wells plated. Typically, 5–15 % of these clones are specific for the immunized antigen.
10. Many of the hybridomas secrete very low amounts of antibodies. It has been reported that supplementing the hybridoma culture with recombinant human interleukin-6 enhances immunoglobulin secretion of rabbit hybridomas [12]. Alternatively, after a clonal population of hybridoma cells is available, the IgH and IgL genes of interest can be PCR-amplified and cloned into mammalian expression vectors for transfection into a cell line that will secrete high amounts of intact immunoglobulin.
11. Sometimes hybridoma cultures overgrow by the time the antibody specificity is determined. With time, these cultures should recover after they have been expanded. However, cultures may also look unhealthy if they are weaned from HAT too quickly. It is important that HT is supplemented in the medium used for cells that are being weaned from HAT. Aminopterin is highly toxic and even trace amounts can kill cells if HT is not provided. After the cells have been in HT medium for several passages, the hybridomas can be cultured in medium alone.
12. Positive hybridomas should be frozen as soon as possible; they should be frozen within days rather than weeks from the time

the supernatants are tested. Since the hybridomas grow slowly, you might not be able to freeze many aliquots with a sufficient number of cells. At least one vial should be frozen immediately, even if there are less than 1×10^6 cells; after thawing this vial, put the cells back in a 12-well plate rather than into a flask. Freeze down more aliquots within the next few days.

13. Alternatively, hybridoma cells can be cloned using fluorescence-activated cell sorting (FACS). Since we have had less success cloning by FACS, we suggest that the rabbit hybridomas are sensitive to manipulation through the cell sorter, resulting in decreased viability. We encourage investigators to try cloning by FACS, but recognize that it may be necessary to resort to the old-fashioned way of cloning, by limiting dilution.

References

1. Raybould TJG, Takahashi M (1988) Production of stable rabbit-mouse hybridomas that secrete rabbit mAb of defined specificity. Science 240:1788–1790
2. Kuo M, Sogn JA, Max EE, Kindt TJ (1985) Rabbit-mouse hybridomas secreting intact rabbit immunoglobulin. Mol Immunol 22: 351–359
3. Verbanac KM, Gross UM, Rebellato LM, Thomas JM (1993) Production of stable rabbit-mouse heterohybridomas: Characterization of a rabbit monoclonal antibody recognizing a 180 kDa human lymphocyte membrane antigen. Hybridoma 12:285–295
4. Spieker-Polet H, Setupathi P, Yam P-C (1995) Rabbit monoclonal antibodies: Generating a fusion partner to produce rabbit-rabbit hybridomas. Proc Natl Acad Sci 92:9348–9352
5. Rossi S, Laurino L, Furlanetto A, Chinellato S et al (2005) Rabbit monoclonal antibodies: a coparative study between a novel category of immunoreagents and the corresponding mouse monoclonal antibodies. Anatomic Pathol 124: 295–302
6. Nunes CB, Rocha RM, Reis-Filho JS, Lambros MB, Rocha GF, Sanches FS, Oliveira FN, Gobbi H (2008) Comparative analysis of six different antibodies against Her2 including the novel rabbit monoclonal antibody (SP3) and chromogenic in situ hybridisation in breast carcinomas. J Clin Pathol 61:934–938
7. Cheuk W, Wong KO, Wong CS, Chan JK (2004) Consistent immunostaining for cyclin D1 can be achieved on a routine basis using a newly available rabbit monoclonal antibody. Am J Surg Pathol 28:801–807
8. Yu Y, Lee P, Ke Y, Zhang Y, Yu Q, Lee J, Li M, Song J, Chen J, Dai J, Do Couto FJ, An Z, Zhu W, Yu GL (2010) A humanized anti-VEGF rabbit monoclonal antibody inhibits angiogenesis and blocks tumor growth in xenograft models. PLoS ONE [Electronic Resource] 5:e9072
9. Zhu W, Yu G-L (2009) Rabbit hybridoma. In: An Z (ed) Therapeutic monoclonal antibodies: from bench to clinic. 889. John Wiley & Sons, Hoboken, NJ
10. Knight KL, Burnett RC, McNicholas JM (1985) Organization and polymorphism of rabbit immunoglobulin heavy chain genes. J Immunol 134:1245–1250
11. Oi VT, Herzenberg LA (eds) (1980) Immunoglobulin-producing hybrid cell lines. W. F. Freeman & Co, San Francisco
12. Liguori MJ, Hoff-Velk JA, Ostrow DH (2001) Recombinant human interleukin-6 enhances the immunoglobulin secretion of a rabbit-rabbit hybridoma. Hybridoma 20:189–198

Chapter 6

Screening Hybridomas for Cell Surface Antigens by High-Throughput Homogeneous Assay and Flow Cytometry

Alejandro Uribe-Benninghoff, Teresa Cabral, Efthalia Chronopoulou, Jody D. Berry, and Cindi R. Corbett

Abstract

Described herein are methods for the successful screening of monoclonal antibodies (mAbs) of the desired specificities via high-throughput (HTP) homogeneous assay and flow cytometry. We present a combination of screening techniques that allow the scientist to efficiently eliminate nontarget-specific antibody as soon as possible. This compilation of protocols will enable researchers with basic immunology skills to make decisions regarding the design of screening algorithms for the generation of mAbs. Although we have provided an informative overview of both HTP homogeneous assay and flow cytometry, it is imperative for the beginner to acquire fundamental knowledge on how both of these technologies work so as to use these screening strategies effectively.

Key words High-throughput homogeneous assay, Hybridoma screening, Flow cytometry, Fluorochromes, Cell surface antigens

1 Introduction

This chapter describes procedures used for the identification of cells producing antigen-specific monoclonal antibodies (mAbs). These procedures are suitable for soluble (cell-free globular proteins), cell surface, or cell-associated antigens. High-throughput (HTP) cell- or bead-based screening has been used as a means to identify mAbs to very closely related and structurally similar targets [1].

1.1 Hybridomas

Hybridomas are generated as a means of creating mAbs to a variety of targets used for research, diagnostic, or therapeutic purposes. A classical hybridoma screening technique includes enzyme-linked immunosorbent assay (ELISA, see the chapter on hybridoma technology for the generation of rodent mAbs via hybridoma fusion). Herein, we describe alternative screening methods including HTP homogeneous assays and flow cytometry against cell surface-expressed antigens or a HTP homogeneous assay for

Vincent Ossipow and Nicolas Fischer (eds.), *Monoclonal Antibodies: Methods and Protocols*, Methods in Molecular Biology, vol. 1131, DOI 10.1007/978-1-62703-992-5_6, © Springer Science+Business Media New York 2014

soluble antigens coated on beads. Providing advantages for each of these approaches will enable a researcher to decide upon the most appropriate method for screening novel mAbs with the desired binding characteristics for their downstream purposes.

1.2 High-Throughput Homogeneous Cell- or Bead-Based Assay

HTP homogeneous cell- or bead-based assays provide an alternative method from ELISA for the primary HTP screening of mAb against either cell-expressed antigens or soluble antigens coupled to polystyrene beads. Advantages to this approach include the following: minimal amount of antigen is used, an important factor to consider when the supply of antigen may be limited or price prohibitive, as well as simultaneous screening against related antigens of interest [1]. In addition, cell surface antigens can provide challenges when screening via ELISA due to the potential loss of conformational epitopes and solubility when they are recombinantly expressed for purification as a soluble protein (i.e., not cell associated). Although suitable, this method can result in loss of important conformational epitopes within the antigenic region, which may be critical for the identification of neutralizing antibodies. Expression and screening on a cell surface can be conducted using cell-based ELISAs; however, this method often yields low signal, high background, and variability. Therefore, HTP screening of surface antigens is recommended via a cell-based assay. Furthermore, a benefit of using this system for both bead- and cell surface-expressed antigens is the convenience of screening hybridomas at an early stage of growth within a 96-well plate, allowing for quick identification of antibody-specific hybridomas.

Bead- and cell-based assays are based on the premise that homogeneous incubation of antigen, antibodies, and a fluorescent detection system provide a highly sensitive method for hybridoma screening. Antigen is either expressed on the cell surface or bound to polystyrene beads. The mixture is incubated from 1 to 4 h allowing the antibodies to bind targets and settle to the bottom of the well. These assays utilize instruments that scan the bottom of a micotiter plate by an excitation source (typically 2 or more lasers). Such instruments include the Applied Biosystems 8200 Cellular Detection System, mirrorball (TTP Labtech), iCyte (Compucyte), and the ImageExpress microXL (TTP Labtech). These microplate cytometers have proprietary laser scanning technology, which scans either 1 mm portions of a well (Applied Biosystems 8200 CDS) or the entire well (mirrorball, iCyte, ImageExpress microXL), both with a fixed depth. Excitation emissions from the laser are captured through photomultiplier tubes (PMT), which are extremely sensitive detectors of light in the visible, UV, and infrared range of the electromagnetic spectrum. This allows for an accurate object scan with a low signal-to-noise ratio. Most instruments are capable of scanning 96-, 384-, and 1,536-well plates (Fig. 1).

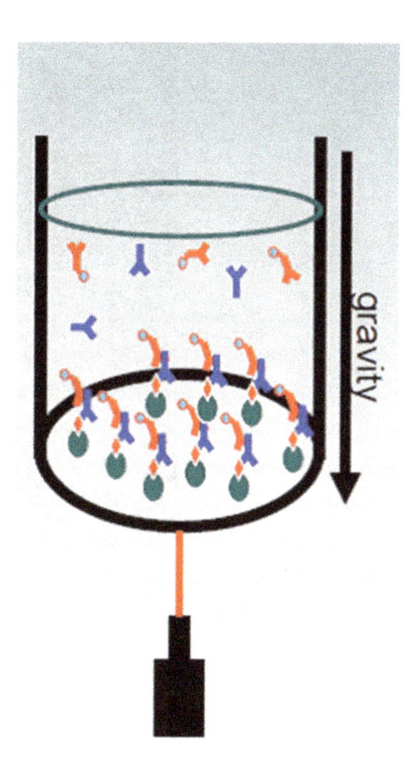

Fig. 1 Schematic representation of a homogeneous assay detection system. Supernatant from hybridoma culture plates is mixed with either antigen-coated beads or antigen-expressing cells together with a fluorescently labeled anti-IgG Fc antibody. Where specific antibody, antigen, and conjugate come together, the beads or the cells settle to the bottom of the well and a positive emission fluorescent is recorded by the imaging software

Prior to starting these screening approaches, the user must determine the appropriate methods to obtain antigen in the desired conformation. In addition, if cells expressing the desired antigen for screening are used in the assay, they need to be assessed for protein expression (for example, via Western blots or HTP homogeneous assay with a known positive control). Preparation work (i.e., protein expression) should be established prior to utilizing the method.

1.3 Data Acquisition and Analysis for HTP Homogeneous Assays

As depicted in Fig. 1, when the hybridoma supernatant contains specific mAbs that bind to the beads or cells they settle on the bottom of the plate, the fluorochrome conjugated to the secondary antibody is excited by the laser, and the emissions are collected by the PMT. The emissions are converted into data representing the number of fluorescent counts and intensity.

For example, captured images enable visualization of the fluorescent activity within a well (Fig. 2). When using cells as the target, it is important to visualize single cells and not clumps. It is up to the user to be familiar with the instrument in order to assess these criterions.

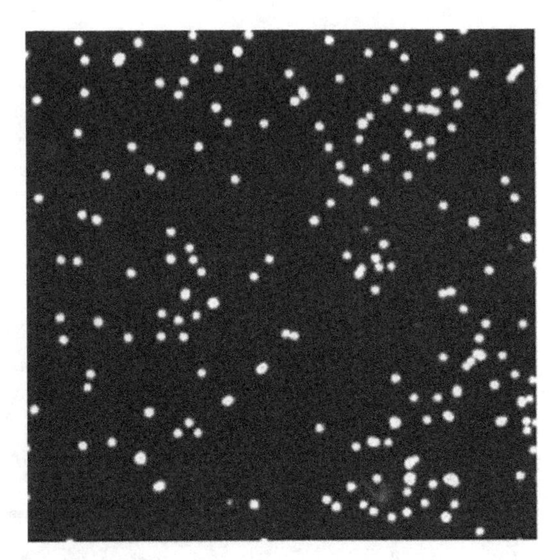

Fig. 2 Single-well optical image captured of a hybridoma screen [1]. The white dots represent fluorescence of a positive clone (image was captured with Applied Biosciences® 8200 Cellular Detection System Analyses Software, version 3.0, 2003, USA)

In the system depicted here, the data can also be represented as 3D histograms of the plate (Fig. 3). The *X*-axis denotes column number, the *Y*-axis fluorescent count observed per well, and the *Z*-axis row on the plate. Viewing the scanned plate in this format allows for the quick identification of potentially positive clones and cross-reference to the captured image (Fig. 2).

Another valuable method of data analysis using the Applied Biosciences® 8200 Cellular Detection System Analyses Software is the spreadsheet (Fig. 4). The data output will vary depending on the system however and is given per well for the number of events measured, the level of fluorescence detected, and the total fluorescence measured by multiplying the events by the fluorescence (*see* **Note 1**). It is important to set a threshold for fluorescent detection to eliminate false positives. For example, the fluorescence threshold presented in Fig. 4 was set to 50; therefore, wells with a fluorescence detection below 50 were assigned a value of zero. Typically positive clones that demonstrate a high increase in total count (this is the number of fluorescent events/well) and total fluorescence are chosen for further analyses. Assessing both of these parameters has proven to be essential within the conditions conducted within our laboratory for successfully picking positive hybridoma clones.

1.4 Screening by Flow Cytometry

Screening by flow cytometry is the most labor-intensive and time-consuming methods to screen mAbs. The majority of the hybridoma cells produced will not produce antibody of the desired specificity, and hybridomas that secrete antigen-specific antibody

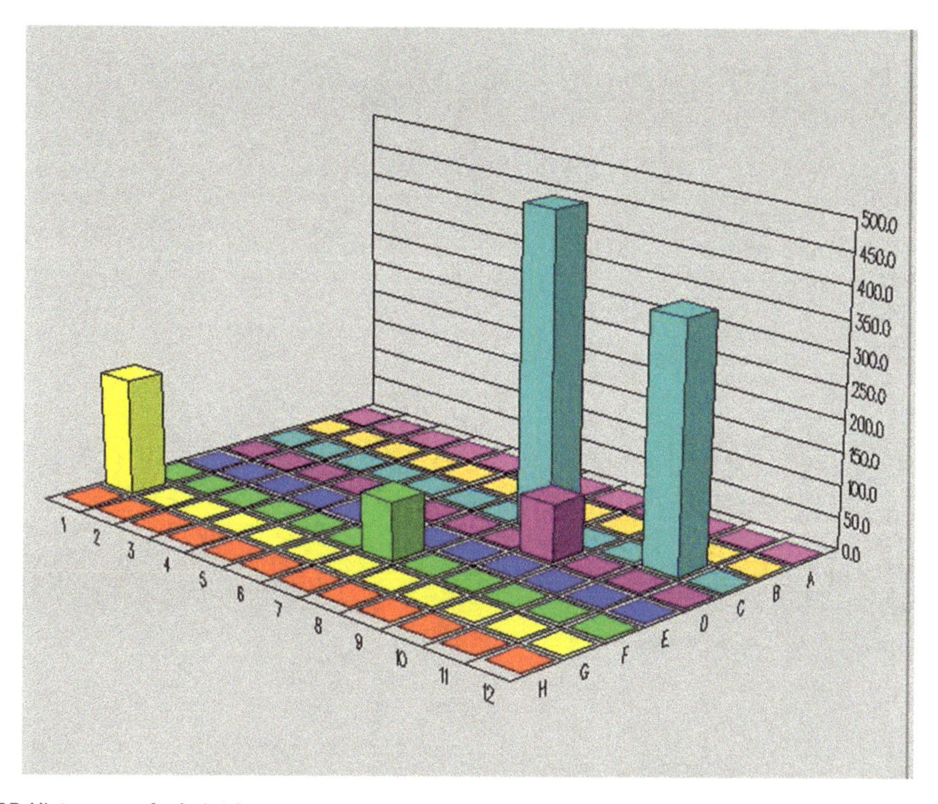

Fig. 3 3D Histogram of a hybridoma screen plate. The optical image of well C8 is seen in Fig. 2 (image is from hybridoma screening data [1] captured on Applied Biosciences® 8200 Cellular Detection System Analyses Software, version 3.0, 2003, USA)

range from 1 to 5 % of the total population. The most common method used to screen hybridomas post fusion is an ELISA (please see the chapter on hybridoma technology for the generation of rodent mAbs via hybridoma fusion). Positive clones selected by a primary fusion screened via ELISA can then be tested for their functionality in flow cytometry or other applications. However, not all clones identified by ELISA screening will be suitable for a flow cytometric application. Clone screening could be performed simultaneously for both ELISA and flow cytometry, but the amount of effort required to pursue both strategies could hinder the researcher. Indeed, it is not recommended to perform initial screening on a flow cytometer due to a higher risk of nonspecific or cross-reactive clone selection. Thus, coupling HTP homogeneous cell assay screening followed by flow cytometry for cell surface antigens provides a researcher with a tangible and robust screening approach. Regardless of the approach chosen, great care must be given when selecting clones, as their specificity should be verified by more than one screening technique. Antibody specificity can be verified by Western blot (WB) or immunoprecipitation, IP/WB; however, these techniques are not of HTP and should only be used

Plate= 1											
Population=	PopA	Parameter=	FL1								
0	0	0	0	0	0	0	0	0	0	0	0
0	0	0	0	0	0	0	0	0	0	0	0
0	0	0	0	0	0	0	495.8	0	0	384.9	0
0	0	0	0	0	0	0	0	88.6	0	0	0
0	0	0	0	0	0	0	0	0	0	0	0
0	0	0	0	0	0	91	0	0	0	0	0
182.1	0	0	0	0	0	0	0	0	0	0	0
0	0	0	0	0	0	0	0	0	0	0	0
Plate= 1											
Population=	PopA	Parameter=	Count								
8	10	10	15	15	15	12	6	6	10	8	5
7	11	23	19	12	12	16	15	13	12	3	13
12	14	5	9	11	7	11	136	8	24	130	6
5	9	10	8	9	11	5	17	79	12	9	6
3	11	9	9	14	7	11	14	11	9	5	5
10	13	8	9	18	7	121	18	8	11	5	9
123	11	12	13	16	15	13	22	11	18	9	11
8	12	14	22	11	7	12	13	7	6	7	6
Plate= 1											
Population=	PopA	Parameter=	EventCount								
8	10	10	16	15	15	12	6	6	10	8	5
7	11	24	20	12	13	16	16	14	13	3	13
12	14	5	9	12	7	11	139	8	25	132	6
5	9	10	8	9	11	5	18	82	12	9	6
3	12	9	9	14	7	11	14	11	9	5	5
10	14	8	9	19	7	123	19	8	12	5	9
127	11	12	13	17	15	14	23	11	18	9	11
8	13	14	24	12	7	13	13	7	6	7	6
Plate= 1											
Population=	PopA	Parameter=	FL1_Total								
0	0	0	0	0	0	0	0	0	0	0	0
0	0	0	0	0	0	0	0	0	0	0	0
0	0	0	0	0	0	0	67436	0	0	50049	0
0	0	0	0	0	0	0	0	7003	0	0	0
0	0	0	0	0	0	0	0	0	0	0	0
0	0	0	0	0	0	11022	0	0	0	0	0
22400	0	0	0	0	0	0	0	0	0	0	0
0	0	0	0	0	0	0	0	0	0	0	0

Fig. 4 Excel spreadsheet generated from the software of the ABI8200 of one plate of a hybridoma screen. The corresponding 3D histogram is seen in Fig. 3, as well as the optical image of well C8 seen in Fig. 2 (Applied Biosciences® 8200 Cellular Detection System Analyses Software, version 3.0, 2003, USA)

for a limited number of clones. The volume of samples to be screened (1,200 on average) can pose a challenge on its own, even if the process is automated. The crux of the problem lies in the sample preparation, model system used to screen, and data analysis and interpretation. Assay results should be available within 1–2 days following screening to ensure early subcloning of hybridoma-secreting antibody of the desired specificity. If subcloning becomes protracted, other co-plated cells, potentially with a growth advantage, could overtake the culture and result in loss of good mAbs from a fusion. Indeed, HTP homogeneous assay screening is desired for these reasons as very small amounts of primary antibody are detected within this method.

1.5 Flow Cytometry

Flow cytometry employs an instrument that scans single cells or particles (0.2–150 μm in size) that flow past an excitation source (usually one or more lasers) in a liquid medium. This technology allows the researcher to perform quantitative multiparameter analysis of biochemical, biophysical, and antigenic characteristics of single cells (both live and dead) at a very high speed [2].

Applications of flow cytometry range from quantitative analysis to fluorescence-activated cell sorting (FACS), in which large numbers of cells are separated for functional studies. Other uses of flow cytometry include gene library preparation, direct cloning of single rare transfected cells, and most recently microbiological and bacteriological identification of pathogens in industrial processes (water quality, brewing, and distillation).

By combining cells with antibodies conjugated to fluorochromes (immunofluorescent labeling), visible and fluorescent light emissions are collected by a coupled optical and electronic system producing measurements that include relative cell size, granularity, and fluorescence intensity (*see* Fig. 5).

All these applications are dependent on two main variables, the first of which is instrument configuration. The lasers (emission wave length) and detectors (wave length range) of the flow cytometer define whether a given fluorochrome can be excited and whether there are enough detectors to properly detect a given combination of fluorochromes [3].

The second is defined by the ability to identify and analyze the cells of interest so that they may be characterized or sorted. Thus, the researcher should pay close attention to the reagent selection, cell preparation, and protocol design for staining cells when performing a flow cytometric experiment (*see* **Note 2**).

1.6 Fluorochromes

Fluorochromes are molecules that can absorb light of a specific wavelength and re-emit light at a longer wavelength. These molecules can be either organic in nature (fluorescent proteins or small molecules) or synthetically made (quantum dots). The assortment of conjugated fluorochromes to mAbs allows the researcher to directly calculate parameters such as the nucleic acid content of a cell, enzyme activity, calcium flux, membrane potentials, and stage of the cell cycle to name a few. Conjugated fluorochromes have also allowed the study and identification of cell surface receptors, intracellular molecules such as cytokines, chemokines, and transcription factors and the discovery of functional subpopulations of cells (*see* **Note 3**).

There are two main methods of immunofluorescent labeling: direct, whereby the antibody against the molecule of interest is

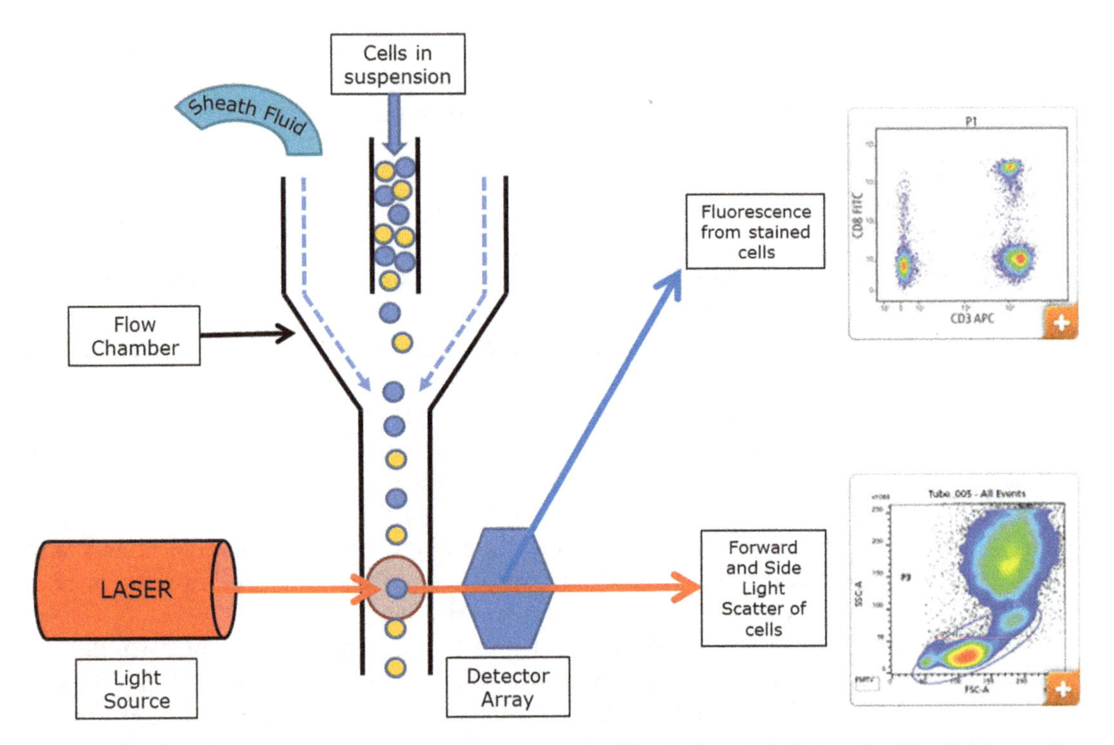

Fig. 5 A flow cytometer consists of three main systems: fluidics, optics, and electronics. The fluidics system transports cells by using hydrodynamic focusing and presenting them to a laser beam. The laser then illuminates each individual cell, and the light emitted is passed through optical filters and then to the appropriate detector. The electronics system then converts the light signal into electronic signals that can be processed in the analysis software

chemically conjugated to a fluorochrome, and indirect where the antibody specific for the molecule of interest (called the primary antibody) is unlabeled and a second anti-immunoglobulin antibody directed toward the constant portion of the first antibody (called the secondary antibody) is tagged with the fluorochrome. It is this last method of labeling that is used for hybridoma screening by flow cytometry.

The first rule to reagent selection is to pick the brightest fluorochrome compatible to the laser setup of the researcher's flow cytometer. The second rule of reagent selection is to minimize spectral overlap when choosing a reagent pair. Once the fluorochromes have been defined, antibody specificities should be matched to the particular fluorescent dyes. The brightest fluorochromes should be reserved for antibodies that are dim and vice versa [4] (*see* Table 1).

Due to the fact that antibody affinity in the hybridoma supernatant being tested is unknown, the researcher should select the brightest fluorochrome that is compatible with the instrument configuration in the lab. This will ensure an adequate detection, even on antibodies that have a low affinity for the target.

Table 1
Fluorochrome reference chart of commonly used dyes identifying fluorescence emission color, maximum levels of excitation and emission, and corresponding laser excitation line

Fluorochrome	Fluorescence Emission Color	Ex-Max (nm)	Excitation Laser Line (nm)	Em-Max (nm)
Alexa Fluor® 405	Blue	401	360, 405, 407	421
Pacific Blue®	Blue	410	360, 405, 407	455
Alexa Fluor® 488	Green	495	488	519
FITC	Green	494	488	519
PE	Yellow	496, 546	488, 532	578
PE-Texas Red®	Orange	496, 546	488, 532	615
Texas Red®	Orange	595	595	615
APC	Red	650	595, 633, 635, 647	660
Alexa Fluor® 647	Red	650	595, 633, 635, 647	668
PE-Cy™5	Red	496, 546	488, 532	667
PerCP	Red	482	488, 532	678
PerCP-Cy™5.5	Far Red	482	488, 532	695
PE-Cy™7	InfraRed	496, 546	488, 532	785
APC-Cy™7	InfraRed	650	595, 633, 635, 647	785

This same strategy should be applied for molecules that are expressed in low levels on the target cells.

Data acquisition and analysis for screening by flow cytometry: As a cell passes through the laser of a flow cytometer, a light pulse is sent to two different detectors: one that measures how the light scatters and another that measures fluorescence. The signal produced is called a voltage pulse, which is then converted into a numerical value. Data generated from each event can be represented in a number of different graphic displays, yet there are two fundamental types that can be used effectively to analyze the results: frequency histograms and bivariate plots [3, 5] (*see* **Note 4**).

Frequency histograms are the simplest mode of data analysis allowing the graphic depiction of a single measured parameter on the *x*-axis and the number of events (i.e., cell count) on the *y*-axis.

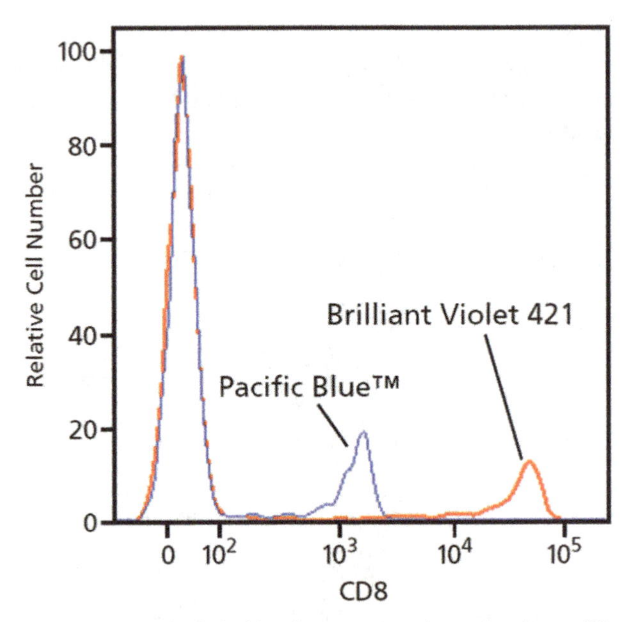

Fig. 6 Histogram of lysed whole blood comparing the stain of two different fluorochromes conjugated to CD8 (image courtesy of www.bdbiosciences.com)

Mean fluorescence and cell distribution can also be measured by defining markers placed on the plotted graph. The graph can be expressed in both linear and logarithmic scales, allowing the researcher to focus either on major characteristics (log scale) or more subtle ones (linear). Multiple histograms can be overlaid on one another so as to determine the percentage of cells that express a positive response when compared to a control sample (*see* Fig. 6). It is this last method of data analysis that is used for hybridoma screening by flow cytometry, when the cell population expressing the target of interest is present at high levels and a high percentage of the cell population.

Two-parameter histograms, commonly known as dot plots, yield more information than their single-parameter counterpart. Similar to a topographical map, the graph represents the measurement of two parameters, one on the *x*-axis and the other on the *y*-axis, compared to a density gradient that represents the number of events (i.e., cell count) (*see* Fig. 3).

Dot plots allow the researcher to visualize populations of cells that may remain hidden when analyzed via the single-parameter histogram, especially if the cells in question are found to be in a low percentage of the population being analyzed or if the expression of the target on the cells of interest is very weak or dim. Cells are stained with two distinct fluorochromes, and the data obtained from these events is then combined with data obtained for the light

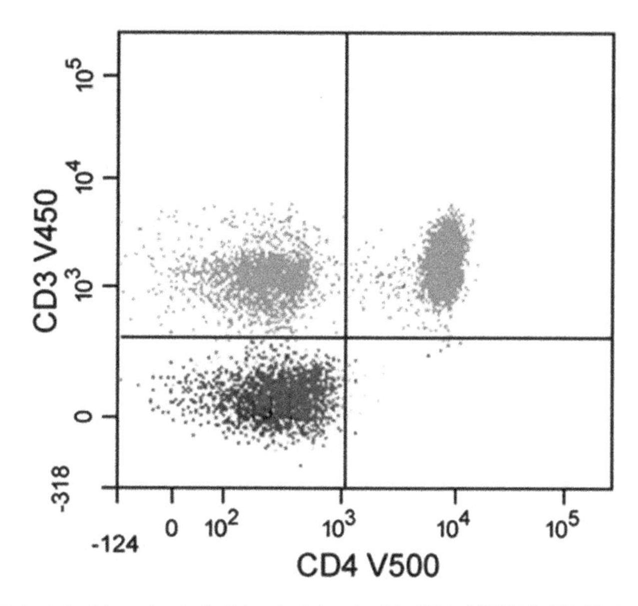

Fig. 7 Dot plot of lysed whole blood stained with CD3 (UCHT1) BD Horizon V450 and CD4 (RPA-T4) BD Horizon V500 (image courtesy of www.bdbiosciences.com)

scattering parameters (FSC and SSC). When the cell population expressing the target of interest is present at low levels or the expression of the target of interest is hard to observe, this is the data analysis method to follow (*see* Fig. 7).

2 Materials

2.1 High-Throughput Homogeneous Assay

1. Six micron polystyrene beads (for example, Spherotech beads or equivalent).
2. Siliconized microfuge tubes (ThermoFisher or equivalent).
3. End-over-end mixer to accommodate microfuge tubes.
4. Benchtop microfuge.
5. Benchtop centrifuge for spinning down cells.
6. 10 ml serological pipettes and pipette aid.
7. Inverted microscope for tissue culture.
8. 96-well assay plate (96-well black clear bottom plate, i.e., Applied Biosystems 8100HTS or equivalent).
9. 8- or 12-multichannel pipette.
10. Reagent reservoir.
11. Cell counter/hemocytometer.

12. Micropipetters and appropriate tips.
13. Biological safety cabinet.
14. 50 ml screw cap tubes.
15. Platform shaker.
16. Sterile flasks.
17. Sterile microfuge tubes (1.5 and 2.0 ml).
18. 1 % BSA in 1× PBS pH 7.2.
19. Dilution buffer (1 % BSA or 1 % skim milk) in 1× PBS pH 7.2.
20. Secondary fluorescent antibody goat anti-mouse IgG (Fc) at 0.4 μg/ml in dilution buffer.
21. 293 freestyle expression media (Invitrogen).
22. 293fectin transfection reagent (Invitrogen).
23. Opti-MEM media (Invitrogen).

2.2 Flow Cytometry

1. Sterile serological pipettes, Pipet-Aid, 50 ml centrifuge tubes (Falcon or Corning sterile), Eppendorf tubes, 96-well U-bottom microtiter plates (Falcon).
2. Multichannel pipettor and appropriate tips, reagent reservoir.
3. Cell-counting device such as the Cellometer™ Auto 2000 or equivalent.
4. Benchtop centrifuge (such as Beckman), tabletop centrifuge with rotor SX4750 that can accommodate microplate carriers such as the Allegra-X14 or equivalent.
5. Inverted microscope for tissue culture.
6. Ice bucket or refrigerator.
7. Accuri ™ C6, BD FACS LSR II with plate loader or equivalent.
8. Staining buffer: PBS (1×), 2 % fetal bovine serum (FBS), 0.1 % sodium azide (NaN3) used for antibody dilution, incubations, and washes (*see* **Notes 5** and **6**).
9. Sheath fluid for flow cytometer.
10. Secondary antibody: Anti-mouse IgG conjugated to phycoerythrin (PE) or allophycocyanin (APC) with minimal cross-reactivity to human, rat, and other species. Fluorochrome-conjugated antibodies should be stored at 4 °C in the dark. Freezing is not recommended (*see* **Note 7**).
11. Cell viability solutions: 7-Amino-actinomycin D (7AA-D) (*see* **Notes 8** and **9**) or 50 μg/ml propidium iodide solution.
12. Sterile 1× PBS pH 7.2.
13. 10 mg/ml BSA in 1× PBS pH 7.2.

3 Methods

3.1 High-Throughput Homogeneous Assay for Screening Hybridoma Supernatants Using Cell Surface-Expressed Antigens or Soluble Antigens Coupled to Beads

Refer **Notes 10–12** prior to running this method.

3.1.1 Bead Preparation/ Coating

1. Prepare 200 μl of polystyrene beads from a 5 % slurry in 1 ml PBS in a siliconized microfuge tube. Vortex the beads to obtain a homogeneous solution.
2. Vortex and microfuge for 1 min at max speed using a benchtop centrifuge.
3. Discard the supernatant, wash beads twice by re-suspension in 1 ml sterile PBS pH 7.2, and repeat **step 2**.
4. Resuspend the washed beads in 1 ml PBS containing 20–200 μg of purified protein and mix end over end at 4 °C overnight (*see* **Note 13**).
5. Centrifuge beads in a microfuge at max speed for 1 min.
6. Remove supernatant containing unbound protein and use for protein quantitation to ensure protein bound to the beads. Protein concentration prior to binding must be known to ensure successful coating of the beads.
7. Wash the beads to remove any unbound protein 3× with PBS pH 7.2 as above.
8. Block the beads by re-suspending in a 10 mg/ml solution of BSA in PBS (w/v) mixing end over end for 2 h at room temperature.
9. Wash the beads twice as in **step 2** and resuspend in 1 ml PBS containing 1 % BSA.
10. Resuspend the washed antigen-coated beads to a final concentration of 2.0×10^6 beads/ml.

3.1.2 Cell Preparation

There are numerous methods for transfecting cells, and so, it is up to the user to determine which method is suitable for their needs. Described here is a simple method utilized for surface expression of protein.

1. Obtain 293 F cells and culture as per manufacturer's instructions (Invitrogen). Transfections are conducted after a minimum of three passages. On the day prior to transfection, cells must be seeded at a density of 0.6 cells/ml.

2. On the day of transfection prepare the following (*see* **Note 14**):
3. In a sterile microfuge tube place 30 μg of vector in OPTI-MEM media to a final vol of 1 ml.
4. In a sterile microfuge tube place 60 μl of 293fectin in DMEM media to a final vol of 1 ml.
5. Incubate both mixtures at room temperature for 5 min.
6. Combine the two mixtures in 1 × 2 ml sterile tube and incubate for 30 min at room temperature.
7. 25 min into the incubation, remove cells from incubator and vortex to break apart cell clumps.
8. At the end of the 30-min incubation, add the mixture to the cells dropwise.
9. Incubate the cells with vector encoding the protein of interest (antigen) for 72 h.
10. Harvest the cells by centrifugation at 1,500 × *g* for 10 min.
11. Resuspend cells in 10 ml of 1 % BSA/PBS, and obtain the cell counts.
12. Wash cells twice using 1 % BSA/PBS.
13. Resuspend the cells to a final concentration of 2.0×10^6 cells/ml (*see* **Note 15**).

3.1.3 Preparing Plates for Screening Assay

For the cells/beads and secondary antibody, prepare 5 ml of each per assay plate.

1. Prepare coated beads or transfected cells at 2.0×10^6 per ml in dilution buffer. Transfer mixture to reservoir, and use a multichannel pipette to transfer 50 μl per well.
2. For detection, prepare 0.4 μg/ml of secondary fluorescent antibody in dilution buffer (*see* **Note 16**). Transfer mixture to reservoir, and use a multichannel pipette to transfer 50 μl per well.
3. Add 5 μl of 3-day hybridoma supernatants to each well using a multichannel pipette (*see* **Note 17**). Ensure to change tips between samples.
4. Incubate bead/cell, primary antibody, and secondary antibody plates for 1–4 h at room temperature.
5. Determine fluorescence on a multiplate reader (for example on ABI8200 or equivalent).
6. Data should be analyzed to ascertain which clones are positive. You need to pick clones that demonstrate and increase in fluorescence over appropriate controls (pre-bleed sera or isotype antibody control) as well as a secondary-only antibody control. Please note that overall fluorescence as well as counts per well should be assessed as this will give you the most accurate indication of positive results.

3.2 Sample Preparation for Screening Hybridoma Supernatants via Flow Cytometry

Screening hybridoma supernatants via flow cytometry relies on the selection and preparation of an optimized cell system where the target of interest is expressed. The cell system used for screening can be cell lines, transductants, transfectants (transient or stable), or primary cells (blood, PBMC, splenocytes, or other cells) (*see* **Notes 18–20**) In some cases the cells are treated with an activator or an inhibitor to upregulate or downregulate the expression of a specific target (e.g., such treatments could induce target phosphorylation or dephosphorylation). In other cases cells are cultured with a variety of factors for extended periods of time that allow them to differentiate to a cell type that expresses the target antigen (e.g., differentiation of naïve T cells to Th17 cells).

The sample preparation protocols vary depending upon the type of cells used for screening (suspension or adherent, blood or other tissue) and the target's cellular localization (surface or intracellular). For example if the target of interest is a surface protein and the cell system used for flow screening is a suspension cell line no special sample treatment is necessary. However, if the cell system is an adherent cell line, special care should be taken while detaching the cells since the use of enzymes or mechanical scrapping for cell detachment could affect the cell surface expression of the target (*see* **Note 21**). In addition, if the target is intracellular, the buffer used for fixation and permeabilization prior to immunofluorescent staining is critical (*see* **Note 22**). Generally for cytoplasmic targets such as cytokines a permeabilization buffer that contains a detergent such as saponin is optimal while for nuclear proteins, e.g., stats, a high methanol-containing buffer is recommended. For detailed protocols of immunofluorescent staining for intracellular targets refer to published protocols [6, 7].

3.3 Screening for mAbs Using Flow Cytometry

This procedure is for screening mAb supernatants against cell surface targets. Refer **Notes 23–30** prior to running this method.

1. Prior to immunofluorescent staining observe the cell suspension(s) used in the process under the inverted microscope to ensure their quality (cell suspensions should be viable and contamination free).
2. Harvest cells in a 50 ml conical tube and pipette up and down gently to break up any cell clumps since best results are obtained with single-cell suspensions. If the target cells are adherent use appropriate methods for their detachment from the tissue culture vessel that would not affect the expression of the target antigen (*see* **Note 21**). Cell suspensions (positive and negative cell target) used for immunofluorescent staining should be maintained on ice throughout the staining process to maintain high cell viability. The staining buffer should also be maintained on ice throughout the process.
3. Pellet cells by centrifugation at $350 \times g$ for 5 min, and resuspend them in cold staining buffer. Take a small aliquot and

perform a cell count and viability assessment using an automated cell counter. Adjust the cell concentration to 1×10^7/ml. Cell viability should be 85 % or higher.

4. Transfer 100 µl/ well of the cell suspension (10^6 cells/well) to a 96-well U-bottom microtiter plate, and pellet them by centrifugation at $350 \times g$ for 5 min.
5. Aspirate wells one by one carefully making sure to avoid touching the bottom of the well where the cells are located. The cell pellet should be visible as a white precipitate on the bottom of each well. Alternatively flick the plate, and paper blot it dry.
6. Transfer 100 µl of hybridoma supernatant in each well, and resuspend the cells on the bottom of each well three to four times. Make sure to use different tip for each well transferred. Incubate the plate at 4 °C in the dark for 15–20 min (*see* **Notes 31** and **32**).
7. Controls for staining must be prepared for compensation of overlapping fluorochromes as well as accounting for cell death. *See* **Notes 33** and **34** for details on these controls.
8. While incubating prepare the second-step reagent by diluting the optimally titrated secondary antibody in cold staining buffer [8]. Make sure to prepare enough second-step reagent for all wells received hybridoma supernatant. Dilute the second-step reagent so that the volume to be transferred per well is 50–100 µl. Best staining results are obtained when the staining volume is kept less or equal to 100 µl. Vortex the secondary antibody and withdraw the amount needed for the experiment in an microfuge tube containing staining buffer. Mix the diluted second step by vortexing gently.
9. Once the incubation is over remove the microtiter plate from 4 °C to wash unbound-to-cell antibodies. With a multichannel pipette transfer 150 µl of cold staining buffer per well making sure to change tips for each well. Pellet cells by centrifugation at $350 \times g$ for 5 min (first wash).
10. Aspirate wells carefully without touching the cell pellet, and refill them with 200 µl of cold staining buffer making sure to use one tip per well. Pellet wells by centrifugation at $350 \times g$ for 5 min (second wash).
11. Vortex the microfuge tube with the diluted second step and transfer 50–100 µl per well so that the optimal amount of the second step is delivered per well. Incubate for 15–20 min at 4 °C in the dark (*see* **Note 35**).
12. Wash microtiter plate with cold staining buffer twice as indicated in **steps 9–10**.
13. Following the second wash resuspend cells in each well with 200 µl of staining buffer and proceed to data acquisition and

analysis. Data acquisition should take place within a couple of hours following completion of the staining process to ensure that the cells are in good condition. Prolonged storage at 4 °C is not recommended unless the samples are fixed.

14. Dead cells can be gated out during acquisition by addition of a viability dye solution just before the samples are acquired [9, 10].

4 Notes

1. Upon analyzing the data, one must be careful interpreting the results. Instances may occur where there are larger clumps of fluorescent cells leading to a high total FL count. The captured image is important for analysis as the true fluorescence value will be depicted.
2. It is imperative to know the emission wavelength of the laser in the instrument that will be used for analysis. This wavelength will determine the fluorochromes selected by the researcher. Understanding that a different range of wavelengths is detected for each fluorescent channel is also important.
3. Common fluorochrome characteristics:
   - *Alexa Fluor® 405* dye, with visible-wavelength excitation, has minimal spectral overlap with green fluorophores.
   - *Alexa Fluor® 488 and 647* conjugates are highly photostable and remain fluorescent over a broad pH range. For more information please visit www.invitrogen.com.
   - *APC* is an accessory photosynthetic pigment found in blue green algae. Its molecular weight is approximately 105 kDa. APC has six phycocyanobilin chromophores per molecule, which are similar in structure to phycoerythrobilin, the chromophore in R-phycoerythrin (R-PE).
   - *APC–Cy™7* is a tandem conjugate system that combines APC and a cyanine dye (Cy7). It is recommended that special precautions be taken with APC–Cy7 conjugates, and cells stained with them, to protect the fluorochrome from long-term exposure to visible light. It is recommended that a 750-nm longpass filter be used along with a red-sensitive detector such as the Hamamatsu R3896 PMT for this fluorochrome.
   - *Fluorescein isothiocyanate (FITC)* is a fluorochrome with a molecular weight of 389 Daltons. The isothiocyanate derivative (FITC) is the most widely used form for conjugation to antibodies and proteins, but other derivatives are available. FITC has a high quantum yield (efficiency of energy transfer from absorption to emission fluorescence),

and approximately half of the absorbed photons are emitted as fluorescent light. The number of FITC molecules per conjugate partner (antibody, avidin, streptavidin, etc.) is usually in the range of three to five molecules.

- *Pacific Blue®*, a UV light excitable dye, is based on the 6,8-difluoro-7-hydroxycoumarin fluorophore and is strongly fluorescent, even at neutral pH. For more information please visit www.invitrogen.com.
- *R-PE* is an accessory photosynthetic pigment found in red algae. In vivo, it functions to transfer light energy to chlorophyll during photosynthesis. In vitro, it is a 240-kDa protein with 34 phycoerythrobilin fluorochromes per molecule. The large number of fluorochromes per PE molecule makes R-PE an ideal pigment for flow cytometry applications.
- *PE–Cy™5* is a tandem conjugate system which combines R-PE (a 240-kDa protein) and a cyanine dye (MW 1.5 kDa). The efficiency of the light energy transfer between the two fluorochromes allows less than 5 % of the absorbed light to be lost as fluorescence at 575 nm by R-PE. An average of one PE–Cy5 molecule is coupled per antibody or protein. Because of its broad absorption range, PE–Cy5 is not recommended for use with dual-laser flow cytometers where excitation by both lasers is possible.
- *PE–Cy™7* is a tandem conjugate system that combines PE and a cyanine dye (Cy7). PE–Cy7 can be used simultaneously with APC with minimal crossbeam compensation.
- *PE–Texas Red®* is a tandem conjugate system which combines R-PE and Texas Red. Special care must be taken when using PE–Texas Red conjugates in conjunction with R-PE as there is considerable spectral overlap in the emission profiles of both fluorochromes.
- *Peridinin chlorophyll protein (PerCP)* is a component of the photosynthetic apparatus found in the dinoflagellate, *glenodinium.* PerCP is a protein complex with a molecular weight of approximately 35 kDa. Due to its photobleaching characteristics, PerCP conjugates are not recommended for use on stream-in-air flow cytometers.
- *PerCP–Cy™5.5* is a tandem conjugate system that combines PerCP with a cyanine dye (Cy5.5). PerCP–Cy5.5 is recommended for use with stream-in-air flow cytometers.
- *PI* (Ex-Max 536 nm/Em-Max 617 nm, when bound to nucleic acid): Propidum iodide is an intercalating agent that binds nonspecifically to nucleic acid with the stoichiometry of 1 dye per 4–5 bases. Because PI nonspecifically binds to all nucleic acids it is necessary to treat the cells

with nucleases to distinguish between DNA and RNA. Once PI is bound to the nucleic acid its fluorescence is enhanced 20- to 30-fold. PI is membrane impermeable and generally excluded from viable cells. PI is commonly used to stain DNA for cell cycle analysis, and the cells must be permeabilized prior to incubation with the dye. PI is also commonly used as a cell viability marker, as it is excluded from healthy cells, but easily penetrates the disrupted membranes of dead/dying cells. PI is a potential mutagen and should be handled with care.

- *Qdot® nanocrystals* are nanometer-scale semiconductor particles comprising a core, shell, and coating. The core is made up of a few hundred to a few thousand atoms of a semiconductor material, often cadmium mixed with selenium or tellurium. The fluorescence properties of Qdot® nanocrystals are different from those of typical dye molecules. The color of light that the Qdot® nanocrystal emits is strongly dependent on the particle size, creating a common platform of labels from green to red, all manufactured from the same underlying semiconductor material. For more information, please visit www.invitrogen.com.
- *Texas Red®* is a sulfonyl chloride derivative of sulforhodamine 101 with a molecular weight of 625 Daltons. Texas Red conjugated to avidin is a useful second step for multicolor analysis. When performing multicolor analysis involving both Texas Red and R-PE, a dual-laser flow cytometer equipped with a tunable dye laser to avoid "leaking" into the PE detector is recommended. If a krypton laser, emitting light at 568 nm, is used, the laser light will "leak" into the R-PE channel. Texas Red can be used in conjunction with APC for multicolor analysis when both dyes are excited in the 595–605 nm range with a dye laser.

4. It is very important for researchers to familiarize themselves with these two fundamental types of graphic displays, as the data analysis step in flow cytometry is paramount to the success of the selection of the clones.
5. If a characterized source of protein in the staining buffer is desirable, the FBS can be replaced with 1 % bovine serum albumin (BSA, fraction V).
6. Wear appropriate protective clothing while handling sodium azide. It is harmful if swallowed and should be disposed with large volumes of water since its reaction with cooper plumbing could lead to explosive conditions.
7. The use of secondary detection reagents against H + L chains enables IgM mAbs to routinely show up in the detection and screening stages. While these can later on be weeded out by

isotyping they have little value as routine reagents due to the fact that they are less simple to purify and have higher background.

8. Dead cells can be excluded during acquisition by adding propidium iodide or 7AA-D to each sample prior to acquisition and analysis.
9. Dissolve 1 mg of 7-AAD powder by adding 50 μl of absolute methanol directly to the vial. Mix well and add 950 μl of 1× PBS with Ca^{2+} and Mg^{2+} to achieve a concentration of 1 mg/ml. Store solution tightly closed and protected from light at 4 °C.
10. Questionable results (samples with a weak positive result) should be rescreened to ensure that all hybridomas are detected. Ensure that the growth of the cells is sufficient for screening, waiting a few more days to rescreen.
11. The fluorescence intensity associated with cells decreases at high primary antibody concentrations. This is due to saturation of cell-binding sites and a higher concentration of primary antibodies in solution. This is why the recommended time for screening is 3 days post hybridoma transfer to 96-well plate.
12. Proper controls must be prepared for each plate. Positive and negative sera should be used in duplicate. In addition, prior to running your experiment, initial screens with secondary antibody only should be done in order to assess background levels.
13. The amount of protein used to coat beads should be assessed prior to running the assay. This can be done by running positive controls with different amounts of protein to assess background levels.
14. The transfection protocol is based on a 30 ml transfection. Transfections can be scaled up as needed.
15. It is important to ensure that the solution is homogeneous. This is essential when using cell-based assays as cells tend to adhere to one another.
16. The secondary antibody used in the assay is dependent on the system that is used. Knowledge of the system and the lasers is needed in order to obtain the proper fluorochrome.
17. It is recommended that supernatants are used 3–4 days post 96-well plate transfer. However, cell density is of the highest priority.
18. Screen on the preferred application early on following fusion to ensure that the mAb would optimally work on that application.
19. For isolating mAbs that work well in flow cytometry an optimized screening system should be developed that has been validated to work prior to fusion.
20. If an overexpressed cell system is used for screening by flow such as transfectants (transient or stable) or transductants the

investigator should confirm that the final clones would work on the preferred application in a primary model system.

21. A variety of methods have been employed to maintain the expression of surface targets following detachment of adherent cells such as the use of versene or accutase instead of trypsin for cell detachment or enzyme-free dissociation buffers. Alternatively incubating cells in 5 mM EDTA in PBS (without Ca or Mg) for 5 min on ice followed with 10–15-min incubation at 37 °C will allow them to dissociate.
22. Paraformaldehyde cross-links proteins and generally preserves well the cell morphology.
23. Saponin is a detergent used for cell permeabilization. Upon interaction with membrane cholesterol it selectively removes it leading to small openings on the cell membrane.
24. Methanol, organic solvent, is used for membrane permeabilization. Its mode of action is to dissolve lipids from the cell membrane making it permeable to antibodies.
25. Autofluorescence is the self-fluorescence or inherent fluorescence of an object [11].
26. Spectral overlap is the spillover between two (or more) fluorescence emission spectra.
27. Fluorescence compensation is a mathematical method to correct for bleed-through emission.
28. Isotype control is an antibody of the same class (isotype) of immunoglobulin as the specific antibody but generated against an antigen that is not present on or inside the cells under study.
29. Median fluorescence intensity (MFI)—The median is the fluorescence value below which 50 % of the events are found; i.e., it is the 50th percentile in fluorescence. In general, the median is a better estimator of the central tendency of a population than the mean, especially for log-amplified data.
30. Signal-to-noise ratio—Ratio of the median fluorescence intensity of the positively stained cell population versus the median fluorescent intensity of the negative cell population.
31. Investigators might have to increase the incubation time or modify the incubation temperature from 4 °C to room temperature to achieve optimal binding of low-affinity antibodies.
32. Certain cell types such as B cells, macrophages, natural killer cells, monocytes, dendritic cells (at low levels), granulocytes, and mast cells carry mouse FcγIII and FcγII receptors on their surface that nonspecifically bind immunoglobulin through its Fc domain. It may be necessary to block those receptors prior to adding the hybridoma supernatants with anti-CD16/CD32 antibodies. Generally 0.5–1 μg of rat anti-mouse CD32/

CD16 antibody per 10^6 cells is sufficient to block nonspecific binding. Cells are preincubated with the blocking reagent for 15 min at 4 °C, and the hybridoma supernatant is added in the presence of the blocking reagent. No wash is necessary. Proper blocking of Fc receptors can be verified with a matching isotype control antibody.

33. Compensation controls: No compensation controls are necessary if the experiment has been designed as a single-color experiment. However, for multicolor experiments, single-color control tubes have to be set up according to the following rules (*see* **Note 36**) [12, 13]:
    - Controls need to be as bright as or brighter than any sample the compensation will be applied to.
    - Background fluorescence should be the same for the positive and negative control.
    - Compensation controls must match the exact experimental fluorochrome.
    - Alternatively the instrument can be compensated using bead polystyrene particles mixed with fluorochrome-conjugated antibodies (e.g., BD™ Comp Bead). This is the method of choice if antibodies conjugated to tandem dyes are used in a multicolor experiment since it allows compensating with the same tandem dye–antibody clone used in the experiment (*see* **Note 37**).
34. Other controls:
    - Unstained cells (lack fluorescent antibodies) should be always used to establish levels of background autofluorescence.
    - Cells stained with a second-step antibody allow the researcher to determine background staining of secondary antibody to the cell population under investigation.
    - Cells are stained with pre-fusion bleed or other antibody of the same specificity that is known to work in flow cytometry.
35. If the screening system by flow is a primary cell and the target of interest is only a small subset of the total population multicolor staining might be necessary to identify the cell population of interest. In those cases, directly conjugated optimally titrated antibodies specific for cell surface markers other than the target of interest should be used to identify the cell population of interest. The fluorochrome selection on those antibodies should follow the following rules:
    - Dim fluorochrome-conjugated antibodies should be used for highly expressed antigens and vice versa
    - Select fluorochromes that are excited by different lasers to minimize spillover issues between fluorochromes.

- Select fluorochromes excited by the red laser for highly autofluorescent cells (e.g., APC, Alexa Fluor 647, APC-H7, PE-Cy7).
- In the case where direct and indirect stainings are performed in the same assay, the indirect staining should be executed first.

36. For screening hybridoma supernatants via flow it is recommended to develop a simple, efficient, and cost-effective system for screening that would allow testing of a large number of clones. If investigators have a choice, they should refrain from using primary cells and multicolor immunofluorescent staining at this stage. Overexpressed cell systems and cell lines and single-color staining should be the cell systems of choice.
37. There is a large lot-to-lot variability observed in antibodies conjugated to tandem dyes.

Acknowledgments

The views and opinions expressed herein are those of the authors only and do not necessarily represent the views and opinions of the Public Health Agency of Canada or the Government of Canada.

References

1. Corbett CR, Elias MD, Simpson LL, Yuan XY, Cassan RR, Ballegeer E, Kabani A, Plummer FA, Berry JD (2007) High-throughput homogeneous immunoassay readily identifies monoclonal antibody to serovariant clostridial neurotoxins. J Immunol Methods 328:128–138
2. Carter NP, Ormerod MG. Introduction to the principles of flow cytometry. In flow cytomstry: A Practical Approcach. 3rd ed. Oxford University Press: 2000. PP 1–22
3. Radcliff G, Jaroszeski MJ (1998) Basics of flow cytometry. Methods Mol Biol 91:1–24
4. Maecker HT, Troter J. Selegng reagents for multicolor flow cytometry with BD LSRII and BD Facscanto systems. [Appplication Notes]. Nat. Methods 2008:5:AN6–AN7
5. Herzenberg LA, Tung J, Moore WA, Herzenberg LA, Parks DR (2006) Interpreting flow cytometry data: A guide for the perplexed. Nat Immund 7:681
6. Camplejohn RS (2011) The measurement of intracellular antigens and DNA by multiparametric flow cytometry. J Microscopy 176(1):1–7
7. Widen RH (1997) Detection of intracellular/lntranuclear antigens: applications in leukemia/lymphoma analysis in. Methods Mol Biol 263:45–55
8. Hulspas R (2010) Titration of Fluorochrome-Conjugated UNIT 6.29 Antibodies for labeling cell surface markers on live cells. Curr Protoc Cytom 54:6.29.1–6.29.9, C_2010 by John Wiley & Sons, Inc
9. Schmid I, Uittenbogaart CH, Krall WJ, Braun J, Giorgi JV (1992) Dead cell discrimination with 7-amino-actinomycin D in combination with dual color immunofluorescence in single laser flow cytometry. Cytometry 13:204–208
10. Sasaki DT, Dumas SE, Dr. Edgar G (1987) Engleman Discrimination of viable and nonviable cells using propidium iodide in two color immunofluorescence. Cytometry 8(4):413–420
11. http://www.flowjo.com/html/statdefinitions.html
12. Hulspas R, O'Gorman MRG, Wood BL, Gratama JW, Robert Sutherland D (2009) Considerations for the control of background fluorescence in clinical flow cytometry. Cytometry Part B (Clinical Cytometry) 76B:355–364
13. Lanier LL, Recktenwald DJ (1991) Multicolor immunofluorescence and flow cytometry in. Methods 2(3):192–199
14. http://www.med.unc.edu/flowcytometry/basic-flow-cytometry/compensation

Chapter 7

Screening and Subcloning of High Producer Transfectomas Using Semisolid Media and Automated Colony Picker

Suba Dharshanan and Cheah Swee Hung

Abstract

Generation of high-producing clones is a perquisite for achieving recombinant protein yields suitable for biopharmaceutical production. However, in many industrially important cell lines used to produce recombinant proteins such as Chinese hamster ovary, mouse myeloma line (NS0), and hybridomas, only a minority of clones show significantly above-average productivity. Thus, in order to have a reasonable probability of finding rare high-producing clones, a large number of clones need to be screened. Limiting dilution cloning is the most commonly used method, owing to its relative simplicity and low cost. However the use of liquid media in this method makes the selection of monoclonal hybridoma and transfectoma colonies to be labor intensive and time consuming, thus significantly limiting the number of clones that can be feasibly screened. Hence, we describe the use of semisolid media to immobilize clones and a high-throughput, automated colony picker (ClonePix FL) to efficiently isolate monoclonal high-producing clones secreting monoclonal antibodies.

Key words High-throughput selection, Mammalian cells, Monoclonal antibodies, Recombinant proteins, Semisolid media

1 Introduction

Theoretically, a population of cells that have been transfected with genes, e.g., genes that produce specific proteins, should be genetically and phenotypically identical; however, in practice, a significant variation in growth rate and specific productivity occurs [1–5]. Thus, cloning is required to identify and isolate the high-producing cells. The conventional method of cloning using liquid media has many drawbacks [6]. With liquid media, it is impossible to be entirely certain that the cell line generated is derived from a single cell, and cell lines can only be said to have a probability of being monoclonal [7].

Monoclonality is important because in a heterogenous population, non-producing transfectomas will overgrow producing

Vincent Ossipow and Nicolas Fischer (eds.), *Monoclonal Antibodies: Methods and Protocols*, Methods in Molecular Biology, vol. 1131, DOI 10.1007/978-1-62703-992-5_7, © Springer Science+Business Media New York 2014

cells, and this presents a barrier to the enhancement of product yields. With liquid media, it is physically impossible to separate individual clones, and low- and high-producing cells growing together will eventually lead to the clones to be dominated by low-producing cells.

To overcome this, semisolid media which has high viscosity has been used to immobilize clones and allow the clonal progeny of a single cell to stay together. Additionally, semisolid media also minimizes diffusion of secreted proteins, and these secreted proteins are retained in the vicinity of the clones [8]. With the inclusion of appropriate probes such as fluorescence-labeled antibodies against the secreted protein in the semisolid media, the rare high producers can be identified among the majority of modest and low producers, isolated and subcloned [9].

In order to quickly isolate the high-producing monoclonal clones in the semisolid media for downstream applications, an automated colony picker, the ClonePix FL system, is used. Using pre-programmed software and micro-pins, the desired clones are selected without contamination from neighboring clones [10]. Hence, the clones which were initially monoclonal remain monoclonal during incubation and also after isolation [11].

2 Materials

2.1 Seeding of Transfectomas in Semisolid Media

1. Antibiotic-selected transfectomas [12–15] (*see* **Note 1**).
2. CloneMedia semi-solid medium (Molecular Devices, California, USA) (*see* **Note 2**).
3. Liquid growth medium: DMEM with 1 % v/v of glutamax, 1 % v/v of antibiotic/antimycotic, 5 % v/v of fetal bovine serum (*see* **Note 3**).
4. CloneDetect, anti-IgG antibody conjugated to fluorescein isocyanate (FITC) (Molecular Devices) (*see* **Note 4**).
5. PetriWell-1 EquiGlass non-tissue culture treated plates (Molecular Devices) (*see* **Note 5**).

2.2 High-Throughput Screening of High-Producing Transfectomas

1. Sterilizing agent Perasafe (Molecular Devices).
2. 70 % ethanol.
3. Eight sterilized F1 pins (Molecular Devices) (*see* **Note 6**).
4. ClonePix FL system (Molecular Devices).

2.3 Selection of High-Producing Transfectomas

1. 96-well plates with liquid growth media.
2. ClonePix FL system.

3 Methods

All work should be done in a level 2 biosafety cabinet with proper clothing and clean environment.

3.1 Seeding of Transfectomas in Semisolid Media

1. Prior to use, thaw CloneMedia and CloneDetect overnight at 4 °C (*see* **Note 7**).
2. The CloneMedia, CloneDetect, and liquid growth media should be brought to room temperature at least 90 min before seeding of cells (*see* **Note 8**).
3. Determine the volume of supernatant required for the desired number of cells (e.g., 10^5 cells) using trypan blue dye exclusion assay.
4. To one bottle of 90 ml of CloneMedia, add 10 ml of liquid growth media (*see* **Note 9**). Mix vigorously for 2 min.
5. Add the whole content of one bottle of CloneDetect (1 ml), and gently mix it into the CloneMedia by inverting and rotating the bottle for 1 min (*see* **Note 10**).
6. Add the appropriate volume of cell culture media to give a final concentration of 1,000 cells/ml, and gently mix it into the CloneMedia by inverting and rotating the bottle for 1 min (*see* **Note 11**).
7. Slowly pipette 9 ml of CloneMedia to each well of PetriWell-1 EquiGlass non-tissue culture-treated plates using 10 ml serological pipettes (*see* **Note 12**).
8. Spread the CloneMedia across the whole well by slowly rotating the plates; remove large bubbles using the end of a sterile pipette tip (*see* **Note 13**).
9. Label one of the plates, and designate this plate as "observation plate" for colony formation. Place all plates in a humidity-controlled incubator and leave undisturbed for at least 4 days (*see* **Note 14**).
10. On the fifth day, colony formation in the observation plate should be verified using an inverted microscope (*see* **Note 15**).

3.2 Sanitization and Calibration of ClonePix FL System

Diligently follow the instructions of the pre-programmed software in the ClonePix FL system.

1. Add the contents of one perasafe sachet (Molecular Devices) to 1 l of lukewarm water (30–37 °C) and leave for 15 min. Invert and rotate the bottle for 1 min to ensure that the powder is fully dissolved.
2. Transfer the perasafe solution into a feed bottle.
3. Soak pins in 70 % ethanol for 15 min prior to use.

4. Perform pinfire test to ensure that the pins are firing correctly and freely.
5. Perform alignment to ensure that the empty source plates are aligned to the camera.
6. Ensure that the ethanol feed bath is filled with 70 % ethanol.
7. Purge the whole system with perasafe with the following settings:
   (a) Number of cycles for the head to purge: 35.
   (b) Number of cycles the pins will be scrubbed in the wash bath: 8.
   (c) Number of seconds the halogen dryer will be on: 8 s.
8. Substitute the feed bottle containing perasafe with feed bottle containing 2 l of sterile water (*see* **Note 16**).
9. Flush out the perasafe from the system by repeating **step 7** with sterile water.
10. Sanitize the system with ultraviolet light for 10 min.

3.3 Selection of High-Producing Transfectomas Using ClonePix FL System

1. Transfer all the plates with CloneMedia to source stacker.
2. Place 96-well plates containing 200 μl of liquid growth media per well in the destination stacker.
3. Image the transfectomas under the white and FITC acquisition, with the former as the prime probe.
4. Adjust the setting of the aspirate, and dispense volume of CloneMedia to 7 and 9 μl, respectively.
5. Adjust the setting of the number of dispersal cycle to 10 and dispersal volume to 20 μl (*see* **Note 17**).
6. The pins should be sanitized after each round of picking (a maximum of eight colonies each time), with the following settings:
   (a) Purge cycle: 2.
   (b) Bath cycles: 4.
   (c) Number of seconds the halogen dryer will be on: 8 s.
7. The plates will then be automatically imaged, and the properties of each colony will be displayed (*see* **Note 18**).
8. Identify the desired colonies with high FITC reading, and the chosen colonies will then be automatically transferred into each well in the destination plates (*see* **Note 19**).
9. Transfer the destination plates to the incubator for the expansion of the high-producing monoclonal transfectomas and subsequently for downstream applications.
10. If a second screening of colonies is necessary, the CloneMedia should be kept in the incubator for 3 more days, before the imaging and picking processes are repeated. Otherwise the plates may be discarded (*see* **Note 20**).

4 Notes

1. To minimize the cost, we suggest that, prior to seeding into semisolid media, the transfectomas be screened at least thrice using conventional ELISA to confirm the stability and the productivity of the transfectomas.
2. The specific CloneMedia should be used for each type of cell line:
   (a) CloneMedia-Chinese hamster ovary (CHO) for CHO-S, CHOK1, and CHO-DG44 cells.
   (b) CloneMedia-HEK for HEK-293 cells.
   (c) CloneMedia-Hyb for NS0 and hybridoma cells.
3. The choice of liquid growth media depends on the media used previously to culture the cells. Ideally, the cells should be cultured in serum-free, protein-free, or chemically defined media in the absence of antibiotics.
4. The CloneDetect used must match the species of secreted monoclonal antibody. Usually CloneDetect anti-mouse is used for hybridoma cells secreting mouse IgG, while CloneDetect anti-human is used for transfectoma cells secreting chimeric, humanized, or human IgG.
5. The use of plates with non-tissue culture surface promotes the growth of spherical suspended colonies.
6. For adherent adapted cell lines, F2 pins should be used instead for the isolation of the transfectomas.
7. Do not shake the contents until completely thawed.
8. Do not pre-warm contents to 37 °C. Thawed CloneMedia MUST be used within the same day.
9. The total volume of CloneMedia used should not exceed 110 ml in order to maintain the viscosity of the semisolid media.
10. The light in the biosafety cabinet should be switched off to minimize photo-bleaching of CloneDetect.
11. It is not advisable to seed more than 1,000 cells/ml of CloneMedia semisolid growth media. This is because this will increase the risk of overlapping of precipitation of secreted proteins among two or more colonies, hence giving a false-positive high fluorescence reading. Also the cells added should be contained in a volume of 100 μl or less.
12. This step should be done immediately after the addition of cells before they have the opportunity to settle.
13. To minimize evaporation during incubation period, add 3 ml sterile PBS into each reservoir of the PetriWell-1 EquiGlass plates.

14. In order to maintain the monoclonality of the transfectomas, the plates should preferably be left undisturbed. Transferring of plates should be done as gently as possible to minimize vibration and movement of CloneMedia. Avoid frequent repeated checking of the plates.
15. The accumulation of secreted monoclonal antibodies from each transfectoma in the CloneMedia could also be verified using inverted fluorescence microscope.
16. Only high-quality filtered sterile water should be used to avoid contamination and clogging of tubing. The transferring of sterile water to the feed bottle should be carried out in the biosafety cabinet.
17. This step is vital to ensure homogenous distribution of aspirated colonies into liquid growth media and also to ensure the high viability of the isolated transfectomas.
18. From the images obtained under white light and FITC acquisitions, it is possible to differentiate among the high-, low-, and non-producing transfectomas (*see* Fig. 1). Precipitations are observed around the high-producing transfectomas and under high magnification; the precipitation around the high producer is visible even under white light (*see* Fig. 1a).
19. To determine transfectomas expressing high amounts of secreted proteins such as monoclonal antibodies, the colonies should be compared statistically using "exterior median intensity" and "sum total intensity." On the other hand, for determining transfectomas expressing high amounts of non-secreted proteins such as green fluorescent proteins (GFP), "interior median intensity" should be used to compare the colonies statistically.
20. It is vital to incubate the clones in semisolid media for at least 7 days but not more than 10 days prior to picking. Incubation for less than 7 days is not recommended as the colonies may not have reached the optimum size and may not be detected. Even if they could be detected, the very low number of cells in the liquid media after selection may result in low viability and subsequently the loss of the high producer clones. However, incubation of more than 10 days is also discouraged as this can lead to overgrowth of low-producing clones and subsequently the loss of high-producing clones. Thus the optimum cell concentration and duration of incubation should be between 800 and 1,000 cells/ml and 7 and 10 days, depending on cell type.

Acknowledgment

This work was supported by Inno Biologics and Ministry of Science, Technology and Innovation, Malaysia.

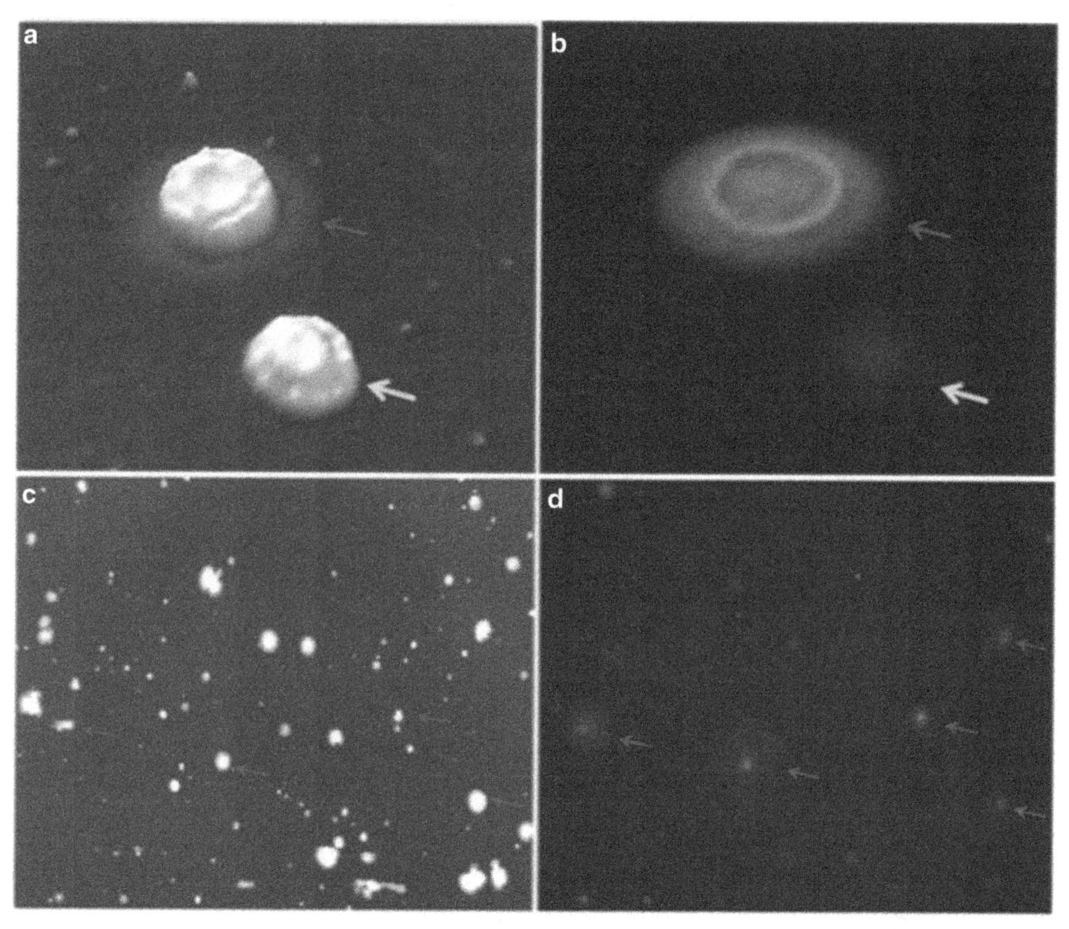

Fig. 1 Screening of colonies under white light and under fluorescence. Cells which were plated in semisolid media and incubated to form discrete colonies were imaged under white light (**a**, **c**) and fluorescence (**b**, **d**), respectively. High and low producer clones are depicted by the *red* and *yellow arrows*, respectively. In monoclonal antibody-producing colonies, precipitations were formed around the respective colonies due to the interaction between secreted humanized anti-C2 monoclonal antibodies from the clones and the capture antibodies conjugated to FITC. Therefore, the greater the quantity of antibodies secreted, the higher the exterior fluorescent intensity displayed (**b**, **d**). In fact, the precipitation around the high producer could be visible even under white light (**a**) (Color figure online)

References

1. Carroll S, Al-Rubeai M (2005) ACSD labelling and magnetic cell separation: a rapid method of separating antibody secreting cells from nonsecreting cells. J Immunol Methods 296: 171–178
2. Browne SM, Al-Rubeai M (2007) Selection methods for high producing mammalian cell lines. Trends Biotechnol 25:425–432
3. Underwood AP, Bean PA (1988) Hazards of the limiting-dilution method of cloning hybridomas. J Immunol Methods 107:119–128
4. Kim NS, Byun TH, Lee GM (2001) Key determinants in the occurrence of clonal variation in humanized antibody expression of CHO cells during dihydrofolate reductase mediated gene amplification. Biotechnol Prog 17:69–75
5. Barnes LM, Dickson AJ (2006) Mammalian cell factories for efficient and stable protein expression. Curr Opin Biotechnol 17:381–386
6. Bailey CG, Tait SA, Sunstrom NA (2002) High-throughput clonal selection of recombinant CHO cells using a dominant selectable and amplifiable metallothionein-GFP fusion protein. Biotechnol Bioeng 80:670–676
7. Staszewski R (1990) Murphy's law of limiting dilution cloning. Stat Med 9:457–461

8. Caron AW, Nicolas C, Gaillet B et al (2009) Fluorescent labeling in semi-solid medium for selection of mammalian cells secreting high-levels of recombinant proteins. BMC Biotechnol 9:42. doi:10.1186/1472-6750-9-42
9. Lee C, Ly C, Sauerwald T et al (2006) High throughput screening of cell lines expressing monoclonal antibodies. J Ind Microbiol Biotechnol 4:32–35
10. Mann CJ (2007) Rapid isolation of antigen-specific clones from hybridoma fusions. Nat Methods 4:i–ii
11. Dharshanan S, Chong H, Cheah SH et al (2011) Rapid automated selection of mammalian cell line secreting high level of humanized monoclonal antibody using ClonePix FL system and the correlation between exterior median intensity and antibody productivity. Electron J Biotechnol. doi:10.2225/vol14-issue2-fulltext-7
12. Dharshanan S (2012) Development and production of humanized monoclonal antibodies. Lap Lambert, Germany
13. Iznaga-Escobar N, Ramos-Suzarte M, Morales-Morales A et al (2004) 99mTc labeled murine ior C5 monoclonal antibody in colorectal carcinoma patients: pharmacokinetics, biodistribution, absorbed radiation doses to normal organs and tissues and tumor localization. Methods Find Exp Clin Pharmacol 26:687–696
14. Mateo C, Lombardero J, Moreno E et al (2000) Removal of amphipathic epitopes from genetically engineered antibodies: production of modified immunoglobulins with reduced immunogenicity. Hybridoma 19:463–471
15. Roque-Navarro L, Mateo C, Lombardero J et al (2003) Humanization of predicted T-cell epitopes reduces the immunogenicity of chimeric antibodies: new evidence supporting a simple method. Hybrid Hybridomics 22: 245–257

Chapter 8

Design and Generation of Synthetic Antibody Libraries for Phage Display

Gang Chen and Sachdev S. Sidhu

Abstract

Highly functional synthetic antibody libraries can be used to generate antibodies against a multitude of antigens with affinities and specificities that rival or exceed those of natural antibodies. Current design and generation of synthetic antibody libraries are dependent on our insights from previous studies of simplified synthetic antibody libraries, in addition to our knowledge of antibody structure and function and sequence diversity of natural antibody repertoires. We describe a detailed protocol for the design and generation of phage-displayed synthetic antibody libraries built on a single framework with diversity restricted to four complementarity-determining regions by using precisely designed degenerate oligonucleotides. This general methodology could be applied to generation of large, functional synthetic antibody libraries using standard supplies, equipment, and molecular biology techniques.

Key words Phage display, Synthetic antibody library, Diversity, Protein engineering

1 Introduction

Since the first monoclonal antibody (mAb) was generated from mouse B-cell hybridomas in 1975 [1], mAbs have proven to be powerful diagnostic tools in both research and clinical applications and also as effective therapeutic reagents against cancers, immunological disorders, and infectious diseases [2]. Indeed, mAbs are the fastest growing category of pharmaceutical proteins with a predicted sale of $50 billion in 2012 [3].

mAbs are traditionally produced by animal immunization and classic hybridoma technology, which is still the dominant source of mAbs. However, advances in in vitro antibody selection technology, especially phage display, enable mAb generation without the need for the tedious and lengthy procedures of hybridoma technology [4, 5]. Phage display technology, the concept of which was originally introduced in 1985 by George P. Smith [6], is now well established as the most commonly used in vitro selection technology owing to its simplicity and robustness.

Vincent Ossipow and Nicolas Fischer (eds.), *Monoclonal Antibodies: Methods and Protocols*, Methods in Molecular Biology, vol. 1131, DOI 10.1007/978-1-62703-992-5_8, © Springer Science+Business Media New York 2014

Antibody development, the most common application of phage display, is particularly attractive for therapeutic applications because human antibodies can be obtained directly without the need for cumbersome humanization of murine antibodies.

The successful selection of antibodies by phage display depends on the following crucial factors: (1) construction of a large functional library with high diversity; (2) expression of functional antibody fragments in the periplasm of *E. coli* followed by display of the antibody fragments on the surfaces of phage particles; (3) accurate and efficient genotype–phenotype coupling that facilitates decoding of identities of selected antibodies; and (4) effective selection pressure and efficient phage amplification.

As a starting point of antibody development by phage display, a functional phage-displayed antibody library of high quality is vital for isolation of antibodies with desired characteristics. Early approaches for library construction relied on natural repertoires obtained from immunized or naïve B cells. While libraries constructed by this approach are still commonly used for antibody discovery, advances in our knowledge of antibody structure and function have enabled a synthetic approach to design and construct antibody libraries entirely from scratch [5]. Diversity in synthetic antibody libraries is introduced with designed degenerate synthetic DNA that targets precisely the regions encoding the complementarity-determining regions (CDRs) that form the antigen-binding sites within defined variable domain frameworks. Besides vast diversity, the feature of independence from natural immune systems makes synthetic antibody libraries devoid of natural biases, which enables selection against self- or toxic antigens that are difficult to target with hybridoma methods. Moreover, unlike natural repertoires that contain multiple frameworks, synthetic antibody libraries can be built with frameworks that are highly stable and, for therapeutic applications, can be chosen to minimize immunogenicity [7]. Last but not least, the defined nature of synthetic libraries enables incorporation of design features to facilitate affinity maturation and automation for high-throughput antibody generation.

Design and construction of a highly functional synthetic antibody library require detailed knowledge of antibody structure and function. In particular, it is crucial to choose optimal frameworks for supporting diversity, determine positions to be diversified, and design appropriate chemical diversity at each position. Extensive research in antibody structure and function has provided a detailed understanding of antibody function at the molecular level [8]. In addition, we have developed simplified synthetic antibody libraries to explore basic principles of antibody–antigen interactions [9–11]. Our work has shown that libraries constructed by introducing diversity into only a subset of positions within four of the six CDRs are highly functional [11]. Moreover, we have been able to restrict

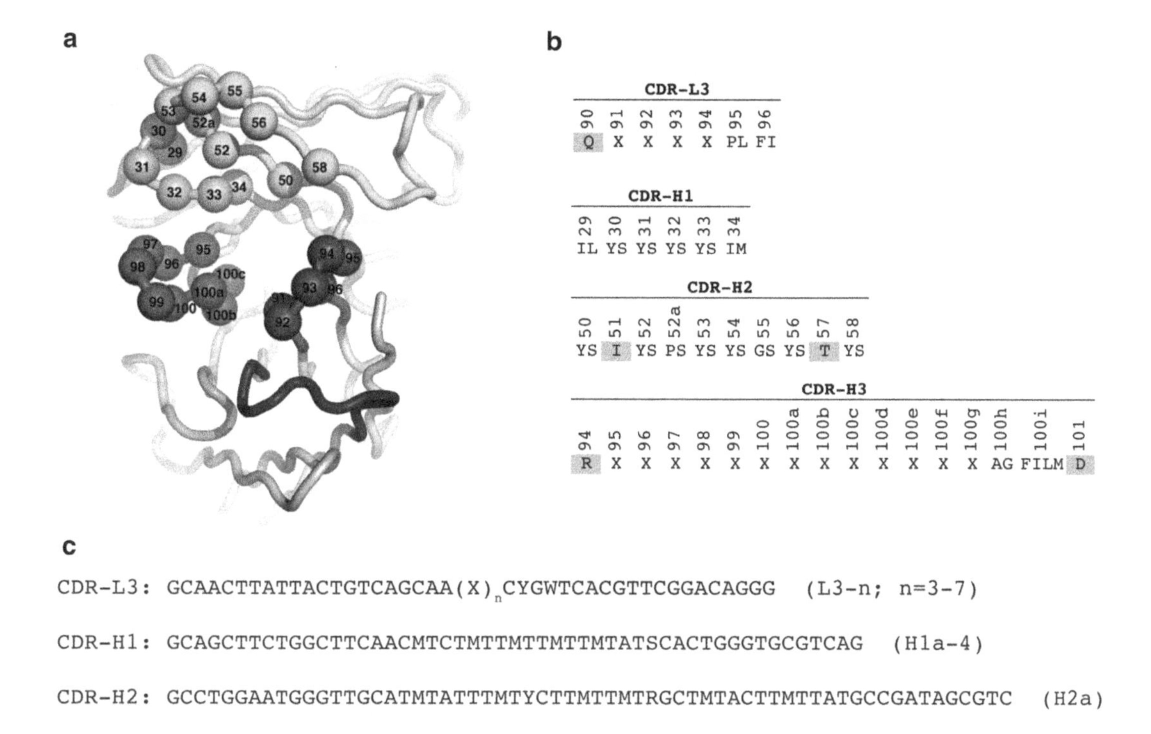

Fig. 1 Phage-displayed synthetic antibody library design. (**a**) Three-dimensional view of the antigen-binding site formed by the heavy- and light-chain variable domains. The backbones are shown as *grey tubes*. The CDR loops defined by Kabat et al. [26] are colored as follows: CDR-L1, *blue*; CDR-L2, *green*; CDR-L3 *magenta*; CDR-H1, *cyan*; CDR-H2, *yellow*; CDR-H3, *red*. *Spheres* represent positions that were diversified and are colored according to the CDR coloring scheme. The figure was generated using PyMOL (http://www.pymol.org/) with crystal structure coordinates (PDB ID: 1FVC). Residue numbering is according to the Kabat nomenclature [26]. (**b**) CDR diversity design. Positions shaded in *grey* were fixed as the parental sequence. At each diversified position, the allowed amino acids are denoted by the single-letter code. X denotes nine amino acids (Y, S, G, A, F, W, H, P, and V) in a ratio of 5:4:4:2:1:1:1:1:1. Additional diversity was introduced by varying the number of positions denoted by X in CDR-L3 and -H3 from 3 to 7 and 1 to 17, respectively. (**c**) Sequences of mutagenic oligonucleotides for library construction using oligonucleotide-directed mutagenesis based on a 4D5-derived anti-maltose-binding protein template. Each oligonucleotide is designed to have at least 15 nucleotides complementary to the template sequence on either side of the region targeted for randomization. Equimolar DNA degeneracies are represented in the IUB code (K, G/T; M, A/C; R, A/G; S, C/G; W, A/T; Y, C/T). X denotes a randomized position where a custom trimer phosphoramidite mixture containing a preset ratio of nine trimers encoding the nine amino acids described above was used in the synthesis of the oligonucleotides. The DNA sequences are shown in 5′–3′ direction, and their names are shown in *brackets* at the ends of the sequences (color figure online)

the chemical diversity introduced into CDRs, and in an extreme case, binary tyrosine and serine diversity has been used to construct highly functional libraries [9, 10, 12, 13].

Valuable insights gained from existing libraries could also be applied to new library designs. In this chapter, we focus on methods for the construction of a comprehensive library (Fig. 1) designed based on our knowledge of natural antibody repertoires and insights from previous studies of simplified synthetic antibody libraries. In this phage-displayed Fab library, extensive tailored diversity designed to mimic the types of amino acids that are abundant in natural binding sites [9, 14, 15] was introduced in the third CDR loops of the heavy and light chains and limited diversity was introduced in CDR-H1 and CDR-H2. The library has proven to be highly functional in generating antibodies with high affinity and specificity against over 100 antigens [16–21]. This protocol outlines the library design and describes procedures to construct the library using degenerate oligonucleotides.

2 Materials

1. 0.2-cm gap electroporation cuvette.
2. 1.0 M Tris–HCl, pH 8.0.
3. 1.0 mM HEPES, pH 7.4 (4.0 ml of 1.0 M HEPES, pH 7.4 in 4.0 l of ultrapure irrigation USP water, filter sterilize).
4. 3 M sodium acetate, pH 5.0.
5. 10 % (v/v) ultrapure glycerol (100 ml ultrapure glycerol in 900 ml ultrapure irrigation USP water, filter sterilize).
6. 10 mM ATP.
7. 10× TM buffer (0.1 M $MgCl_2$, 0.5 M Tris–HCl, pH 7.5).
8. 40 mM dNTP mix (solution containing 10 mM each of dATP, dCTP, dGTP, dTTP).
9. 96-well microtubes.
10. 100 mM Dithiothreitol (DTT).
11. 200 mM phenylmethylsulfonyl fluoride (PMSF) in ethanol.
12. 500 mM ethylenediaminetetraacetic acid (EDTA), pH 8.0.
13. 2YT medium (10 g bacto-yeast extract, 16 g bacto-tryptone, 5 g NaCl; add water to 1.0 l; adjust pH to 7.0 with NaOH; autoclave).
14. 2YT top agar (16 g tryptone, 10 g yeast extract, 5 g NaCl, 7.5 g granulated agar; add water to 1.0 l, adjust pH to 7.0 with NaOH, heat to dissolve, autoclave).
15. 2YT/carb/cmp medium (2YT, 100 μg/ml carbenicillin, 5 μg/ml chloramphenicol).
16. 2YT/carb/kan medium (2YT, 100 μg/ml carbenicillin, 25 μg/ml kanamycin).
17. 2YT/carb/kan/uridine medium (2YT, 100 μg/ml carbenicillin, 25 μg/ml kanamycin, 0.25 μg/ml uridine).
18. 2YT/kan medium (2YT, 25 μg/ml kanamycin).

19. 2YT/kan/tet medium (2YT, 25 μg/ml kanamycin, 10 μg/ml tetracycline).
20. 2YT/tet medium (2YT, 10 μg/ml tetracycline).
21. ECM® 630 electroporation system (BTX, Holliston, MA).
22. *E. coli* CJ236 (New England Biolabs, Ipswich, MA).
23. *E. coli* SS320 (Lucigen, Middleton, WI).
24. *E. coli* OmniMAX™ 2 T1 Phage-Resistant ($T1^R$) (Invitrogen, Grand Island, NY).
25. LB/carb plates (LB agar, 50 μg/ml carbenicillin).
26. LB/kan plates (LB agar, 25 μg/ml kanamycin).
27. LB/tet plates (LB agar, 5 μg/ml tetracycline).
28. M13K07 helper phage (New England Biolabs, Ipswich, MA).
29. Magnetic stir bars (2 in.) soaked in ethanol.
30. UT-MLB buffer (1 M sodium perchlorate, 30 % (v/v) isopropanol).
31. Phosphate-buffered saline (PBS) (137 mM NaCl, 3 mM KCl, 8 mM Na_2HPO_4, 1.5 mM KH_2PO_4; adjust pH to 7.4 with HCl, autoclave).
32. PB buffer (PBS, 0.5 % bovine serum albumin (BSA)).
33. PEG/NaCl (20 % PEG-8000 (w/v), 2.5 M NaCl; mix and filter sterilize).
34. QIAprep Spin M13 Kit (Qiagen, Valencia, CA).
35. QIAquick Gel Extraction Kit (Qiagen, Valencia, CA).
36. SOC medium (5 g bacto-yeast extract, 20 g bacto-tryptone, 0.5 g NaCl, 0.2 g KCl; add water to 1.0 l, adjust pH to 7.0 with NaOH, autoclave; add 5.0 ml of autoclaved 2.0 M $MgCl_2$ and 20 ml of filter-sterilized 1.0 M glucose).
37. Superbroth medium (12 g tryptone, 24 g yeast extract, 5 ml glycerol; add water to 900 ml, autoclave; add 100 ml of autoclaved 0.17 M KH_2PO_4 and 0.72 M K_2HPO_4).
38. Superbroth/tet/kan medium (Superbroth medium, 10 μg/ml tetracycline, 25 μg/ml kanamycin).
39. SYBR® Safe DNA gel stain (Invitrogen, Grand Island, NY).
40. T4 polynucleotide kinase.
41. T4 DNA ligase.
42. T7 DNA polymerase.
43. TAE buffer (40 mM Tris–acetate, 1.0 mM EDTA; adjust pH to 8.0; autoclave).
44. TAE/agarose gel (TAE buffer, 1.0 % (w/v) agarose, 1:10,000 (v/v) SYBR® Safe DNA gel stain).

45. TE buffer (10 mM Tris–HCl, 1 mM EDTA, pH 8.0).
46. Ultrapure glycerol.
47. Ultrapure irrigation USP water.
48. Uridine (0.25 mg/ml in water, filter sterilize).

3 Methods

The following sections describe optimized protocols for the construction of phage-displayed synthetic Fab libraries based on a single antibody framework. The use of a single framework facilitates library design and construction and also downstream characterization of isolated antibodies. Introduction of diversity in library construction is achieved by oligonucleotide-directed mutagenesis, which allows for precise control of the type and site of diversity to be introduced in the library construction. The protocol described here is an optimized scale-up version of the mutagenesis method developed by Kunkel et al. [22]. A highly diverse library containing more than 10^{10} unique Fabs can be constructed by introducing appropriate diversity into the CDRs with precisely designed synthetic degenerate oligonucleotides. This genetic library is then passed through an *E. coli* host, thereby converting it into a phage-displayed Fab library that can be used for biopanning against any antigen of interest (Fig. 2).

3.1 Design of a Phage-Displayed Synthetic Antibody Library

Template. In this example, the library template encodes an anti-maltose-binding protein Fab with a framework derived from the humanized anti-HER2 antibody 4D5 [11]. This framework contains variable domains from subgroups V_H3 and $V\kappa1$ that are highly prevalent in natural human antibodies. Moreover, this framework is highly stable and displays well on phage particles.

Phagemid design. A phagemid for displaying library template as bivalent Fabs on phage particles is outlined in Fig. 3. As with other phagemids, the vector utilized for phage display contains a double-stranded DNA origin of replication (dsDNA ori), which enables production of the phagemid as a plasmid in *E. coli*, and a single-strand DNA (ssDNA) filamentous phage origin of replication (f1 ori), which allows single-stranded DNA replication and packaging into phage particles. In addition, the vector contains a β-lactamase

Fig. 2 (continued) mutated strand (*solid circle*) is preferentially preserved and converted into the replicative form of phagemid DNA. Mutations can be introduced into different regions simultaneously by using multiple oligonucleotides as long as sequences within the oligonucleotides targeting different regions do not overlap with each other. Here, two oligonucleotides are shown for descriptive purpose. As a result, the library pool contains members in which one or two regions are replaced by the mutagenic oligonucleotides

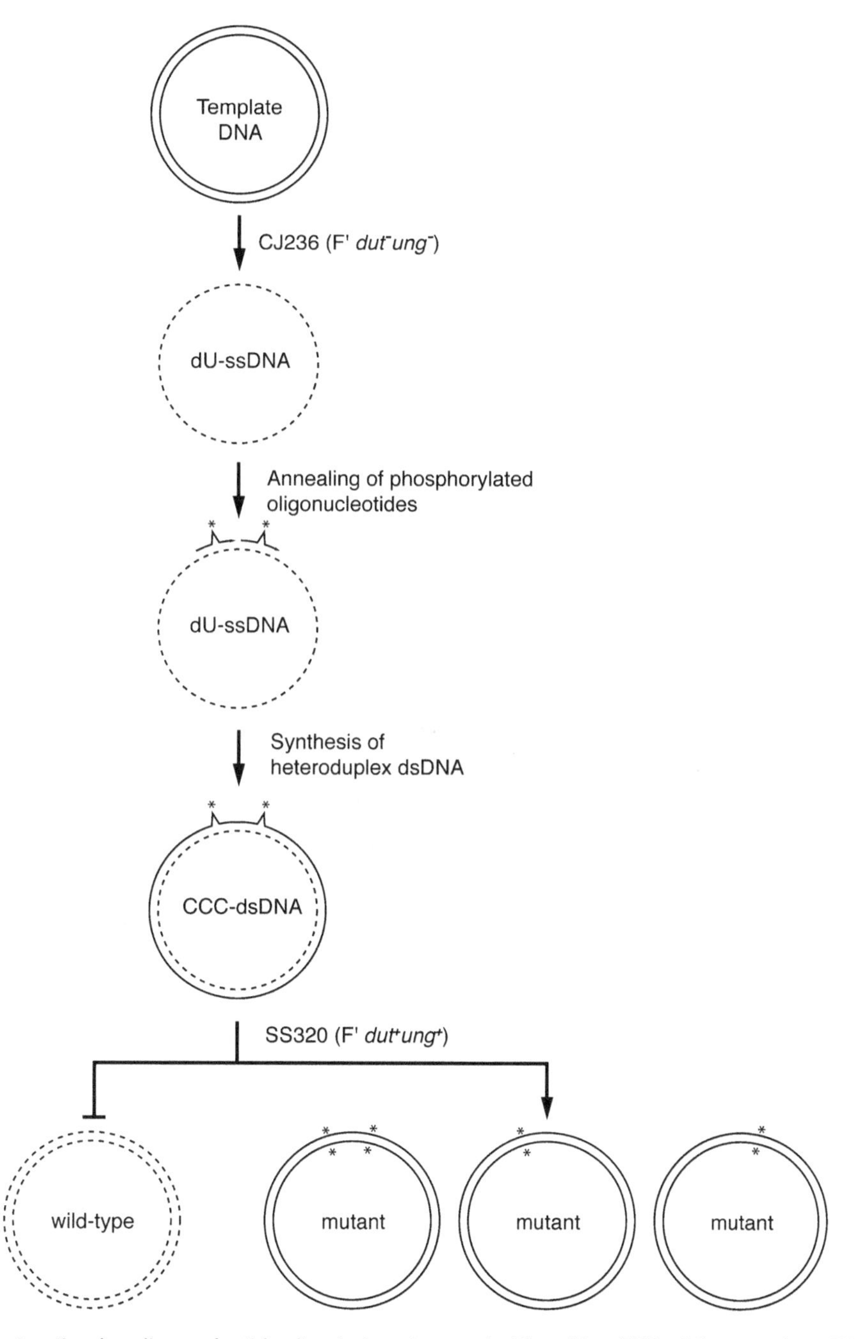

Fig. 2 Library construction by oligonucleotide-directed mutagenesis. The dU-ssDNA of the template DNA is prepared from phage particles produced by template phagemid-carrying CJ236 superinfected with M13K07 helper phage. Phosphorylated synthetic oligonucleotides (*arrows*) designed to introduce mutations (*asterisk*) in regions of interest are annealed to the dU-ssDNA template (*dashed circle*) owing to the complementary flanking sequences on both ends of the oligonucleotides. Heteroduplex covalently closed circular, double-stranded DNA (CCC-dsDNA) is enzymatically synthesized by T7 DNA polymerase and T4 DNA ligase in the presence of dNTPs and ATP. The resulting CCC-dsDNA is then introduced into *E. coli* SS320 (or other *dut*$^{+}$*ung*$^{+}$ *E. coli* hosts) where the mismatched regions are repaired to either the wild-type sequence or the mutant sequence. The uracil-containing parental strand (*dashed circle*) is degraded in *dut*$^{+}$*ung*$^{+}$ host, and the

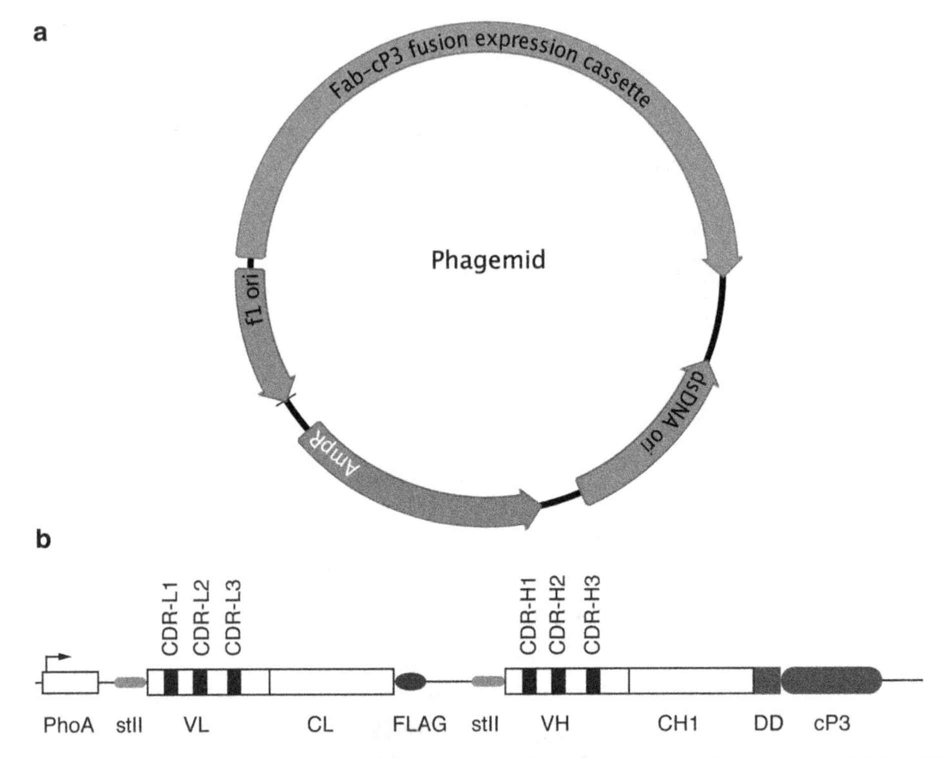

Fig. 3 A phagemid vector designed for Fab display. (**a**) In addition to an expression cassette of Fab-cP3 fusion, the phagemid vector contains origins of single-stranded (f1 ori) and double-stranded (dsDNA ori) DNA replication as well as a β-lactamase gene conferring resistance to ampicillin and carbenicillin (AmpR). (**b**) Fab-cP3 expression cassette for bivalent display of Fabs on phage particles. A PhoA promoter drives transcription of a bicistronic operon encoding the light chain (VL-CL) and the variable and first constant domains of the heavy chain (VH-CH1) fused to a truncated gene-3 minor coat protein (cP3). The three CDRs in the VL and VH domains are also shown. A FLAG tag is fused to the C-terminus of the light chain, whereas a dimerization domain (DD) is placed in-frame between CH1 and cP3. Both polypeptides contain the stII secretion signal at their N-termini. Upon expression in *E. coli*, secretion signals direct the proteins to the periplasm, where light and heavy chains associate to form a functional Fab anchored on the inner membrane via cP3. Infection of the phagemid-harboring *E. coli* host with M13 helper phage results in assembly of phage particles with Fabs displayed on the surface as bivalent dimers

gene, conferring resistance to ampicillin and carbenicillin. The display of heterodimeric Fab fragments is achieved by the use of a bicistronic operon under the control of the same alkaline phosphatase promoter (PhoA). The first open reading frame encodes the light chain of the framework followed by a FLAG tag (sequence: DYKDDDDK) [23], and the second encodes the VH domain and first constant domain (CH1) of the heavy chain fused to a dimerization domain (DD) containing a disulfide linkage for bivalent format Fab display [24], followed by the C-terminal domain of the M13 gene-3 minor coat protein (cP3). Both polypeptides are secreted into the periplasm, directed by stII secretion signals [25], where they assemble into Fab protein displayed on phage surfaces once the phagemid-harboring *E. coli* host is superinfected with M13 helper phage.

Library design. The library described here, Lib F (*see* Fig. 1), is designed similarly to the previously reported Library D [11], with modifications based on our further understanding of natural antibody repertoires and study of our previous designed synthetic antibody libraries. Lib F uses a 4D5-derived Fab as the scaffold, and diversity was restricted to positions within four of the six CDRs. Within CDR-H1 and -H2, solvent-accessible paratope positions were restricted to binary diversity consisting of Tyr and Ser. Greater chemical diversity was introduced within CDR-L3 and -H3. The tentative paratope positions within these two CDRs were randomized using tailored degenerate codons that biased in favor of four amino acids (Tyr, Ser, Gly, and Ala) and with lesser quantities of five other amino acids (Phe, Trp, His, Val, and Pro) in a ratio of 5:4:4:2:1:1:1:1:1. Further diversity was introduced by varying the length of both CDR-L3 and CDR-H3. Based on loop lengths found within these regions of natural antibodies, 5 and 17 different lengths were allowed in CDR-H3 and -L3, respectively. In addition, two non-paratope positions in each of the four CDR regions were subjected to limited randomization using binary or quaternary degenerate codons encoding residues commonly found at their corresponding positions of natural antibody sequences [26], which potentially present different conformations of the CDRs.

3.2 Preparing dU-ssDNA of the Template Phagemid Vector (See Note 1)

1. Transform the phagemid vector carrying the template Fab sequence to be diversified into competent *E. coli* CJ236 (or another *dut⁻/ung⁻* strain). Plate onto LB agar plate containing an antibiotic for selection of the vector (carbenicillin in this case) and grow overnight at 37 °C.
2. Pick a single colony of *E. coli* CJ236 harboring the template phagemid vector to inoculate 1 ml of 2YT medium supplemented with M13K07 helper phage (10^{10} pfu/ml) and appropriate antibiotics to maintain the host F′ episome and the phagemid. For example, 2YT/carb/cmp medium contains carbenicillin to select for β-lactamase gene-carrying phagemid and chloramphenicol to select for the F′ episome of *E. coli* CJ236.
3. Incubate with shaking at 200 r.p.m. and 37 °C for 2 h.
4. Add kanamycin to a final concentration of 25 μg/ml to select clones coinfected with M13K07, which carries a kanamycin resistance gene.
5. Incubate with shaking at 200 r.p.m. and 37 °C for 6 h; then transfer the culture to 30 ml of 2YT/carb/kan/uridine medium.
6. Grow the culture with shaking at 200 r.p.m. and 37 °C for 20 h.
7. Centrifuge the culture at 28,880 × *g* for 10 min to remove the bacterial cells. Precipitate phage particles by adding 1/5 final volume of phage precipitation solution PEG/NaCl into the

phage-containing supernatant and incubate for 5 min at room temperature.

8. Centrifuge at 28,880 × *g* and 4 °C for 10 min. Decant the supernatant, centrifuge briefly at 4,000 × *g*, and aspirate the remaining supernatant (*see* **Note 2**). Make sure that barrier tips are used when handling phage to avoid contamination.
9. Resuspend the phage pellet in 0.5 ml of PBS and transfer to a 2-ml conical bottom microcentrifuge tube (*see* **Note 3**). Centrifuge again at 20,000 × *g* for 5 min in a benchtop microcentrifuge to remove any remaining bacterial debris. Transfer the supernatant to a fresh 1.5-ml microcentrifuge tube.
10. Add 7 μl of buffer MP (Qiagen) and mix. Incubate at room temperature for at least 2 min. Phage particles are precipitated from the culture medium, and thus, a cloudy solution should be visible at this point.
11. Apply the sample to a QIAprep spin column in a 2-ml microcentrifuge tube. Centrifuge at 20,000 × *g* for 30 s in a benchtop microcentrifuge. Discard the flow-through. The phage particles remain bound to the column matrix.
12. Add 0.7 ml of UT-MLB buffer to the column. Centrifuge at 20,000 × *g* for 30 s, and discard the flow-through.
13. Add another 0.7 ml of UT-MLB buffer and incubate at room temperature for at least 1 min.
14. Centrifuge at 20,000 × *g* for 30 s. Discard the flow-through. The DNA is separated from the protein coat and remains bound to the column matrix.
15. Add 0.7 ml of buffer PE (Qiagen). Centrifuge at 20,000 × *g* for 30 s, and discard the flow-through.
16. Repeat **steps 13–15**. Residual proteins and salt are removed.
17. Centrifuge the column at 20,000 × *g* for 1 min to remove residual PE buffer.
18. Transfer the column to a clean 1.5-ml microcentrifuge tube.
19. Add 100 μl of buffer EB (Qiagen; 10 mM Tris–HCl, pH 8.5) to the center of the column membrane. Incubate at room temperature for 10 min.
20. Centrifuge at 20,000 × *g* for 1 min to elute the dU-ssDNA.
21. Analyze the DNA by electrophoresing 1 μl on a TAE/agarose gel. The DNA should appear as a predominant single band, but faint bands with lower electrophoretic mobility are often visible. These are likely caused by secondary structure in the dU-ssDNA.
22. Determine the DNA concentration by measuring absorbance at 260 nm (A_{260} = 1.0 for 33 ng/μl of ssDNA). Typical DNA concentrations range from 200 to 500 ng/μl.

Table 1
Generation of oligonucleotide pools

Oligonucleotide pool	Oligonucleotides
L3-37	L3-3, L3-4, L3-5, L3-6, L3-7
H3-13	H3-1, H3-2, H3-3
H3-46	H3-4, H3-5, H3-6
H3-79	H3-7, H3-8, H3-9
H3-1012	H3-10, H3-11, H3-12
H3-1315	H3-13, H3-14, H3-15
H3-1617	H3-16, H3-17

In each oligonucleotide pool, the oligonucleotides encoding sequences of different lengths are pooled at equimolar ratios

3.3 Generating Oligonucleotide Pools

In order to minimize the potential differences in bacterial growth and/or phage production that might be caused by differences in CDR-H3 lengths, six oligonucleotide pools for CDR-H3 are generated (Table 1). As the length diversity introduced in CDR-L3 is not as extreme as that in CDR-H3, one oligonucleotide pool with all possible CDR-L3 lengths is generated. Accordingly this entails six mutagenesis reactions in which each CDR-H3 oligonucleotide pool is individually mixed with oligonucleotides H1a-4, H2a, and oligonucleotide pool L3-37 at a 1:1:1:1 ratio; thus, Lib F consists of six sub-libraries.

3.4 In Vitro Synthesis of Heteroduplex CCC-dsDNA

A three-step procedure is used to incorporate the mutagenic oligonucleotides into heteroduplex covalently closed, circular, double-stranded DNA (CCC-dsDNA), using dU-ssDNA as the template (*see* **Note 4**). Six synthetic antibody sub-libraries are generated.

1. Oligonucleotide phosphorylation with T4 polynucleotide kinase: Combine 0.6 μg of each of the six mutagenic oligonucleotide pools designed to mutate CDR-H3 (Fig. 1b and Table 1) with 2 μl of 10× TM buffer, 2 μl of 10 mM ATP, 1 μl of 100 mM DTT, and 2 μl of T4 polynucleotide kinase (10 U/μl). Add water to a total volume of 20 μl in a 1.5-ml microcentrifuge tube. Each mutagenic oligonucleotide pool requires a separate phosphorylation reaction. For mutagenic oligonucleotides (or oligonucleotide pool in the case of CDR-L3) designed to mutate CDR-H1, CDR-H2, or CDR-L3, the kinase reactions are scaled up to 6× as there are six sub-libraries to be generated according to the number of mutagenic oligonucleotide pools for CDR-H3.
2. Incubate the reactions for 1 h at 37 °C. Place on ice before use for annealing (*see* **Note 5**).

3. Annealing of the oligonucleotides to the library template: To 20 μg of dU-ssDNA template, add 25 μl of 10× TM buffer, 20 μl of each phosphorylated oligonucleotide or oligonucleotide pool, and water to a final volume of 250 μl (*see* **Note 6**). As four CDRs are to be mutated in one Kunkel mutagenesis reaction, in total 80 μl of phosphorylated mutagenic oligonucleotides are added simultaneously, including oligonucleotide pool L3-37 for CDR-L3, H1a-4 for CDR-H1, H2a for CDR-H2, and one of the six oligonucleotide pools for CDR-H3: H3-13, H3-47, etc. (Table 1).
4. Incubate at 90 °C for 3 min, 50 °C for 5 min, and room temperature for 5 min.
5. Synthesis of CCC-dsDNA by fill-in reaction with T7 DNA polymerase and T4 DNA ligase: For each reaction, add 10 μl of 10 mM ATP, 25 μl of 40 mM dNTP mix, 15 μl of 100 mM DTT, 5 μl of T4 DNA ligase (400 U/μl), and 3 μl of T7 DNA polymerase (10 U/μl) to the 250 μl mixture of annealed oligonucleotide/template. Mix and incubate overnight at room temperature.
6. Analyze the mutagenesis products by electrophoresing 1 μl aliquot of each mutagenesis reaction mixture alongside the dU-ssDNA template on a TAE/agarose gel (*see* **Note 7**).
7. Purify and desalt the DNA using a modified protocol based on the protocol for the QIAquick PCR purification kit (Qiagen). Add 1 ml of buffer QG (Qiagen) to each reaction mixture (*see* **Note 8**).
8. To bind DNA, transfer half volume of each sample to each of the two QIAquick spin columns placed in 2-ml microcentrifuge tubes. Centrifuge at 20,000 ×*g* for 1 min in a microcentrifuge.
9. Discard the flow-through. Add 750 μl buffer PE (Qiagen) to each column, let the column stand for 2–5 min (*see* **Note 9**), and then centrifuge at 20,000 ×*g* for 1 min.
10. Discard the flow-through. Centrifuge the columns at 20,000 ×*g* for an additional 1 min to remove excess buffer PE.
11. Transfer the spin column to a clean 1.5-ml microcentrifuge tube, and add 35 μl of ultrapure irrigation USP water to the center of the membrane. Incubate for 5–10 min at room temperature.
12. To elute the purified DNA from the columns, centrifuge at 20,000 ×*g* for 1 min. Combine the eluants from the two columns, and determine DNA concentration by measuring absorbance at 260 nm (A_{260} = 1.0 for 50 ng/μl of dsDNA). The total recovery should be at least 20 μg. The DNA can be used immediately for *E. coli* electroporation or frozen at −20 °C for future use.

3.5 Preparation of Electrocompetent SS320 (See Note 10)

1. Inoculate 25 ml 2YT/tet medium with a single colony of *E. coli* SS320 from a fresh LB/tet plate. Grow at 37 °C with shaking at 200 r.p.m. to mid-log phase (OD_{600} = 0.6–0.8).
2. Make 10 tenfold serial dilutions of M13K07. Add 90 μl of PBS to each 1.5-ml centrifuge tube. Transfer 10 μl of M13K07 stock to the first tube and mix well. Make tenfold serial dilution by taking 10 μl of M13K07 dilution from a single tube and adding it to the next (use a new pipette tip for each dilution).
3. Mix 500 μl aliquots of mid-log phase *E. coli* SS320 with 90 μl of each M13K07 dilution and 4 ml of 2YT top agar.
4. Pour the mixtures onto pre-warmed LB/tet plates and shake gently to distribute the top layer. Incubate overnight for growth at 37 °C.
5. Pick a well-separated single plaque and inoculate into 1 ml of 2YT/kan/tet medium. Grow for 8 h at 37 °C with shaking at 200 r.p.m.
6. Transfer the culture to 250 ml of 2YT/kan medium in a 2-l baffled flask. Grow overnight at 37 °C with shaking at 200 r.p.m.
7. Inoculate six 2-l baffled flasks containing 900 ml of superbroth/tet/kan medium with 5 ml of the overnight culture. Incubate at 37 °C with shaking at 200 r.p.m. to mid-log phase.
8. Centrifuge at 5,000 ×*g* and 4 °C for 10 min in six 1-l centrifuge bottles to harvest the cells.
9. The following steps should be done on ice in a cold room with pre-chilled solutions and equipment.
10. Decant the supernatant.
11. Fill the bottles with 500 ml of 1.0 mM HEPES, pH 7.4, and add sterile magnetic stir bars to facilitate pellet resuspension. Swirl to dislodge the pellet from the tube wall, and stir at a moderate rate to resuspend the pellet completely.
12. Centrifuge at 5,000 ×*g* and 4 °C for 10 min. Decant the supernatant, being careful to retain the stir bar. To avoid disturbing the pellet, maintain the position of the centrifuge tube when removing from the rotor.
13. Repeat **steps 11** and **12**.
14. Resuspend each pellet in 150 ml of 10 % ultrapure glycerol. Use magnetic stir bars, and do not combine the pellets.
15. Remove the stir bars from centrifuge bottles. Centrifuge at 5,000 ×*g* and 4 °C for 15 min. Decant the supernatant.
16. Spin briefly, and remove the remaining traces of supernatant with a pipette.

17. Add 3 ml of 10 % ultrapure glycerol to one bottle, and resuspend the pellet by pipetting. Transfer the suspension to the next bottle, and repeat until all of the pellets are resuspended.
18. Transfer 350 μl aliquots into 1.5-ml microcentrifuge tubes.
19. Flash freeze with liquid nitrogen, and store at −80 °C (*see* **Note 11**).

3.6 Conversion of CCC-dsDNA into a Phage-Displayed Antibody Library

1. Chill the purified, desalted CCC-dsDNA (20 μg in a minimum volume) and a 0.2-cm gap electroporation cuvette on ice.
2. Thaw a 350 μl aliquot of electrocompetent *E. coli* SS320 on ice. Add the cells to the DNA and mix gently by pipetting several times (avoid introducing bubbles).
3. Transfer the cells/CCC-dsDNA mixture to the cuvette and electroporate. For electroporation, follow the manufacturer's instructions. We use a BTX ECM® 630 electroporation system with the following settings: 2.5 kV field strength, 125 Ω resistance, and 50 μF capacitance.
4. Immediately rescue the electroporated cells by adding 1 ml pre-warmed SOC medium (*see* **Note 12**) and transferring to 25 ml pre-warmed SOC medium in a 250-ml baffled flask. Rinse the cuvette twice with 1 ml pre-warmed SOC medium, and transfer the rinsing medium to the same flask.
5. Incubate for 30 min at 37 °C with shaking at 200 r.p.m.
6. Make serial dilutions to determine the library diversity. Add 90 μl of 2YT medium to each well of a single column of a 96-microwell plate. Transfer 10 μl from the culture flask, and add it to the first well of the plate. Make 8 tenfold serial dilutions by taking 10 μl from a single well and adding it to the well in the next row of the same column. Plate 5 μl of the serial dilutions on LB/carb plates (select for the phagemid) to determine the library diversity. For quality control, plate 5 μl of the same serial dilutions on LB/tet and LB/kan plates (select for SS320 cells and M13K07 helper phage, respectively) to check the titers after electroporation (*see* **Note 13**). Incubate the plates overnight at 37 °C.
7. Transfer the culture to a 2-l baffled flask containing 500 ml of 2YT/carb/kan medium. The antibiotics are supplemented for phagemid and M13K07 helper phage selection, respectively.
8. Incubate overnight at 37 °C with shaking at 200 r.p.m.
9. On the same day for phage harvesting, inoculate 25 ml 2YT/tet medium with a single colony of *E. coli* OmniMax™ 2 T1^{R} from a fresh LB/tet plate. Grow at 37 °C with shaking at 200 r.p.m. to mid-log phase (OD_{600} = 0.6–0.8).

10. Centrifuge the overnight culture for 10 min at 20,000×*g* and 4 °C.
11. Transfer the supernatant to a fresh 1-l centrifuge bottle, and add 1/5 final volume of PEG/NaCl solution to precipitate the phage particles. Incubate for 20 min on ice.
12. Centrifuge for 20 min at 20,000×*g* and 4 °C. Decant the supernatant. Spin briefly, and remove the remaining supernatant with a pipette.
13. Resuspend each phage pellet in 1/20 volume (25 ml to each bottle) of pre-chilled TE buffer supplemented with 0.5 mM PMSF by gentle pipetting. Transfer the resuspended phage solution to a clean 50-ml Falcon tube.
14. Pellet insoluble matter by centrifuging for 10 min at 20,000×*g* and 4 °C. Transfer the supernatant to a clean tube. Add 15 ml ice-cold TE supplemented with 0.5 mM PMSF to each tube.
15. Add 1/5 volume (10 ml to each tube) of PEG/NaCl to the supernatant. Mix gently, and incubate on ice for 20 min.
16. Repeat **steps 12** and **13**, but resuspend each phage pellet with 5 ml of PB buffer by gentle pipetting.
17. Centrifuge again for 10 min at 20,000×*g* and 4 °C to remove any debris. Transfer each supernatant to a clean tube.
18. Make serial dilutions to determine the phage concentrations (*see* **Note 14**). Add 90 μl of 2YT medium to each well of a 96-microwell plate, using a single row per sub-library. Transfer 10 μl from the individual phage tubes, and add separately to 90 μl of 2YT medium in individual wells of the first column of the plate, that is, each row should contain a different phage sub-library. Make 10 tenfold serial dilutions by taking 10 μl from a single well and adding it to the well in the next column of the same row. Add 90 μl of mid-log-phase *E. coli* OmniMax™ 2 T1^R to each well of another 96-microwell plate, using the same single row for the same sub-library as that in phage dilution. Transfer 10 μl of each serial diluted phage to the plate pre-added with OmniMax™ 2 T1^R. Incubate for 30 min at 37 °C with shaking at 200 r.p.m. Transfer 5 μl of infected culture on LB/tet, LB/carb, and LB/kan plates to determine phage concentrations (*see* **Note 15**).
19. Mix six sub-libraries according to their phage titers. The combined library (or individual sub-library) can be used immediately for selection experiments. Alternatively, the library can be frozen and stored at −80 °C, following the addition of glycerol to a final concentration of 10 % and EDTA to a final concentration of 2 mM (*see* **Note 16**).

4 Notes

1. We recommend using Qiagen QIAprep Spin M13 Kit for dU-ssDNA purification. The protocol provided herein is a modified version based on the kit manual. The MLB buffer originally provided in this kit was replaced with PB buffer (Qiagen) by Qiagen, but the lysis efficiency of PB buffer (Qiagen) is pretty lower, leading to much lower yield of dU-ssDNA. Our homemade UT-MLB buffer is as efficient as discontinued MLB in lysing the phage particles. By using this buffer, we consistently obtain dU-ssDNA with comparable yield and quality to the preparation with discontinued MLB.
2. A beige or light grey pellet should be visible. Thorough removal of the PEG containing precipitation solution is the key to preparation of dU-ssDNA with high yield.
3. Conical bottom centrifuge tubes are recommended in this step, as it is easier for removal of the remaining bacterial debris.
4. The first step of a complete mutagenesis protocol involves a kinase reaction, in which oligonucleotides are 5′-phosphorylated. Then the phosphorylated oligonucleotides are annealed to a dU-ssDNA library template. This annealing reaction is followed by a fill-in reaction in which the oligonucleotides are enzymatically extended and ligated to form heteroduplex CCC-dsDNA (Fig. 2). The protocol produces ~20 μg of highly pure, low-conductance CCC-dsDNA after purification and desalting. This is sufficient for the construction of a library containing more than 10^{10} unique members.
5. Immediate use of phosphorylated oligonucleotides for synthesis of CCC-dsDNA is recommended for achieving highest mutagenesis frequencies. However, the phosphorylated mutagenic oligonucleotide can be stored at −20 °C for a month without significantly compromising its performance.
6. Normally the length ratio of the oligonucleotide and library template is approximately 1:100. Thus the DNA quantities used in this protocol provide an oligonucleotide:template molar ratio of 3:1.
7. Complete conversion of ssDNA to dsDNA results in a lower electrophoretic mobility when samples are resolved on the agarose gel. No ssDNA template should remain in the reaction after overnight incubation, and usually at least two product bands are visible. The product band with higher electrophoretic mobility represents the desired mutated product: correctly extended and ligated CCC-dsDNA. The product band with lower electrophoretic mobility is a strand-displaced product resulting from an intrinsic, unwanted activity of T7 DNA polymerase [27]. Sometimes a third band is visible, with an

electrophoretic mobility between the other two product bands. This intermediate band is correctly extended but contains unligated dsDNA (nicked dsDNA) resulting from either insufficient T4 DNA ligase activity or from incomplete oligonucleotide phosphorylation. Desired CCC-dsDNA provides a high mutation frequency (~80 %) and transforms *E. coli* efficiently, whereas strand-displaced product provides a low mutation frequency (~20 %) and transforms *E. coli* at least 30-fold less efficiently than CCC-dsDNA. Therefore, a successful reaction should result in complete conversion of ssDNA to dsDNA and CCC-dsDNA being the major product.

8. The adsorption of DNA to the QIAquick column matrix is efficient only at pH $\leq$7.5, under which the pH indicator in buffer QG is yellow. If the mixture turns orange or violet upon addition of reaction mixture, adjust the pH of the mixture by adding 10 μl of 3 M sodium acetate, pH 5.0.
9. Successful electroporation depends greatly on the quality of the plasmid solution, especially on its salt content. Low salt concentrations help protect cells from potential electrical discharge. Incubation for 2–5 min after addition of buffer PE helps remove salt in DNA solution.
10. The unique feature of the competent *E. coli* SS320 cells used for library construction lies in the cells being pre-infected with M13K07 helper phage. Competent *E. coli* SS320 cells harboring M13K07 retain the high-efficiency transformation quality. Compared with the traditional approach by which electroporation of DNA is normally followed by helper phage superinfection, our method simplifies the whole transformation procedure by bypassing the need for helper phage superinfection. Therefore this novel approach not only improves the overall efficiency of transformation/infection but also facilitates library construction.
11. The titer of SS320 should be 2–6 $\times$ 10^{11} cfu/ml on both LB/carb and LB/kan plates.
12. In addition to DNA purity, temperature and time gap between electroporation and addition of pre-warmed SOC medium are also key for successful library construction. Always keep electroporation cuvette and cells/CCC-dsDNA mixture on ice before electroporation. Adding pre-warmed SOC medium to electroporated cells as fast as possible helps cell recovery from the electric shock.
13. Approximately 50 % cells survive after electroporation. The titers from LB/tet and LB/kan plates should be approximately the same.
14. Alternatively, estimate the phage concentration spectrophotometrically (OD_{268} = 1.0 for a solution of 5 $\times$ 10^{12} phage/ml).

15. The expected phage concentration is 10^{12}–10^{13} cfu/ml.
16. Alternatively, the phage-displayed antibody libraries can be stored at −20 °C with addition of glycerol to a final concentration of 50 % to avoid freezing. However, it is best to use libraries immediately, as levels of displayed proteins can be reduced over time due to denaturation or proteolysis.

References

1. Kohler G, Milstein C (1975) Continuous cultures of fused cells secreting antibody of predefined specificity. Nature 256:495–497
2. Nelson AL, Dhimolea E, Reichert JM (2010) Development trends for human monoclonal antibody therapeutics. Nat Rev Drug Discov 9:767–774
3. Morrow KJ Jr (2012) The new generation of antibody therapeutics: current status and future prospects. Cambridge Healthtech Institute, Needham, MA
4. Bradbury ARM, Sidhu S, Dübel S et al (2011) Beyond natural antibodies: the power of *in vitro* display technologies. Nat Biotechnol 29:245–254
5. Sidhu SS, Fellouse FA (2006) Synthetic therapeutic antibodies. Nat Chem Biol 2:682–688
6. Smith GP (1985) Filamentous fusion phage: novel expression vectors that display cloned antigens on the virion surface. Science 228:1315–1317
7. Fellouse FA, Sidhu SS (2006) Making antibodies in bacteria. In: Howard GC, Kaser MR (eds) Making and using antibodies: a practical handbook. CRC, Boca Raton, FL, pp 157–180
8. Johnson G, Wu TT (2000) Kabat database and its applications: 30 years after the first variability plot. Nucleic Acids Res 28:214–218
9. Fellouse FA, Li B, Compaan DM et al (2005) Molecular recognition by a binary code. J Mol Biol 348:1153–1162
10. Fellouse FA, Wiesmann C, Sidhu SS (2004) Synthetic antibodies from a four-amino-acid code: a dominant role for tyrosine in antigen recognition. Proc Natl Acad Sci U S A 101: 12467–12472
11. Fellouse FA, Esaki K, Birtalan S et al (2007) High-throughput generation of synthetic antibodies from highly functional minimalist phage-displayed libraries. J Mol Biol 373: 924–940
12. Birtalan S, Zhang Y, Fellouse FA et al (2008) The intrinsic contributions of tyrosine, serine, glycine and arginine to the affinity and specificity of antibodies. J Mol Biol 377:1518–1528
13. Fisher RD, Ultsch M, Lingel A et al (2010) Structure of the complex between HER2 and an antibody paratope formed by side chains from tryptophan and serine. J Mol Biol 402: 217–229
14. Sidhu SS, Li B, Chen Y et al (2004) Phage-displayed antibody libraries of synthetic heavy chain complementarity determining regions. J Mol Biol 338:299–310
15. Knappik A, Ge LM, Honegger A et al (2000) Fully synthetic human combinatorial antibody libraries (HUCAL) based on modular consensus frameworks and CDRs randomized with trinucleotides. J Mol Biol 296:57–86
16. Persson H, Ye W, Wernimont A et al (2013) CDR-H3 diversity is not required for antigen recognition by synthetic antibodies. J Mol Biol 425:803–811
17. Karauzum H, Chen G, Abaandou L et al (2012) Synthetic human monoclonal antibodies toward staphylococcal enterotoxin B (SEB) protective against toxic shock syndrome. J Biol Chem 287:25203–25215
18. Koellhoffer JF, Chen G, Sandesara RG et al (2012) Two synthetic antibodies that recognize and neutralize distinct proteolytic forms of the Ebola virus envelope glycoprotein. Chembiochem 13:2549–2557
19. Colwill K, Gräslund S, Persson H et al (2011) A roadmap to generate renewable protein binders to the human proteome. Nat Methods 8:551–561
20. Laver JD, Ancevicius K, Sollazzo P et al (2012) Synthetic antibodies as tools to probe RNA-binding protein function. Mol Biosyst 8: 1650–1657
21. Qazi O, Rani M, Gnanam AJ et al (2011) Development of reagents and assays for the detection of pathogenic burkholderia species. Faraday Discuss 149:23–36
22. Kunkel TA, Roberts JD, Zakour RA (1987) Rapid and efficient site-specific mutagenesis without phenotypic selection. Methods Enzymol 154:367–382
23. Hopp TP, Prickett KS, Price VL et al (1988) A short polypeptide marker sequence useful for

recombinant protein identification and purification. Nat Biotechnol 6:1204–1210

24. Lee CV, Sidhu SS, Fuh G (2004) Bivalent antibody phage display mimics natural immunoglobulin. J Immunol Methods 284:119–132
25. Picken RN, Mazaitis AJ, Maas WK et al (1983) Nucleotide sequence of the gene for heat-stable enterotoxin II of *Escherichia coli*. Infect Immun 42:269–275
26. Kabat EA, Wu TT, Perry HM et al (1991) Sequences of proteins of immunological interest, 5th edn. National Institutes of Health, Bethesda, MD
27. Lechner RL, Engler MJ, Richardson CC (1983) Characterization of strand displacement synthesis catalyzed by bacteriophage T7 DNA polymerase. J Biol Chem 258: 11174–11184

Chapter 9

Selection and Screening Using Antibody Phage Display Libraries

Patrick Koenig and Germaine Fuh

Abstract

Phage display is a powerful tool to isolate specific binders from a large and diverse combinatorial library. Here we provide a step-by-step protocol in how to set up a successful phage panning experiment in order to isolate useful antibodies. The protocol includes testing antigens for their suitability in the phage panning procedure and optimizing the panning conditions and alternative screening methods to minimize nonspecific binding. We describe example phage panning experiments starting from the library transformation to the phage clone screening.

Key words Antibody discovery, Phage display, Combinatorial antibody library, Phage panning, Phage ELISA

1 Introduction

Since the early pioneer work of George Smith [1] panning phage display libraries has been proven to be an efficient and effective way of isolating binders from a diverse set of libraries [2]. To date, there are roughly 20 therapeutic antibodies that are derived from phage display libraries approved for clinical use or in various stages of clinical trials [3]. One advantage of phage display is that fully human antibodies can be isolated. Furthermore, phage display gives greater control over the selection process than using animal immunization. This allows the selection of binders with special properties such as species cross-reactivity [4], pH dependency [5], conformation specificity [6], or dual specificity [7]. It further allows the selection of binders from alternative scaffold libraries [8]. Finally, compared to other in vitro selection methods such as cell display, phage display is more readily accessible as it does not require expensive equipment like fluorescence-activated cell sorter (FACS).

Here, we describe a simple workflow to isolate binders from a phage library. We start by describing how to find the right panning conditions for each antigen and then continue with a description

Vincent Ossipow and Nicolas Fischer (eds.), *Monoclonal Antibodies: Methods and Protocols*, Methods in Molecular Biology, vol. 1131, DOI 10.1007/978-1-62703-992-5_9, © Springer Science+Business Media New York 2014

of an actual phage panning process. We finish with a protocol for the initial screening procedure to identify selected binders. The expression of the binders and the determination of the actual K_d as well as subsequent affinity maturation are not within the scope of this chapter.

2 Materials

2.1 Testing Antigen Suitability and Optimization of Panning Condition

1. 10 μg of antigen (*see* Subheading 3.1).
2. 96-microwell MaxiSorp plate, flat bottom (Nunc).
3. 96-microwell plate, flat bottom (Nunc).
4. TMB microwell peroxidase substrate system: A two-component horseradish peroxidase substrate (KPL).
5. Mouse monoclonal anti-M13, Horseradish peroxidase conjugated (GE Healthcare).
6. M13KO7 helper phage (NEB) or any other M13 phage, which displays a binder against an antigen, unrelated to the one, which will be tested (irrelevant phage).
7. Phosphate-buffered saline (PBS) pH 7.4: Dissolve 0.2 g KCl, 8.0 g NaCl, 0.2 g KH_2PO_4, 1.13 g Na_2HPO_4 in 1 L of deionized H_2O, sterile filtered.
8. 10 % (v/v) Tween 20 in H_2O stock solution.
9. Bovine serum albumin (Sigma-Aldrich).
10. 5 M NaCl stock solution.
11. ELISA blocking buffer: 1× PBS, 0.5 % (w/v) BSA, 15 ppm ProClin. Sterile filtered. Store at 4 °C.
12. ELISA assay buffer: 1× PBS, 0.5 % (w/v) BSA, 0.05 % (v/v) Tween 20, 15 ppm ProClin. Sterile filtered. Store at 4 °C.
13. 1.0 M phosphoric acid, pH 3.0.
14. Microplate shaker.
15. ELISA plate reader.
16. ELISA plate washer.

2.2 Library Transformation

1. 15 μg of purified library phagemid DNA (*see* **Note 1**).
2. SS320 cells (*see* **Note 2**).
3. M13KO7 helper phage (NEB).
4. LB Agar plate with 5 μg/mL tetracycline.
5. 2YT top agar: Dissolve 10 g yeast extract, 16 g tryptone, 5 g NaCl, 7.5 g granulated agar, in 1 L deionized water. Heat to dissolve and autoclave.
6. 2YT medium: Dissolve 16 g tryptone, 10 g yeast extract, 5 g NaCl, in 1 L deionized water and autoclave. Store at 4 °C.

7. Superbroth: Component 1: Dissolve 12 g tryptone, 5 mL glycerol, 24 g yeast extract in 900 mL deionized water. Component 2: Dissolve 125 g K_2HPO_4 and 38 g KH_2PO_4 in 1 L of deionized water. Autoclave both components separately, and then mix 900 mL component 1 with 100 mL component 2.
8. SOC medium: Dissolve 20 g bactotryptone, 5 g yeast extract, 0.6 g NaCl, and 0.2 g KCl in 1 L deionized water; adjust the pH to 7.0; and autoclave. Cool solution, and add 10 mL 1 M $MgSO_4$, 10 mL 1 M $MgCl_2$, and 7.2 mL 50 % glucose. Filter sterilize.
9. Glycerol.
10. 1 mM Hepes: 4 L of deionized, sterile water with 4 mL HEPES 1 M, pH 7.5.
11. 10 % glycerol in ddH_2O, sterile filtered.
12. PEG precipitation solution: 20 % (w/v) PEG 8000, 2.5 M NaCl, autoclaved.
13. 5 mg/mL tetracycline, store at −20 °C.
14. 5 mg/mL carbenicillin, store at 4 °C.
15. 5 mg/mL kanamycin, store at 4 °C.
16. One large autoclaved magnetic stirrer.

2.3 Phage Selection

1. High-quality antigen (at least 5 μg per panning round).
2. *E. coli* XL1-Blue cells (Agilent).
3. LB agar plate with 50 μg/mL carbenicillin.
4. LB agar plate with 5 μg/mL tetracycline.
5. Flat bottom 96-well tissue culture plate (BD).
6. 96-microwell MaxiSorp plate, flat bottom (Nunc).
7. 96-microwell plate, flat bottom (Nunc).
8. PEG precipitation solution: 20 % (w/v) PEG 8000, 2.5 M NaCl, autoclaved.
9. M13KO7 helper phage (NEB).
10. 2YT medium: *See* Subheading 2.2 for recipe.
11. PBS, sterile filtered. Store at room temperature.
12. Phage blocking buffer 1: 1 % BSA, 0.05 % (v/v) Tween 20 in PBS.
13. Phage blocking buffer 2: 1 % Casein, 0.05 % (v/v) Tween 20 in PBS.
14. Plate blocking buffer 1: 1 % BSA in PBS.
15. Plate blocking buffer 2: 1 % Casein in PBS.
16. Washing buffer 0.05 % (v/v) Tween 20 in PBS.

17. 10 % (v/v) Tween 20 in H_2O stock solution.
18. 1 M Tris base, pH 11.
19. 0.1 N HCl.
20. pH paper.
21. 1.8 mL 96-deep-well plate.
22. 96 single 0.65 mL microtube rack (National Scientific).
23. Autoclaved wooden toothpicks.
24. 5 mg/mL tetracycline, store at −20 °C.
25. 5 mg/mL carbenicillin, store at 4 °C.
26. 5 mg/mL kanamycin, store at 4 °C.
27. Blocker solution: 1 % (w/v) casein in PBS (Thermo).

2.4 Screening of Phage Selection Output

1. High-quality antigen.
2. 96-microwell MaxiSorp plate, flat bottom (Nunc).
3. 96-microwell plate, flat bottom (Nunc).
4. 96-well tissue culture cell, flat bottom (BD).
5. 96 single 0.65 mL microtube rack (National Scientific).
6. TMB microwell peroxidase substrate system: A two-component horseradish peroxidase substrate (KPL).
7. Mouse monoclonal anti-M13, Horseradish peroxidase conjugated antibody (GE Healthcare).
8. M13KO7 helper phage (NEB).
9. Bovine serum albumin (Sigma-Aldrich) or another nonrelated protein.
10. ELISA blocking buffer: 1× PBS, 0.5 % (w/v) BSA, 15 ppm ProClin. Sterile filtered. Store at 4 °C.
11. ELISA buffer: 1× PBS, 0.5 % (w/v) BSA, 0.05 % (v/v) Tween 20, 15 ppm ProClin. Sterile filtered. Store at 4 °C.
12. Phosphoric acid, pH 3.0.
13. 96 well mini prep kit (Qiagen).
14. 2YT medium: *See* Subheading 2.2 for recipe.
15. Terrific broth component 1: Dissolve 12 g tryptone, 4 mL glycerol, 24 g yeast extract in 900 mL deionized water. Component 2: Dissolve 12.54 g K_2HPO_4 and 2.31 g KH_2PO_4 in 100 mL of deionized water. Autoclave both components separately, and then mix 900 mL of component 1 with 100 mL of component 2.
16. LB agar plate, supplemented with 50 μg/mL carbenicillin.
17. 5 mg/mL carbenicillin, store at 4 °C.
18. Competing protein (5 mL at 1 μM).

19. ELISA plate washer.
20. ELISA plate reader.
21. Microplate shaker.

3 Methods

3.1 Testing Antigen Suitability and Optimization of Panning Condition

Antigens used for the phage display experiment should be of high quality. Antigen purity, homogeneity, and monodispersity should be optimized. If possible, the antigen should be tested in functional assays (e.g., cell assays or receptor binding assays) prior to phage panning to ensure that antigen used in the experiment is functional and binders selected from the panning will bind to endogenous protein. Also, since panning experiments can stretch over several weeks, the antigen should be stored under conditions, which minimize degradation and precipitation.

Some antigens, although of high quality, may bind nonspecifically to phage particles. In order to identify the appropriate antigen preparation used for phage panning, one needs to test for nonspecific binding of the available antigen constructs or batches, by a phage titration ELISA. Phage, which displays a Fab or an scFv against an unrelated antigen and/or M13KO7 helper phage, can be used as a probe to detect nonspecific binding of phage particles to the antigen. As a negative control, use BSA or an antigen that has been successfully used previously in a panning experiment.

1. For each antigen to test coat 8 wells (one column) of a 96-well MaxiSorp plate with 100 μL/well of 2 μg/mL antigen in PBS. As negative control coat a column with 2 μg/mL BSA or another irrelevant antigen that has been used previously successfully in phage panning. Incubate the plate overnight at 4 °C.
2. The next day, block the plate for 1 h at room temperature using 100 μL/well of ELISA blocking buffer.
3. Meanwhile, dilute your phage (the irrelevant and/or M13K07 helper phage) to a final optical density at 268 nm (OD_{268}) of 0.1 using the ELISA buffer. You will need 180 μL of phage. Use a fresh 96-microwell plate to generate a serial dilution of the prepared phage solution. Pipette the 180 μL into the first well of a column. Add 120 μL of ELISA buffer to the other seven wells of the column. Now pipette 60 μL of the phage dilution from the first well into the buffer of the second well. Mix the plate for a few seconds, and then pipette 60 μL from the second well into the third well of the column. Mix again. Proceed until you added phage to the second last well of the column. No phage is added to the last well, as it will serve as a negative control.

4. After 1 h of blocking, pour off the blocking solution from the coated plate, wash the plate using an ELISA plate washer, and transfer 100 μL of the serially diluted phage solution to the blocked wells. Shake the plate for 15 min at room temperature at 200 rpm.
5. Pour off the phage solution, and wash the plate with an ELISA plate washer.
6. Add 100 μL of anti-M13 antibody diluted 1:5,000 in ELISA buffer to each of the wells. Shake the plate for 30 min at room temperature at 200 rpm.
7. Pour off the antibody solution, and wash the plate using an ELISA plate washer.
8. Add 100 μL of peroxidase substrate solution to each of the wells, and shake the plate for 3–5 min, depending on the signal strength.
9. Stop the ELISA with 100 μL/well 1 M phosphoric acid, pH 3.0.
10. Use a plate reader to read the optical density at 450 nm.

For data evaluation, plot for each antigen tested the common logarithm of the phage concentration (OD_{268}) versus the respective ELISA signal. Use the antigen preparation that gives you the lowest signal over the background for your phage panning experiment. In some cases all of the antigen constructs tested may give you a high signal over background. One possible reason could be that the antigen contains positively charged patches on the surface, which can result in charge–charge interactions of the antigen to the negatively charged phage particles [9]. Salts such as sodium chloride or other additives can minimize these nonspecific interactions. We found that sodium chloride is a useful additive to reduce the nonspecific background. In order to find the optimal sodium chloride concentrations repeat the phage-ability ELISA in the presence of no additional salt or with addition of 250, 500, and 750 mM sodium chloride to the ELISA assay buffer. A typical result is shown in Fig. 1. In this instance, 500 mM of additional NaCl drastically reduced the nonspecific phage binding to the antigen. This buffer condition was later used for phage panning and yielded well-behaving, high-affinity antibodies which bound specifically to the antigen.

3.2 Library Transformation

This protocol consists of two parts. We describe in the first half (**steps 1–9**) how to generate SS320 cells (*see* **Note 2**) that harbor M13K07 helper phage, and the second half (**steps 10–18**) explains the library transformation process. Alternatively, you can skip the first half of the protocol and add M13KO7 helper phage after the library transformation; however, the advantage of using

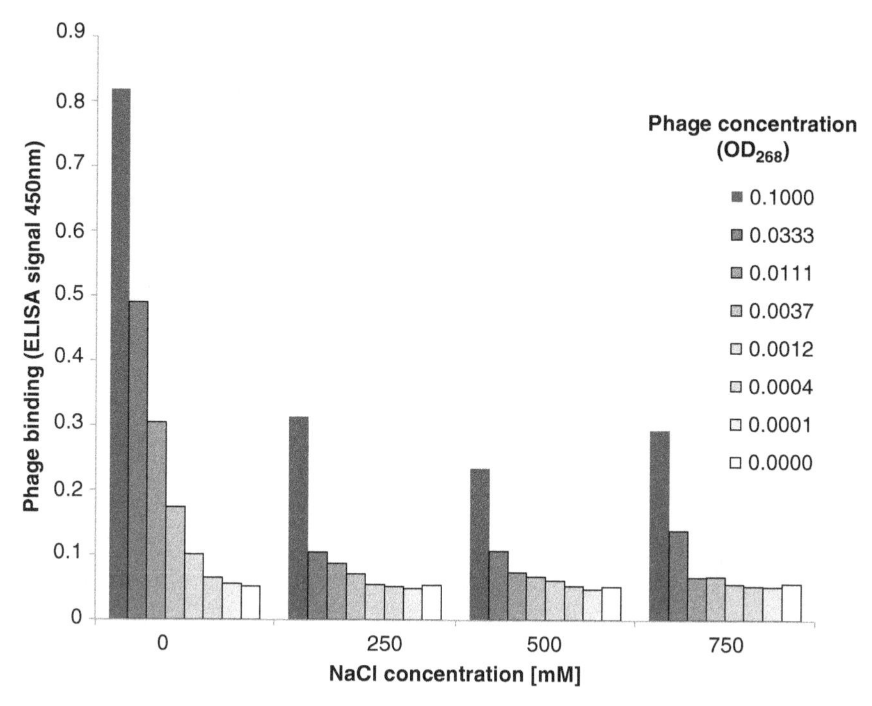

Fig. 1 Testing the suitability of antigen X for phage panning and the effect of NaCl on nonspecific antigen binding of phage particles. The binding of irrelevant phage to the antigen X is measured by ELISA assay. Shown here is an example where, under regular ELISA condition ("0") with no additional NaCl added to the ELISA buffer, there is a significant amount of nonspecific phage binding to the antigen X in a concentration-dependent manner, which would preclude the use of this antigen in phage panning experiments. However, adding 250–500 mM NaCl reduced this nonspecific binding and allowed panning against antigen X

SS320-harboring M13KO7 cells is that it ensures that every transformed clone will produce phage particles.

1. Inoculate 25 mL of 2YT medium supplemented with 10 μg/mL tetracycline with a colony of *E. coli* SS320 cells, and incubate the culture shaking at 37 °C until it reaches mid-log-phase growth.
2. Infect SS320 cells with serially diluted M13KO7 phage (starting dilution 1×10^9). Mix dilutions in 2YT top agar to cover LB agar plates supplemented with 10 μg/mL tetracycline. Incubate plates overnight at 37 °C.
3. On the next day, pick an individual plaque which should contain SS320 cells infected with helper phage and inoculate in 1 mL of 2YT medium supplemented with 10 μg/mL tetracycline and 50 μg/mL kanamycin. Shake the culture for 4–6 h at 37 °C until it reaches mid-log-phase growth.
4. Use the 1 mL culture to inoculate 25 mL of 2YT medium containing 10 μg/mL tetracycline and 50 μg/mL kanamycin. Shake culture overnight at 37 °C.

5. On the next day inoculate two 4 L shake flask containing 900 mL of superbroth with 10 μg/mL tetracycline and 50 μg/mL kanamycin with 5 mL of the overnight culture. Incubate the cultures at 37 °C with shaking at 200 rpm until they reach OD_{600} of 0.6–0.8.
6. Cool the cultures down for 10 min by gently shaking them in an ice bath.
7. Spin the cultures for 5 min at 5,000 × *g* at 4 °C, and discard the supernatant.
8. Re-suspend the pellet in 150 mL of 1 mM HEPES pH 7.5 until it is completely dissolved (*see* **Note 3**).
9. Centrifuge again, and repeat the wash once more with 150 mL of 1 mM HEPES and then once with 150 mL of 10 % glycerol.
10. Add 1 mL of 10 % glycerol to the cells. The total volume will be usually 10–11 mL. Aliquot the cells into precooled Eppendorf tubes. Add 400 μL (~10^{11} cells) to each tube, and freeze them down in dry ice or liquid nitrogen. Store cells at −80 °C.
11. To generate a library electroporate 10^{11} SS320-harboring M13KO7 cells with 15 μg of library phagemid DNA. Recover the cells in 25 mL of pre-warmed SOC media, and shake them for 40 min at 37 °C.
12. Take 100 μL from the culture and determine the titer by serial dilution (*see* **Note 4**).
13. If you use SS320 cells without M13KO7, this is the point to add 10^{11} plague-forming unit (pfu)/mL of M13KO7 to the culture. Continue culturing for another 30 min at 37 °C.
14. Transfer the culture to 350 mL of 2YT medium supplemented with 50 μg/mL carbenicillin and 50 μg/mL kanamycin and grow with shaking for 4 h at 37 °C and then overnight at 30 °C.
15. On the next day, centrifuge the overnight culture for 10 min at 7,500 × *g*.
16. Transfer the supernatant to 70 mL of PEG precipitation solution, mix, and incubate it for 5 min on ice.
17. Spin down the phage for 20 min at 20,000 × *g*, and discard the supernatant. To dry the pellet, spin it again for 2 min at 3,000 × *g*, and remove the remaining supernatant.
18. Re-suspend the pellet in 5 mL of PBS, and spin it down at 16,000 × *g* for 10 min. Transfer the supernatant to a fresh centrifugation tube, and spin it down again for 10 min at 16,000 × *g*. Transfer the supernatant again to a fresh tube.
19. Measure the OD_{268} to determine the phage concentration. Add 50 % (final) glycerol and store at −20 °C.

3.3 Phage Selection

Phagemids that express an antibody molecule as N-terminal fusion to the C-terminal fragment of protein 3 of the M13 phage can be used for the panning procedure [10]. The Fab expression may be under the control of a PhoA promoter [11,12] that is induced under low-phosphate conditions [13]. Only leaky expression from the promoter is used to generate phage particle as sufficient levels of Fab or scFv for selection are displayed on phage. The media and incubation condition might vary from the conditions established for the widely used phagemids with Lac Z promoter [14,15]. If you are planning on using a LacZ phagemid, please refer for the propagation steps in the protocol to the condition described for LacZ phagemids, e.g., [16]. Although this protocol was developed for the use of a library with diversified heavy-chain CDRs [17], it has been tested with a variety of libraries and should therefore be applicable to a wide variety of libraries and various antibody-displayed formats. Also, this protocol covers the panning against antigens, which are coated to the surface of wells of a plate either directly or indirectly as biotinylated antigen captured on wells coated with neutravidin (plate panning). If you are interested in switching during the selection process to a panning procedure, with higher stringency, you can easily convert to a solution-based panning strategy. We have previously published a detailed protocol on solution panning for affinity maturation [18].

Please note the following antibiotic resistance: *E. coli* XL1-Blue cells carry tetracycline resistance on the F′ episome, which is required for the phage infection process. The phagemid in our protocol carries ampicillin/carbenicillin and the M13KO7 helper phage kanamycin resistance. If your phagemid carries a different resistance you should adapt the protocol accordingly.

3.3.1 Phage Panning

1. On the day before starting the panning experiment, plate out some *E. coli* XL1-Blue cells fresh from a frozen stock on an LB agar tetracycline plate. Incubate the plate overnight at 37 °C.
2. Also, coat a 96-microwell MaxiSorp plate using 5 μg/mL antigen in PBS. Add 100 μL per well. The number of wells you have to coat will depend on the size of your library and the number of sub-libraries it is made up. For example, for libraries comprising 12 sub-libraries with each roughly a size of 10^9, we recommend preparing one coated well for each sub-library for the first round of panning. Incubate the plate overnight at 4 °C.
3. The next day, use some colonies of the *E. coli* XL1 agar plate to inoculate about 20 mL of 2YT medium containing 10 μg/mL tetracycline. We will need about 4 mL of *E. coli* XL1 cells at 0.6 OD_{600}. Shake the cells at 200 rpm at 37 °C (*see* **Note 5**).
4. For each sub-library, pipet 100 μL of 3 OD_{268} of phage from the library glycerol stock into a 1.5 mL Eppendorf tube.

Adjust the volume to 1 mL using PBS, and add 200 μL of the PEG precipitation solution. Incubate for 5 min on ice to precipitate the phage. The solution should become cloudy.

5. Spin down the phage at 16,000 × *g* in a benchtop centrifuge. Carefully discard the supernatant, and re-suspend the pellet in 100 μL of blocking solution containing 1 % (w/v) BSA and 0.05 % (v/v) Tween 20 in PBS. Incubate the phage for 1 h at room temperature with slow shaking.
6. Pour out the coating solution from the 96-well plate, and add 65 μL 1 % BSA in PBS per coated well. Cover the plate with plastic wrap, and incubate it with gentle shaking (50–100 rpm) for 30 min at room temperature.
7. After 30 min add 40 μL/well 10 % (v/v) Tween 20 and incubate for another 30 min gently shaking at room temperature.
8. Discard the blocking solution, and dry the plate by tapping it on a pile of paper tissues. Wash wells once more with 100 μL of blocking solution.
9. Apply 100 μL of the blocked phage solution to each coated well on the 96-well plate. Cover the plate with plastic wrap and incubate with gentle shaking for 3 h at room temperature.
10. Pour out the phage solution, and wash the plate with 0.05 % (v/v) Tween 20 in PBS by rinsing the wells of the plate ten times with approximately 150 mL washing solution per washing step (*see* **Note 6**). After washing, tap the plate dry.
11. To elute phage, add to each panned well of the plate 100 μL of 100 mM HCl, and incubate it for 20 min with gentle shaking at room temperature.
12. Add to the eluted phage solution one-tenth volume of 1 M Tris base at pH 11 and one-tenth volume of 1 % BSA in PBS. For example if you were panning on ten wells, you will collect about 1,000 μL of phage solution, so add 100 μL of 1 M Tris base and 100 μL of 1 % BSA in PBS. Check the pH by using pH paper. The solution should have a pH of 7.
13. Proceed to titer determination and propagate the phage (*see* Subheadings 3.3.2 and 3.3.3) for the next round of panning. It is best to do this on the same day, especially in the early rounds.
14. The eluted phage can be stored at 4 °C. They should keep infectivity for at least 4 weeks.

3.3.2 Titer Determination

The next step is to determine the titer of the eluted phage. This provides an overview of how many infectious phage particles are present in the eluate. We will use the titer data in later rounds to determine the enrichment; however, for the first round it is sufficient to determine the overall titer.

1. Pipet 90 μL of an *E. coli* XL1-Blue culture at 0.6–0.8 OD_{600} to the first well of a column of a fresh 96-well tissue culture plate and add 10 μL of eluted phage.
2. Cover the plate, and shake it at 200 rpm at 37 °C for 30 min.
3. Determine the titer by serial dilution (*see* **Note 4**).

3.3.3 Phage Amplification and Purification

In parallel to determining the titer, start a propagation culture to amplify the phage for the next round of panning.

1. Add 400 μL of eluted phage, which is approximately half of the total elution, to 4 mL *E. coli* XL1-Blue cells at mid-log phase in a 100 mL shaking culture flask. Shake the culture for 30 min at 37 °C.
2. Add 50 μL KO7 helper phage to the culture and incubate for an additional 1 h at 37 °C.
3. Add 25 mL 2YT containing 50 μg/mL kanamycin, 50 μg/mL carbenicillin, and 10 μg/mL tetracycline.
4. Incubate the culture first for 4 h at 37 °C, and then continue to incubate the culture overnight at 30 °C.
5. The next day start purifying the phage: Transfer the overnight culture into a 35 mL centrifugation tube, and spin it down for 10 min at 7,500 × *g* at 4 °C.
6. Prepare a fresh centrifugation tube with 5 mL of the PEG precipitation solution and incubate on ice.
7. Transfer the supernatant of the centrifugation into the PEG precipitation solution containing tube and incubate for 10 min on ice.
8. After incubation, centrifuge the tube for 15 min at 20,000 × *g* at 4 °C.
9. This time discard the supernatant, and spin the tube again for 1 min at 3,000 × *g*. Discard the remaining liquid.
10. Re-suspend the pellet into 1 mL PBS and transfer into a 1.5 mL Eppendorf tube.
11. Use a tabletop centrifuge to spin the sample down for 10 min at 16,000 × *g*.
12. Transfer the supernatant to a fresh 1.5 mL Eppendorf tube, and add 200 μL of the PEG precipitation solution. Incubate the mix for 5 min on ice.
13. Spin down the phage by centrifugation at 16,000 × *g* for 10 min.
14. Remove the supernatant, and re-suspend the pellet in 500 μL of PBS.

Table 1
Overview of the conditions used in different rounds of a typical phage panning experiment

Selection round	Concentration of phage for panning (OD_{268})	Blocking reagent for plate and phage	Washing protocol (0.05 % Tween 20 in PBS)
1	3.0	Plate: 65 μL 1 % BSA in PBS, add 40 μL 0.05 % Tween 20 after 30 min of blocking Phage: 1 % BSA, 0.05 % Tween 20 in PBS	Wash/rinse 10× with 150 mL
2	2.0	Plate: 65 μL 1 % Casein in PBS, add 40 μL 0.05 % Tween 20 after 30 min of blocking Phage: 1 % Casein in PBS, 0.05 % Tween 20	Wash/rinse 15× with 150 mL
3	1.5	Plate: 65 μL 1 % BSA in PBS, add 40 μL 0.05 % Tween 20 after 30 min of blocking Phage: 1 % BSA, 0.05 % Tween 20 in PBS	Wash/rinse 10× with 150 mL Incubate the wells with 100 μL washing buffer for 10 min, shaking Wash/rinse 10× with 150 mL
4	1.0	Plate: 65 μL 1 % Casein in PBS, add 40 μL 0.05 % Tween 20 after 30 min of blocking Phage: 1 % Casein in PBS, 0.05 % Tween 20	Wash/rinse 10× with 150 mL Incubate the wells with 100 μL washing buffer for 10 min, shaking Wash/rinse 10× with 150 mL

To increase stringency the phage concentration is reduced in each successive round. In addition, the stringency of the washing protocol is increased. To avoid the selection of binders against the blocking reagent, BSA and casein blocking solutions are alternately used for blocking

15. Spin the tube one more time at 16,000 × *g* for 3 min, and transfer the supernatant to a fresh tube.
16. Determine the phage concentration by measuring the OD_{268}. The purified phage can be stored at 4 °C. It should be stable for at least 6 months.

3.3.4 Further Rounds of Panning

The protocol for the next rounds of selection is similar to the one for the first round. However there are some important differences, as listed here:

1. First of all, in order to increase the stringency of the selection the input amount of phage as well the incubation time can be decreased. Furthermore the plate is washed more stringently.
2. Secondly, to avoid selecting binders against the blocking reagent BSA, BSA and casein are alternated as blocking reagents. A typical experiment is shown in Table 1.

To determine the enrichment factor for each of the rounds we have to determine the background binding:

3. For each panning condition, coat one column (eight wells) overnight with 5 μg/mL antigen in PBS.
4. To determine the amount of nonspecific binders, block in addition to the eight coated wells two additional wells (background). Treat these wells the same as the coated plates. Incubate them with blocked phage obtained from the previous round, wash them, and elute the bound phage. Determine the titer of the phage. You can calculate the enrichment factor by dividing the titer per well of the antigen-coated plate by the titer per well of the background wells. Also, when serial diluting your phage culture for determining the titer, store the tissue culture plate with the dilutions at 4 °C. They are used for plating out the colonies for phage spot ELISA (*see* Subheading 3.4.1, **step 1**).

In our experience you should expect enrichment in rounds 3 or 4. However this will depend on several factors: the library you are using, the antigen you are panning against, and the stringency of your panning protocol. In the event you are not seeing enrichment after five rounds of panning, proceed to pick a plate full of clones and screen them for binding (see below). If the screen shows no binders, you will have to repeat your panning. Since the antigen as well as the phage library have the biggest influence on the successful outcome of the panning experiment, you should consider using alternative antigens (e.g., different constructs, expression hosts, or purification protocols) and phage libraries for panning.

3.4 Screening of Phage Selection Output

If your panning experiment goes well you will see robust enrichment of your antigen track over the respective background (typical between 5- and 100-fold). The fastest way to get a quick overview on how many binders are there in your current panning track is to perform a phage spot ELISA. It is also important to determine if your phage is just "sticky" and binds nonspecifically to the antigen. We therefore recommend also measuring the binding of the phage to an unrelated protein (e.g., BSA or some arbitrary IgG).

3.4.1 Phage Spot ELISA

1. Take the dilution plate from the round you plan to screen. Plate each of the cell suspension from the wells with the three highest dilutions, that still gave you colonies on the titer plate (*see* Subheading 3.3.4, **step 4**), onto a fresh LB agar carbenicillin plate. Incubate the plates overnight at 37 °C.
2. The next day, fill the tubes of a 96-microtube rack with 400 μL 2YT media containing 50 μg/mL carbenicillin and 1×10^9 pfu/mL M13KO7.
3. Use toothpicks to inoculate each tube of the rack with a single colony from the plated dilutions. Shake the plate overnight at 37 °C.
4. Coat one 96-well MaxiSorp plate with 100 μL of 2 μg/mL antigen in PBS. Coat another 96-well MaxiSorp plate with

100 μL of 2 μg/mL stickiness control protein in PBS. Incubate the plate overnight at 4 °C.

5. The next day, pour out the coating solutions, and block the MaxiSorp plates with 100 μL/well ELISA blocking buffer.
6. Transfer 50 μL of phage and cell suspension from the 96-microtube rack to a 96-well-covered tissue culture plate. Store the tissue culture plate at 4 °C. It will be used later to start cultures for DNA miniprep and sequencing.
7. Spin down the 96-microtube rack at 1,100×*g* for 10 min. Transfer the resulting supernatant which contains the phage soup to a fresh 96-microtube rack. Take care not to disturb the cell pellet. Discard the rack with the cell pellets.
8. Dilute the phage supernatant 1:3 by transferring 70 μL of supernatant per well to a fresh 96-well plate containing 140 μL/well of ELISA assay buffer.
9. After 1 h of blocking, pour off the blocking solutions from the coated plates, and fill the wells with 100 μL of the 1:3 diluted phage supernatant. Cover the plate with adhesive plate cover, and incubate the plates with gentle shaking for 15 min at room temperature.
10. Pour off the phage soup, and wash the plates using an ELISA plate washer.
11. Add 100 μL of 1:5,000 in assay buffer diluted anti-M13 antibody HRP conjugate to each well of the ELISA plates. Shake the plates for 30 min at room temperature at 200 rpm.
12. Pour off the antibody solution, and wash the plates using an ELISA plate washer.
13. Add 100 μL of peroxidase substrate solution to each of the wells, and shake the plates for 3–5 min, depending on the required signal strength.
14. Stop the ELISA reaction with 100 μL/well of 1 M phosphoric acid, pH 3.0.
15. Use a plate reader to read the optical density at 450 nm.

3.4.2 Phage Spot Competition ELISA

You can design this screen in a way that you get information if your phage clones target the desired epitope. One example is if you panned against a ligand (protein 1) and you now want to determine if your binders block ligand from binding the respective receptor (protein 2). To do so, modify Subheading 3.4.1 in the following ways:

1. In **step 4**, for screening 96 clones, coat two plates with your antigen (protein 1). Block and wash as described above.
2. Instead of **step 8**, prepare 5 mL of a 1 μM solution of your competing protein (protein2) in ELISA assay buffer (*see* **Note 7**).

Add 50 μL/well of the competing protein solution to every well of one plate while adding 50 μL/well of ELISA assay buffer to the wells of your second plate. Incubate both plates shaking at room temperature for 30 min. Meanwhile, dilute the phage soup 1:2 with ELISA assay buffer. Prepare 110 μL diluted phage soup from each well. After 30 min, add 50 μL of diluted phage soup to all the wells. And continue the shaking incubation for another 15 min.

3. Perform **steps 10–15** from Subheading 3.4.1.
4. For evaluation, compare the ELISA signal of the well with competitor with the ELISA signal of the well without competitor. If a binder competes with competitor for antigen binding then the ELISA signal will be reduced on the competitor ELISA plate compared to the non-competitor plate (*see* **Note 8**).

3.4.3 Sequencing of Individual Phage

In order to check how diverse your phage binder population is you will have to sequence the isolated phagemids:

1. Fill each well of a 1.8 mL 96-deep-well plate with 1.5 mL terrific broth containing 50 μg/mL carbenicillin. Inoculate each well with 20 μL of the cell phage suspension plate saved during the screening ELISA (*see* Subheading 3.4.1, **step 6**). Shake the plate overnight at 37 °C.
2. The next day, spin down the cells at 1,100 × *g* for 10 min. Discard the supernatant.
3. The cell pellet can be used to isolate phagemid DNA using a miniprep kit for 96-well plates. Alternatively, you can also amplify the region containing the CDRs by PCR and sequence the PCR product.
4. Sequence the isolated phagemid DNA. Depending on the format of your library you will have to submit for two sequencing reactions for each well if you, for example, seek both the heavy- and light-chain CDR DNA sequences.
5. Align the obtained sequences to determine how many of them are clonal and carry the same CDR sequences. Use this information together with your spot ELISA result to pick positive binding clones for further characterization.

3.5 Further Characterization

The next step is to confirm the spot ELISA results using purified phage. Propagate and purify the positive binding phage clones (use Subheading 3.3.3) that you decided to characterize further. Confirm binding by performing a phage titration ELISA (similar as in Subheading 3.1). In addition, you can also determine the approximate K_d by using a phage IC_{50} ELISA [18]. Once you have confirmed your results with the purified phage, express your binder as an IgG. Confirm binding again by ELISA and other methods such as through Octet or Biacore assays. Note that occasionally, some clones lose their affinity towards their antigen following their reformatting into IgG. IgG is most often the best format to characterize in biological assays the function and impact of these antibodies on activities such as cell binding or cell growth inhibition.

4 Notes

1. Different protocols exist for the construction of antibody libraries. In addition, a variety of antibody fragments such as single-chain Fv and Fab and even full-length IgGs can be displayed on phage [19]. Finally, even the oligomeric state of the molecule displayed can be controlled [20]. A protocol for the construction of an antibody library is out of the scope of this chapter and is therefore not described here.
2. SS320 is an *E. coli* strain, which exhibits high transformation efficiency. The strain was generated by mating the XL1-Blue strain, which carries the F′ episome required for phage infection, with the MC1061 strain. Electrocompetent SS320 cells are available from Lucigen.
3. Add a large autoclaved magnetic stirrer to the centrifugation tube, and dissolve the pellet by shaking the tube gently.
4. For serial dilution it is convenient to use a 96-well tissue culture plate. For each sample to titrate fill the wells of a column with 90 μL 2YT buffer. Add 10 μL of the cells you would like to titrate to the first well of the column, and shake the plate shortly. Serial dilute by transferring 10 μL from the first well column to second well. Shake the plate shortly, and then transfer 10 μL from the second to third well of the column. Shake again, and continue until the end of the column is reached. Use a multichannel pipette to spot 10 μL from each well on a pre-warmed agar plate supplemented with the appropriate antibiotic. Let the spots dry at room temperature, and incubate the plate overnight at 37 °C. The next day, count the colonies of the spot with the highest dilution at which colonies are still visible. To determine the titer per well multiply the colony count times the dilution factor.
5. Check the cell growth on a regular basis during the panning procedure. If the cells reach OD_{600} 0.6 before the panning procedure is finished, dilute the cells in fresh 2YT media containing 10 μg/mL tetracycline and continue culturing.
6. You can use a plate washer for washing, but ensure that it has been rinsed thoroughly to avoid carryover phage from previous experiments.
7. The concentration of your competitor depends on the affinity of your competitors towards the antigen. In our experience 0.5–1 μM is a good starting point for a competition ELISA.

You can use a serial dilution of the competitors to observe a trend of decreasing phage binding levels in the presence of increasing concentration of competitors, provided that the binding between the competitor and phage clone is competitive.

8. Please note that the phage spot competition assay does not distinguish between steric and allosteric competition.

Acknowledgment

We thank Alessandro Palumbo, Sarah Sanowar, Jack Lin, and C. Vivian Lee for contributing to the development of protocols.

References

1. Smith GP (1985) Filamentous fusion phage: novel expression vectors that display cloned antigens on the virion surface. Science 228: 1315–1317
2. Hoogenboom HR (2005) Selecting and screening recombinant antibody libraries. Nat Biotechnol 23:1105–1116
3. Tohidkia MR, Barar J, Asadi F et al (2012) Molecular considerations for development of phage antibody libraries. J Drug Target 20: 195–208
4. Liang WC, Wu X, Peale FV et al (2006) Cross-species vascular endothelial growth factor (VEGF)-blocking antibodies completely inhibit the growth of human tumor xenografts and measure the contribution of stromal VEGF. J Biol Chem 281:951–961
5. Murtaugh ML, Fanning SW, Sharma TM et al (2011) A combinatorial histidine scanning library approach to engineer highly pH-dependent protein switches. Protein Sci 20:1619–1631
6. Haque A, Andersen JN, Salmeen A et al (2011) Conformation-sensing antibodies stabilize the oxidized form of PTP1B and inhibit its phosphatase activity. Cell 147:185–198
7. Bostrom J, Yu SF, Kan D et al (2009) Variants of the antibody herceptin that interact with HER2 and VEGF at the antigen binding site. Science 323:1610–1614
8. Koide S, Koide A, Lipovsek D (2012) Target-binding proteins based on the 10th human fibronectin type III domain ((1)(0)Fn3). Methods Enzymol 503:135–156
9. Marvin DA, Hale RD, Nave C et al (1994) Molecular models and structural comparisons of native and mutant class I filamentous bacteriophages Ff (fd, f1, M13), If1 and IKe. J Mol Biol 235:260–286
10. Sidhu SS, Li B, Chen Y et al (2004) Phage-displayed antibody libraries of synthetic heavy chain complementarity determining regions. J Mol Biol 338:299–310
11. Kikuchi Y, Yoda K, Yamasaki M et al (1981) The nucleotide sequence of the promoter and the amino-terminal region of alkaline phosphatase structural gene (phoA) of Escherichia coli. Nucleic Acids Res 9:5671–5678
12. Garrard LJ, Yang M, O'Connell MP et al (1991) Fab assembly and enrichment in a monovalent phage display system. Biotechnology (N Y) 9:1373–1377
13. Hsieh YJ, Wanner BL (2010) Global regulation by the seven-component Pi signaling system. Curr Opin Microbiol 13:198–203
14. Jacob F, Monod J (1961) Genetic regulatory mechanisms in the synthesis of proteins. J Mol Biol 3:318–356
15. Qi H, Lu H, Qiu HJ et al (2012) Phagemid vectors for phage display: properties, characteristics and construction. J Mol Biol 417:129–143
16. Viti F, Nilsson F, Demartis S et al (2000) Design and use of phage display libraries for the selection of antibodies and enzymes. Methods Enzymol 326:480–505
17. Lee CV, Liang WC, Dennis MS et al (2004) High-affinity human antibodies from phage-displayed synthetic Fab libraries with a single framework scaffold. J Mol Biol 340:1073–1093
18. Bostrom J, Lee CV, Haber L et al (2009) Improving antibody binding affinity and specificity for therapeutic development. Methods Mol Biol 525:353–376, xiii
19. Mazor Y, Van Blarcom T, Carroll S et al (2010) Selection of full-length IgGs by tandem display on filamentous phage particles and Escherichia coli fluorescence-activated cell sorting screening. FEBS J 277:2291–2303
20. Lee CV, Sidhu SS, Fuh G (2004) Bivalent antibody phage display mimics natural immunoglobulin. J Immunol Methods 284:119–132

Chapter 10

Yeast Surface Display for Antibody Isolation: Library Construction, Library Screening, and Affinity Maturation

James A. Van Deventer and Karl Dane Wittrup

Abstract

Antibodies play key roles as reagents, diagnostics, and therapeutics in numerous biological and biomedical research settings. Although many antibodies are commercially available, oftentimes, specific applications require the development of antibodies with customized properties. Yeast surface display is a robust, versatile, and quantitative method for generating these antibodies and is accessible to single-investigator laboratories. This protocol details the key aspects of yeast surface display library construction and screening.

Key words Yeast surface display, Antibody engineering, Antibody fragment, Affinity maturation, Homologous recombination

1 Introduction

In contemporary biological and biomedical research settings, monoclonal antibodies occupy a privileged position as reagents, diagnostics, and—increasingly—therapeutics. The frequent use of antibodies in these settings will likely continue far into the future, as both academic and industrial researchers possess familiarity and expertise in the areas of monoclonal antibody isolation, engineering, and production. Despite the scientific community's increasing adeptness with antibodies, applications involving these proteins are still sometimes limited by specific assay or therapeutic requirements. Each experiment has its own unique considerations, and "one-antibody-fits-all" solutions can be difficult to find.

As a result, methods for isolating and engineering monoclonal antibodies continue to be critical in advancing many lines of research and therapeutic development. The invention of in vitro display methodologies presents single-investigator laboratories with an accessible, powerful alternative to traditional animal immunization and hybridoma-based monoclonal antibody isolation. Most work with display methodologies can be performed at the benchtop or with the aid of increasingly ubiquitous flow cytometry

Vincent Ossipow and Nicolas Fischer (eds.), *Monoclonal Antibodies: Methods and Protocols*, Methods in Molecular Biology, vol. 1131, DOI 10.1007/978-1-62703-992-5_10, © Springer Science+Business Media New York 2014

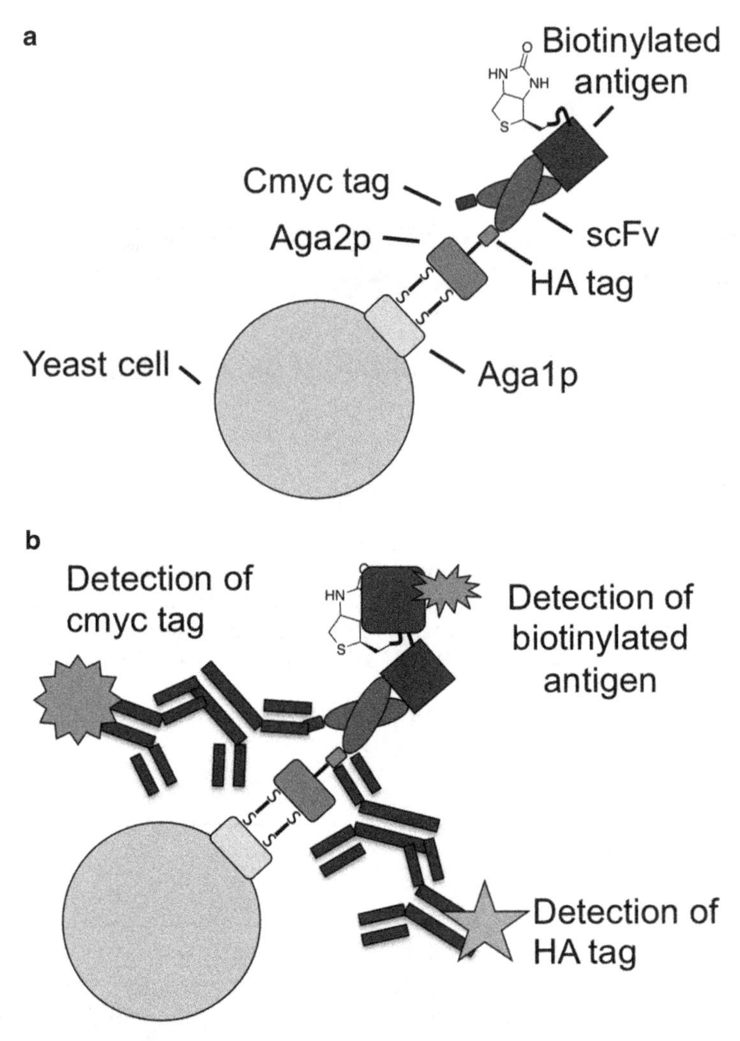

Fig. 1 Schematic representation of yeast surface display and detection strategies. (**a**) Heterologous protein display. A protein of interest such as an scFv is encoded between two epitope tags and fused in-frame with Aga2p. During processing and secretion, Aga2p forms disulfide bonds with Aga1p, resulting in a physical linkage between the protein of interest and the genetic information encoding the protein. (**b**) Detection of protein display and antigen binding. Yeast can be probed for the presence of full-length protein and antigen binding using fluorescent antibodies and streptavidin reagents

instrumentation. In display methodologies, a collection, or a library, of proteins is physically linked to the genetic material encoding these constructs of interest (usually on the surfaces of cells [1, 2] or phage [3, 4], but in some cases directly physically linked to RNA or DNA [5]). Most of these methods are well suited for isolating and engineering antibodies or antibody fragments.

Over the years, yeast surface display (Fig. 1) [6] has proven to be a robust, versatile, quantitative methodology for isolating and engineering antibody fragments [7]. The technique is amenable to screening naïve antibody fragment libraries for binders to novel antigens as well as the affinity maturation of existing antibody fragments. Some key strengths of yeast surface display are the ease with which a variety of libraries can be constructed and the quantitative, real-time assessment of candidate clones made possible using flow cytometric analysis. Most implementations of yeast surface display utilize proteins containing single domains, although recent reports of the display of proteins such as the dimeric antibody Fc region and intact IgGs suggest that the single-domain constraint can be overcome with the appropriate protein and display constructs [7]. Yeast surface display is now routinely used to screen libraries of antibody fragments with a high degree of success.

This protocol covers the basics of yeast-based construction and screening of antibody fragment libraries for novel binding proteins and high-affinity fragments to a specified target. Previous protocols describe the key methodologies of yeast surface display admirably, and some of the methods described here remain essentially unchanged from these earlier technical references [8, 9]. Other descriptions representing new, straightforward methodologies for constructing libraries, screening large populations of yeast, and reducing the likelihood of finding antibodies recognizing secondary reagents are integrated within the existing tried and true methodologies. The main topics to be described in detail in this protocol are the construction of naïve antibody fragment libraries, bead-based scFv screening of naïve libraries, flow cytometry-based scFv library screening, DNA preparation and evaluation of individual clones, and affinity maturation using error-prone PCR libraries.

2 Materials

2.1 Library Construction and Cell Growth

1. pCTCON2 (yeast surface display shuttle vector available from the authors upon request; contains tryptophan marker for use in yeast and ampicillin marker for propagation in *E. coli*).
2. Restriction enzymes: SalI, NheI, and BamHI.
3. Phusion DNA polymerase, Phusion buffer, and DMSO (provided as a kit; New England Biolabs).
4. dNTPs (10 mM concentration of each deoxynucleotide).
5. Template DNA for library amplification (*see* **Note 1**).
6. Primers for template amplification and homologous recombination. Each scFv library will contain different 5′ and 3′ ends and will need to be designed to accommodate the specific library design. However, for use with pCTCON2, recommended primers for homologous recombination are forward: 5′-GGAGGCGGTAGCGGAGGCGGAGGGTCGGCTAGC (start scFv DNA)-3′ and reverse: 5′-GTCCTCTTCAGAAA TAAGCTTTTGTTCGGAT (end reverse complement of scFv DNA)-3′.

7. Forward and reverse sequencing primers (forward: 5′-GTT CCAGACTACGCTCTGCAGG-3′; reverse: 5′-GATTTTGTT ACATCTACACTGTTG-3′).
8. DNA gel electrophoresis equipment.
9. Nucleic acid gel extraction kit.
10. Spectrophotometer for measuring DNA concentrations.
11. Sterile, doubly distilled water (ddH_2O) chilled to 4 °C.
12. Pellet Paint co-precipitant (EMD Millipore).
13. 70 % ethanol.
14. 100 % ethanol.
15. 100 mM sterile lithium acetate.
16. 1 M dithiothreitol, freshly dissolved in water and sterile filtered immediately prior to use.
17. Sterile 30 % glycerol.
18. 2 mm electroporation cuvettes (chilled on ice prior to use).
19. Biorad Gene Pulser XCell Total System (Biorad).
20. Benchtop vortexer.
21. *Saccharomyces cerevisiae* yeast surface display strain EBY100 (available from the authors upon request).
22. YPD plates: Mix 20 g peptone, 10 g yeast extract, and 15 g agar in 900 mL ddH_2O. Separately, make a solution of 100 mL 20 % glucose (20 g in 100 mL). Autoclave both solutions, cool the solutions to 55–60 °C with stirring, mix them together, and pour plates. The entire mixture can be prepared and autoclaved together, but this tends to lead to caramelization of the glucose.
23. YPD media: Mix 20 g peptone and 10 g yeast extract in 900 mL ddH_2O. Separately, prepare a solution of 100 mL 20 % glucose. Autoclave both solutions, let them cool, and combine the two to make the final product (*see* **Note 2**).
24. Penicillin/streptomycin (Pen/Strep; 10,000 IU/mL and 10,000 μg/mL, respectively, in 100× solution).
25. Sterile glass culture tubes.
26. Sterile baffled flasks for liquid culture growth (*see* **Note 3**).
27. SDCAA medium, pH 4.5: Dissolve 20 g glucose, 6.7 g yeast nitrogen base without amino acids, 5 g casamino acids, and citrate buffer salts (10.4 g sodium citrate, 7.4 g citric acid monohydrate) in 1 L ddH_2O. Filter sterilize the solution and store in autoclaved bottles (*see* **Note 4**).
28. SDCAA plates, pH 6.0: Mix phosphate buffer salts (5.4 g sodium phosphate dibasic, anhydrous, and 8.56 g sodium phosphate monobasic monohydrate), 15 g agar, and 182 g

sorbitol in a final volume of 900 mL with ddH_2O. Autoclave the mixture and cool with stirring at room temperature. At the same time, dissolve 20 g glucose, 6.7 g yeast nitrogen base without amino acids, and 5 g casamino acids in a final volume of 100 mL using vigorous stirring. Once the autoclaved solution has cooled to approximately 60 °C, filter sterilize the glucose/yeast nitrogen base/casamino acid mixture directly into the autoclaved solution, mix briefly, and pour plates.

29. SGCAA medium, pH 6.0: Dissolve 20 g galactose, 2 g glucose, 6.7 g yeast nitrogen base without amino acids, 5 g casamino acids, and phosphate buffer salts (5.4 g sodium phosphate dibasic, anhydrous, and 8.56 g sodium phosphate monobasic monohydrate) in 1 L ddH_2O. Filter sterilize the solution and store in autoclaved bottles (*see* **Note 5**).
30. 50 mL conical tubes.
31. Refrigerated centrifuge for spinning 50 mL conical tubes.
32. 1.7 mL microcentrifuge tubes.
33. Stationary incubator at 30 °C (for yeast plate incubation).
34. Shaking incubator at 30 °C, 250 rpm (for yeast liquid culture growth).
35. Shaking incubator at 20 °C, 250 rpm (for induction of liquid cultures).
36. Floor centrifuge or other centrifuge for pelleting large volumes (1 L or greater).
37. Autoclavable centrifuge bottles.
38. Microscope (for inspecting bacterial contamination).
39. Zymoprep Yeast Plasmid Miniprep II kit (Zymo Research).
40. Luria-Bertani medium (for *E. coli* growth) (available as premixed powder or use the following recipe: for 1 L, mix 10 g tryptone, 5 g yeast extract, and 10 g sodium chloride in 1 L ddH_2O; autoclave).
41. Ampicillin (1,000× stock: 100 mg/mL dissolved in water and sterile filtered).
42. Luria-Bertani plates (recipe: for 1 L, mix 10 g tryptone, 5 g yeast extract, 10 g sodium chloride, and 15 g agar in 1 L ddH_2O; autoclave, allow media to cool with stirring to 55–60 °C, mix in ampicillin stock, and pour plates).
43. Stationary incubator at 37 °C (for bacterial plate incubation).
44. Shaking incubator at 37 °C (for bacterial liquid culture growth).
45. Competent *E. coli* (*see* **Note 39**).
46. Taq polymerase and PCR buffer.
47. 8-oxo-dGTP.
48. dPTP.

2.2 Bead Sorting

1. Refrigerated benchtop centrifuge for spinning microcentrifuge tubes.
2. Refrigerated centrifuge for spinning plates in swinging bucket rotors.
3. 2.0 mL microcentrifuge tubes (2 mL tubes).
4. 1× PBS, pH 7.4: Mix 8 g sodium chloride, 0.2 g potassium chloride, 1.44 g sodium phosphate dibasic (anhydrous), and 0.24 g potassium phosphate monobasic (anhydrous) in 1 L ddH_2O. Use hydrochloric acid or sodium hydroxide to adjust the pH to 7.4.
5. Sterile PBS + 0.1 % bovine serum albumin (BSA), pH 7.4 (PBSA): Add 1 g BSA to 1 L 1× PBS, pH 7.4, dissolve, and sterile filter. Store in autoclaved bottles.
6. Dynabeads Biotin Binder (Life Technologies).
7. Biotinylated antigen (*see* **Note 6**).
8. Rotary wheels stationed at 4 °C and room temperature.
9. Dynamag-2 magnet with tube holder (Life Technologies).

2.3 Flow Cytometry

1. Primary antibodies.
   (a) Chicken anti-cmyc (Gallus Immunotech).
   (b) Mouse anti-HA antibody 16B12 (Covance).
2. Secondary antibodies.
   (a) Goat anti-chicken Alexa Fluor 647.
   (b) Goat anti-mouse Alexa Fluor 488.
   (c) Streptavidin–Alexa Fluor 488.
   (d) Mouse anti-biotin PE.
3. Flow cytometry tubes compatible with available flow cytometer.
4. 96-well plates (compatible with flow cytometer of choice; U or V bottom preferred for enhanced pelleting).
5. Shaking platforms stationed at 4 °C and room temperature.

3 Methods

3.1 Preparation of Yeast Surface Display Libraries via Homologous Recombination

Steps 1–4 and **steps 5** and **6** can be performed in parallel. **Steps 10** and **11** should be performed concurrently.

1. *Digest pCTCON2 using SalI, NheI, and BamHI over the course of approximately 48 h.* To set up the first portion of the digest, prepare a 100 μL-scale reaction as follows: mix 10 μL 10× restriction buffer, 1 μL 100× BSA, 10 μg backbone vector, and 3 μL SalI. Bring the volume of the reaction to 100 μL using doubly distilled water, mix well, and incubate at 37 °C overnight.

The next day, add 3 μL each of NheI and BamHI to the reaction mixture, mix well, and incubate at 37 °C overnight. The following morning, add 0.5 μL of all three restriction enzymes to the reaction mixture, mix well, and incubate for one final hour (this serves as insurance that the vector has been completely digested). Upon completion of the digest, no further preparation of the vector DNA is required. Store the reaction mixture at −20 °C until it is needed.

2. *Amplify scFv insert DNA using polymerase chain reaction (PCR)*. To preserve the initial library design and diversity, use Phusion polymerase and set up 200 μL PCR reactions to prepare insert DNA (*see* **Note 1**). Prepare reaction mixtures according to the manufacturer's recommendations using 20–200 ng template DNA per reaction. Cycle the reaction as follows:

   1 cycle of 98 °C for 30 s; 30 cycles of 98 °C for 30 s, 58 °C for 30 s, and 72 °C for 30 s; 1 cycle of 72 °C for 10 min; and 1 cycle of 4 °C forever.

   One 200 μL reaction will yield approximately 4–8 μg of DNA. Primers for amplification should contain sufficient homology to the backbone vector to enable efficient recombination in yeast (see recommended primers in Subheading 2.1).

3. *Run a DNA gel to extract properly sized insert DNA*. Run a 1 % gel at 100 V until the insert band (approximately 700–800 base pairs) is well separated from any contaminating bands. Gel extract the DNA and measure its concentration using a spectrophotometer.

4. *Pellet DNA samples for transformation*. Mix 4 μg insert DNA and 1 μg triply digested pCTCON2 DNA (if the digestion protocol of **step 1** was followed, 10 μL ≈ 1 μg) for each transformation in 1.7 mL microcentrifuge tubes and pellet the DNA using Pellet Paint according to the manufacturer's protocol (*see* **Note 7**). In parallel, pellet a sample of digested vector only as a negative control. At the end of the procedure, allow the pelleted DNA to air-dry for approximately 20–30 min. Confirm that the pellet is dry by tapping the tube firmly against the benchtop. A completely dry pellet should be readily dislodged from the bottom of the tube and move upon repeated tapping. Once the pellet is dry, resuspend it in 10 μL sterile ddH_2O. Repeated pipetting will be necessary to completely resolubilize the DNA. If extended storage of pelleted and resuspended DNA is required, the DNA can be stored at −20 °C.

5. *Prepare growing, healthy EBY100 cells to be used for electroporation*. Use a cryogenic stock of EBY100 to streak a YPD plate (*see* **Note 8**). Large colonies should be visible after 2 days of growth at 30 °C in a stationary incubator. Pick a single colony

of EBY100 from the plate and inoculate a 5 mL liquid YPD culture. Grow the cells at 30 °C with shaking overnight. The following day, passage the EBY100 into 5 mL fresh YPD by transferring 100 μL of saturated culture into 5 mL fresh YPD (*see* **Note 9**). Grow the cells at 30 °C with shaking overnight. The next morning, measure the OD_{600} of the EBY100 culture. Dilute the culture to an OD_{600} of 0.2 in a large volume of YPD and grow at 30 °C with shaking until the cells reach an OD_{600} of approximately 1.5 (*see* **Note 10**). In this protocol, 50 mL EBY100 at an OD_{600} of approximately 1.5 will yield enough electrocompetent cells for two transformations; more transformations can be performed if the volume of EBY100 prepared is scaled appropriately.

6. *Prepare EBY100 for electroporation.* Upon reaching an OD_{600} of 1.5, use sterile technique to transfer the culture to 50 mL conical tubes. Pellet the cells by spinning them at 2,000 × *g* for 5 min. Decant the supernatant and resuspend each tube in 25 mL sterile 100 mM lithium acetate (half of the pelleted culture volume) by shaking the tubes vigorously. Add 0.25 mL freshly prepared, sterile-filtered 1 M DTT to each tube. Loosen the cap on each tube, tape the lid on to ensure adequate oxygenation, and then incubate the cells at 30 °C in the shaking incubator for 10 min. Following the incubation, the cells should remain at 4 °C or on ice for the remainder of the preparation. After incubation, tighten the lids on the tubes and pellet the cells at 2,000 × *g* for 5 min. Decant the supernatant and resuspend the cells in 25 mL/tube chilled, sterile ddH_2O with vigorous shaking. Pellet the cells at 2,000 × *g* for 5 min. Decant the supernatant and resuspend the cells in 0.25 mL/tube sterile, chilled ddH_2O using repeated pipetting. Be sure to examine the bottom of the tube to confirm that all of the cells are resuspended. The cells are now electrocompetent.
7. *Transform electrocompetent EBY100 with library DNA.* Take 250 μL electrocompetent cells and add them to the 10 μL preparations of library DNA (or control). Mix cells and DNA well using a pipette and transfer the cells to prechilled 2 mm electroporation cuvettes. Dry the outside of a single electroporation cuvette, place the cuvette in the shock pad of the Biorad Gene Pulser XCell, and shock the cells using a square wave protocol with the following parameters: 500 V, 15 ms pulse duration, 1 pulse only, and 2 mm cuvette. Immediately after the shock is completed, remove the cuvette from the shock pad, add 1 mL YPD to the cuvette (this does not need to be performed under sterile conditions), and mix the YPD and cells multiple times. Dump the contents of the cuvette into a sterile glass test tube, then add 1 mL YPD to the cuvette, rinse the cuvette, and dump the contents into the glass tube.

Repeat the electroporation for all cuvettes and then incubate the glass tubes at 30 °C without shaking for 1 h. At this time, warm SDCAA plates to 30 °C (one plate per transformation).

8. *Take samples of transformed cells to estimate the number of transformants in each tube.* Remove the glass tubes containing the transformed cells from the incubator after the incubation is complete, vortex each tube (Gently! Use the lowest vortex setting on a benchtop vortexer), remove 10 μL, and dilute the cells into 990 μL SDCAA medium (100× dilution). Set the dilutions aside briefly at room temperature while completing **step 9**.
9. *Prepare transformed cells for overnight growth in selective media.* Take the tubes of transformed cells and pellet them in glass tubes at 900 × *g* for 5 min. After aspirating the supernatant, resuspend the cells in the tube with approximately 5 mL SDCAA. Dilute the resuspended cells into a total of 100 mL SDCAA for each set of transformed cells. If numerous transformations of the same genetic diversity have been performed, these transformed cells can be pooled and grown in the same, larger amount of media (100 mL/transformation is a good rule of thumb). Grow the cells at 30 °C in baffled culture flasks for approximately 24 h.
10. *Dilute, plate, and grow cells to estimate the total number of transformants from each electroporation.* Take the samples obtained in **step 8** (100× dilutions), vortex, and transfer 10 μL into Eppendorf tubes containing 90 μL SDCAA (1,000× dilutions). Repeat dilutions into fresh media twice more to obtain 10,000× and 100,000× dilutions, using a fresh tip for each dilution to avoid the possibility of cells sticking to tips in one dilution and becoming dislodged in future dilutions. Remove the prewarmed SDCAA plates from the incubator and divide each plate into quadrants using a permanent marker. Take each diluted sample from a transformation, vortex vigorously, remove 20 μL, and deposit into one quadrant of the plate. Use the pipette tip to spread the cells evenly throughout the area. Repeat the process for each dilution, using a fresh tip for each deposition and spreading. Incubate plates at 30 °C for 3 days, and count colonies to determine the number of transformants. A single colony in the 100×, 1,000×, 10,000×, and 100,000× quadrants corresponds to 1×10^4, 1×10^5, 1×10^6, and 1×10^7 transformed cells, respectively.
11. *Passage the library and prepare frozen stocks.* After 24 h of growth, check the OD_{600}s of the growing library cultures. Typically, saturation OD_{600}s in SDCAA range from 8 to 12. If the cultures are not yet saturated, continue growing the cells until saturation is reached. Once the cultures are saturated, pellet the cultures in conical tubes (to remove cellular debris) at 2,000 × *g* for 5 min.

Decant the supernatants and resuspend the pelleted cells in SDCAA to a final OD_{600} of approximately 1. To passage a 100 mL culture, resuspend the entire pellet in 1 L SDCAA. For passaging pooled transformations with a number of transformants totaling less than or equal to 1×10^9, pellet 1×10^{10} cells and resuspend the cells in 1 L SDCAA (*see* **Note 11**). Allow the passaged yeast to grow to saturation. For immediate induction and bead-based screening, use a portion of the saturated culture in Subheading 3.2, **step 1** (*see* **Note 12**). Take a small sample (2–3 μL), place on a microscope slide with a cover slip, and examine the culture under a microscope to confirm the absence of bacterial contamination (*see* **Note 13**). Transfer the remainder of the culture into sterile centrifuge bottles and pellet the cells using a floor centrifuge. Decant the supernatant and resuspend the pellet in a total of 10 mL 30 % sterile glycerol. The combination of the resuspended pellet and glycerol should lead to an approximate total volume of 20 mL and final glycerol concentration of 15 %. Aliquot the cells into 2 mL cryogenic vials and freeze the vials at −80 °C. These library fractions can be stored indefinitely (*see* **Note 14**).

12. *Perform library quality control to sample library genetic diversity and verify that the majority of clones are of full length (optional but recommended)*. Using the methods described in Subheading 3.4, **steps 1–3**, isolate the DNA from individual library members and submit these clones for sequencing (10 gives a good coarse-grain assessment of quality). Verify that the sequences are consistent with the library construction, the clones isolated are distinct, and most clones are free of stop codons (the method of library construction should dictate the probability of encountering a stop codon in a given sequence). Using the small-scale induction method described in Subheading 3.2, **step 11**, and the flow cytometry preparation and methods from Subheading 3.2, **steps 12–14**, and Table 1, determine the percentage of clones in the library displaying full-length scFvs (*see* **Note 15**).

3.2 Bead-Based Enrichments of Yeast Surface Display Libraries

This section assumes a library size of no greater than 1×10^8. Notes in the section explain how to scale the screening procedure to accommodate libraries of larger size.

1. *Thaw a library vial and induce cells for bead sorting*. Remove a vial of cells from the freezer and let it thaw on the benchtop. Once thawed, place the contents of the vial in 1 L SDCAA in a baffled flask and grow the cells overnight at 30 °C (*see* **Note 16**). The next morning, measure the OD_{600} of the culture and pellet 1×10^9 yeast cells (e.g., with an OD_{600} of 10, 10 mL) at $2{,}000 \times g$ for 5 min (in case of flocculation, *see* **Note 12**).

Table 1
Yeast surface display specimen labeling

Specimen name	Cells used	Primary label(s)	Dilution/ concentration	Secondary label(s)	Dilution
Negative	Any	None		None	
488 primary and secondary	Library fraction	Mouse anti-HA	1:200	Goat anti-mouse 488	1:100
488 primary and secondary	Single clone	Biotinylated antigen	Varies	Streptavidin 488	1:100
647 primary and secondary	Any displaying cells	Chicken anti-cmyc	1:250	Streptavidin 647	1:100
PE primary and secondary	Single clone	Biotinylated antigen	Varies	Mouse anti-biotin PE	1:100
Library validation sample	Portion of naïve library	(1) Mouse anti-HA (2) Chicken anti-cmyc	(1) 1:200 (2) 1:250	(1) Goat anti-mouse 488 (2) Goat anti-chicken 647	(1) 1:100 (2) 1:100
Library secondary binder analysis sample	Sorted library	Chicken anti-cmyc	1:250	(1) Streptavidin 488 (2) Goat anti-chicken 647	(1) 1:100 (2) 1:100
Library secondary binder analysis sample	Sorted library	Chicken anti-cmyc	1:250	(1) Mouse anti-biotin PE (2) Goat anti-chicken 647	(1) 1:100 (2) 1:100
Library analysis/sorting sample	Sorted library	(1) Biotinylated antigen (2) Chicken anti-cmyc	(1) Varies (2) 1:250	(1) Mouse anti-biotin PE (2) Goat anti-chicken 647	(1) 1:100 (2) 1:100
Library analysis/sorting sample	Sorted library	(1) Biotinylated antigen (2) Chicken anti-cmyc	(1) Varies (2) 1:250	(1) Streptavidin 488 (2) Goat anti-chicken 647	(1) 1:100 (1) 1:100

Resuspend the cells in 100 mL SDCAA and grow the cells at 30 °C with shaking. This step results in a culture at an initial OD_{600} of 1. Grow the cells until the OD_{600} of the culture is between 2 and 5, the logarithmic growth phase of yeast (*see* **Note 17**). Pellet 1×10^9 cells at $2,000 \times g$ for 5 min in 50 mL conical tubes, decant the supernatant, and resuspend the cells in 100 mL SGCAA (again, resulting in a culture at an OD_{600} of 1). Incubate the cells at 20 °C in a shaking incubator for 18–24 h to induce the cells (*see* **Note 18**).

2. *Prepare antigen-coated beads for positive enrichment.* Mix together 10 μL resuspended biotin binder beads (coated with streptavidin), 100 μL PBSA, and 33 pmol biotinylated antigen in a 2 mL tube (*see* **Notes 19** and **20**). Incubate at 4 °C for 2 h to overnight on a rotary wheel. After the completion of incubation, add 1 mL of ice-cold PBSA to the tube and place the tube on a Dynamag-2 magnet for 2 min. Remove the supernatant completely with a pipette, and resuspend the beads in 1 mL ice-cold PBSA to wash the beads, pipetting the beads up and down. Place the beads on the magnet for 2 min and remove the supernatant. Remove the washed beads from the magnet, resuspend the beads in 100 μL ice-cold PBSA, and place them on ice until needed.
3. *Prepare cells for negative and positive sorts.* Measure the OD_{600} of the induced culture and remove a culture volume containing 2×10^9 cells (*see* **Note 21**). Pellet the cells in 50 mL conical tubes at $2,000 \times g$ for 5 min, decant the supernatant, and resuspend the cells in a total volume of 1 mL sterile, ice-cold PBSA using repeated pipetting. Transfer the cells to a 2 mL tube and pellet the cells at $12,000 \times g$ for 1 min. Remove the supernatant with a pipette tip and resuspend the cells in 1 mL sterile, ice-cold PBSA. Repeat the pelleting, supernatant removal, and resuspension. The cells are now ready for magnetic bead sorting (*see* **Notes 22** and **23**).
4. *Perform a negative bead sort on the cells* (*see* **Note 24**). All sorting should be performed at 4 °C. Add 100 μL resuspended biotin binder beads to the cells. Incubate cells on a rotary wheel at 4 °C for 2 h (longer incubation times are also acceptable). Upon completion of the incubation, place the tube on the magnet, taking care to transfer any liquid lodged in the cap of the tube to the bottom portion of the tube. After 2 min, carefully remove the supernatant from the tube and transfer it into a fresh 2 mL tube. The supernatant will serve as the input for the next sort. Resuspend the beads in 1 mL ice-cold PBSA with a pipette (at this stage, do not invert the tube) and place the tube on the magnet for 2 min. Remove the supernatant of the washed beads and discard it. Resuspend the beads in 1 mL ice-cold PBSA and set aside the sample to enable estimation of the number of cells captured in the negative sort in **step 8**.

5. *Perform a second negative sort on the cells.* Repeat **step 4** using the supernatant from the previous step and then proceed to the following step with the depleted supernatant.
6. *Perform a positive sort on the cells.* This step should also be performed at 4 °C. Mix the supernatant from **step 5** with the antigen-coated beads from **step 2** and incubate on the rotary wheel at 4 °C for 2 h. Upon completion of the incubation, place the tube on the magnet, taking care to transfer any liquid lodged in the cap of the tube to the bottom portion of the tube. Incubate the cells and the beads on the magnet for 2 min, then carefully remove the supernatant from the tube and discard it. Resuspend the beads in 1 mL ice-cold PBSA using a pipette and place the beads on the magnet for 2 min. Remove the supernatant and discard it. The beads should contain a population enriched for binding to the target.
7. *Rescue the enriched population.* Resuspend beads and cells from the previous step in 1 mL SDCAA and transfer the cells to a sterile glass culture tube containing 4 mL SDCAA (5 mL SDCAA + beads + cells in total). Vortex the tube gently and then remove a sample of 5 μL from the culture. Dilute the sample into 995 μL SDCAA (200× dilution) and set aside for **step 8**. Grow the cells and beads at 30 °C overnight with shaking.
8. *Plate fractions of beads from negative and positive sorts to estimate the number of cells recovered in each step.* Vortex the saved beads from the negative sorts and transfer 100 μL beads into 400 μL fresh SDCAA. Take these diluted samples, vortex, and transfer 5 μL of each sample into 995 μL SDCAA (200× dilution). Vortex the 200× dilutions of the negative sort and the positive sort (from the previous step) and transfer 10 μL from each population into 190 μL SDCAA (4,000× dilution). Divide an SDCAA plate into six regions using a permanent marker. Vortex each dilution and plate 20 μL. Grow the plate at 30 °C for 3 days and count the colonies. One colony in the 200× and 4,000× dilutions represents 5×10^4 and 1×10^6 cells recovered, respectively.
9. *Prepare sorted cells for further sorts by removing the beads from the first sort, expanding the cells, and inducing a portion of the cells.* After the overnight growth, measure the OD_{600} of the cells. If the OD_{600} is still low, continue to allow the cells to grow; another day of growth is acceptable in case of especially low OD_{600}s (*see* **Note 25**). Once the culture approaches saturation, pellet the cells (at $900 \times g$ for 5 min) and aspirate the supernatant. Resuspend the pellet in 1 mL SDCAA and transfer the cells to a 2 mL tube. Place the tube on the magnet for 2 min. Recover the supernatant and dilute the cells into two cultures for further expansion. Dilute 2.5×10^8 cells into 25 mL SDCAA medium for growth and induction and dilute the

remaining cells into 25 mL SDCAA for overnight growth and temporary storage at 4 °C in case the first sorted population needs to be induced and sorted again (for revival recommendations, *see* **Note 14**). Once the 2.5×10^8 cell culture reaches an OD_{600} between 2 and 5, pellet 5×10^8 cells and resuspend in 50 mL SGCAA. Incubate the cells at 20 °C with shaking for 18–24 h to induce.

10. *Perform an additional series of negative and positive bead sorts.* Prepare more antigen-coated beads according to **step 2**. Once the cells are induced, take an OD_{600} to enable estimation of the total number of induced cells. Take between 0.5 and 2×10^9 cells and perform **steps 3–8**, yielding beads and cells growing in a 5 mL SDCAA culture.

11. *Perform a small-scale induction to enable initial flow cytometry analysis.* Once the twice-enriched population from **step 10** reaches saturation, pellet the cells and remove the beads as in **step 9**. Take 5×10^7 cells and dilute them to an OD_{600} of 1 in 5 mL SDCAA. Grow the cells at 30 °C until the OD_{600} is between 2 and 5. Take 5×10^7 cells, transfer them to a new glass culture tube, pellet them at $900 \times g$ for 5 min, remove the supernatant, and resuspend the cells in 5 mL SGCAA. Grow the cells at 20 °C with shaking for 18–24 h to obtain induced cells.

12. *Prepare the twice-sorted population for flow cytometry analysis.* For this step, the preparation of control samples from a single clone known to display a full-length, functional binding protein is highly encouraged and will be required for fluorescence-activated cell sorting in Subheading 3.3 (*see* **Note 26**). The labeling procedures described in this step assume that only the sorted sample is available, but if the control construct is available, use the induced control construct for negative, 488 primary and secondary, and 647 primary and secondary samples. Measure the OD_{600} of the induced cells and remove 5×10^6 cells for the following samples: negative, 488 primary and secondary, 647 primary and secondary, streptavidin detection only, and detection of antigen. Pellet the cells at $12{,}000 \times g$ for 30 s, aspirate the supernatant, and wash the cells by resuspending the cells in 1 mL sterile PBSA, pelleting ($12{,}000 \times g$ for 30 s), and aspirating the supernatant. Resuspend the washed cells in 250 μL PBSA and aliquot 50 μL into each of the five 1.7 mL microcentrifuge tubes. Label the tubes and set the negative sample aside on ice. Add primary labels to the cells according to Table 1, vortex the cells briefly (*see* **Note 27**), and incubate the cells at room temperature on a rotary wheel for 30 min. For the sample in which antigen will be detected, be sure to use a high concentration of antigen (1 μM is ideal) to ensure that cells displaying weak binding proteins appear positive on the flow cytometer. After completion of the incubation,

pellet the cells using a refrigerated tabletop centrifuge at 4 °C (12,000 × *g* for 30 s for the remainder of this step), aspirate the supernatant, resuspend the cells in 1 mL ice-cold PBSA, pellet, aspirate, then resuspend the cells in 50 μL ice-cold PBSA and add secondary labeling reagents according to Table 1. Vortex the cells briefly, place the cells on ice, and protect them from light while incubating with the secondary reagents for 10 min (up to 30 min in the secondary reagents is acceptable). Pellet the cells, wash them once in ice-cold PBSA, and leave the aspirated pellets on ice until immediately prior to analysis.

13. *Perform flow cytometry analysis on labeled cells.* Resuspend the labeled cell populations and transfer them to tubes compatible with the flow cytometer to be used (*see* **Note 28**). Run samples one by one, starting with the negative, 488 primary and secondary, and 647 primary and secondary samples first. Adjust detector voltages so that stained cells appear on scale for the corresponding detector. If some cells have especially high fluorescence values, continue to decrease the voltage at the expense of the negative cells, which may appear below the detection limit. Finish running all samples on the cytometer, making sure to record data on all samples after the detector voltages have been finalized. Figure 2 depicts typical histograms of a bead-sorted library population and a control clone (*see* **Note 29**).

14. *Analyze data and determine future experiments.* Examine the flow cytometry data using two-dimensional dot plots comparing display and antigen binding levels as in Fig. 2. Sorted library populations will vary greatly in the number of antigen-positive cells present after two rounds of bead sorting;

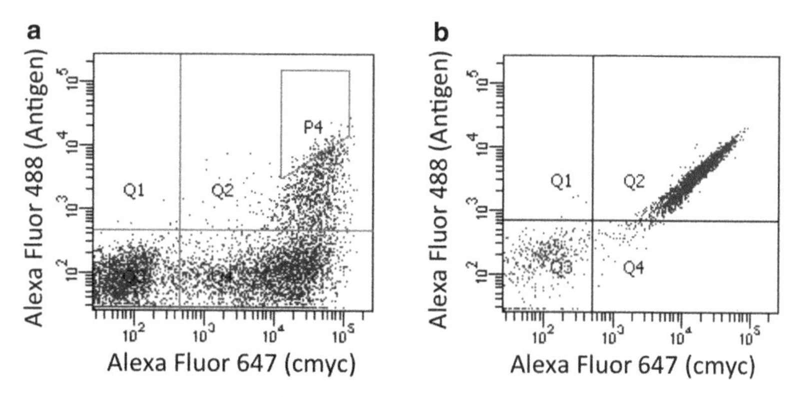

Fig. 2 Flow cytometry setup. (**a**) Gating strategy for enrichment of high-affinity antigen binders. A library subjected to two rounds of bead sorting appears enriched for antigen binders. Full-length clones binding only the highest amounts of antigen will be collected in an aggressive attempt to isolate the highest affinity binders. (**b**) The typical fluorescence profile of a single clone. Note the high degree of correlation between display levels and antigen binding levels

this will be largely dependent upon library quality. The "secondary only" labeling will reveal the presence of streptavidin-binding clones. Enriched populations should be carefully monitored for secondary binders due to their occasional persistence even after repeated library depletion. If the sorted population reveals even a small percentage of antigen-positive, full-length clones (as low as 0.1 %), use flow cytometry to enrich the population for antigen-specific clones using Subheading 3.3. If more than 1–2 % of cells are antigen positive, perform additional flow cytometry experiments to determine if a substantial antigen-binding population persists at lower concentrations of antigen (e.g., 100 nM, 10 nM). If the data reveal no evidence for antigen-positive clones, consider performing additional rounds of bead-based enrichments or constructing a library with a different diversification strategy.

3.3 Flow Cytometric Enrichments Using Equilibrium Conditions

Steps 1 and **2** can be performed concurrently.

1. *Prepare a control population of induced cells to be used in setting up sorting conditions.* Using a plasmid encoding a functional, surface-displayed binding protein (*see* **Note 26**), transform zymocompetent EBY100 according to the manufacturer's protocol. Plate the cells on SDCAA plates and allow the cells to grow for 2–3 days at 30 °C. Pick a single colony from the plate and use it to inoculate a 5 mL SDCAA culture. Grow the culture at 30 °C with shaking until saturation is reached (1–2 days). Perform a 5 mL-scale induction using the steps outlined in Subheading 3.2, **step 11**. Measure and record the OD_{600} of the induced cells and store the cells at 4 °C. These induced cells can be used as controls for flow cytometry experiments for at least 1 month without significant loss of the quality of display or the physical properties of the cells. Uninduced cells can also be stored at 4 °C for at least 1 month without loss of viability; these cells can be used to prepare more induced cells when necessary or proliferate yeast containing the construct for future use.
2. *Prepare populations for flow cytometric enrichment using antigen labeling under equilibrium conditions.* Use the instructions of Subheading 3.2, **step 1** or **11**, to grow and induce cells for flow cytometric cell sorting. The choice of large- or small-scale induction will depend on the estimated diversity of the population to be sorted (*see* **Note 30**). Upon completing induction, label cells displaying the control construct and cells displaying the population to be sorted. For populations to be sorted, label at least 1×10^7 cells per sample (regardless of population diversity) and try as best as possible to oversample the population diversity by tenfold. Also, bear in mind that flow cytometers can usually sort $0.5–1 \times 10^8$ cells in an hour;

do not expect to process billions of cells in a day unless several machines are available. All cells can be labeled using the general procedure outlined in Subheading 3.2, **step 12**, with three minor modifications: (1) For the first round of flow cytometric enrichment, use the secondary label mouse anti-biotin PE in place of streptavidin 488 in order to avoid enrichment for any remaining streptavidin binders in the population (*see* Table 1 for labeling of samples and controls). (2) The sorted population of cells will likely need to be labeled in a larger volume. When antigen is present at very high concentrations, the highest recommended cell density is 1×10^7 cells per 50 μL volume. When cells are exposed to lower concentrations of antigen, ensure that the antigen concentration exceeds ten times the effective concentration of binding protein in the solution. Assume that each cell displays approximately 5×10^4 scFvs (*see* **Note 31**). (3) Ensure that the antigen labeling approaches equilibrium prior to moving to secondary labeling using the formula $\tau = (k_{on}[\text{Ag}]_0 + k_{off})^{-1}$, where τ is the equilibrium time constant; k_{on} and k_{off} are the on and off rates of antigen–scFv binding, respectively; and $[\text{Ag}]_0$ is the initial concentration of antigen in solution. The binding has reached 95 % of equilibrium at 3τ and 99 % equilibrium at 4.6τ. A typical scFv k_{on} is $1 \times 10^5\ \text{M}^{-1}\ \text{s}^{-1}$, and the k_{off} can be determined using the relationship $K_d = k_{off}/k_{on}$, where K_d is the equilibrium dissociation constant (*see* **Note 32**) [8]. In all cases, perform labeling with primary antigen for at least 30 min. Upon completion of labeling, leave cells to be sorted as pellets on ice until immediately prior to sorting. The labeling will remain stable for long periods of time under these conditions. In cases where a long sort (>10–20 min) or labile antigen binding is anticipated, divide the labeled population into several tubes and prepare each tube as the sorting requires in the following step (1×10^7 cells/sample is a good place to start).

3. *Run flow cytometry controls and sort populations of interest.* Run the negative, PE-only, and Alexa Fluor 647-only control populations, adjusting voltages as in Subheading 3.2, **step 13** (*see* **Note 33**). Use these samples to set up a two-dimensional plot with Alexa Fluor 647 and PE as the axes and draw a quadrant that places negative, PE-positive, and Alexa Fluor 647-positive cells in separate quadrants. Resuspend the first sample to be sorted and obtain data from a few thousand cells. Draw a polygon gate that captures full-length, antigen-positive cells (*see* **Note 34**). Set up a glass culture tube containing 1 mL SDCAA in the collection area, specify the population to be collected, and initiate the sort. Be sure to examine the fluorescence values as a function of time. If the fluorescence of the population is changing rapidly, frequently adjust the gating or change tubes. In a single collection tube, collect no more than 1 mL

volume of sorted cells to avoid flocculation issues associated with most phosphate-based flow cytometry buffers (*see* **Note 12**). When a collection tube contains 1 mL sorted cells or a new population will be sorted, remove and cap the collection tube and vortex briefly. Leave the collection tube on ice, and proceed to **step 4** once all sorting is complete. Be sure to record the number of cells collected in each tube to ensure that population diversity can be preserved in subsequent steps by carrying on an appropriate number of cells.

4. *Rescue sorted cells*. Take the collection tubes and wash down the sides of the tubes with 4 mL additional SDCAA. Grow the cells at 30 °C with shaking until the cells reach saturation (if speed is an issue, *see* **Note 25** regarding induction of cells prior to saturation). The number of collected cells will determine the time needed to reach saturation. Tubes containing only a few thousand collected cells will almost certainly take 2 days to reach saturation OD_{600}, while tubes containing higher numbers will reach saturation in 1–1.5 days.
5. *Assess sort quality and continue sorting library until sufficiently enriched*. Using Subheading 3.2, **steps 11–13**, perform inductions and flow cytometry analysis on the sorted populations (*see* **Note 35**). If a high percentage of antigen binding is anticipated, prepare samples in which cells are labeled with decreasing concentrations of antigen to determine the concentration of antigen to be used for future sorting. Also, be sure to prepare samples in which binding to mouse anti-biotin PE is assessed (Table 1). Run the samples on the flow cytometer and determine the percentage of cells displaying full-length clones that recognize the antigen and the percentage recognizing secondary reagents. If the population appears to be enriched for antigen-specific binding, continue with flow cytometry-based enrichment until a majority of cells exhibit binding to the antigen, decreasing antigen concentration if higher affinity antibodies appear to be present in the population. If the population exhibits some antigen recognition and some mouse anti-biotin PE recognition, use the methods of this section to sort the population again using streptavidin–Alexa Fluor 488 for secondary detection (*see* **Note 36**). Do not use the same secondary reagent for antigen detection for more than two rounds of sorting in a row to minimize the possibility of isolating secondary reagent binders. Continue sorting the library until the population has become enriched for binders of the highest affinity possible (*see* **Note 37**). If the population appears to contain clones with the desired affinities, or if individual clones are to be affinity matured, evaluate the properties of individual clones using the methods of Subheading 3.4. If the entire population is to be affinity matured, use Subheading 3.4, **steps 1** and **2**, to prepare DNA suitable for use in Subheading 3.5.

3.4 DNA Preparation and Evaluation of Individual Clones

1. *Isolate plasmid DNA from yeast population.* Take up to 1×10^8 yeast from a growing culture with an OD_{600} below approximately 2 and perform a zymoprep according to the manufacturer's recommendations with the following modifications: (1) use double the amount of zymolyase recommended and increase the digestion time to at least 2 h with mixing every hour; (2) following the pelleting of cellular debris, decant the supernatant into a silica column with high binding capacity and complete the procedure (*see* **Note 38**). In the last step, elute the DNA into 40 μL ddH_2O.
2. *Transform E. coli with zymoprepped DNA, grow transformed E. coli, and perform minipreps.* Using chemically competent or electrocompetent *E. coli* (*see* **Note 39**) according to the manufacturer's instructions, transform the cells with 1–10 μL of the zymoprepped DNA population of interest. Rescue the cells for 1 h at 37 °C in 0.5–2 mL SOC medium. If individual clones are to be isolated, plate 20–200 μL of the rescued cells on prewarmed LB plates containing 100 μg/mL ampicillin (LB-amp) and incubate the plates at 37 °C for 12–18 h. Use single colonies from the plates to inoculate 5 mL cultures in LB media containing 100 μg/mL ampicillin (*see* **Note 40**). If the DNA from the entire population is needed in bulk (e.g., for affinity maturation), dilute the rescued cells into 50 mL LB medium containing 100 μg/mL ampicillin (*see* **Note 41**). Grow liquid cultures for 12–18 h or until the cultures reach saturation. Pellet the cells and perform minipreps according to the manufacturer's instructions. If DNA is to be sequenced, elute the DNA using ddH_2O, not buffer. If DNA will be immediately mutagenized, proceed to Subheading 3.5.
3. *Sequence individual clones.* Because highly enriched populations are oftentimes dominated by a few clones, identifying unique clones can cut down on labor in subsequent evaluation of scFv properties. Submit DNA for sequencing using the forward sequencing primer (*see* Subheading 2). In many cases, the sequencing read quality will remain high throughout the length of the scFv. If a reverse sequence is needed, use the reverse primer (*see* Subheading 2). Confirm that clones of interest are of full length and select unique clones for further evaluation.
4. *Transform zymocompetent EBY100 with individual clones and prepare induced samples.* Use the instructions of Subheading 3.3, **step 1**, and Subheading 3.2, **step 11**, to prepare induced samples of EBY100 displaying the clones of interest. Note that more than one transformation can be plated on the same SDCAA plate.
5. *Confirm antigen-specific binding using flow cytometry (optional, but recommended).* Before thoroughly evaluating the binding properties of individual clones, verify that each clone recognizes

the antigen of interest. Perform flow cytometry analysis on cells displaying individual clones stained with secondary antigen detection reagents only and antigen plus secondary (Table 1) to validate the antigen specificity of each clone.

6. *Titrate antigen with individual clones and perform flow cytometry.* With antigen-specific clones in hand, titrate the antigen concentration on a series of cell samples and perform flow cytometry to enable estimation of the equilibrium dissociation constant (K_d). For crude estimations, fourfold dilutions two orders of magnitude above and below the anticipated K_d are sufficient, while fine K_d determination can be obtained with twofold dilutions in triplicate. The large numbers of samples necessary to obtain K_ds are most efficiently obtained by performing staining in 96-well plates (*see* **Note 42**). For these experiments, a greatly reduced number of cells and lower concentrations of anti-cmyc and secondary labels can be used. Prepare the plate samples in parallel with negative and single-color controls in tubes (as described in Subheading 3.2, **step 12**) for voltage adjustment on the flow cytometer. Wash induced cells in PBSA and resuspend the cells to a final concentration of 1.5×10^6 cells/mL in PBSA. Remove a large enough volume to enable distribution of 10 μL of cells (1.5×10^4) into each of 8 wells (fourfold dilutions) or 16 wells (twofold dilutions) and place in a new 1.7 mL microcentrifuge tube. Add chicken anti-cmyc to a final dilution of 1:200 and vortex the cells briefly. If sufficiently low concentrations of antigen will be reached, the antibody-displaying cells will need to be reduced in order to maintain a tenfold excess of effective antibody concentration. In these cases, use bare EBY100 to dilute the antibody-displaying cells to a lower concentration while maintaining a sufficient number of cells to establish a pellet. Prepare bare EBY100 in the same way as the antibody-displaying cells (including addition of chicken anti-cmyc) and mix with cells displaying scFvs in appropriate ratios (e.g., 4:1 or 9:1 bare:displaying). Set aside the cells in tubes while preparing the titrations on the plate. Add 40 μL PBSA to each well of a 96-well plate to be used in the experiment except the well that will contain the highest concentration of antigen in each titration. Add antigen diluted in PBSA to the first well for each titration at a concentration of 1.25 times the final highest desired concentration and a volume of 53.33 μL (fourfold dilutions) or 80 μL (twofold dilutions). Complete the titration by diluting 13.33 μL (fourfold) or 40 μL (twofold) antigen into the well intended to have the next highest concentration of antigen and so forth until 7 out of 8 (fourfold) or 15 out of 16 (twofold) wells to be used for a sample have had antigen diluted into them. Remove 13.33 μL (fourfold) or 40 μL (twofold) from the 7th or 15th wells, respectively, and discard.

Leave the 8th or the 16th wells blank to serve as a secondary-only staining reference. Add 10 μL cmyc-labeled cells to each well, taking care to add cells diluted with EBY100 to wells that contain especially low concentrations of antigen as required. This lowers the antigen concentration in each well by a factor of 1.25. Seal the plate with adhesive foil and incubate the cells on a shaking platform at 150 rpm until near-equilibrium conditions are reached as described in Subheading 3.3, **step 2**. Upon completion of the primary incubation, add 150 μL/well ice-cold PBSA to the plate and spin the cells at 1,300 × *g* for 5 min in a swinging bucket rotor. Dump the supernatant of the pelleted wells out by inverting the plate and using quick shaking motions, then place the plate on ice. Resuspend each well in 50 μL of secondary labeling reagent diluted in ice-cold PBSA. In this case, 1:500 dilutions of each detection reagent are sufficient to conserve labeling reagents. Seal the plate with adhesive foil and incubate it at 4 °C on a shaking platform for 10 min. Add 150 μL/well ice-cold PBSA and pellet the cells. Dump the supernatant and resuspend the cells in 200 μL/well ice-cold PBSA. Run the samples on the flow cytometer, starting with controls (prepared in tubes) to set detector voltages, followed by the plate.

7. *Process flow cytometry data and estimate* K_d*s of individual clones.* Take the raw flow cytometry data and gate out the population of cells positive for the cmyc label. Process the data to obtain the median fluorescence values corresponding to antigen display in all samples using FlowJo or BD FACSDiva software. Fit the fluorescence data as a function of antigen concentration to the equation $MF_{high} = MF_{low} + (MF_{observed} \times [Ag])/([Ag] + K_d)$ using a least squares approach in a program such as Microsoft Excel or Matlab. In the equation, MF_{high} equals the mean fluorescence value of the sample exposed to the highest antigen concentration, MF_{low} equals the mean fluorescence obtained in the absence of antigen, and $MF_{observed}$ equals the fluorescence observed at a specific antigen concentration. To average multiple experiments, average the K_ds obtained in individual experiments and report the average and standard deviations or 95 % confidence intervals of the values. These data can be used to determine if further affinity maturation is desired. An approach to affinity maturation is described in Subheading 3.5.

3.5 Affinity Maturation Using Error-Prone PCR Libraries

1. *Introduce mutations into DNA encoding scFvs to be affinity matured.* Using the DNA encoding an individual clone, several pooled clones, or an entire population (*see* **Note 43**), perform error-prone PCR at 50 μL scale using the following recipe:

   1.0 μL 0.7 ng/μL template DNA (roughly 1×10^8 plasmids).

   2.5 μL 10 μM forward amplification primer (*see* **Note 44**).

2.5 μL 10 μM reverse amplification primer.

5.0 μL 10× ThermoPol buffer.

1.0 μL Taq enzyme.

1.0 μL 10 mM dNTP mix.

5.0 μL 20 μM 8-oxo-dGTP.

5.0 μL 20 μM dPTP.

27 μL ddH_2O.

Cycle the reaction as follows:

1 cycle of 95 °C for 3 min; 12–16 cycles of 95 °C for 45 s, 60 °C for 30 s, and 72 °C for 90 s; 1 cycle of 72 °C for 10 min; and 1 cycle of 4 °C forever.

Running the PCR with the mutagenic dNTPs (8-oxo-dGTP and dPTP) under these conditions will introduce, on average, one or more amino acid mutations per gene [10].

2. *Purify the mutagenized DNA, amplify the DNA further, and purify the amplified DNA*. Run a DNA gel and purify the full-length PCR product using a gel extraction kit. Elute the DNA in 40 μL ddH_2O. Amplify the DNA further in a 200 μL PCR reaction using Taq polymerase according to the following recipe:

   10 μL template (from previous PCR).

   2 μL 100 μM forward primer.

   2 μL 100 μM reverse primer.

   20 μL 10× ThermoPol buffer.

   4 μL Taq enzyme.

   4 μL 10 mM dNTP mix.

   158 μL ddH_2O.

Use the following cycling:

1 cycle of 95 °C for 3 min; 30 cycles of 95 °C for 45 s, 55 °C for 30 s, and 72 °C for 90 s; 1 cycle of 72 °C for 10 min; and 1 cycle of 4 °C forever.

Run a DNA gel of the amplified product, extract the properly sized band, and purify it. This PCR reaction will usually yield between 4 and 8 μg of purified product (*see* **Note 45**).

3. *Create a library*. Using the techniques outlined in Subheading 3.1, construct a library in yeast using a single electroporation cuvette per library. Given the limited throughput of flow cytometers, no benefit will be obtained from having libraries larger than approximately 1×10^7 clones, although there is no harm in having a library of larger size, either. Induce the library using the techniques described in Subheading 3.1.
4. *Screen the library using equilibrium screening (if appropriate)*. If the K_ds of the majority of the clones in the library remain of

low to medium affinity (weaker than 5–10 nM), equilibrium screening remains the most feasible and straightforward method to search for clones with improved affinity for the antigen. To maximize the probability of finding clones with higher affinities, label the induced library with a concentration of antigen resulting in weak but perceptible binding for the majority of clones in the library (*see* **Note 46**). Large volumes (50–100 mL) may be required to enable sorting with antigen in the low nanomolar range. If this is the case, do not label the cells with cmyc during incubation with the antigen. Instead, after completing antigen labeling, pellet the cells, wash once, and resuspend cells in ice-cold PBSA containing a 1:250 dilution of chicken anti-cmyc, keeping the concentration of cells at 1×10^7 cells/50 μL or lower. Vortex the cells briefly and incubate on ice for 10 min before proceeding with washing and secondary labeling as described in Subheading 3.2, **step 12**. Sort the library using the techniques described in Subheading 3.3 until no further enrichment can be obtained. If the enriched population remains bright upon labeling with subnanomolar concentrations of antigen, screen the library using the kinetic screening approach described in the next step. Otherwise, isolate the DNA from the enriched population, evaluate individual clones, and initiate construction of a second error-prone PCR library.

5. *Screen the library using kinetic sorting*. If the library contains clones with K_ds at or below the low nanomolar range, equilibrium screening becomes impractical due to the large volumes required to label cells with antigen. Kinetic sorting can be used to circumvent this issue. To set up kinetic sorting, label cells with a concentration of biotinylated antigen that will allow all of the antibody-displaying cells to become saturated with antigen (i.e., tenfold higher than the K_d of the lowest affinity antibody in the population). Pellet the cells and wash them once in 1 mL ice-cold PBSA. After pelleting a second time, resuspend the cells in a large excess of nonbiotinylated antigen (at least 10–100-fold higher than the effective antibody concentration in solution) and allow biotinylated antigen to dissociate from the cells at room temperature. Rebinding of biotinylated antigen will be prevented by the presence of the nonbiotinylated antigen. In an analogous situation to equilibrium screening, allow the biotinylated antigen to dissociate from cells until the majority of the clones in the library have a weak but perceptible amount of antigen binding when analyzed on the flow cytometer (*see* **Note 47**). After a sufficient length of time, pellet the cells and wash them once in ice-cold PBSA. As in **step 4**, resuspend the cells with 1:250 cmyc in ice-cold PBSA to provide a primary label for the detection of scFv display, and then proceed with secondary labeling and

sorting as in Subheading 3.3. Sort the library until sufficient enrichment is obtained and then determine whether to characterize individual mutants or move directly on to more affinity maturation.

4 Notes

1. Numerous methods exist for preparing antibody fragment libraries or for obtaining DNA for construction of a library. For example, the Sidhu laboratory at the University of Toronto has described one accessible approach to the de novo assembly of minimal antibody fragment libraries in a recent book chapter [11]. Many antibody libraries have been made for phage display and can easily be adapted for yeast surface display. Some existing antibody fragment libraries include Tomlinson I and J libraries [12] and the ETH-2 human antibody library [13]. Construction of nonimmune antibody libraries from human antibody repertoires is also a well-established approach [14].
2. Typically, YPD media should not be supplemented with antibiotics, especially when preparing libraries. However, in cases where contamination poses a significant threat, supplement the media with 100× Pen/Strep solution.
3. Whenever possible, grow liquid cultures in flasks in which no more than 40 % of the culture volume is occupied by liquid. One exception: 2 L flasks seem to work well for growing 1 L cultures.
4. Growth of yeast at pH 4.5 discourages bacterial growth. Addition of Pen/Strep to all SDCAA and SGCAA medium is also recommended to prevent contamination. If contamination appears, supplement the culture with 50 μg/mL kanamycin and continue growing the culture to attempt to eliminate the bacteria.
5. A pH of 6.0 increases the probability that antibodies will remain properly folded when displayed on the surface of yeast.
6. Some proteins are sold commercially as biotinylated proteins. A good positive control antigen is lysozyme, sold in biotinylated form by Sigma (product #L0289). Biotinylation of antigens can also be accomplished via chemical [15] or enzymatic [16] means.
7. Each transformation performed using this protocol will yield roughly 1×10^7–1×10^8 transformants. Thus, if a billion-member library is to be constructed, preparations should be made for approximately 20 transformations.
8. A freshly streaked plate of cryogenically frozen EBY100 helps to ensure that the cells retain their tryptophan auxotrophy.

Cryogenic stocks can be prepared by mixing 1 mL saturated EBY100 culture with 1 mL sterile 30 % glycerol in a 2 mL cryogenic vial and freezing the vial at −80 °C.

9. Two passages should be performed prior to electroporation to ensure that the EBY100 are as healthy as possible. For especially large-scale electroporation efforts, the second passage should be performed into a larger volume of YPD. This volume can be estimated based on the observation that the saturation OD_{600} of EBY100 in YPD is typically between 8 and 16.
10. The doubling time of EBY100 in YPD media is approximately 1.5–2 h.
11. The OD_{600} measurement of liquid yeast cultures allows for an estimate of the number of yeast in a given culture. An OD_{600} of 1 corresponds to approximately 1×10^7 cells in 1 mL media. At the time of passaging, the number of transformants in the library (also known as the library size) will not yet be known due to the slow rate of colony formation on SDCAA plates. Thus, the library diversity must be preserved by assuming the highest possible number of transformants in the library and passaging sufficient numbers of cells. The electroporation procedure described here typically yields 1×10^7–1×10^8 transformants per electroporation. Thus, if five electroporations are pooled, the maximum expected library size would be approximately 5×10^8. Allowing for an additional safety factor of 2, the library should be assumed to contain 1×10^9 members. In order to preserve library diversity in this case, a number of cells equal to or greater than ten times the maximum library size should be passaged (based on Poisson statistical calculations [17]).
12. Occasionally, when yeast cells are grown to saturation or experience other conditions of stress, they become flocculent, forming clumps of cells that tend to sediment quickly. This does not appear to affect plasmid retention, but flocculation can impact the ability to accurately read OD_{600}s, aliquot cells for different experiments, or screen cells individually in a flow cytometry experiment. Therefore, if a culture is observed to be flocculent, pellet a portion of the offending culture and resuspend the pellet to an OD_{600} of 1 in SDCAA medium containing 4 % glucose instead of 2 %. Dilution of the cells and growth in richer medium tend to reverse flocculation.
13. Under a microscope, yeast (~3–5 μm in diameter) will appear stationary, while any contaminating bacteria (~1 μm in diameter) will be quite motile.
14. For libraries in which repeated use of frozen fractions is anticipated, expand the culture to a greater volume, prepare a large number of stocks (40–100), and deem this collection of vials

the master cell bank. To obtain the maximum possible number of stocks out of the library, upon thawing a member of the master bank (to be described in Subheading 3.2, **step 1**), expand the culture to a large volume and create 10–100 frozen vials of a working bank. Use all of the vials in the working bank, and when it is spent, make a new working bank from a single vial of the master bank. For libraries to be used for short-term purposes only (≤1 month), these libraries can be stored at 4 °C. Revival can be accomplished by resuspending the stored culture, pelleting 1×10^{10} cells, resuspending the pellet in fresh medium, and initiating sorting following the steps laid out in Subheading 3.2 starting with **step 1**.

15. In this analysis any cell exhibiting positive staining for both cmyc and HA tags will be assumed to be of full length. Given the limited throughput of the flow cytometer, only a small amount of the total library can be investigated practically. Thus, while a small-scale induction of the library will not contain the entirety of the library diversity, it will contain enough of the library to determine the percentage of full-length clones.
16. In general, approximately 70 % of frozen cells will be viable upon thawing. For especially large libraries, multiple aliquots may need to be thawed in order to adequately preserve diversity. To validate a freezing procedure, plate serial dilutions of the thawed cells on SDCAA plates and count colonies to estimate the number of viable cells in the vial.
17. The doubling time of yeast in SDCAA is approximately 3–4 h.
18. Larger libraries (up to 1×10^9 members) can be thawed and induced on a similar scale. After thawing the cells and letting the initial culture grow to saturation, carry a number of cells equal to ten times the library diversity on to each subsequent step. Freshly diluted cultures should have an OD_{600} of no greater than 1, a criteria that can be satisfied by using the rule of thumb that 100 mL media should be used per 1×10^9 cells (or per 1×10^8 library members).
19. The incubation step can be performed as induction of the yeast cells is occurring. Final washing of the beads for positive enrichment should be performed within 1–2 h of adding the beads to the cells after negative selections. The value of 33 pmol represents sufficient antigen to saturate the available streptavidin-binding sites present in 10 μL biotin binder beads according to the manufacturer's information. If biotinylated antigen is especially precious, as few as 6.7 pmol antigen can be used. However, in this case, the negative selections are especially important to avoid enriching the library for streptavidin or magnetic particle binders.

20. In the case of larger libraries, use 10 μL beads for every 2×10^9 cells to be sorted for up to 1×10^{10} cells. For cell numbers between 1×10^{10} and 1.5×10^{10}, 50 μL beads can be used.
21. In general, the use of more cells in a bead sort will lead to a lower probability of nonspecific bead binding, with the ratio of 2×10^9 cells to 10 μL biotin binder beads providing a high level of consistency. Induction of 1×10^9 cells will almost always yield more than two billion cells after 18–24 h. Typically, induced cultures reach an OD_{600} between 2.5 and 4.
22. Strict sterile technique is not necessary during bead preparation and sorting. Minimal instances of contamination will be observed as long as all reagents and pipettes are sterile and tubes are open for a short time.
23. Resuspend cells to be used in sorts in ice-cold PBSA. For cell numbers less than 2×10^9, use 1 mL PBSA. For cell numbers in between 2×10^9 and 1×10^{10}, use 1 mL PBSA per 2×10^9 cells. For cell numbers between 1×10^{10} and 1.5×10^{10}, use 5 mL.
24. For libraries that will be used for routine binder isolation, consider depleting the library of bead binders repeatedly and storing this depleted library in cryogenic vials. This will result in a library that can be immediately enriched for binding to antigens. To deplete, perform three negative depletions on the first depletion day, recover the supernatant from the third depletion, pellet the cells, and resuspend them in 1 L SDCAA. Grow, passage, and induce the cells, perform a second day of three depletions, and expand the cells for pelleting and cryogenic freezing as described in Subheading 3.1. To perform bead sorts with the library, thaw and induce the cells as outlined in Subheading 3.2, **step 1**; prepare antigen-coated beads as in Subheading 3.2, **step 2**; and proceed immediately to Subheading 3.2, **step 6**.
25. When speed is especially crucial, cells can be induced as soon as they reach logarithmic phase after a bead sort. Remove the beads from the cells, and resuspend the cells in SGCAA at a final OD_{600} of 1 and induce at 20 °C for 18–24 h. The one drawback to this accelerated approach is that fewer cells are used in the second round of sorting, which can potentially lead to more nonspecific binding and lower enrichment factors. In general, the bead sorting seems to work efficiently when at least 5×10^8 cells are used as input.
26. Numerous plasmids encoding functional, high-affinity binders to different targets now exist. Antibody fragment and fibronectin lysozyme binders are especially easy to use given the ready availability of biotinylated lysozyme and include the scFv D1.3 [18] and the fibronectin L7.5.1 [19].

Many other displayed binding proteins suitable as controls have also been described in the literature [7].

27. Vortex all samples after flow cytometry reagent additions in order to improve the uniformity of labeling.
28. Preferred tubes will be different for distinct models of flow cytometers, but most instrument manufacturers recommend sample filtration with devices such as tubes topped with a mesh filtration cap such as Falcon 12 × 75 mm polystyrene test tubes (Falcon #352235, VWR #21008-948).
29. When a population of yeast is induced, a certain percentage (roughly 30–50 %) fails to display the protein of interest, even in a genetically uniform population.
30. In general, populations smaller than 1×10^7 members should be induced at the 5 mL scale.
31. For example, in a volume of 50 μL, 1×10^7 cells displaying approximately 5×10^4 antibodies per cells will have an effective concentration of $(10^7$ cells/0.00005 L$) \times (5 \times 10^4$ antibodies/cell$) \times (1$ mol/6.02×10^{23} antibodies$) = 16.6$ nM. Thus, cell labeling with antigen concentrations lower than approximately 170 nM should be performed in larger volumes.
32. For example, for a population in which the highest expected binders have K_ds of roughly 10 nM and the solution is being labeled at 100 nM, 95 % equilibrium will be reached in $3\tau = 3((1 \times 10^5\ M^{-1}\ s^{-1}) \times (100 \times 10^{-9}\ M) + 1 \times 10^{-3}\ s^{-1})^{-1} = 273$ s or roughly 5 min. For most naïve libraries, the highest K_ds encountered will not fall below 1 nM. The apparent K_ds of populations of cells can also be determined using the titration methods outlined in Subheading 3.4.
33. Because detection of antigen binding tends to yield lower levels of cell staining than the detection of N- or C-terminal display (Fig. 1), the signal of antigen-binding cells may not be easily maximized without raising the apparent fluorescent values of negative cells to inappropriate levels. In this case, use the negative cells as a guide to set the PE voltage so that these cells fall within the lowest decade of fluorescence values.
34. In cases where only high-affinity binders are deemed important, this gate can be quite stringent, encompassing a region substantially separated from the borders of the quadrants (Fig. 2) and as few as 0.1 % or less of the total events. Regardless of antigen binding stringency, be sure to gate only on cells that are clearly of full length to minimize the isolation of truncated clones.
35. This analysis can be performed in parallel with additional sorting.
36. If streptavidin binders were present after bead enrichments, perform analysis of the population using streptavidin–Alexa

Fluor 488 to ensure that these binders are now absent from the population. Furthermore, if secondary reagent binding becomes a severe problem at any point, sort the population for full-length clones that do not recognize the secondary reagents before sorting again for antigen binders.

37. Most naïve libraries can be enriched using equilibrium sorting methods alone, but some special cases (library with exceptional properties, bi- or multivalent antigens) may require kinetic sorting as described in Subheading 3.5, **step 5**.
38. The columns accompanying the Zymoprep kit do not always bind the plasmid DNA efficiently, so alternatives such as Qiagen Qiaprep columns or Epoch Life Sciences EconoSpin columns are preferred.
39. Either chemically competent or electrocompetent *E. coli* can be transformed with zymoprepped DNA. In some cases, the zymoprep procedure gives low yields of DNA. In these cases, use electrocompetent cells for transformation. The *E. coli* strains DH10B and DH5α are excellent cloning strains, and competent versions of these cell strains are readily available from companies such as Life Technologies (DH10B: electrocompetent: ElectroMAX DH10B Cells; chemically competent: MAX Efficiency DH10B Competent Cells; DH5α: electrocompetent: ElectroMAX DH5α-E Competent Cells; chemically competent: MAX Efficiency DH5α Competent Cells).
40. For routine evaluations, ten colonies per transformed DNA population are sufficient to determine basic information about the genetic diversity of a population such as the number of distinct clones.
41. If bulk DNA is to be used for affinity maturation purposes only, the number of transformants is not critical as long as the population contains only a few dominant clones. On the other hand, if library coverage is crucial, take a portion of the freshly inoculated 50 mL culture and plate it on an LB-Amp plate and grow the plate at 37 °C for 12–18 h. Use the number of colonies that appear on the plate to estimate the total number of transformants; ensure that this number is greater than or equal to ten times the number of yeast cells recovered in the final sort.
42. Tube-based titration setup has been described previously [8].
43. If multiple scFvs are used, this enables recombination within scFv sequences to occur, leading to potentially faster affinity maturation [20]. DNA shuffling [21] or other approaches to sequence diversification can also be considered for the rapid affinity maturation of antibodies.
44. Use the same primers in the previous amplification of library DNA for homologous recombination. This will exclude roughly

nine to ten residues at the start and end of the protein from mutagenesis. If these residues are suspected to play an important role in the structure or the function of the scFv of interest, use primers that bind outside of the coding sequence of the scFv [8].

45. If between 1 and 4 μg DNA is obtained from the reaction, proceed on to Subheading 3.5, **step 3**, and create a library. The resulting library will likely contain enough diversity (1×10^6 or more transformants) to reliably yield improved scFvs.

46. If the library was created from a single clone, experiments with the single clone can be used to empirically determine an appropriate antigen concentration. If the K_d of the clone is known, an optimal value can be determined by using theories described in the literature [22].

47. As in the case of equilibrium screening, if the library was created from a single clone, the single clone can be used to empirically determine an appropriate length of competition prior to preparing library samples for sorting. If the dissociation rate (k_{off}) of the clone is known, theory can also be used to determine suitable lengths of competition [22].

Acknowledgments

We thank Alessandro Angelini and Tiffany Chen for providing comments on the manuscript. This work was supported by CA96504, CA101830, and postdoctoral fellowship support to J.A.V. from the National Institutes of Health (Ruth L. Kirschstein National Research Service Award #1 F32 CA168057-01) and the Ludwig Fund for Cancer Research.

References

1. Gai SA, Wittrup KD (2007) Yeast surface display for protein engineering and characterization. Curr Opin Struct Biol 17:467–473
2. van Bloois E, Winter RT, Kolmar H et al (2011) Decorating microbes: surface display of proteins on Escherichia coli. Trends Biotechnol 29:79–86
3. Hoogenboom HR (2005) Selecting and screening recombinant antibody libraries. Nat Biotechnol 23:1105–1116
4. Winter G, Griffiths AD, Hawkins RE et al (1994) Making antibodies by phage display technology. Annu Rev Immunol 12:433–455
5. Lipovsek D, Plückthun A (2004) In-vitro protein evolution by ribosome display and mRNA display. J Immunol Methods 290:51–67
6. Boder ET, Wittrup KD (1997) Yeast surface display for screening combinatorial polypeptide libraries. Nat Biotechnol 15:553–557
7. Boder ET, Raeeszadeh-Sarmazdeh M, Price JV (2012) Engineering antibodies by yeast display. Arch Biochem Biophys 526:99–106
8. Chao G, Lau WL, Hackel BJ et al (2006) Isolating and engineering human antibodies using yeast surface display. Nat Protoc 1:755–768
9. Colby DW, Kellogg BA, Graff CP et al (2004) Engineering antibody affinity by yeast surface display. Methods Enzymol 388:348–358
10. Zaccolo M, Williams DM, Brown DM et al (1996) An approach to random mutagenesis of DNA using mixtures of triphosphate derivatives of nucleoside analogues. J Mol Biol 255:589–603

11. Rajan S, Sidhu SS (2012) Simplified synthetic antibody libraries. Methods Enzymol 502:3–23
12. de Wildt RMT, Mundy CR, Gorick BD et al (2000) Antibody arrays for high-throughput screening of antibody-antigen interactions. Nat Biotechnol 18:989–994
13. Pini A, Viti F, Santucci A et al (1998) Design and use of a phage display library. J Biol Chem 273:21769–21776
14. Feldhaus MJ, Siegel RW, Opresko LK et al (2003) Flow-cytometric isolation of human antibodies from a nonimmune Saccharomyces cerevisiae surface display library. Nat Biotechnol 21:163–170
15. Hermanson GT (2008) Bioconjugate techniques. Elsevier, Amsterdam
16. Beckett D, Kovaleva E, Schatz PJ (1999) A minimal peptide substrate in biotin holoenzyme synthetase-catalyzed biotinylation. Protein Sci 8:921–929
17. Wittrup KD (2002) Directed evolution of binding proteins by cell surface display: analysis of the screening process. In: Brakmann S, Johnsson K (eds) Directed molecular evolution of proteins. Wiley, New York, pp 111–126
18. VanAntwerp JJ, Wittrup KD (2000) Fine affinity discrimination by yeast surface display and flow cytometry. Biotechnol Prog 16: 31–37
19. Hackel BJ, Kapila A, Wittrup KD (2008) Picomolar affinity fibronectin domains engineered utilizing loop length diversity, recursive mutagenesis, and loop shuffling. J Mol Biol 381:1238–1252
20. Swers JS, Kellogg BA, Wittrup KD (2004) Shuffled antibody libraries created by in vivo homologous recombination and yeast surface display. Nucleic Acids Res 32:e36
21. Stemmer WPC (1994) Rapid evolution of a protein in vitro by DNA shuffling. Nature 370:389–391
22. Boder ET, Wittrup KD (1998) Optimal screening of surface-displayed polypeptide libraries. Biotechnol Prog 14:55–62

Chapter 11

Human B Cell Immortalization for Monoclonal Antibody Production

Joyce Hui-Yuen, Siva Koganti, and Sumita Bhaduri-McIntosh

Abstract

Infection of primary B lymphocytes with Epstein–Barr virus gives rise to growth-transformed and immortalized lymphoblastoid cell lines (LCL) in vitro. Among their many applications is the use of LCL to present antigens in a variety of immunologic assays and to generate human monoclonal antibodies. This chapter describes a method to generate LCL from donor peripheral blood with rapid immortalization and cryopreservation times.

Key words Epstein–Barr virus, Lymphoblastoid cell lines, B cells, Proliferation, Immortalization, Transformation, Monoclonal antibody

1 Introduction

Isolation and expansion of human B cells in vitro have been previously described using a variety of methods. Lymphoblastoid cell lines (LCL) provide a model system to investigate Epstein–Barr virus (EBV) latency and persistence, EBV-driven B cell proliferation, and tumorigenesis [1]. LCL have also been used as antigen-presenting cells in a variety of immunologic assays [2, 3], as the source of human monoclonal antibodies [4, 5], and as a potentially unlimited source of genetic material when access to primary biologic material is limited [6, 7]. The efficiency of EBV-mediated immortalization has been investigated using different mitogens [8, 9] and/or immunosuppressive agents to inhibit T cell-mediated killing of infected B cells [7, 10–12].

This chapter describes a method which results in rapid and reliable EBV-driven B cell growth transformation. This method uses a small amount of peripheral blood mononuclear cells (PBMC) that are infected in vitro with EBV in the presence of a T cell immunosuppressant, FK506. Through the combination of high titres of infectious EBV and FK506, one is able to consistently infect, promote proliferation, and transform EBV-infected B cells in an

Vincent Ossipow and Nicolas Fischer (eds.), *Monoclonal Antibodies: Methods and Protocols*, Methods in Molecular Biology, vol. 1131, DOI 10.1007/978-1-62703-992-5_11, © Springer Science+Business Media New York 2014

efficient manner, thus resulting in generation of rapidly expanding LCL for downstream applications, including monoclonal antibody production. Proliferating cells appear as clusters of cells. Cell clusters can be seen under the microscope approximately 1 week after infection with EBV, and larger clumps of LCL can be visualized by the naked eye a few weeks thereafter.

2 Materials

2.1 Components for Preparing EBV Stock

1. Complete RPMI 1640 medium: To RPMI 1640, add heat-inactivated fetal bovine serum (FBS) to a final concentration of 10 % (v/v), penicillin at 100 U/ml, streptomycin at 100 μg/ml, and amphotericin B at 0.5 μg/ml.
2. B95-8 cells (ATCC #CRL 1612).
3. Tetradecanoyl phorbol acetate (TPA) (Calbiochem).

2.2 Components for Isolating and Immunosuppressing Peripheral Mononuclear Cells

1. Incomplete RPMI 1640 medium.
2. Complete RPMI 1640 medium (as above).
3. Phosphate-buffered saline.
4. Ficoll Hypaque lymphocyte separation medium (Mediatech).
5. FK506 (Fisher Scientific).

2.3 Components for Cryopreservation

1. Ice-cold freezing medium: 90 % heat-inactivated FBS + 10 % dimethylsulfoxide.

3 Methods

Carry out all procedures within a level II biosafety cabinet at room temperature unless otherwise specified.

3.1 EBV Stock Preparation (See Note 1)

1. Using sterile technique, exponentially growing B95-8 cells (ATCC # CRL 1612; 13) are subcultured at 3×10^5 cells/ml in a 75 cm^2 tissue culture flask (*see* **Note 5**). Add complete RPMI 1640 (*see* **Note 2**), and place cells at 37 °C in the presence of 5 % CO_2 (CO_2 incubator).
2. After 48 h, resuspend cells in fresh complete RPMI 1640 at 1×10^6 cells/ml. Add TPA to a final concentration of 20 ng/ml to induce virus production. Place the flask in CO_2 incubator.
3. After 1 h, wash cells three times with RPMI 1640 to remove TPA.
4. Use complete RPMI 1640 to resuspend cells in the original volume, and place flask in a CO_2 incubator for 96 h.
5. To separate EBV-containing culture supernatant from cells, centrifuge at 2,000 rpm ($600 \times g$) $\times$ 10 min at 4 °C.

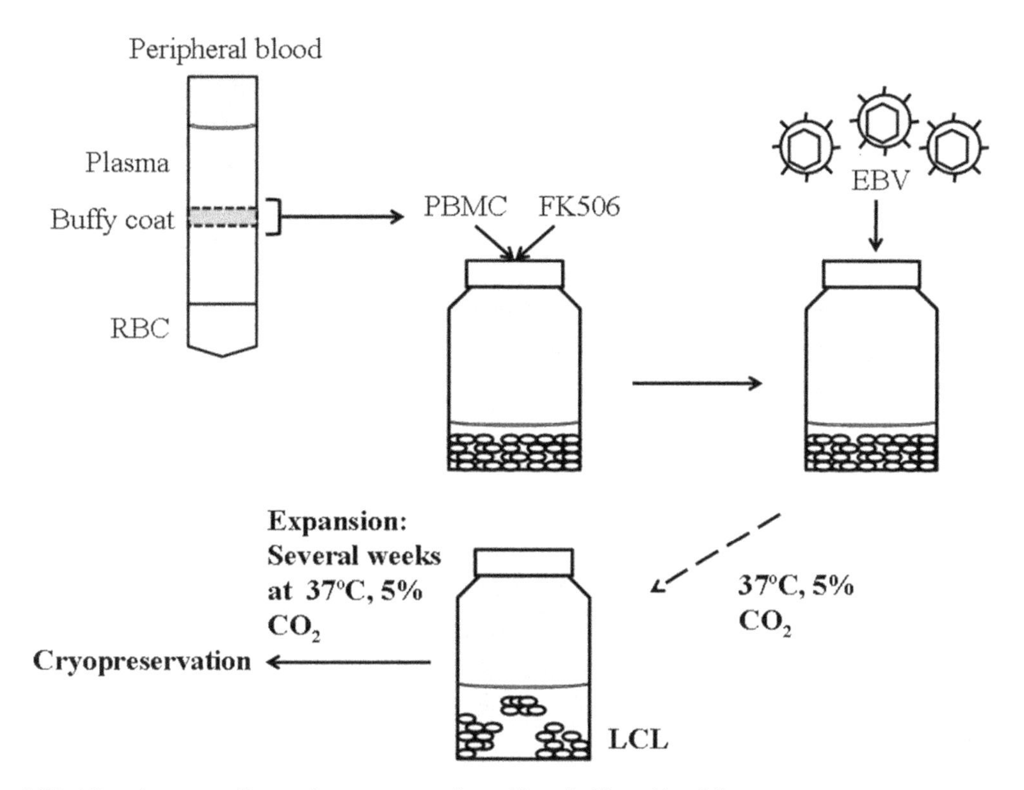

Fig. 1 Workflow for generation and cryopreservation of lymphoblastoid cell lines

6. Use a 0.45-μM filter to filter supernatant. Then aliquot and store at −70 °C. This procedure produces high titres of infectious EBV particles which should be viable for over a year.

3.2 Isolation of Peripheral Blood Mononuclear Cells

1. Draw 10 ml blood from a human donor into a heparinized syringe or heparinized blood tube.
2. Dilute 1:2 by adding 20 ml incomplete RPMI at room temperature in a 50 ml conical tube using sterile technique.
3. Carefully underlay diluted blood with 15 ml Ficoll Hypaque lymphocyte separation medium. To establish gradient, spin at 1,200 rpm (225 × *g*) without brake at room temperature for 30 min.
4. Remove buffy coat (*see* **Notes 3** and **4**) and transfer into a new 50 ml conical tube; see the schematic for generation of LCL (Fig. 1).
5. Raise volume to 50 ml with incomplete RPMI and spin at 2,000 rpm (600 × *g*) at room temperature for 10 min (use of brake permitted). A pellet of cells should be visible at the end of the spin.
6. Gently pour off supernatant. Wash cells by resuspending pellet in 50 ml incomplete RPMI and centrifuging at 2,000 rpm at room temperature for 10 min.

7. Wash once more as above.
8. Resuspend washed cells in 1 ml of complete RPMI 1640.
9. Prepare cells for counting: Add 5 μL of cells to 45 μL trypan blue. Use this 1:10 dilution of cell suspension to count live cells using a hemocytometer.
10. Obtain a cell concentration of 2×10^6/ml by adjusting the volume using complete RPMI. Transfer cells to a 25 cm^2 tissue culture flask.

3.3 Infection with EBV

1. Add FK506 to the cell suspension from Subheading 3.2, **step 10**, to a final concentration of 20 nM.
2. Place flask in a CO_2 incubator at 37 °C for 1 h.
3. Rapidly thaw an aliquot of EBV from −70 °C. Remove the flask from the CO_2 incubator, and add EBV to the cells at a 1:10 dilution.
4. Gently swirl the flask to mix, and place it upright in the CO_2 incubator with the cap on but slightly loose.

3.4 Expansion and Cryopreservation

1. Visualization of cells by light microscopy: By 5–7 days after EBV infection, clusters of cells are visible by light microscopy, indicating successful proliferation (Fig. 2b). Lack of cell clusters in uninfected control cells is shown for comparison (Fig. 2a).
2. Within the next week or two, continued proliferation of cells leads to larger microscopic clusters of cells (Fig. 2c) that also become visible to the naked eye.
3. Feeding cells: On day 12, double the volume of supportive medium with fresh complete RPMI.
4. The rate of cell growth will help determine the periodicity of feeding cells. A general rule of thumb is to triple the culture medium with fresh complete RPMI when the cell culture medium turns yellow, typically once per week. Some cell lines may need to be fed more or less frequently.
5. Using this method, expand the cell culture to 75–100 ml over the next few weeks.
6. Cryopreservation: 48 h after feeding cells, centrifuge cells at 2,000 rpm ($600 \times g$), using brakes, for 10 min at room temperature. Resuspend cells in complete RPMI to count live cells. Centrifuge cells to pellet again. Remove supernatant, and resuspend the cell pellet in ice-cold freezing medium at 1×10^7 cells/ml. Place tube in a liquid nitrogen-containing cryopreservation tank.
7. Of note: If LCL are to be used for generation of monoclonal antibodies, cells would need to be screened for antibody production by standard immunologic techniques. Cultures producing desired antibodies would need to be subcloned by limiting dilution followed by amplification of clonal populations of interest.

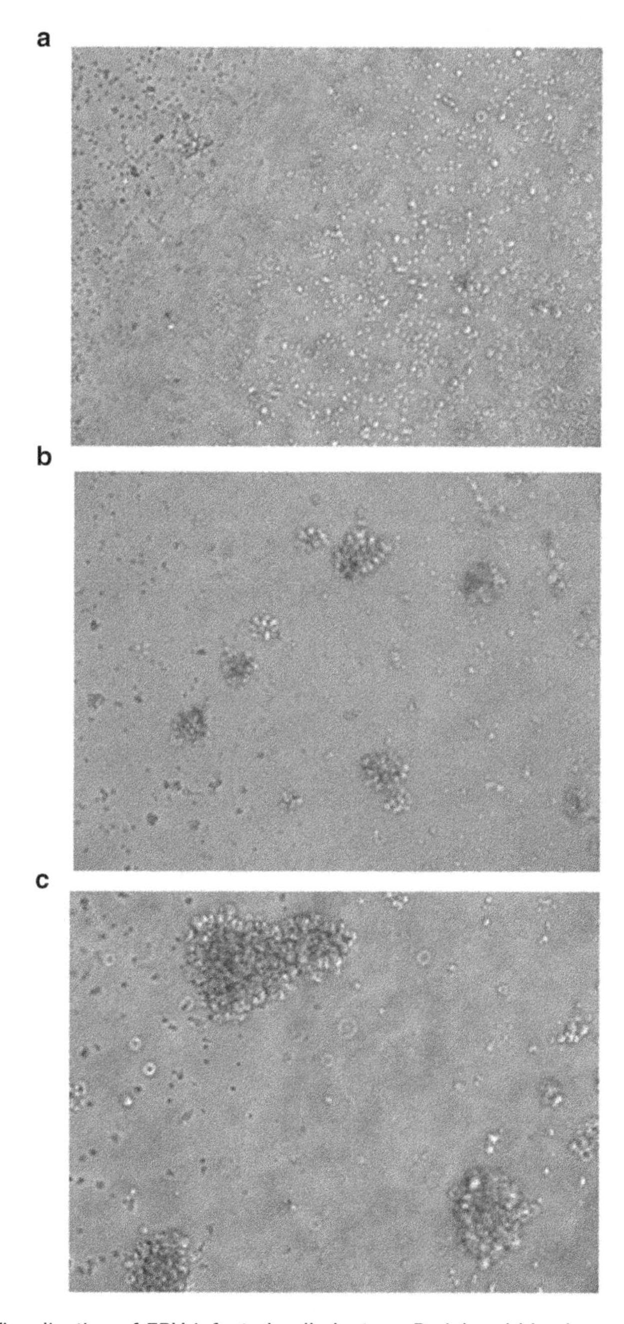

Fig. 2 Visualization of EBV-infected cell clusters. Peripheral blood mononuclear cells were infected with EBV in the presence of FK506. Uninfected cells placed in culture as control (**a**) and EBV-exposed cells (**b**) were visualized 1 week later by phase contrast microscopy (10× magnification). Larger clusters of EBV-infected cells (LCL) visualized microscopically at 5 weeks are shown in (**c**)

Special attention should be given to their storage to prevent cross-contamination with other cell lines. If downstream applications include therapeutic use, it would also be important to inactivate carryover virus.

4 Notes

1. An alternative to preparing EBV stock is to obtain EBV-containing B95-8 cell culture supernatant from the ATCC (VR-1492) and use at the dilution recommended by the manufacturer.
2. Components of complete RPMI 1640 should be mixed at 37 °C. Complete and incomplete RPMI 1640 should be warmed to 37 °C prior to use.
3. Optimal isolation of PBMC: Remove buffy coat with automated pipette with 50 ml conical tube held at eye level, at 45° angle to increase available surface area. Care should be taken to avoid contamination of PBMC by not piercing through the buffy coat to the lymphocyte separation medium below.
4. Work as efficiently as possible to minimize the time spent by cells in lymphocyte separation medium, as prolonged exposure can be toxic to cells.
5. Maintenance of sterile technique in every step is critical to the success of generating LCL.

Acknowledgments

This research was funded by the NIH grants K08 AI062732, K12 HD001401, and 1UL1RR024139-02; a Child Health Research Grant from the Charles H. Hood Foundation; and the Research Foundation for the State University of New York at Stony Brook to S.B.-M.

References

1. Thorley-Lawson DA, Gross A (2004) Persistence of the Epstein-Barr virus and the origins of associated lymphomas. N Engl J Med 350:1328–1337
2. Kubuschok B et al (2002) Use of spontaneous Epstein-Barr virus lymphoblastoid cell lines genetically modified to express tumor antigen as cancer vaccines: mutated p21 ras oncogene in pancreatic carcinoma as a model. Hum Gene Ther 13:815–827
3. Kuppers R (2003) B cells under influence: transformation of B cells by Epstein-Barr virus. Nat Rev Immunol 3:801–812
4. Traggiai E et al (2004) An efficient method to make human monoclonal antibodies from memory B cells: potent neutralization of SARS coronavirus. Nat Med 10:871–875
5. Bernasconi N, Traggiai E, Lanzavecchia A (2002) Maintenance of serological memory by polyclonal activation of human memory B cells. Science 298:2199–2202
6. Oh H-M et al (2003) An efficient method for the rapid establishment of Epstein-Barr virus immortalization of human B lymphocytes. Cell Prolif 36:191–197
7. Ventura M et al (1988) Use of a simple method for the Epstein-Barr virus transformation of lymphocytes from members of large families of Reunion Island. Hum Hered 38:36–43
8. Henderson E et al (1977) Efficiency of transformation of lymphocytes by Epstein-Barr virus. Virology 76:152–163
9. Bird AG et al (1981) Characteristics of Epstein-Barr virus activation of human B lymphocytes. J Exp Med 154:832–839
10. Neitzel H (1986) A routine method for the establishment of permanent growing lymphoblastoid cell lines. Hum Genet 73:320–326

11. Pelloquin F, Lamelin JP, Lenoir GM (1986) Human B lymphocytes immortalization by Epstein-Barr virus in the presence of cyclosporin A. In Vitro Cell Dev Biol 22:689–694
12. Pressman S, Rotter JI (1991) Epstein-Barr virus transformation of cryopreserved lymphocytes: prolonged experience with technique. Am J Hum Genet 49:467
13. Miller G et al (1972) Epstein-Barr virus: transformation, cytopathic changes, and viral antigens in squirrel monkey and marmoset leukocytes. Proc Natl Acad Sci U S A 69:383–387

Chapter 12

Using Next-Generation Sequencing for Discovery of High-Frequency Monoclonal Antibodies in the Variable Gene Repertoires from Immunized Mice

Ulrike Haessler and Sai T. Reddy

Abstract

Historically, isolation of antigen-specific monoclonal antibodies has relied on screening-based approaches. Here we describe a simple and rapid method for antibody isolation without screening, which capitalizes on next-generation DNA sequencing and bioinformatic analysis of antibody variable region (V) gene repertoires from the bone marrow plasma cells of immunized mice. The highest frequency antibody variable heavy (V_H) and variable light (V_L) gene sequences are paired based on their relative frequency, and their respective antibody genes are constructed by DNA synthesis and recombinantly expressed, purified, and validated for antigen-binding specificity.

Key words Plasma cells, Immune repertoire, Systems immunology, Variable region CDR3, Fab

1 Introduction

Mammalian humoral immune responses are capable of generating a diverse repertoire of antibodies in response to stimuli (i.e., immunization), which has been substantially exploited for biotechnology applications, most notably for monoclonal antibody (mAb) discovery. There are ~30 clinically approved mAbs and ~300 in clinical development [1]. With this in mind, there is a substantially high demand for antibody discovery platforms that rapidly provide panels of mAbs with desired specificity and affinity. Traditionally, the discovery of mAbs relies on screening-based approaches such as B cell immortalization (i.e., hybridoma) and in vitro surface display of recombinant antibody libraries (e.g., phage and microbial display) [2–6]. Such methods are heavily dependent on the high-throughput screening of large clonal libraries to identify antibodies of desired antigenic specificity. For example, the low efficiency for hybridoma fusions makes it necessary to extensively screen for antigen-specific mAbs by limiting dilution into microtiter plates.

Vincent Ossipow and Nicolas Fischer (eds.), *Monoclonal Antibodies: Methods and Protocols*, Methods in Molecular Biology, vol. 1131, DOI 10.1007/978-1-62703-992-5_12,

Surface display also requires screening of large libraries often approaching ~10^9 variants, primarily due to combinatorial diversity introduced from random variable light- and heavy-chain (V_L and V_H) pairing during library generation and cloning. Screening is accomplished through successive rounds of affinity panning or flow cytometric sorting that allow for enrichment of antigen-specific antibodies [6].

For both B cell immortalization and display techniques, murine immunized antibody repertoires have mainly been derived from the splenic B cell population, which have focused on mainly memory B cells. However, there are several other stages to B cell differentiation, including plasmablasts and mature antibody-secreting plasma cells [7]. B cell maturation and homing culminate in the terminal, non-proliferative stage of B cell development—the formation of plasma cells, which serve as immunoglobulin production factories [8]. Plasma cells represent less than 1 % of all lymphoid cells and yet are responsible for the majority of antibodies in circulation [7, 9]. The bone marrow constitutes the major compartment where plasma cells take residency and produce antibodies for prolonged periods of time. In mice, a highly enriched antigen-specific bone marrow plasma cell population of ~10^5 cells (10–20 % of all bone marrow plasma cells) appears 6 days following secondary immunization and persists for prolonged periods [10, 11]. In contrast, the splenic plasma cell population is highly transient, as it peaks at day 6 and rapidly declines to $<10^4$ cells by day 11. Furthermore, bone marrow plasma cells likely represent the source of the most abundant circulating antibodies and thus are likely to be relevant for protective humoral immunity.

Here we describe a simple and rapid method for mAb isolation without the laborious need for traditional screening [12]. We capitalize on a systems immunology-based approach, specifically by utilizing next-generation DNA sequencing and bioinformatic analysis of antibody V gene repertoires from the bone marrow plasma cells of protein antigen-immunized mice. The bioinformatics analysis is followed by synthetic DNA construction of V_H and V_L genes that are highly represented in the repertoire, likely to encode antigen-specific antibodies. High-frequency antibody clones are then expressed and purified from a recombinant expression system (e.g., HEK293 cells) for specificity and affinity characterization to the antigen of interest.

2 Materials

2.1 Immunization and Serum Titer Measurement Components

1. Female BALB/c mice, 6–8 weeks old (Charles River Laboratories).
2. $CaCl_2$- and $MgCl_2$-free sterile phosphate-buffered saline (PBS).
3. Complete Freund's Adjuvant (CFA, Pierce Biotechnology).
4. Incomplete Freund's Adjuvant (IFA, Pierce Biotechnology).

5. Purified or recombinant protein antigen resuspended at 1 mg/ml in PBS (e.g., ovalbumin).
6. 26- and 21-G needles.
7. 1 ml syringes.
8. Surgical blades.
9. Mouse restrainer.
10. MaxiSorp ELISA plate (Nunc).
11. ELISA washing buffer: 0.05 % (ml/ml) Tween-20 in PBS.
12. ELISA blocking buffer: 2.5 % (g/ml) casein in PBS.
13. Anti-Mouse Fc-HRP antibody (Biorad; 170-6516).
14. TMB Substrate Solution (Thermo Scientific).
15. ELISA stop solution: 2 N H_2SO_4 in H_2O.

2.2 Plasma Cell Isolation

1. Surgical scissors, forceps.
2. Washing buffer: 2 mM EDTA, 0.5 % (g/ml) bovine serum albumin (BSA) in PBS.
3. 70 μm cell strainer.
4. Red blood cell lysis buffer (eBioscience).
5. Purification antibodies: Biotinylated rat anti-mouse CD45R (eBioscience, 13-0452, 0.5 mg/ml), biotinylated rat anti-mouse CD138 (BD, 553713).
6. Magnetic isolation: M-280 streptavidin-coated magnetic beads (Dynabeads) and magnet (for instance DynaMag2, both from Invitrogen).

2.3 RNA Isolation, First-Strand cDNA Synthesis, and V_L/V_H PCR

1. TRIzol Reagent, PureLink RNA Mini Kit (both from Ambion).
2. Retroscript kit (Ambion).
3. Taq DNA polymerase with Thermopol buffer 10× (New England Biolabs; NEB).
4. dNTP Mix 10 mM (Thermo Scientific).
5. EconoSpin All-in-One Mini Spin Columns (Epoch Life Science) with Buffer WS (wash buffer; 10 mM Tris–HCl, pH 7.5 with 80 % ethanol), Buffer PEX (DNA absorption buffer; 5.5 M guanidine hydrochloride, 20 mM Tris–HCl, pH 6.6).
6. TAE buffer: Dissolve 24.2 g Tris base in water, add 5.7 ml glacial acetic acid and 10 ml EDTA at 0.5 M, and adjust to 1 l with water, final pH 8.0.
7. Agarose 1 % (g/ml) in TAE buffer.
8. DNA purification reagents: 6× DNA Loading Dye, Gene Ruler 100 bp, Sybr Safe DNA gel stain (Invitrogen), Zymoclean gel DNA recovery kit (Zymo Research).
9. Nanodrop (Thermo Scientific).

2.4 Recombinant Antibody Expression, Purification, and Antigen-Binding Analysis

1. Human constant region IgG1 expression plasmids pFUSE-CHIg-hG1 (VH) and pFUSE2-CLIg-hk/pFUSE2-CLIg-hl2 (VL) (Invivogen).
2. Suspension FreeStyle 293-F cells cultured in FreeStyle 293 Expression Medium (both from Invitrogen) within a 125 ml Erlenmeyer culture flask on a shaker at 250 rpm.
3. FreeStyle MAX Reagent (Invitrogen).
4. OptiPro SFM (for DNA and transfection reagent dilution, Invitrogen).
5. 12-well plate.
6. Protein A/G agarose for IgG purification with Protein A/G IgG Binding Buffer, IgG Elution Buffer, and Disposable Polypropylene Columns (all from Thermo Scientific).
7. Neutralization buffer 1 M TRIS (pH 8.0).
8. 96-well cell culture round-bottom plates and high-binding 96-well ELISA plate.
9. Blocking buffer (e.g., 0.5 % (g/ml) BSA/PBS, 1 % (g/ml) gelatin/PBS, or 0.5 % (g/ml) nonfat dry milk/PBS).
10. Anti-human Fc-HRP antibody (Thermo Scientific).

3 Methods

All procedures are carried out at room temperature if not stated otherwise.

3.1 Immunization and Serum Titer Measurements

1. Keep mice in conventional barrier space under specific pathogen-free (SPF) conditions.
2. The day of injection (day 1), mix 25 μl of CFA with 25 μl protein antigen (e.g., ovalbumin) and 50 μl PBS per injected mouse using a 21-G needle. Adjust for dead volume in the syringe by making a little extra volume. Mix the white emulsion vigorously for several minutes and leave on ice (*see* **Note 1**).
3. Before immunization, restrain mice and isolate tail using mouse restrainer. Wipe tail clean with alcohol. Collect blood (~25–50 μl) from the tail vain by making a small incision with a surgical blade and pipetting pooled blood. Blood is collected in a 1.5 ml Eppendorf tube.
4. Let blood clot at room temperature for 20–30 min. Centrifuge blood tubes at 1,500 × *g* for 15 min, and transfer the upper clear phase (serum) into a new tube. Store serum at −20 °C for later analysis.

5. Inject 100 μl of CFA–antigen mixture with a 26-G needle subcutaneously into the back pad.
6. Perform the secondary immunization on day 21. Mix 25 μl IFA with 25 μl of antigen solution and 50 μl of PBS per mouse (as described before for the CFA). Inject the mixture intraperitoneally (*see* **Note 2**).
7. Six days after booster injection (~day 27), collect blood and serum as previously described (*see* **Note 3**).
8. Verify antigen-specific antibody titer by ELISA: Coat an ELISA plate with 100 μl/well of antigen solution (for OVA: 5 μg/ml in PBS). Incubate overnight at 4 °C.
9. Wash plates 3× with ELISA washing buffer.
10. Add 200 μl/well of ELISA blocking buffer and incubate for 2 h.
11. Wash plates 3× with ELISA washing buffer.
12. Add a serial dilution of serum in blocking buffer, starting with a dilution of $1:10^2$ up to $1:10^7$ (total volume of 100 μl/well). Add PBS only to one of the wells to determine your background absorbance.
13. Wash plates 3× with ELISA washing buffer.
14. Add 100 μl/well of HRP–secondary antibody, diluted 1:2,500 in blocking buffer, and incubate for 1 h.
15. Wash plate 3× with ELISA washing buffer.
16. Add 100 μl of TMB substrate solution to each well and keep protected from light.
17. Let substrate react for 10–30 min, and check periodically for a positive signal (substrate–HRP reaction—wells change color to blue).
18. When it is possible to visually (by eye) detect a positive signal, stop the reaction with 100 μl H_2SO_4 stop solution.
19. Detect absorbance of wells at an optical density wavelength of 450 nm (using a standard microplate reader). Serum dilutions with a positive signal above the background well represent a positive titer (presence of antigen-specific antibodies).
20. If mice are observed to have a low titer (positive signal only at serum dilutions $<1:10^5$) repeat booster immunization (*see* **Note 4**).
21. When mice have been confirmed to have a high titer (signal at serum dilutions $>1:10^5$), euthanize ~7 days following final booster immunization. Remove femurs and tibia and keep in washing buffer on ice. Also collect blood and serum (as previously described).

3.2 Plasma Cell Isolation

All the following steps concerning cell isolation are performed on ice unless stated otherwise. Always centrifuge cells at 335 × *g* at 4 °C for 5 min.

1. Remove soft tissue from harvested femurs and tibiae and store in washing buffer (*see* **Note 5**).
2. Cut off the ends of the bones, and flush out the bone marrow with a syringe filled with washing buffer and connected to a 26-G needle.
3. Filter cell suspension through a 70 μm cell strainer into a 50 ml tube.
4. Pellet cells by centrifugation, and decant supernatant.
5. Resuspend cells in 3 ml of red blood cell lysis buffer and keep at room temperature for 3 min.
6. Add washing buffer up to 50 ml to neutralize red blood cell lysis buffer (*see* **Note 6**).
7. Centrifuge cells, and decant supernatant.
8. Resuspend cells in 200 μl washing buffer. To determine total bone marrow cell yield, perform a cell count. Typical yields are $5–9 \times 10^7$ total bone marrow cells per mouse.
9. Add 2.5 μl anti-CD45R-biotin to resuspended cells. Incubate for 25 min on ice.
10. While cells are incubating wash 50 μl Dynabeads per mouse in washing buffer twice using a magnetic stand.
11. Following incubation of 25 min and labeling with anti-CD45R-biotin antibody, wash cells twice by centrifuging and resuspend in 1 ml washing buffer.
12. Add 25 μl Dynabead suspension per sample and incubate with constant rotation at 4 °C for 30 min (*see* **Note 7**).
13. Place cells on magnet for 3 min, and transfer cell suspension into a new tube; these are negatively selected cells (include plasma cells). Cells remaining on the magnet (CD45R+) can be discarded.
14. Centrifuge negative selection cells and resuspend in 200 μl washing buffer.
15. Add 4 μl of anti-CD138-biotin antibody and incubate for 25 min.
16. Wash 2× by centrifuging and resuspend in 1 ml washing buffer.
17. Add 25 μl of bead suspension and incubate for 30 min with rotation at 4 °C (*see* **Note 7**).
18. Place cells on the magnet and wash 2×. Cells bound to the beads are plasma cells. Keep cells on ice until proceeding to the RNA isolation. Perform a cell count; a typical number of plasma cells to expect is $0.5–2 \times 10^6$ cells per mouse (*see* **Note 8**) (Fig. 1).

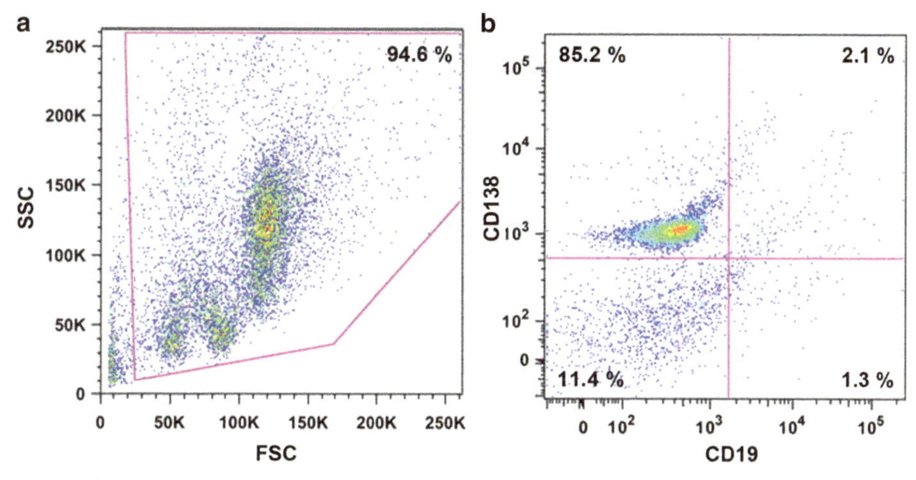

Fig. 1 Plasma cell isolation with magnetic beads. (**a**) Forward- and side-scatter plot of a typical bone marrow population. Gated cells are used for further analysis. (**b**) CD19-positive cells (B cells) and CD138-positive cells. *Upper left quadrant* are plasma cells, which are enriched to 85 %

3.3 RNA Isolation, First-Strand cDNA Synthesis, and V_L/V_H PCR

All steps are performed at room temperature unless stated otherwise.

1. Resuspend magnetic bead-bound plasma cells and in 1 ml TRIzol reagent. Follow the manufacturer's protocol for the PureLink RNA Mini Kit. Elute twice with 30 μl with RNase-free H_2O.
2. Measure total RNA concentration (e.g., on a Nanodrop; *see* **Note 9**). The expected amount of total RNA is in the range of 0.5–2 μg per mouse (*see* **Note 10**). Store the isolated RNA at −80 °C until proceeding with first-strand cDNA synthesis.
3. Perform first-strand cDNA synthesis according to the manufacturer's protocol for the Retroscript kit using oligo(dT) primers with 200 ng plasma cell total RNA per reaction.
4. Perform PCR amplification of V_L and V_H separately (see next step); typically parallel reactions (at least four reactions) for each V_L and V_H should be performed. A complete list of primers is found in Table 1 (*see* **Note 11**).
5. For a 50 μl PCR reaction use (*see* **Note 12**):
   (a) 2 μl unpurified cDNA template.
   (b) 5 μl Thermopol buffer 10×.
   (c) 1 μl dNTPs.
   (d) 1 μl Taq DNA polymerase.
   (e) 1 μl forward V_H or V_L primer mix.
   (f) 1 μl reverse V_H or V_L primer mix.
   (g) 39 μl of double-distilled water.

Table 1
Primers containing restriction sites for cloning purposes

	V_L	V_H
Forward		
1	GAY ATC CAG CTG ACT CAG CC	GAK GTR MAG CTT CAG GAG TC
2	GAY ATT GTT CTC WCC CAG TC	GAG GTB CAG CTB CAG CAG TC
3	GAY ATT GTG MTM ACT CAG TC	CAG GTG CAG CTG AAG SAS TC
4	GAY ATT GTG YTR ACA CAG TC	GAG GTC CAR CTG CAA CAR TC
5	GAY ATT GTR ATG ACM CAG TC	CAG GTY CAG CTB CAG CAR TC
6	GAY ATT MAG ATR AMC CAG TC	CAG GTY CAR CTG CAG CAG TC
7	GAY ATT CAG ATG AYD CAG TC	CAG GTC CAC GTG AAG CAG TC
8	GAY ATY CAG ATG ACA CAG AC	GAG GTG AAS STG GTG GAA TC
9	GAY ATT GTT CTC AWC CAG TC	GAV GTG AWG YTG GTG GAG TC
10	GAY ATT GWG CTS ACC CAA TC	GAG GTG CAG SKG GTG GAG TC
11	GAY ATT STR ATG ACC CAR TC	GAK GTG CAM CTG GTG GAG TC
12	GAY RTT KTG ATG ACC CAR AC	GAG GTG AAG CTG ATG GAR TC
13	GAY ATT GTG ATG CBC AGK C	GAG GTG CAR CTT GTT GAG TC
14	GAY ATT GTG ATA ACY CAG GA	GAR GTR AAG CTT CTC GAG TC
15	GAY ATT GTG ATG ACC CAG WT	GA AGT GAA RST TGA GGA GTC
16	GAY ATT GTG ATG ACA CAA CC	CAG GTT ACT CTR AAA GWG TST G
17	GAY ATT TTG CTG ACT CAG TC	CAG GTC CAA CTV CAG CAR CC
18	GAR GCT GTT GTG ACT CAG GAA TC	GAT GTG AAC TTG GAA GTG TC
19		GAG GTG AAG GTC ATC GAG TC
Reverse		
1	ACG TTT GAT TTC CAG CTT GG	GAG GAA ACG GTG ACC GTG GT
2	ACG TTT TAT TTC CAG CTT GG	GAG GAG ACT GTG AGA GTG GT
3	ACG TTT TAT TTC CAA CTT TG	GCA GAG ACA GTG ACC AGA GT
4	ACG TTT CAG CTC CAG CTT GG	GAG GAG ACG GTG ACT GAG GT
5	ACC TAG GAC AGT CAG TTT GG	
6	ACC TAG GAC AGT GAC CTT GG	

6. Perform PCR with a thermocycle program of the following:
   (a) 92 °C for 3 min.
   (b) Four cycles (92 °C for 1 min, 50 °C for 1 min, 72 °C for 1 min).
   (c) Four cycles (92 °C for 1 min, 55 °C for 1 min, 72 °C for 1 min).
   (d) 20 cycles (92 °C for 1 min, 63 °C for 1 min, 72 °C for 1 min).
   (e) 72 °C for 7 min, 4 °C until removal.
7. Perform a PCR cleanup with the EconoSpin Columns pooling parallel reactions of V_L and V_H. Elute with 15 μl double-distilled H_2O.

8. Prepare a DNA gel with 1 % agarose in TAE buffer with SybrSafe staining.
9. Add 4 μl of 6× DNA loading dye to PCR samples and load onto the gel together with a 100 bp DNA ladder.
10. Run gel at 100 V for ~20 min until loading buffer is across ~¾ of the gel length.
11. Visualize DNA bands on a gel imager (preferentially blue light), and perform gel extraction and purification of the bands for V_L and V_H; typically these bands will be at ~370 bp for V_L and ~400 bp for V_H (*see* **Note 13**).
12. Measure DNA concentrations and quality with an accurate DNA-measuring device, such as Bioanalyzer (Agilent) or NanoQuant (Tecan).
13. Submit samples to a facility or a service provider for next-generation DNA sequencing (*see* **Note 14**) (Fig. 2).

3.4 Bioinformatics and Selection of Synthetic Antibody Genes

1. Submit the DNA sequences of V_H and V_L obtained by next-generation DNA sequencing to IMGT/HighV-Quest at the IMGT website (http://www.imgt.org/HighV-QUEST/index.action). HighV-Quest is an open-access software platform that does extensive analysis of antibody sequences [13]; importantly HighV-quest will provide amino acid identification of V(D)J junctional recombination sites, which represent complementarity-determining region 3 (CDR3) and the clonal identity of V regions. Based on IMGT data analysis, sort and create a list according to the frequency of each V gene CDR3 amino acid sequence (CDRH3 and CDRL3). The highest ranking CDR3s (typically above 1 %) represent potential antigen-binding clones (*see* **Note 15**).

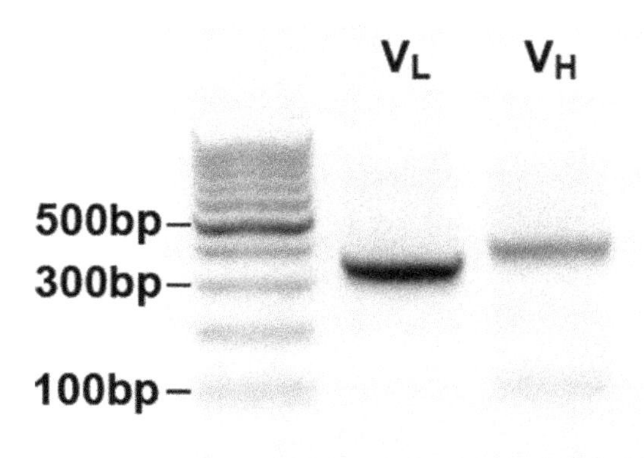

Fig. 2 DNA gel of the variable region light (V_L) and heavy (V_H) chain with the respective DNA ladder (*left lane*)

2. Following identification of high-frequency CDR3 regions, use a pairing strategy based on relative frequency to identify potential V_L and V_H gene combinations (e.g.., #1 ranked CDRL3 with #1 ranked CDRH3; #2 CDRL3 with #2 CDRH3). Typically pairing is reliable for the top four to five CDRL3 and CDRH3 clones per mouse.
3. Use the full-length V region amino acid sequences of the high-frequency CDR3 regions to perform a multiple-sequence alignment (for each respective CDR3 clone; *see* **Note 16**).
4. Use the consensus amino acid sequences of each full-length V gene (from high-frequency CDR3 clones; *see* **Note 17**) and a de novo DNA synthesis service, such as GenScript. To avoid cloning, request the synthetic genes to be directly incorporated into an antibody expression vector (e.g., pFUSE-CHIg-hG1 (V_H) and pFUSE2-CLIg-hk/pFUSE2-CLIg-hl2 (V_L)).

3.5 Recombinant Antibody Expression, Purification, and Antigen-Binding Analysis

Always spin cells at 335 × *g* at 4 °C for 10 min.

1. Transiently co-transfect 293-F cells with the retrieved V_H and V_L antibody expression vectors, following the FreeStyle MAX expression system manufacturer's protocol.
2. Seed cells 2 days before transfection at 3×10^5 cells/ml in FreeStyle media.
3. On the day of transfection, spin cells down and resuspend at 1×10^6 cells/ml in FreeStyle media.
4. Transfer 1 ml/well of cell suspension into a 12-well plate.
5. Dilute 20 μg of V_H and 20 μg of V_L plasmid with OptiPro SFM to 600 μl.
6. In a separate tube, dilute 40 μl of FreeStyle MAX Reagent with OptiPro SFM to 600 μl.
7. Mix the two solutions described in **steps 5** and **6** containing plasmid and MAX Reagent; incubate for 10 min at room temperature.
8. Slowly add 40 μl of the transfection mixture to each well with cell suspension by manually swirling the plate.
9. Put plate back onto an orbital shaker (135 rpm) inside an incubator.
10. Harvest supernatant after 72–96 h of cell culture by centrifuging cells.
11. Mix supernatant 1:1 with column binding buffer.
12. Add 2 ml of A/G agarose slurry to the column.
13. Wash column with 5 ml of binding buffer.

14. Run the diluted supernatant through the column, and keep the flow through to test for the remaining IgG.
15. Wash with 15 ml of binding buffer.
16. Elute antibodies with 5 ml of elution buffer, and collect 0.5 ml fractions.
17. Neutralize samples immediately by adding 50 μl neutralization buffer per fraction.
18. Measure the absorbance of each fraction at 280 nm, and pool the ones with the highest recombinant antibody concentration for ELISA measurements.
19. Prepare an ELISA plate as described under Subheading 3.1, **step 8**, with 10 μg/ml antigen in PBS.
20. Incubate overnight at 4 °C.
21. Block ELISA plate with 200 μl/well blocking buffer for 2 h.
22. Wash ELISA plate three times with washing buffer.
23. Mix recombinant antibodies 1:1 with blocking buffer.
24. Transfer 100 μl of recombinant antibody mixtures (concentrations may vary; typically operate dilutions within a range of 1–500 nM).
25. Incubate for 1 h on a shaker at room temperature.
26. Wash ELISA plate three times with washing buffer.
27. Add 100 μl of anti-human Fc-HRP detection antibody at 1:2,500 dilution in blocking buffer.
28. Wash ELISA plate three times with washing buffer.
29. Add 50 μl TMB substrate to each well and keep protected from light.
30. Incubate for 10–30 min, and check periodically for positive signal.
31. Stop reaction by adding 50 μl H_2SO_4 to each well.
32. Detect absorbance of wells at an optical density wavelength of 450 nm.

4 Notes

1. Immunization can also be done with other alternative adjuvants such as Alum, poly I:C, CpG oligonucleotides, monophosphoryl lipid A (all available from Invivogen; consult the manufacturer's protocol for details).
2. Booster immunizations with an adjuvant other than IFA can also be performed subcutaneously.

3. In order to collect bigger volumes of blood and serum, a cardiac puncture can be performed with a 25-G needle terminally (on the day the mice are sacrificed).
4. It is possible with some mice and some antigens that a multiple (two to four) booster immunization (hyperimmunization) protocol will be necessary before detection of high-serum titer.
5. Alternatively, populations of plasma cells can be isolated from spleen and lymph nodes.
6. Make sure to use $CaCl_2$- and $MgCl_2$-free PBS to avoid aggregation of cells.
7. As magnetic beads settle quickly due to gravitation it is important to rotate samples thoroughly in a sufficient volume.
8. Purity of plasma cell populations can be evaluated by flow cytometry following labeling with fluorescence antibodies (e.g., anti-mouse CD19 or anti-mouse CD45R/B220 and anti-mouse CD138). As Dynabeads have some autofluorescence in most of the channels, preferentially use antibodies conjugated to Pacific Blue or some equivalent fluorescent dye.
9. If available, use a Nanodrop to get a quick estimate on RNA concentration. More precise measurements and information on RNA integrity can be acquired using, e.g., a Bioanalyzer.
10. Exact yields of total RNA from bone marrow plasma cells may substantially vary depending on experimental conditions (i.e., size of mouse, quality of bone marrow isolation).
11. Depending on the downstream procedure it may be best to increase the number of PCR reactions (e.g., 8–12 reactions). It is common for PCR to amplify V_L at higher quantities than V_H.
12. Any standard MMLV reverse transcriptase (e.g., Superscript II, Invitrogen) can be used for cDNA synthesis. Also oligo(dT) primers can be substituted for gene-specific primers corresponding to antibody constant regions for different light (kappa or lambda) or heavy (e.g., IgG or IgM) isotypes. The reverse primer mixes (specific for J regions) to PCR amplify V_L and V_H genes can also be substituted with primers specific for constant regions.
13. Gel extraction is highly recommended as even low amounts of primer dimers occupy space on the sequencing chip and therefore reduce the number of high-quality reads. For this purpose, any standard gel extraction recovery kit can be used (e.g., Zymoclean Kit, Zymo).
14. Several next-generation DNA sequencing platforms are capable of sequencing full-length antibody variable genes. Most commonly used is the 454 GS-FLX pyrosequencing platform (Roche); it is now possible to also use Illumina miSeq paired-end (PE300) sequencing. Depending on which platform is

chosen, sample preparation of variable gene libraries may be slightly altered.

15. Alternatively, a custom CDR3 identification program can be used based on searching for the conserved flanking motifs on the N- and C-terminal sides of CDR3 (for more details *see* ref. 12).
16. Multiple-sequence alignment can be performed with a standard ClustalW alignment using a BLOSUM62 scoring matrix.
17. If bioinformatic sequence analysis results in ambiguous pairing, co-transfect cells with different V_H and V_L combinations.

References

1. Walsh G (2010) Biopharmaceutical benchmarks 2010. Nat Biotechnol 28:917–924
2. Clackson T, Hoogenboom HR, Griffiths AD, Winter G (1991) Making antibody fragments using phage display libraries. Nature 352: 624–628
3. Feldhaus MJ, Siegel RW, Opresko LK, Coleman JR, Feldhaus JMW, Yeung YA, Cochran JR, Heinzelman P, Colby D, Swers J, Graff C, Wiley HS, Wittrup KD (2003) Flow-cytometric isolation of human antibodies from a nonimmune Saccharomyces cerevisiae surface display library. Nat Biotechnol 21:163–170
4. Harvey BR, Georgiou G, Hayhurst A, Jeong KJ, Iverson BL, Rogers GK (2004) Anchored periplasmic expression, a versatile technology for the isolation of high-affinity antibodies from Escherichia coli-expressed libraries. Proc Natl Acad Sci U S A 101:9193–9198
5. Mazor Y, Van Blarcom T, Mabry R, Iverson BL, Georgiou G (2007) Isolation of engineered, full-length antibodies from libraries expressed in Escherichia coli. Nat Biotechnol 25:563–565
6. Hoogenboom HR (2005) Selecting and screening recombinant antibody libraries. Nat Biotechnol 23:1105–1116
7. Radbruch A, Muehlinghaus G, Luger EO, Inamine A, Smith KGC, Dörner T, Hiepe F (2006) Competence and competition: the challenge of becoming a long-lived plasma cell. Nat Rev Immunol 6:10
8. Calame KL (2001) Plasma cells: finding new light at the end of B cell development. Nat Immunol 2:1103–1108
9. Höfer T, Muehlinghaus G, Moser K, Yoshida T, E Mei H, Hebel K, Hauser A, Hoyer B, Luger O, Dörner T, Manz RA, Manz F, Hiepe F, Radbruch A (2006) Adaptation of humoral memory. Immunol Rev 211: 295–302
10. Manz RA, Thiel A, Radbruch A (1997) Lifetime of plasma cells in the bone marrow. Nature 388:133–134
11. Manz RA, Hauser AE, Hiepe F, Radbruch A (2005) Maintenance of serum antibody levels. Annu Rev Immunol 23:367–386
12. Reddy ST, Ge X, Miklos AE, Hughes RA, Kang SH, Hoi KH, Chrysostomou C, Hunicke-Smith SP, Iverson BL, Tucker PW, Ellington AD, Georgiou G (2010) Monoclonal antibodies isolated without screening by analyzing the variable-gene repertoire of plasma cells. Nat Biotechnol 28:957–961
13. Alamyar E, Giudicelli V, Li S, Duroux P, Lefranc M-P (2012) IMGT/HighV-QUEST: the IMGT® web portal for immunoglobulin (IG) or antibody and T cell receptor (TR) analysis from NGS high throughput and deep sequencing. Immunome Res 8:26

Part II

Antibody Expression and Purification

Chapter 13

Cloning, Reformatting, and Small-Scale Expression of Monoclonal Antibody Isolated from Mouse, Rat, or Hamster Hybridoma

Jeremy Loyau and François Rousseau

Abstract

The hybridoma technology, first described in 1975 by Milstein and Köhler, is still to date one of the most commonly used approaches to produce monoclonal antibodies. However, despite many advantages, this approach suffers from limitations like limited antibody productivity. Here, we describe a method for efficient cloning of antibody VH and VL produced by mouse, rat, or hamster hybridoma before reformatting in full-length IgG and small-scale expression in mammalian cell line.

Key words Hybridoma, Cloning, Reformatting, Isotypes, Small-scale expression

1 Introduction

The hybridoma technology allows the isolation and production of monoclonal antibody with high affinity and specificity against soluble or membrane-bound antigens using animal immune systems [1]. Several species have been used as host to isolate B cells expressing antibodies after antigen immunization, and before fusion with myeloma cells, like mouse, rat, or hamster. Recently, transgenic mice modified to express the human antibody repertoire have also been developed and can be used to generate hybridoma expressing fully human antibodies [2].

Despite these advantages, the hybridoma technology has some limitations. In some cases, hybridoma cells are poor antibody producers. Therefore, the protein of interest must be isolated and expressed in a different mammalian host cell line [3]. In addition, the isotype of monoclonal antibody expressed by hybridoma may have specific Fc-mediated functions which are

Vincent Ossipow and Nicolas Fischer (eds.), *Monoclonal Antibodies: Methods and Protocols*, Methods in Molecular Biology, vol. 1131, DOI 10.1007/978-1-62703-992-5_13, © Springer Science+Business Media New York 2014

not suitable for antibody application [4, 5]. This situation often leads to the isolation of the variable heavy (VH) and light (VL) domains to switch antibody isotype by molecular engineering. Finally, mouse, rat, or hamster monoclonal antibodies, which are of particular clinical interest, are usually not good drug candidates because of their potential immunogenicity in humans. Cloning of the binding antigen domains of nonhuman antibody is usually the first step before reformatting as a chimeric antibody with human constant regions to increase its drugability [6, 7]. Moreover, once the VH and VL sequences have been isolated, complementarity-determining regions (CDRs) of the mouse, rat, or hamster antibody can be grafted on human VH and VL scaffolds by humanization technologies [8].

For these reasons, cloning of mouse, rat, or hamster VH and VL sequences from hybridoma is often a prerequisite to facilitate the production and development of an antibody of interest. In this chapter, we present a straightforward and robust method, described schematically in Fig. 1, to isolate VH and VL sequences from hybridoma by PCR cloning. The first step consists in the purification of mRNA from the hybridoma of interest followed by reverse transcription to generate cDNA. Then, we suggest using commercially available set of primers to screen hybridoma cDNA and amplify VH and VL nucleotide sequences. For cloning of antibodies from hamster hybridoma, we suggest to use RACE-PCR to isolate hamster VH and VL. After cloning of these sequences into plasmids, VH and VL sequences can be conveniently analyzed by sequencing. We also describe the reformatting of these variable regions into human or mouse IgG of different isotypes. For this second step, we recommend to use the pFUSE-CLIg (cloning of the variable light chain) and pFUSE-CHIg plasmids (cloning of the variable heavy chain), which allow the production of either mouse or human chimeric antibodies by co-transfection in mammalian cells. Furthermore, these vectors are designed to allow convenient selection of different antibody isotypes to generate antibodies with different effector functions. Finally, small-scale expression in mammalian cells and antibody purification is described. We perform a transient expression in HEK-293 cells by co-transfecting the recombinant pFUSE-CLIg and pFUSE-CHIg vectors encoding the light chain and the heavy chain, respectively.

This method relies on commercially available kits, plasmids, and common lab equipment. This approach should facilitate the isolation and recombinant expression of an antibody of interest using the constant regions of the most suitable species and isotypes before testing in bioassays.

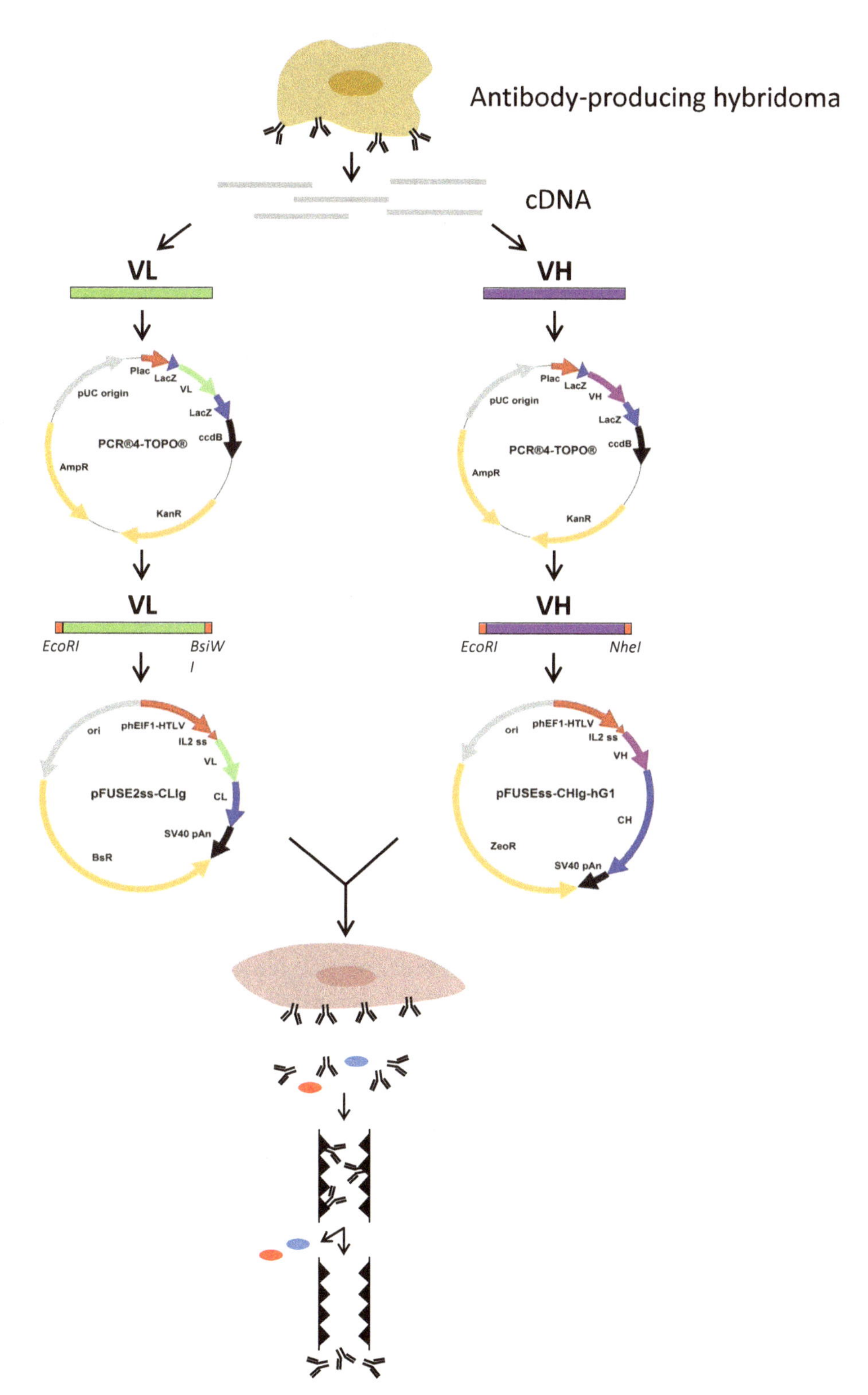

Fig. 1 Flow chart of the method. Monoclonal antibody cloning from hybridoma, reformatting, and small-scale expression steps are schematically described

2 Material

2.1 Antibody Variable Sequence Cloning and Reformatting

1. PCR thermocycler.
2. Nanodrop® spectrophotometer (Thermo Scientific) or equivalent.
3. Bacterial culture equipment: Incubator shaker, static incubator.
4. Thermomixer Comfort (Eppendorf).
5. UV table.
6. Scalpel.
7. RNase-free water (Qiagen).
8. Isopropanol (Sigma).
9. Ethanol (Sigma).
10. β-Mercaptoethanol (Sigma).
11. Dulbecco's Phosphate Buffered Saline, PBS (Sigma).
12. RNase*Zap*® Wipes (Invitrogen).
13. RNeasy® Plus Mini Kit (Qiagen).
14. QIAshredder® (Qiagen).
15. ThermoScript™ RT-PCR System for First-Strand cDNA Synthesis (Invitrogen).
16. Oligo$(dt)_{20}$ (Invitrogen).
17. RNaseOUT® (Invitrogen).
18. RNase H (Invitrogen).
19. SMARTer™ RACE cDNA Amplification Kit (Clontech/Takara).
20. GoTaq® Green Master Mix (Promega).
21. Progen Mouse IgG library Primer set (Progen).
22. E-Gel® 1 Kb Plus DNA ladder (Invitrogen).
23. E-Gel® PowerBase™ v.4 (Invitrogen).
24. E-Gel® 1.2 % with SYBR Safe™ (Invitrogen).
25. MinElute® PCR purification Kit (Qiagen).
26. MinElute® Gel Extraction Kit (Qiagen).
27. PfuUltra II Hotstart PCR master Mix (Agilent Technologies).
28. NEBNext® dA-Tailing Module.
29. Topo® TA cloning kit for sequencing (Invitrogen).
30. Bacterial growth medium (LB).
31. Ampicillin (Sigma).
32. QIAprep® Spin Miniprep Kit (Qiagen).

33. Rapid DNA Dephos & Ligation Kit (Roche).
34. pFUSE-CLIg: Cloning plasmid with constant region of the light chain (Invivogen).
35. pFUSE-CHIg: Cloning plasmid with constant region of the heavy chain (Invivogen).
36. Restriction Enzymes EcoRI (New England BioLabs® Inc.).
37. Restriction Enzymes NheI (New England BioLabs® Inc.).
38. Restriction Enzymes BsiWI (New England BioLabs® Inc.).
39. NEBuffer 2 (New England BioLabs® Inc.).
40. NEBuffer 4 (New England BioLabs® Inc.).
41. Purified BSA 100× (New England BioLabs® Inc.)
42. XL1-blue competent cells (Agilent).
43. Blasticidin S HCL (Invitrogen).
44. Zeocin™ (Invitrogen).
45. LB AGAR Fast-Media® Amp (Invivogen) or LB agar plates with ampicilin at 100 μg/ml.
46. LB AGAR Fast-Media® Zeo (Invivogen) or LB agar plates with zeocin at 25 μg/ml.
47. LB-AGAR Fast-Media® Blas (Invivogen) or LB agar plates with blasticidin at 100 μg/ml.
48. Endofree Plasmid Maxi kit (Qiagen).
49. Oligonucleotide primers (design described in Subheading 3).
50. 3-Hamster-Lambda-out: GCTCTTCTCCACAGTTTCCCC TTCATGGGTAACTTGGCAGGAAACC.
51. 3-Hamster-Lambda-in: CCTTCCAGGTCACTGTTGCAGA ACCCGGG.
52. 3-Hamster-Kappa-out: CATTCCTGTTCAGGGTCTTGACA ATG.
53. 3-Hamster-Kappa-in: CTGCTTTGGTCAGCGAGAGGGTG CTGC.
54. 3-Hamster-HC-out: CTTGTCCACCTTGGTGCTGCTGG.
55. 3-Hamster-HC-in: CTACGTTGCAGGTGATGGGCTGCTTG.
56. Universal M13 forward primer (-20): GTAAAACGAC GGCCAG.
57. Universal M13 reverse primer: CAGGAAACAGCTATGAC.
58. HTLV 5′UTR forward primer: TGCTTGCTCAACTCTA CGTC.
59. Human Fc (IgG1, IgG2, IgG3, IgG4) reverse primer: CTCACGTCCACCACCACGCA.

60. Mouse IgG1 Fc reverse primer: GTGAGCACATCCTTG GGCTT.
61. Mouse IgG Fc (IgG2a, IgG2b, IgG3) reverse primer: GGAAGATGAAGACGGATGGTC.

2.2 Transient Antibody Expression in Mammalian Cells

1. Cell culture hood (laminar flow).
2. CO_2 cell culture incubator.
3. Microscope.
4. Cell counting device, Vi-CELL (Beckman Coulter, Inc.), or equivalent.
5. Vacuum pump.
6. FastELISA IgG quantification kit (R&D Biotech).
7. T75 Flask Nunclon™ Δ Surface (Nunc).
8. 6-Well plates Multidishes Nunclon™ Δ Surface (Nunc).
9. Dulbecco's Modified Eagle Medium, DMEM (Sigma).
10. Dulbecco's PBS (Sigma).
11. Fetal Bovine Serum, FBS (Sigma).
12. l-Glutamine 200 mM (Sigma).
13. SFM 293 II (Invitrogen).
14. *Trans*IT®-LT1 Transfection Reagent (Mirus).
15. HEK-293 cells (ATCC).
16. Stericup® Filter Units (Millipore).
17. Complete HEK-293 growth medium: DMEM, 4 mM l-glutamine, 10 % of decomplemented FBS (must be heated at 56 °C during 35 min). Filter the medium using a Stericup Filter Unit.
18. Production medium: SFM 293 II medium, 6 mM l-glutamine. Filter the medium using a Stericup Filter Unit.

2.3 Material for Small-Scale Antibody Purification

1. Nanodrop® (Thermo Scientific).
2. Rotator (Stuart®).
3. Mabselect Sure (GE healthcare) or equivalent protein A beads.
4. Protein G Sepharose™ 4 Fast Flow (GE healthcare).
5. SigmaPrep Spin column (Sigma).
6. illustra™ NAP™-5 Columns (GE healthcare).
7. Amicon® Ultra-0.5 Centrifugal Filter Devices, 50K (Millipore).
8. Dulbecco's PBS (Sigma).
9. Elution buffer: 50 mM Glycine-HCl, pH 3.

3 Methods

3.1 Mouse or Rat IgG mRNA Isolation from Hybridoma and cDNA Preparation for PCR

1. Perform purification of total RNA from mouse or rat hybridoma using the RNeasy® Plus Mini Kit as described below.
2. Carefully pellet 2×10^6 hybridoma cells by centrifugation at $300 \times g$ for 5 min, and gently resuspend the cells in 10 ml PBS. Repeat this washing step twice, and after the third wash, pellet the cells and carefully remove the supernatant.
3. To avoid RNase contamination and unwanted degradation of RNA, we recommend using RNase*Zap*® Wipes to clean your bench, pipettes, and gloves before starting RNA purification.
4. Disrupt the cells with 350 μl of RLT plus buffer complemented with β-mercaptoethanol (10 μl of β-mercaptoethanol for 1 ml of RLT plus buffer). For homogenization, put the lysate in QIAshredder column and centrifuge at $14{,}000 \times g$ during 2 min.
5. Transfer the flow-through corresponding to the homogenized lysate in gDNA eliminator column provided with the RNeasy® Plus Mini Kit to remove contaminant genomic DNA and spin for 1 min at $14{,}000 \times g$. Recover the flow-through, and mix well with 350 μl of 70 % ethanol.
6. Transfer this sample to RNeasy spin column, spin for 1 min at $14{,}000 \times g$, and discard the flow-through.
7. Wash the column with 700 μl of RW1 buffer, spin for 1 min at $14{,}000 \times g$, and discard the flow-through.
8. Wash two times the column with 500 μl of reconstituted RPE buffer, spin for 1 min at $14{,}000 \times g$, discard the flow-through, and centrifuge the column at $14{,}000 \times g$ for 1 additional minute to dry the filter membrane.
9. Transfer the column to 1.5 ml Eppendorf tube, add 50 μl of RNase-free water to the filter membrane, wait for 1 min, and centrifuge the column for 1 min at $14{,}000 \times g$ to elute RNA. The usual amount of total RNA for 2×10^6 hybridoma cells is about 75 μg.
10. cDNA can then be prepared using ThermoScript™ RT-PCR System for First-Strand cDNA Synthesis kit. The procedure of cDNA synthesis is briefly described below, but if supplementary information is needed, refer to the product information sheet.
11. In a nuclease-free microcentrifuge tube, mix 1 μl of 50 μM Oligo$(dt)_{20}$, 10 pg to 5 μg of RNA, and 2 μl of 10 mM dNTP mix in a final volume of 12 μl adjusted with RNase-free water. Denature RNA and primer by incubating at 65 °C for 5 min and then place on ice.

12. Prepare a master reaction mix on ice mixing 4 μl of 5× cDNA synthesis buffer, 1 μl of 0.1 M DTT (provided with the ThermoScript kit), 1 μl of RNaseOUT™ at 40 U/μl, 1 μl of RNase-free water, and 1 μl of ThermoScript™ RT at 15 U/μl.
13. Add 8 μl of the master reaction mix to the previous one.
14. Perform the reaction using a PCR thermocycler applying the following thermal cycling conditions: 30 min at 50 °C and 30 min at 65 °C and then terminate the reaction by incubating for 5 min at 85 °C (*see* **Note 1**).
15. Add 1 μl of RNase H and incubate for 20 min at 37 °C (optional).
16. cDNA can be stored at −20 °C or use for PCR immediately.

3.2 Hamster mRNA Isolation from Hybridoma and cDNA Preparation for RACE-PCR

1. Perform purification of total RNA from hamster hybridoma using the RNeasy® Plus Mini Kit, as described in Subheading 3.1. Prepare 5′ RACE-Ready cDNA using the SMARTer™ RACE cDNA amplification kit, using the procedure described below.
2. Prepare buffer mix 1 for all your reactions (all the components are provided with the kit). For one reaction, mix 2 μl of 5× first-strand buffer with 1 μl of DTT and 1 μl of dNTP mix.
3. Mix 2.75 μl of hamster hybridoma total RNA with 1 μl 5′-CDS Primer A, then incubate at 72 °C during 3 min, and cool the tube at 42 °C during 2 min in a PCR thermocycler. Spin the tube to collect the content.
4. Mix 4 μl of buffer mix 1 with 3.75 μl of denatured RNA mix from the previous step, and add 1 μl of the SMARTer IIA oligo, 0.25 μl of RNase inhibitor (provided with the kit), and 1 μl of SMARTScribe Reverse Transcriptase for a final volume of 10 μl. Mix well by pipetting.
5. Incubate at 42 °C during 90 min and then at 70 °C during 10 min in a PCR thermocycler.
6. You can then dilute your 5′-RACE-Ready cDNA sample with Tricine–EDTA buffer (provided in the kit). Depending on the total amount of total RNA that you used, add 20 μl of Tricine–EDTA buffer if you started with 200 ng of total RNA or less and add 100 μl of Tricine–EDTA buffer if you started with more than 200 ng of total RNA.

3.3 Mouse or Rat VH and VL DNA Isolation by PCR

1. The VH and VL nucleotide sequences are amplified by PCR using the cDNA prepared in Subheading 3.1 as template. Mix 20 μl of GoTaq® Green Master Mix 2×, 1 μl of forward and reverse primer (from Progen, described below), and 1 μl of cDNA, obtained previously, in a final volume of 40 μl adjusted with sterile water (*see* **Note 2**).

2. Perform 25 separate reactions with the first set of primers. Each variable heavy-chain forward primer (1A-L) has to be combined with the constant region IgG reverse primer (1 M). Similarly, each κ- or λ-light-chain forward primer (1N-W, 1Y) has to be combined with the corresponding constant region reverse primer (1X or 1Z), respectively.
3. Perform the reaction using a PCR thermocycler applying the following thermal cycling conditions: 2 min at 95 °C; 45 s at 95 °C, 30 s at 55 °C, and 30 s at 72 °C, 30 cycles; and 5 min at 72 °C (*see* **Note 3**). After the end of the cycle, immediately cool down to 4 °C.
4. Load 10 μl of the PCR products and 10 μl of E-Gel® 1 Kb Plus DNA ladder on an E-Gel® 1.2 %. After migration, analyze the PCR products on a UV table and estimate their sizes using the DNA ladder. The expected size of both variable regions is approximately 400 bp (*see* **Note 4**).
5. Purify the PCR product (~30 μl) using the MinElute® PCR purification as described below.
6. Mix 5 volumes of PB buffer to 1 volume (30 μl) of the PCR reaction.
7. Apply the sample to MinElute column, and centrifuge for 1 min at full speed. Discard flow-through.
8. Wash the column with 750 μl of reconstituted PEB buffer. Centrifuge for 1 min at full speed. Discard flow-through. Centrifuge the column at full speed for 1 additional minute to dry the filter membrane.
9. Transfer the column to 1.5 ml Eppendorf tube, add 10 μl of EB buffer to the filter membrane, wait for 1 min, and centrifuge the column for 1 min at full speed to elute DNA.
10. Measure DNA concentration using Nanodrop® or equivalent spectroscopic methods, and go to Subheading 3.5.

3.4 Hamster VH and VL DNA Isolation by RACE-PCR

1. The hamster VH and VL nucleotide sequences are amplified by RACE-PCR using the cDNA prepared in Subheading 3.2 as template.
2. Prepare two reaction tubes, one for the first amplification of VH sequence and the other for the first amplification of the VL sequence. For each tube, mix 25 μl of PfuUltra II Hotstart PCR master Mix 2× with 2.5 μl of 5′-RACE-Ready cDNA sample and 5 μl of universal primer mix 10× (UPM, provided with SMARTer™ RACE cDNA amplification kit). Then, add 1 μl of 3-Hamster-HC-out (10 μM) in the VH tube and 1 μl of either 3-Hamster-Lambda-out or 3-Hamster-Kappa-out depending on antibody light chain in the VL and complete with PCR-grade water to a volume of 50 μl.

3. Perform the first PCR reaction using a PCR thermocycler applying the following thermal cycling conditions: 2 min at 95 °C; 20 s at 95 °C, 20 s at 55 °C, and 15 s at 72 °C, 30 cycles; and 5 min at 72 °C. After the end of the cycle, immediately cool down to 4 °C.
4. Load 10 μl of the first PCR products and 10 μl of E-Gel® 1 Kb Plus DNA ladder on an E-Gel® 1.2 %.
5. After migration, analyze the PCR products on a UV table and estimate their sizes using the DNA ladder. The expected size of both variable regions amplified by RACE-PCR can be variable, and bands can be observed between 600 and 800 bp. If no clear bands can be identified, go to **step 6**. If several bands are present go to **step 11**.
6. If no clear bands can be identified, purify the PCR products according to the protocol provided with the MinElute® PCR purification kit which was described in Subheading 3.3.
7. Prepare two new reaction tubes, one for the amplification of VH sequence and the other for VL sequence. For each tube, mix 25 μl of PfuUltra II Hotstart PCR master Mix 2× with 2.5 μl of purified PCR VH or VL products and 5 μl of universal primer mix 10× (UPM, provided with SMARTer™ RACE cDNA amplification kit). Then, add 1 μl of 3-Hamster-HC-in (10 μM) in the VH tube and 1 μl of either 3-Hamster-Lambda-in or 3-Hamster-Kappa-in depending on antibody light chain in the VL and complete with PCR-grade water to volume of 50 μl.
8. Perform the second PCR reaction using a PCR thermocycler applying the following thermal cycling conditions: 2 min at 95 °C; 20 s at 95 °C, 20 s at 55 °C, and 15 s at 72 °C, 30 cycles; and 5 min at 72 °C. After the end of the cycle, immediately cool down to 4 °C.
9. Load 10 μl of the second PCR products and 10 μl of E-Gel® 1 Kb Plus DNA ladder on an E-Gel® 1.2 %.
10. After migration, analyze the PCR products on a UV table and estimate their sizes using the DNA ladder. The expected size of both variable regions amplified by RACE-PCR can be variable, and bands can be observed between 600 and 800 bp. It is possible to observe at this step multiple bands.
11. Purify the DNA fragments using the MinElute® Gel Extraction Kit as follows.
12. Cut the bands from the gel using scalpel and add 3 volumes of QG buffer for 1 volume of gel.
13. Incubate at 50 °C during 10 min under agitation with a thermomixer to dissolve gel slice.
14. Add 1 volume of isopropanol to the sample, mix it, and apply it to a MinElute column. Centrifuge for 1 min at full speed. Discard the flow-through.

15. Wash the column with 500 μl of QG buffer. Discard the flow-through.
16. Wash the column with 750 μl of reconstituted PE buffer. Discard the flow-through, and centrifuge the column at full speed for 1 additional minute to dry the filter membrane.
17. Transfer the column to 1.5 ml Eppendorf tube, add 10 μl of EB buffer to the filter membrane, wait for 1 min, and centrifuge the column for 1 min at full speed to elute DNA.
18. The purified DNA is blunt as it was amplified with PFU enzyme, a proofreading polymerase. Proceed to dA-tailing experiment to incorporate a non-templated dAMP on the 3′ end of a blunt DNA fragment using NEBNext® dA-Tailing Module. Briefly, mix 10 μl of purified DNA from **step 16** with 5 μl of NEBNext dA-Tailing Reaction Buffer 10× and 3 μl of Klenow Fragment provided with the kit. Complete to 50 μl with PCR-grade water. Incubate in a thermal cycler for 30 min at 37 °C, and purify DNA sample according to the protocol provided with the MinElute® PCR purification kit as described in Subheading 3.3.

3.5 Mouse, Rat, or Hamster VH and VL DNA Cloning

1. The purified PCR VH and VL obtained with GoTaq® enzyme or after dA tailing of blunt PCR products are compatible with and can be cloned using the TOPO TA Cloning® Kit for sequencing. The procedure is briefly described below, but if supplementary information is needed, refer to the product information sheet.
2. Mix 2 μl of purified PCR product and 1 μl of salt solution in a final volume of 6 μl adjusted with sterile water. Then add 1 μl of TOPO® vector and incubate the reaction for 15 min at room temperature.
3. Transform the ligated product by adding 2 μl of the ligation product to a vial of chemically competent One Shot TOP10 *Escherichia coli* provided with TOPO TA Cloning® Kit for Sequencing. Then, gently mix and incubate for 30 min on ice. Proceed to the heat shock by putting the vial in a 42 °C water bath for 30 s, and then transfer immediately the cells on ice for 2 min. Add 250 μl of pre-warmed S.O.C medium and incubate for 1 h at 37 °C under agitation using a thermomixer. Finally, spread the bacteria on LB agar plates with ampicillin.
4. Incubate the plates overnight at 37 °C in static incubator.

3.6 Clone Screening

1. Pick several colonies (*see* **Note 5**) from the agar plate using a tip, and inoculate a culture of 5 ml LB medium containing 50 μg/ml ampicillin (*see* **Note 6**).
2. Incubate overnight at 37 °C in incubator shaker.
3. Perform plasmid purification according to the protocol provided with the QIAprep® Spin Miniprep Kit. Perform the sequencing of purified pCR4 plasmids containing VH and VL inserts using M13 reverse and M13 forward primers.

3.7 Sequence Analysis

1. Identify the heavy-chain and light-chain germlines using the IMGT/V-QUEST program (http://www.imgt.org/IMGT_vquest/share/textes/). The procedure is briefly described below, but if supplementary information is needed, refer to the documentation available on the IMGT website (http://www.imgt.org/IMGT_vquest/share/textes/imgtvquest.html) and in Chapter 21 of this book.
2. Depending on your antibody, select Mouse or Rat link. For hamster VH and VL, the number of sequences present in databases is limited. We recommend to compare identified sequence against both rat and mice databases. We also suggest performing BLAST research (http://blast.ncbi.nlm.nih.gov/) to verify the sequences.
3. Copy/paste, in Fasta format, the nucleotide sequences corresponding to your VH or VL sequences determined by sequencing in the text area. You can also give the path access to a local file containing your sequences.
4. Your results can be displayed either in a detailed or a synthesis view: choose your preferred view, and start the analysis. Your antibody VH and VL sequences will be aligned to their closest germlines according to the IMGT nomenclature.

3.8 Oligonucleotide Primer Design

1. Design a couple of oligonucleotide primers to amplify the isolated VH sequence with EcoRI as the 5′ cloning site and NheI as the 3′ cloning site (*see* **Note 7**); an example is provided in Fig. 2a.
2. Similarly, design a couple of oligonucleotide primers to amplify the isolated VL sequence with the EcoRI as the 5′ cloning site and BsiWI as the 3′ cloning site. An example is provided in Fig. 2b.

3.9 Selection of the pFUSE Vector

Several pFUSE vectors are available from Invivogen. Depending on your application, choose the most appropriate species, isotypes, and light-chain constant region for the production of your antibody (*see* **Note 8**). Moreover, select the pFUSEss vectors because they contain a multiple cloning site upstream of the antibody constant heavy or light regions and downstream of the IL-2 leader sequence for efficient production and secretion of antibody from mammalian cells. Effectively, the method described in Subheading 3.3 does not allow the isolation of the VH and VL domains with their native leader sequences.

3.10 Amplification of VH and VL Inserts

1. The VH and VL are amplified by PCR. The pCR4-TOPO® containing the variable region of interest is used as the DNA template.
2. Mix 25 μl of PfuUltra II Hotstart PCR master Mix 2×, 0.2 μM of forward and reverse primers designed according to Subheading 3.8, and 10 ng of the DNA template in a final volume of 50 μl adjusted with sterile water.

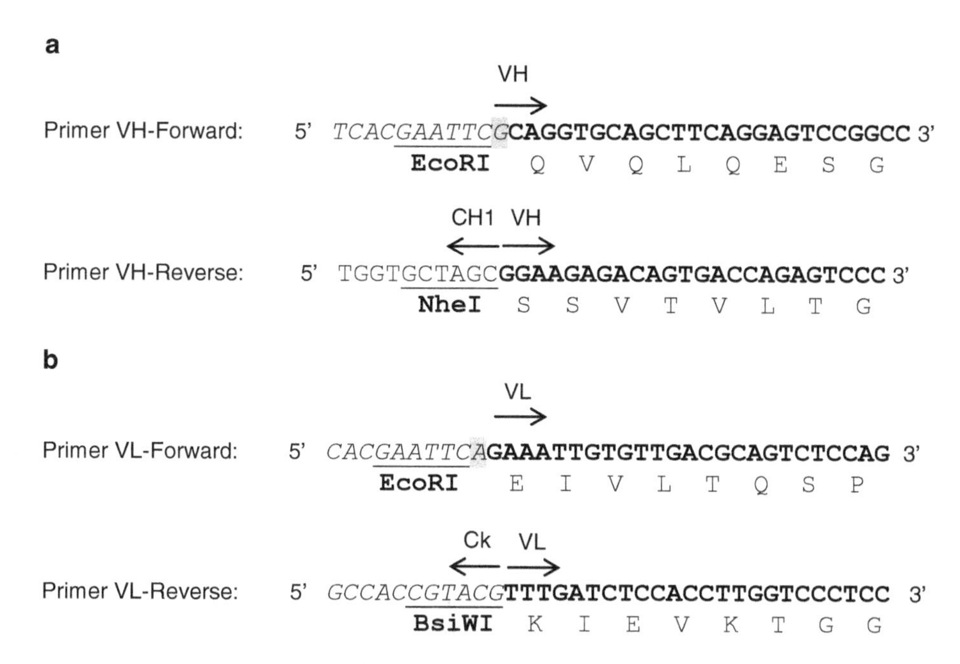

Fig. 2 Design of primers for antibody reformatting. (**a**) Example of reverse and forward primers for reformatting of VH chain in pFuse-CHIg plasmid. The nucleotide stretch that must be added in 5′ part of the primer is in *italic* with the restriction enzyme sites *underlined*. In the forward primer, one nucleotide highlighted in *grey* is inserted after the restriction site to allow the expression of the VH sequence in frame with the hIL-2 signal peptide. Nucleotide sequence encoding an example of VH sequence is in *bold* text and translated in its corresponding amino acid sequence. *Arrows* indicate the start (forward primer) and the end (reverse primer) of the sequence encoding the VH domain and the start of the sequence encoding the CH1 domain (reverse primer). (**b**) Example of reverse and forward primers for reformatting of VL chain in pFuse-CLIg. The same format as in (**a**) was used. *Arrows* indicate the start (forward primer) and the end (reverse primer) of the sequence encoding the VL domain and the start of the sequence encoding the constant κ (*C*κ) domain (reverse primer)

3. Perform the reaction using a PCR thermocycler applying the following thermal cycling conditions: 2 min at 95 °C; 20 s at 95 °C, 20 s at 55 °C, and 15 s at 72 °C, 30 cycles; and 5 min at 72 °C. After the end of the cycle, immediately cool down to 4 °C.
4. Load the PCR products and 10 μl of E-Gel® 1 Kb Plus DNA ladder on an E-Gel® 1.2 %.
5. After migration, analyze the PCR products on a UV table and estimate their sizes using the DNA ladder. The expected size of both variable regions is approximately 400 bp.
6. Purify the band according to the protocol provided with the MinElute® Gel Extraction Kit which is described in Subheading 3.3.
7. Measure the concentration of DNA using a Nanodrop® or equivalent spectroscopic methods.

3.11 VH Insert Digestion

1. The VH insert and pFUSE-CHIg vector are digested using the EcoRI and NheI restriction enzymes.
2. Mix 5 μl of NEBuffer 4 10×, 0.5 μl of purified BSA 100×, 1 μg of VH insert DNA, and 10 U of each enzyme in a final volume of 50 μl adjusted with sterile water.
3. Incubate the reaction for 2 h at 37 °C in a thermomixer.
4. Purify the digested VH using the MinElute® PCR Purification Kit as described in Subheading 3.3.
5. Measure the concentration of the purified DNA using a Nanodrop® or equivalent spectroscopic methods.

3.12 pFUSE-CHIg Vector Digestion

1. Mix 5 μl of NEBuffer 4 10×, 0.5 μl of purified BSA 100×, 1 μg of pFUSE-CHIg vector DNA, and 10 U of each enzyme in a final volume of 50 μl adjusted with sterile water.
2. Incubate the reaction for 2 h at 37 °C in a thermomixer.
3. Add 1 μl of alkaline phosphatase (provided with the Rapid DNA Dephos & Ligation Kit) into the tube containing the digested pFUSE-CHIg vector and incubate during 15 additional minutes at 37 °C (do not exceed this incubation time as it will decrease your ligation efficiency).
4. Load the digested vector product and 10 μl of E-Gel® 1 Kb Plus DNA ladder on an E-Gel® 1.2 %.
5. After migration, extract the band corresponding to the digested vector (the expected size is approximately 4,400 bp).
6. Purify the band containing vector DNA using the MinElute® Gel Extraction Kit as described in Subheading 3.3.
7. Measure the concentration of the purified DNA using a Nanodrop® or equivalent spectroscopic methods.

3.13 VL Insert Digestion

1. The VL and a pFUSE-CLIg are digested using the EcoRI and BsiWI restriction enzyme.
2. Mix 5 μl of NEBuffer 2 10×, 0.5 μl of purified BSA 100×, 1 μg of VL DNA insert, and 10 U of BsiWI enzyme in a final volume of 50 μl adjusted with sterile water.
3. After 1-h incubation at 55 °C into a thermomixer, decrease the temperature at 37 °C and add 10 U of EcoRI enzyme.
4. Incubate the reaction for 2 h at 37 °C.
5. Purify the digested VL using the MinElute® PCR Purification Kit as described in Subheading 3.3.
6. Measure the concentration of the purified DNA using a Nanodrop® or equivalent spectroscopic methods.

3.14 pFUSE-CLIg Vector Digestion

1. Mix 5 μl of NEBuffer 2 10×, 0.5 μl of purified BSA 100×, 1 μg of pFUSE-CLIg, and 10 U of BsiWI enzyme in a final volume of 50 μl adjusted with sterile water.
2. After 1-h incubation at 55 °C into a thermomixer, decrease the temperature at 37 °C and add 10 U of EcoRI enzyme.
3. Incubate the reaction for 2 h at 37 °C.
4. Then, perform the vector dephosphorylation step with the alkaline phosphatase, purify the DNA, and measure its concentration following the procedure described in Subheading 3.12.

3.15 Ligation and Transformation

1. The VH is ligated into a pFUSE-CHIg, and the VL is ligated into a pFUSE-CLIg. The Rapid DNA Dephos & Ligation Kit is used for the ligation.
2. Mix 2 μl of DNA dilution buffer 5×, 50 ng of vector, and 15 ng of insert (corresponding to a 1:3 molar ratio of vector DNA to insert DNA) in a final volume of 10 μl adjusted with sterile water.
3. Also perform a negative ligation control with vectors only (using the same conditions) to determine the self-ligation background of the digested plasmids.
4. After mixing thoroughly, add 10 μl of T4 DNA ligation buffer 5× and 1 μl of T4 DNA ligase.
5. Incubate the reaction mix for 15 min at room temperature.
6. Transform the ligated products and your negative ligation controls into chemically competent XL1-blue *Escherichia coli* by adding directly 50 μl of cells to the ligation mix.
7. Gently mix and incubate for 30 min on ice. Heat the cells in a 42 °C water bath for 45 s, and then replace them for 2 min on ice.
8. Add 500 μl of bacterial growth medium and incubate for 1 h at 37 °C under agitation using a thermomixer. Finally, spread the bacteria transformed with recombinant pFUSE-CHIg, and the corresponding negative control, on LB agar plates containing zeocin as well as the bacteria transformed with recombinant pFUSE-CLIg, and the corresponding negative control, on LB agar plates containing blasticidin.
9. Incubate the plates overnight at 37 °C in a static incubator.

3.16 Colony Screening

1. Determine the ratio of colonies observed on the insert-plus-vector plate versus vector-only plate to control the ligation efficiency. A ratio superior or equal to 2/1 indicates that the ligation was successful (*see* **Note 9**).

2. Pick a single colony from the agar plate using a tip and inoculate a culture of 5 ml LB medium containing the appropriate selective antibiotic. Incubate overnight at 37 °C with vigorous shaking.
3. Perform plasmid purification according to the protocol provided with the QIAprep® Spin Miniprep Kit.
4. Control the sequence of the cloned VL sequence in pFUSE-CLIg with HTLV 5′ UTR forward primer. Depending on your pFUSE-CHIg, use HTLV 5′ UTR forward primer and the most appropriate reverse primer (check the species and isotypes; human Fc reverse primer, mouse IgG1 Fc reverse primer, mouse IgG Fc reverse primer, or rat IgG1 Fc reverse primer) for sequencing. Control that the VH and VL sequences were inserted in frame with the IL-2 leader sequences and the constant part of IgG and that no mutations were introduced during the reformatting process according to the VH and VL templates.
5. Once the final clones of both VH and VL inserted into their respective vectors have been controlled, purify large amount of plasmids using Endofree plasmid maxi kit to get DNA of good quality for transfection (optional).

3.17 Cell Growth and Maintenance

1. The subculturing procedure of HEK-293 cells is briefly described below. Remove and discard culture medium from a T75 flask with confluent HEK-293 cells, rinse the cell layer with 10 ml of PBS, and detach the cells from the flask surface by gentle pipetting with 10 ml of DMEM medium.
2. Add 1 ml of a 1 in 10 dilution of the cell suspension to a new T75 containing 25 ml of complete HEK-293 growth medium. Incubate culture at 37 °C in a CO_2 incubator. Cells become confluent 3 days after the subculturing and reach around 2.5×10^7 cells per T75 flask. Subculture the cells twice a week.

3.18 Cell Transfection

1. *Day 0*: Detach cells from a flask (*see* **Note 10**) as described above, perform a cell count using ViCell or equivalent, and plate 4×10^5 cells in 2 ml of complete medium per well in a 6-well plate.
2. *Day 1*: Prepare a mix with 97 µl of DMEM and 3 µl of *Trans*IT®-LT1 Transfection Reagent for one well in a tube and incubate for 5 min at room temperature. During the incubation, mix 1 µg of pFUSE-CHIg plasmid (encoding the heavy chain) and 1 µg of pFUSE-CLIg plasmid (encoding the light chain) in another tube (*see* **Note 11**). Add the transfection reagent mix drop by drop to the DNA and incubate for 30 min at room temperature. Then, add the preparation drop by drop to the cells and incubate overnight at 37 °C in a CO_2 incubator.

Table 1
Relative binding strengths of IgG for protein A and protein G

Species	Subclass	Protein A binding	Protein G binding
Human	IgG_1	++++	++++
	IgG_2	++++	++++
	IgG_3	–	++++
	IgG_4	++++	++++
Mouse	IgG_1	+	++++
	IgG_{2a}	++++	++++
	IgG_{2b}	+++	+++
	IgG_3	++	+++
Rabbit	IgG	++++	+++

(++++), (+++), (++), (+), and (–) indicate strong binding, medium-to-strong binding, medium binding, weak binding, and no binding to protein A or protein G, respectively

3. *Day 2*: Remove and discard complete medium containing the transfection mix, rinse carefully the cell layer with 2 ml of production medium, and then add 2 ml of production medium (*see* **Note 12**).
4. *Day 5*: Add 500 μl of production medium to each well.
5. *Day 7*: Harvest the supernatant. Centrifuge the supernatant for 5 min at 500 × *g* to remove cells debris. Optional: Filter the supernatant to avoid any contamination. Store the supernatant at 4 °C until purification.
6. Quantify the concentration of antibody in the supernatant using the Fast ELISA kit (optional).

3.19 Antibody Purification

1. For antibody purification, choose the affinity resin according to antibody species and isotypes (Table 1). The binding capacity of the Mabselect Sure resin (protein A) is approximately 30 mg of human IgG/ml of resin. The binding capacity of the protein G resin is approximately 17 mg of human IgG/ml of resin and 6 mg of mouse IgG/ml of resin.
2. Add 200 μl of appropriate resin slurry into a microcentrifuge tube. The resin beads are supplied as 50 % slurry; therefore, pipetting of 200 μl results in 100 μl of total resin.
3. Centrifuge for 2 min at 200 × *g* in a microcentrifuge.
4. Carefully remove the supernatant, and discard it.
5. Add 1 ml of PBS, cap the tube, and resuspend the resin by inverting the tube.
6. Centrifuge for 2 min at 200 × *g*; carefully remove the supernatant, and discard it.

7. Repeat the wash three times.
8. Add the resin to the supernatant from the production.
9. Incubate for 1 h at room temperature or overnight at 4 °C on rotator at 15 rpm.
10. Centrifuge for 1 min at 200 × *g*, and keep the supernatant (*see* **Note 13**).
11. Add 500 μl of PBS, and transfer the resin pellet to a spin column.
12. Centrifuge for 1 min at 200 × *g*; discard liquid from the collection tube.
13. Wash three times with 500 μl of PBS.
14. Place the column in a new collection tube, and elute the antibody by adding 500 μl of elution buffer.
15. Discard the column, and measure the concentration of IgG using a Nanodrop® or equivalent spectroscopic methods (*see* **Note 14**).

3.20 Sample Desalting

1. The eluted sample is desalted using illustra™ NAP™-5 Columns.
2. Remove top and bottom caps of the column.
3. Allow excess liquid to drain by gravity flow.
4. Equilibrate the column with 10 ml of PBS.
5. Add 500 μl of eluted IgG.
6. Allow excess liquid to drain by gravity flow.
7. Place a collection tube under column.
8. Elute with 1 ml of PBS.
9. Collect the eluate by gravity flow.
6. Measure the concentration of IgG using a Nanodrop® or equivalent spectroscopic methods.

3.21 Sample Concentration (Optional)

1. Add 500 μl of sample to an Amicon® Ultra-0.5 Centrifugal Filter Devices 50K, and cap it.
2. Centrifuge at 14,000 × *g* during 10 min to concentrate the sample. During centrifugation, check your sample every minute to avoid protein over-concentration which can lead to antibody precipitation and sample loss.
3. Place the Amicon Ultra filter device upside down in a clean microcentrifuge tube.
4. Centrifuge at 1,000 × *g* for 2 min to transfer the concentrated sample from the device to tube.
7. Measure the concentration of IgG using a Nanodrop® or equivalent spectroscopic methods.
5. Store purified sample at 4 °C or −20 °C.

4 Notes

1. This protocol has been optimized to transcribe RNA with difficult secondary structure.
2. Usually, GoTaq polymerase is able to amplify most VH and VL sequences, and, as this is a non-proofreading enzyme, obtained PCR products are compatible with TA cloning. If you encounter problem with VH or VL sequence amplification, you can use proofreading enzyme like PfuUltra II fusion HS DNA polymerase, but you need to perform dA tailing of the obtained PCR products (using, for example, the NEB dA tailing kit) to facilitate TA cloning.
3. This is a standard PCR cycle protocol which is usually successful with Progen Mouse IgG library primer set; however, the annealing temperature can be optimized using temperature gradient. The Progen primers are also suited for rat antibody variable gene amplification.
4. It is frequent to observe the amplification of several VH and VL sequences with different VH or VL primers provided by the Progen Mouse IgG library primer set. This is usually due to the amplification of the same VH or VL by different primers. Therefore, using PCR temperature gradient, you can find the optimum annealing temperature to decrease the numbers of positive amplification and isolate the correct nucleotide sequence. It is also possible that the hybridoma used to isolate the nucleotide sequences of the antibody was not monoclonal leading to the isolation of multiple VH and VL sequences. To identify the correct VH and VL sequence of the antibody of interest, the different variable domains have to be reformatted into IgG format and the different HC/LC pairs have to be expressed (as described in Subheadings 3.10–3.20) and tested to identify the correct pairing.
5. It is important to perform the sequencing of several clones to determine VH and VL sequence consensus as mutations may have been introduced during the cloning process.
6. The pCR4-topo plasmid, provided with the TOPO TA Cloning® Kit for Sequencing, contains the lethal *E. coli* gene *ccdB* which is disrupted when your DNA sequence is inserted. Therefore, only positive recombinants are theoretically selected after transformation in TOP10 cells. However, we recommend performing PCR screen to select positive clones. Briefly pick several clones and resuspend them individually in Gotaq PCR mix with the M13 reverse and M13 forward primers. Perform PCR amplification with 20–30 cycles, and identify positive clones by gel visualization.

Table 2
ADCC, ADCP, CDC, and half-life of human and mouse IgG isotypes

Species	Subclass	ADCC	ADCP	CDC	Half-life
Human	IgG_1	+++	+++	+++	21
	IgG_2	–	+	+	21
	IgG_3	+++	+++	+++	7
	IgG_4[a]	+	–	–	21
Mouse	IgG_1	+	N/A	–	6–8
	IgG2a	+++	N/A	+++	6–8
	IgG2b	++	N/A	+++	4–6
	IgG3	–	N/A	++	6–8

(+++), (++), (+), and (–) indicate strong function, medium function, weak function, and no function, respectively. (N/A) indicates that this information is not available in the literature. The average half-life in human or mouse serum is expressed in days

[a]IgG4 molecules can exchange half-molecules in a dynamic process termed Fab-arm exchange [9, 10]. This phenomenon can occur between therapeutic antibodies and endogenous IgG4. The S228P mutation has been shown to prevent this recombination process allowing the design of less unpredictable therapeutic IgG4 antibodies. We recommend to use the pFUSE-hIgG4e1-Fc2 from Invivogen which contains this stabilizing mutation to avoid this phenomenon

7. For the design of primers, we recommend to add several nucleotides before the restriction enzyme site to ensure good digestion of the PCR product. For the design of forward primers, one nucleotide after the EcoRI restriction enzyme site (G or A for VH forward primer or VL forward primer, respectively, *see* Fig. 2a, b) has to be conserved to ensure that the chain is expressed in frame with the pFUSE's hIL2 signal sequence. Moreover, the Progen Mouse IgG library primer set can introduce amino acid mutations in the N-terminal parts of VH and VL during PCR amplification. With the sequence analysis provided by the IMGT V-QUEST server, you can easily determine residue changes by direct VH or VL sequence comparison with their closest germline sequences. We therefore suggest to correct these mutations by designing forward primers which correspond to the germline.
8. Several pFUSE-CHIg are available which encode different human and mouse isotypes. Carefully choose the right plasmid according to your application. Table 2 recapitulates the antibody-dependent cell-mediated cytotoxicity (ADCC), antibody-dependent cell-mediated phagocytosis (ADCP), complement-dependent cytotoxicity (CDC), and half-life characteristics of these different isotypes. Moreover, Invivogen provides plasmids with engineered human IgG1 where mutations have been introduced in the Fc region to increase antibody half-life, enhance or disrupt ADCC/CDC, increase ADCP, and increase ADCC or CDC only.
9. A large number of colonies on both plates could be due to a weak digestion of the vector leading to a low-ratio recombinant

Table 3
Transfection conditions in functions of culture vessel

Culture vessel	T175	T75	6-well plate
Surface area	175 cm^2	75 cm^2	9.6 cm^2
Complete growth medium	50 ml	25 ml	2 ml
Nb of cells for *plated* transfection	5.10^6	2.10^6	4.10^5
TransIT®-LT1 reagent	45 μl	18 μl	3 μl
Serum-free medium	1.5 ml	600 μl	100 μl
DNA	30 μg	12 μg	2 μg

The quantity of DNA and TransIT®-LT1 transfection reagent are indicated in function of the size of the culture vessel. The number of cells that have to be plated the day before the transfection is also specified

vector versus empty vector. In this case, modify the conditions of the vector digestion (increase the incubation times). A low number of colonies on both plates could be due to a poor digestion of the insert.

10. This method described antibody production using a 6-well plate, but larger culture vessel can be used to increase the amount of antibodies as described in Table 3.
11. The heavy chain has usually a lower expression compared to the light chain; we suggest to increase the ratio of pFUSE-CHIg:pFUSE-CLIg vectors (for example 2:1) which can lead to a better antibody production.
12. To extend the antibody production in 293 cells, you can also generate semi-stable cell line using zeocin and blasticidin to select cells co-transfected with both pFUSE-CHIg and pFUSE-CLIg.
13. Keep the supernatant to perform a second capture step with the resin if the binding capacity of the resin was exceeded by the sample.
14. The transfection protocol described in this chapter usually leads to the production of 10–30 μg of antibody per milliliter of cell culture supernatant.

References

1. Kohler G, Milstein C (1975) Continuous cultures of fused cells secreting antibody of predefined specificity. Nature 256:495–497
2. Lonberg N (2008) Fully human antibodies from transgenic mouse and phage display platforms. Curr Opin Immunol 20:450–459
3. Chadd HE, Chamow SM (2001) Therapeutic antibody expression technology. Curr Opin Biotechnol 12:188–194
4. Jefferis R, Lund J, Pound JD (1998) IgG-Fc-mediated effector functions: molecular definition of interaction sites for effector ligands and

the role of glycosylation. Immunol Rev 163: 59–76

5. Strohl WR (2009) Optimization of Fc-mediated effector functions of monoclonal antibodies. Curr Opin Biotechnol 20: 685–691
6. Boulianne GL, Hozumi N, Shulman MJ (1984) Production of functional chimaeric mouse/ human antibody. Nature 312:643–646
7. Morrison SL, Johnson MJ, Herzenberg LA, Oi VT (1984) Chimeric human antibody molecules: mouse antigen-binding domains with human constant region domains. Proc Natl Acad Sci U S A 81:6851–6855
8. Riechmann L, Clark M, Waldmann H, Winter G (1988) Reshaping human antibodies for therapy. Nature 332:323–327
9. Aalberse RC, Schuurman J (2002) IgG4 breaking the rules. Immunology 105:9–19
10. Labrijn AF, Buijsse AO, van den Bremer ET, Verwilligen AY, Bleeker WK, Thorpe SJ et al (2009) Therapeutic IgG4 antibodies engage in Fab-arm exchange with endogenous human IgG4 in vivo. Nat Biotechnol 27:767–771

Chapter 14

Cloning of Recombinant Monoclonal Antibodies from Hybridomas in a Single Mammalian Expression Plasmid

Nicole Müller-Sienerth, Cécile Crosnier, Gavin J. Wright, and Nicole Staudt

Abstract

Antibodies are an integral part of biological and medical research. In addition, immunoglobulins are used in many diagnostic tests and are becoming increasingly important in the therapy of diseases. To express antibodies recombinantly, the immunoglobulin heavy and light chains are usually cloned into two different expression plasmids. Here, we describe a method for recombinant antibody expression from a single plasmid.

Key words Recombinant antibodies, Monoclonal antibodies, Hybridomas, Single expression plasmid, Fusion PCR

1 Introduction

Antibodies were first described in the late nineteenth century [1]. Today, they have become an important part of many molecular techniques such as protein assays or immunohistochemical protocols. Therapeutic antibodies are used to treat illnesses ranging from autoimmune diseases, such as rheumatoid arthritis [2], to cancer [3].

B lymphocytes can produce antibodies against any antigen the immune system recognizes as "nonself" [4]. Additionally, in autoimmune diseases antibodies against an individual's own antigens are generated [5].

Antibodies are predominantly Y-shaped molecules consisting of four covalently linked polypeptides: two 50 kDa "heavy" and two 25 kDa "light" chains. Two regions of the heavy and light chain dimer are highly variable (at the tips of the "Y") and mediate specific binding to the antigen (Fig. 1). This is achieved by recombination of distinct variable modules in the genomic locus [6], mutation by receptor editing [7], and somatic hypermutation [8].

Vincent Ossipow and Nicolas Fischer (eds.), *Monoclonal Antibodies: Methods and Protocols*, Methods in Molecular Biology, vol. 1131, DOI 10.1007/978-1-62703-992-5_14,

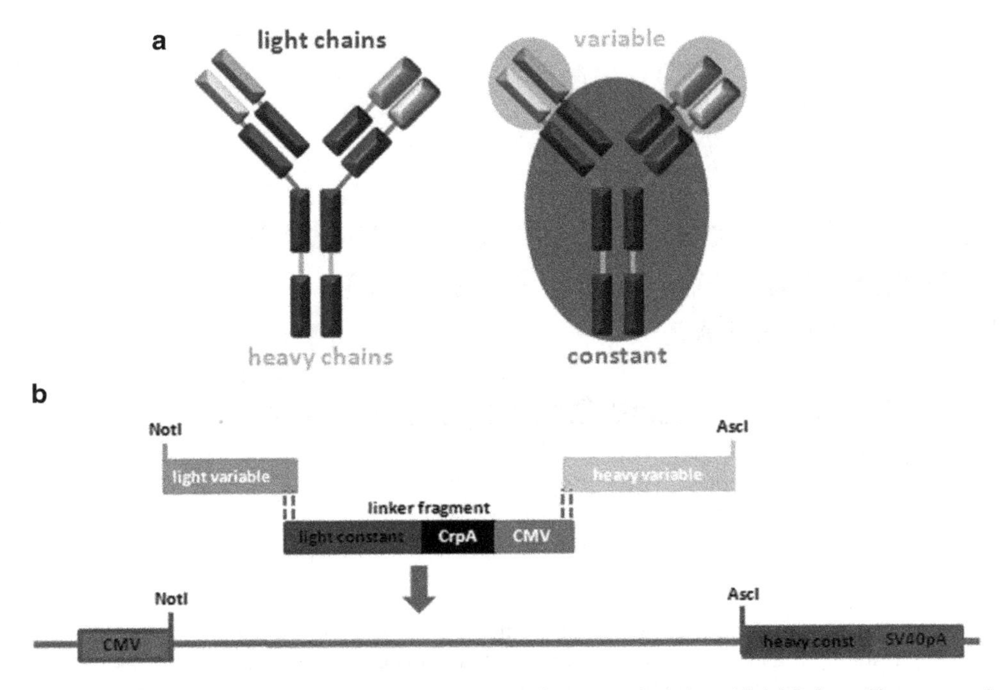

Fig. 1 Cloning strategy for recombinant antibodies in a single expression plasmid. (**a**) Schematic representation of a Y-shaped antibody molecule consisting of two heavy and two light chains. The antigen binding site is located at the tips of the "Y", where the amino acid sequence is unique to the antibody and specifically binds antigen. (**b**) Three PCR products (linker fragment, variable light, and variable heavy) are combined by fusion PCR (*dotted red lines* indicate overlap between single PCR products). The product of the fusion PCR and the plasmid are digested with NotI and AscI to allow a directed insertion into a mammalian expression plasmid backbone (Color figure online)

To raise antibodies, animals are immunized with the antigen of choice, and antibodies are collected from the serum of these animals. The amount of antibody produced is finite and consists of a polyclonal mix of antibodies recognizing different regions of the same antigen, which might not always be desirable. When more antibodies are needed, animals have to be re-immunized and the elicited antibodies may vary in quantity and quality.

The production of monoclonal antibodies is more time consuming but can offer infinite supply of an antibody species recognizing a single epitope. Typically, the antibody-producing B cells of the immunized animals, usually mice, are immortalized by cellular fusion to myeloma cells [9]. The resulting "hybridomas" are kept in culture, and the secreted antibodies are harvested from the cell culture supernatant. One disadvantage of monoclonal antibodies is that the generation and maintenance of hybridoma lines demands a large amount of laborious cell culture work. To establish a monoclonal line, a positive culture usually needs to go through at least two rounds of cellular cloning which can take several weeks. Ideally, this cloning process needs to be repeated regularly to guarantee that all hybridomas secrete the antibody.

An alternative to the maintenance of hybridoma lines is to clone the variable regions of the specific antibodies and to express them as recombinant proteins. This has the added advantage of enabling the introduction of additional protein tags to manipulate the antibody and changes to the sequence or isotype [10]. After the generation of hybridomas, the variable heavy and variable light chain regions of specific antibodies produced in these cells are amplified by RT-PCR. These are then combined with the sequence of the constant regions of heavy and light chains, typically in two separate expression plasmids [11]. To produce a functional antibody, these two plasmids are co-transfected into mammalian cells [12]. We have developed a system that allows the expression of variable heavy and light chains from a single expression plasmid (ref. 13, Fig. 1).

After RNA extraction from antibody-producing hybridomas, the regions encoding both the variable heavy and light chains are amplified by RT-PCR using a set of degenerate primers that have been designed to amplify the vast majority of known mouse antibody regions. In a subsequent PCR reaction, a third "linker" fragment containing the constant region and 3′ UTR of the light chain and a promoter driving the expression of the heavy chain is amplified (Fig. 1). All three PCR products are then combined in a single PCR using overlapping regions and primers containing restriction enzyme sites. After digestion, the PCR product is inserted into a mammalian expression plasmid containing the constant heavy chain region and 3′ UTR (Fig. 1). The final product is a single expression plasmid containing all the sequence information needed for the expression of a recombinant antibody in mammalian cells.

2 Materials

Use sterilized tips, tubes, and ultrapure water for RNA isolation and PCR steps.

1. RNA isolation: Trizol (Invitrogen), chloroform, isopropanol, prepare 70 % ethanol with ultrapure water.
2. cDNA synthesis: Superscript III (200 U/μl, Invitrogen), 10× RT buffer (Invitrogen), RNAse Inhibitor (40 U/μl, Promega), 10 mM dNTPs (Invitrogen), 50 μM Oligo dT (Invitrogen), 0.1 M DTT, 25 mM $MgCl_2$.
3. Amplification of variable heavy, variable light and linker fragment: Pfu Ultra Polymerase (2.5 U/μl, Agilent), 10× Pfu buffer (Agilent), DMSO, 10 mM dNTPs (Fermentas), 5 μM oligonucleotides (Table 1).
4. Fusion PCR: KOD Hot Start polymerase (1 U/μl, Novagen), 10× KOD buffer (Novagen), 2 mM dNTPs (Novagen), 25 mM $MgSO_4$, DMSO, 5 μM oligonucleotides (Table 1).

Table 1
Sequences (5′–3′) of primers used for different PCR steps of the cloning process

Degenerate forward primers for rearranged κ light chains	
1	TCAATAGTTGAACATAGCGGCCGCASAAAWTGTKCTCACCCAGTC
2	TCAATAGTTGAACATAGCGGCCGCAGAWATTGTGCTMACTCAGTC
3	TCAATAGTTGAACATAGCGGCCGCAGACATTGTGCTRACACAGTC
4	TCAATAGTTGAACATAGCGGCCGCAGACATTGTGATGACMCAGTC
5	TCAATAGTTGAACATAGCGGCCGCAGAYATCMAGATRAMCCAGTC
6	TCAATAGTTGAACATAGCGGCCGCAGAYATCCAGATGAYTCAGTC
7	TCAATAGTTGAACATAGCGGCCGCAGATATCCAGATGACACAGAC
8	TCAATAGTTGAACATAGCGGCCGCAGACATTGTGCTGACCCAATC
9	TCAATAGTTGAACATAGCGGCCGCAGACATYSTRATGACCCARTC
10	TCAATAGTTGAACATAGCGGCCGCAGATRTTKTGATGACYCARAC
11	TCAATAGTTGAACATAGCGGCCGCAGAYATTGTGATGACBCAGKC
12	TCAATAGTTGAACATAGCGGCCGCAGATATTGTGATAACCCAGGA
13	TCAATAGTTGAACATAGCGGCCGCAGACATCYTGCTGACYCAGTC
14	TCAATAGTTGAACATAGCGGCCGCAGAAAWTGTGYTGACCCAGTC
15	TCAATAGTTGAACATAGCGGCCGCAGAAACAACTGTGACCCAGTC
16	TCAATAGTTGAACATAGCGGCCGCAGACATTRTGATGWCACAGTC
17	TCAATAGTTGAACATAGCGGCCGCAGACATCCAGMTGACMCARTC
Reverse primer for rearranged κ light chains	
18	GGATACAGTTGGTGCAGCATCAGCCC
Degenerate forward primers for rearranged heavy chains	
19	TTCACGAGTCCAGCCTCAAGCAGTGAKRTRCAGCTTMAGGAGTC
20	TTCACGAGTCCAGCCTCAAGCAGTGAGKTYCAGCTBCAGCAGTC
21	TTCACGAGTCCAGCCTCAAGCAGTCAGGTGCAGMTGAAGSAGTC
22	TTCACGAGTCCAGCCTCAAGCAGTGAGRTCCAGCTGCAACARTC
23	TTCACGAGTCCAGCCTCAAGCAGTCAGGTYVAGCTGCAGCAGTC
24	TTCACGAGTCCAGCCTCAAGCAGTCAGGTYCARCTGCAGCAGTC
25	TTCACGAGTCCAGCCTCAAGCAGTGAGGTGMAGCTGGTGGAATC
26	TTCACGAGTCCAGCCTCAAGCAGTGAVGTGMWGCTSGTGGAGTC
27	TTCACGAGTCCAGCCTCAAGCAGTGARGTGCAGCTGKTGGAGWC
28	TTCACGAGTCCAGCCTCAAGCAGTGAGGTGAAGCTGATGGAATC
29	TTCACGAGTCCAGCCTCAAGCAGTGAGGTGCAGCTTGTTGAGTC

(continued)

Table 1
(continued)

30	TTCACGAGTCCAGCCTCAAGCAGTGAGGTGAAGCTTCTCRAGTC
31	TTCACGAGTCCAGCCTCAAGCAGTGAAGTGAARMTTGAGGAGTC
32	TTCACGAGTCCAGCCTCAAGCAGTCAGGTTACTCWGAAAGWGTCTG
33	TTCACGAGTCCAGCCTCAAGCAGTCAGGTCCAACTGCAGCAGCC
34	TTCACGAGTCCAGCCTCAAGCAGTGATGTGAACCTGGAAGTGTC
35	TTCACGAGTCCAGCCTCAAGCAGTCAGATCCAGTTSGTRCAGTC
36	TTCACGAGTCCAGCCTCAAGCAGTGAGGTRCAGCTKGTAGAGAC
Reverse primers for rearranged heavy chains	
37	CTATTCTAGCTAATCTAGGCGCGCCGAGGAGACGGTGACCGTGGTCC
38	CTATTCTAGCTAATCTAGGCGCGCCGAGGAGACTGTGAGAGTGGTGC
39	CTATTCTAGCTAATCTAGGCGCGCCGCAGAGACAGTGACCAGAGTCC
40	CTATTCTAGCTAATCTAGGCGCGCCGAGGAGACGGTGACTGAGGTTC
Primers for linker fragment	
41	GGGCTGATGCTGCACCAACTGTA
42	ACTGCTTGAGGCTGGACTCGTGAACAATAGCAGC
Primers for fusion PCR	
43	TCAATAGTTGAACATAGCGGCCGC
44	CTATTCTAGCTAATCTAGGCGCGCCG
Primers for colony PCR	
45	GACTGATCTAGATCATTTACCAGGAGAGTGGGAGAG
46	CAATAAAGTGAGTCTTTGCACTTGATAGTTATTAATAGTAATCAATTACG
47	GGATGCTGCAACCTATTACTGTCAGCACATTAGGGAGCTTACACG

Degenerate base code: M = A + C; R = A + G; W = A + T; S = C + G; Y = C + T; K = G + T; V = A + C + G; B = C + G + T. A mixture of degenerate primers for light and heavy chains containing 5 μM of each oligo is prepared

5. Agarose gel purification of PCR products: QIAquick gel elution kit (Qiagen).
6. Restriction of fusion PCR product and plasmid (*see* **Note 10**): NotI (NEB), AscI (NEB), restriction buffer 4 (NEB), PCR purification kit (Qiagen).
7. Ligation: T4 Ligase (1 U/μl, Roche), 10× ligase buffer (Roche).
8. Transformation: XL1 blue supercompetent cells (Agilent Technologies), SOC medium (Invitrogen), Agar plates with ampicillin.
9. Colony PCR: Red Taq mix (Sigma), 5 μM oligonucleotides (Table 1).

10. Mini plasmid preparation: LB medium, Qiaprep spin mini prep kit (Qiagen).
11. Transfection: HEK293F cells (Invitrogen), Freestyle 293 medium (Gibco), linear polyethylenimine (25 kDa, Polysciences), Metafectene (Biontex), 50 ml conical cell culture flasks (Corning), 24-well cell culture plates (Costar), gas-permeable adhesive seals (135 × 80 mm, KBioscience).
12. ELISA: Streptavidin-coated 96-well plates (Nunc), PBS 0.1 % Tween20, anti-mouse alkaline phosphatase conjugate, phosphatase substrate (Sigma), ELISA buffer (10 % diethanolamine 0.5 mM $MgCl_2$, pH 9.2).

3 Methods

To express an antibody recombinantly, the unique mRNA sequence of the specific antibody needs to be amplified. RNA is therefore isolated from antibody expressing hybridoma cells and reverse-transcribed into cDNA.

3.1 RNA Isolation from Hybridomas

1. To pellet hybridoma cells (carefully aspirate 0.5–1 × 10^6 cells in 1 ml culture from a 24-well plate) centrifuge at 1,677 × *g* (5,000 rpm in a tabletop centrifuge with a 6 cm rotor) for 5 min.
2. Add 0.2 ml Trizol (*see* **Notes 1** and **2**) to the cell pellet and homogenize (*see* **Note 3**).
3. Add 40 μl chloroform and vortex for 15 s.
4. Incubate at room temperature for 5 min.
5. Centrifuge at 11,337 × *g* (13,000 rpm in a tabletop centrifuge with a 6 cm rotor) for 15 min.
6. Transfer the upper aqueous phase into a new tube (*see* **Note 4**).
7. Add 500 μl isopropanol, invert the tubes to mix, and precipitate overnight at −80 °C.
8. Centrifuge at 11,337 × *g* (13,000 rpm in a tabletop centrifuge with a 6 cm rotor) for 30 min at 4 °C.
9. Wash the RNA pellet with 500 μl 70 % ethanol.
10. Centrifuge at 11,337 × *g* (13,000 rpm in a tabletop centrifuge with a 6 cm rotor) for 10 min.
11. Air-dry the pellet for 5 min (*see* **Note 5**).
12. Dissolve the pellet in 50 μl RNAse-free water.

Store RNA at −20 °C until proceeding with the cDNA synthesis.

3.2 cDNA Synthesis

1. Mix 5 μl (1–5 μg) total RNA, 1 μl 10 mM dNTPs, 1 μl oligodT, and 3 μl H_2O.
2. Incubate the mixture at 65 °C for 5 min (*see* **Note 6**).
3. Transfer the mixture to ice and leave for 1 min.
4. Add 4 μl 5× RT buffer, 2 μl DTT, 1 μl RNAse Inhibitor, 4 μl 25 mM $MgCl_2$, and 1 μl Superscript III. Mix by pipetting up and down.
5. Incubate the mixture at 50 °C for 50 min (*see* **Note 6**).
6. Inactivate the reverse transcriptase by incubation at 85 °C for 5 min (*see* **Note 6**).
7. Add 1 μl RNAse H and incubate at 37 °C for 20 min (*see* **Note 6**).
8. Store cDNA at −20 °C until use.

3.3 Cloning of Variable Heavy and Variable Light into a Mammalian Expression Plasmid

After the isolation of RNA from antibody expressing hybridoma cells and reverse transcription to cDNA, the variable regions from both heavy and light chains are amplified by PCR. Because we do not know the exact sequence of these variable regions, a set of degenerate primers is required. Each of these primers hybridizes to one of the known variable regions and by using a mixture of these primers the amplification of the vast majority of possible variable regions is achieved. After amplification of the variable regions, we combine both PCR products with a third "linker" fragment by fusion PCR to introduce the constant region and 3′ UTR of the light chain and a promoter for the heavy chain. The sequence encoding the heavy constant region is present in the expression plasmid backbone. Once made, this single plasmid permits expression of both heavy and light chains in mammalian cells (Fig. 1).

The myeloma cells we use (SP2/0 cells) for the generation of hybridomas do not express any heavy chain, but do transcribe a non-functional "aberrant" light chain. This SP2/0 light chain is amplified in our PCR protocol along with the functional light chain that encodes the specific antibody. To identify the unwanted plasmids that contain the aberrant light chain, we perform a diagnostic colony PCR after transformation of the construct into bacterial cells. The plasmids containing the antibody specific light chains are selected and the antibodies are produced by mini-scale transfections for functional testing.

1. To amplify variable heavy and variable light by PCR, prepare two mixes: 5 μl 10× Pfu buffer, 5 μl dNTPs, 5 μl cDNA (*see* **Note 7**), 6 μl light (oligonucleotides 1–18, Table 1) or heavy chain (oligonucleotides 19–40, Table 1) primer mix (5 μM), 2.5 μl DMSO, 2.5 μl Pfu Ultra DNA polymerase, 24 μl H_2O (*see* **Note 7**). Put mixes in PCR machine and use program "heavy/light" (Table 2).

Table 2
Details of the PCR programs for the different steps of the cloning protocol

Step	Temperature (°C)	Time	Repeat
Program: heavy/light			
1	95	2 min	1×
2	95	45 s	9×
3	62	30 s	
4	72	1 min	
5	95	45 s	24×, after every 4× drop temp in step 6 by 1 °C
6	61–56	30 s	
7	72	1 min	
8	72	10 min	1×
Program: linker			
1	95	2 min	1×
2	95	50 s	29×
3	60	50 s	
4	72	2 min	
5	72	10 min	1×
Program: fusion			
1	95	2 min	1×
2	95	45 s	10×
3	63	30 s	
4	72	5 min	
5	95	45 s	6×, drop temp of step 6 by 1 °C in each cycle
6	62–56	30 s	
7	72	5 min	
8	95	45 s	
9	56	30 s	25×
10	72	5 min	
11	72	10 min	1×
Program: colony			
1	95	2 min	1×
2	95	1 min	29×
3	60	1 min	
4	72	3 min	
5	72	10 min	1×

2. Amplify linker fragment: mix 5 μl 10× Pfu buffer, 5 μl dNTPs (2 mM), 1 μl (50–100 ng/μl) DNA template (*see* **Note 8**), 3 μl 5′ primer 41 (Table 1), 3 μl 3′ primer 42 (Table 1), 1 μl Pfu Ultra DNA polymerase, 32 μl H_2O and use program "linker" (Table 2).

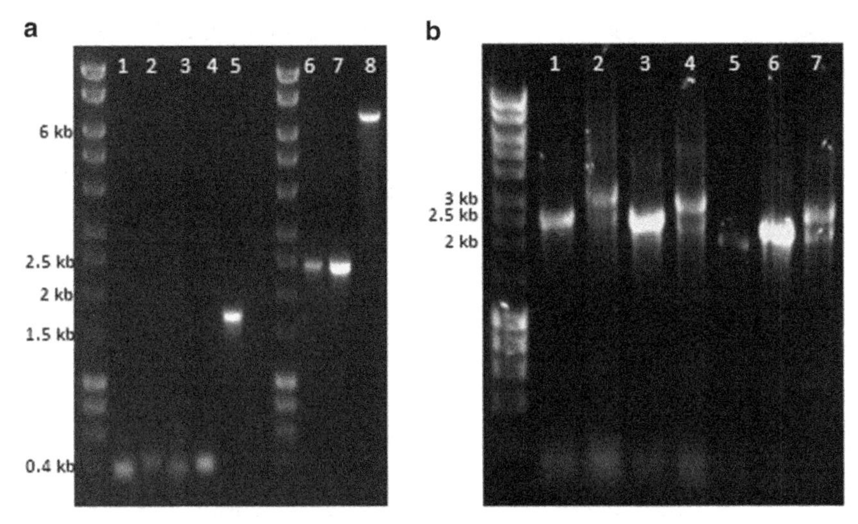

Fig. 2 Examples of gels exemplifying the different stages of the cloning process. (**a**) Heavy (*1*, *3*) and light (*2*, *4*) chain RT-PCR products, linker fragment (*5*), fusion PCR products (*6*, *7*), and linearized expression plasmid (*8*). (**b**) Examples for colony PCR products: recombinant antibodies with specific light chain (*1*, *3*) or myeloma aberrant light chain (*2*, *4*), negative control without insert (*5*), positive control (*6*), myeloma control (*7*). Primer 45 (*see* Table 1) hybridizes in the plasmid (3′ end of the heavy constant region) and primer 46 within the linker fragment (light constant 3′ UTR + CMV promoter for heavy chain) resulting in a PCR product of 2.4 kb. In the presence of the myeloma aberrant chain primer 47 hybridizes as well resulting in an additional 3 kb fragment. In clones without any insert, a 2.1 kb product is produced due to hybridizing of primer 46 to the CMV promoter region present in the plasmid

Resolve PCR products on an agarose gel (Fig. 2), cut out bands sized 400 bp for variable heavy/variable light chains and 1.6 kb for the linker fragment. Purify DNA from gel slices (*see* **Note 9**).

3. Join the three DNA fragments by fusion PCR: mix 5 μl 10× KOD buffer, 5 μl dNTPs (2 mM), 3 μl $MgSO_4$ (25 mM), 3 μl 5′ primer 43 (Table 1), 3 μl 3′ primer 44 (Table 1), 400 ng variable heavy, 400 ng variable light, 100 ng linker fragment, 2.5 μl DMSO, 1 μl KOD Hot Start polymerase, add H_2O to 50 μl. Use program "fusion" (Table 2).

4. After electrophoresis, excise and extract 2.4 kb band from the gel. Set up a restriction digest of purified PCR product and mammalian expression plasmid (*see* **Note 10**) with NotI and AscI for 60 min. Purify restriction mix using spin column (for plasmid after gel electrophoresis, *see* **Note 10**). Ligate the insert and plasmid in a 5:1 ratio in a total volume of 10 μl overnight at 4 °C. Transform 1 μl of the ligation mix into 100 μl XL1-blue cells (*see* **Note 11**). Incubate plates at 37 °C overnight.

5. Pick colonies and perform colony PCR with 20 μl RedTaqMix and program "colony PCR" (*see* **Note 12** and Table 2). Use a mix of two forward (46, 47) and one reverse (45) primer (Table 1).
6. Inoculate 5 ml LB medium with positive colonies (*see* **Note 12** and Fig. 2) and incubate shaking at 37 °C overnight. Isolate the plasmids from 1.5 ml of the bacterial culture.

3.4 Transfection and Testing of Recombinant Antibodies

After isolation of plasmids that do not contain the SP2/0 aberrant light chain (Fig. 2) from mini cultures, these are transfected into mammalian cells on a small scale and supernatants are tested by ELISA. Plasmids of functionally positive recombinant antibodies are amplified in bacteria enabling the production of the recombinant antibody on a larger scale. Finally antibodies can be purified from the cell culture supernatant via protein G.

1. Small scale transfection in 24-well plates: Split HEK293 F cells the day before transfection to a density of 2.5×10^5/ml and leave in a flask overnight at 37 °C in a shaking incubator with 5 % CO_2. Next day, distribute 1 ml of the cells into one well of a 24-well plate. Mix 2.5 μg DNA in 100 μl medium and 2.5 μl Metafectene in 100 μl medium. Combine both mixes and incubate at room temperature for 10 min. Transfer the mix to 1 ml of cells, cover the 24-well plate with a gas-permeable seal and incubate at 37 °C in a shaking incubator with 5 % CO_2.
2. Harvest supernatant after 3 days and test for secretion of a specific antibody by ELISA.
3. For positives, isolate more plasmids from bacteria and perform larger scale transfections in 50 ml (*see* **Note 13**).
4. Purify antibody via a protein G column (*see* **Note 14**) to use in downstream applications.

4 Notes

1. Trizol contains phenol (toxic and corrosive) and guanidine isothiocyanate (an irritant). Carry out all work with Trizol under a fume hood. Dispose of tips and tubes appropriately.
2. The volume of Trizol can be increased for the preparation of RNA from a larger culture sample. In that case, take care to increase the amount of chloroform and isopropanol as well. The volume of 70 % ethanol and RNAse free water should be adjusted accordingly.
3. Small amounts of cells can be homogenized by vortexing. For larger samples use a pestle.
4. Be careful not to transfer anything from the middle and lower phase.

5. Do not let the RNA pellet dry for too long as it gets difficult to dissolve in water.
6. Incubation steps can be done in a water bath, a heat block, or a PCR cycler.
7. The amount of cDNA can be adjusted according to the sample size used for the RNA preparation. Always add water to a total volume of 50 μl.
8. Amplify the linker fragment from one of our recombinant antibodies (e.g., SI3-Flrt3, Addgene ID: 28217; http://www.addgene.com). The same construct can be used to obtaining the backbone plasmid and for use as a positive transfection control (*see* **Note 10**).
9. Gel purification of the right-sized band is critical at this point.
10. For the expression plasmid containing the constant part of the heavy chain use one of our recombinant antibodies from (e.g., SI3-Flrt3, Addgene ID: 28217; http://www.addgene.com). After restriction with NotI/AscI and separation from the insert via agarose gel electrophoresis, the plasmid can be used for ligation with the fusion PCR product.
11. We use ultracompetent XL1-blue cells which are quite sensitive to the ligation buffer; therefore, we use just 1 μl of the ligation mix. If the number of colonies is a problem we suggest to use the entire ligation mix after purification, or to try different competent cells. If the cells grow very slowly, leave in the incubator for a longer time.
12. Positive colonies are detected by colony PCR: Make a numbered grid on an agar plate, pick a colony from the transformation plate with a pipette tip and transfer to the labelled agar "master" plate by carefully touching the agar surface. Put the tip in PCR mix in a tube that corresponds to the number on the master plate and leave for 5 min. After positive clones have been identified by PCR, liquid mini cultures can be inoculated with the positive colony and cultured overnight. When isolating the plasmids from the bacterial culture, keep some of the liquid culture (for longer storage, transfer bacteria to an agar plate) at 4 °C which can be used to inoculate a larger culture in the eventuality that the plasmid encodes a functional recombinant antibody.
13. Bacterial midi or maxi size cultures are inoculated with 100 μl of the mini culture and incubated overnight at 37 °C. After isolation of the plasmid, a 50 ml culture of mammalian cells is transfected. The supernatant is harvested after 5 days.
14. We pack columns with a protein G/agarose slurry (Invitrogen).

References

1. Lindenmann J (1984) Origin of the terms 'antibody' and 'antigen'. Scand J Immunol 19(4):281–285
2. Feldmann M, Maini RN (2001) Anti-TNF alpha therapy of rheumatoid arthritis: what have we learned? Annu Rev Immunol 19: 163–196
3. Plosker GL, Figgitt DP (2003) Rituximab: a review of its use in non-Hodgkin's lymphoma and chronic lymphocytic leukaemia. Drugs 63(8):803–843
4. Medzhitov R, Janeway CA Jr (2002) Decoding the patterns of self and nonself by the innate immune system. Science 296(5566): 298–300
5. Salinas GF, Braza F et al (2012) The role of B lymphocytes in the progression from autoimmunity to autoimmune disease. Clin Immunol 146(1):34–45
6. Market E, Papavasiliou FN (2003) V(D)J recombination and the evolution of the adaptive immune system. PLoS Biol 1(1):E16
7. Pelanda R, Torres RM (2006) Receptor editing for better or for worse. Curr Opin Immunol 18(2):184–190
8. Maul RW, Gearhart PJ (2010) Controlling somatic hypermutation in immunoglobulin variable and switch regions. Immunol Res 47(1–3):113–122
9. Tomita M, Tsumoto K (2011) Hybridoma technologies for antibody production. Immunotherapy 3(3):371–380
10. Kipriyanov SM, Little M (1999) Generation of recombinant antibodies. Mol Biotechnol 12(2): 173–201
11. Gritzmacher CA (1988) Producing novel antibodies by expression of recombinant immunoglobulin genes. Year Immunol 3:260–274
12. Humphreys DP, Glover DJ (2001) Therapeutic antibody production technologies: molecules, applications, expression and purification. Curr Opin Drug Discov Devel 4(2):172–185
13. Crosnier C, Staudt N et al (2010) A rapid and scalable method for selecting recombinant mouse monoclonal antibodies. BMC Biol 8:76

Chapter 15

Monoclonal Antibody Purification by Ceramic Hydroxyapatite Chromatography

Larry J. Cummings, Russell G. Frost, and Mark A. Snyder

Abstract

Hydroxyapatite chromatography is shown to be an excellent method for chromatographically purifying monoclonal antibodies (Mab). Mab contained in eluates from Protein A columns was partially purified on ceramic hydroxyapatite (CHT™) Type I, 40 μm ceramic hydroxyapatite using two scouting methods which provide milligram amounts of Mab typical at laboratory scale. The result from one of the scouting methods was optimized to obtain a high concentration of purified Mab with acceptable clearance of cell culture impurities. Several techniques (linear phosphate screening, linear alkaline salt screening, and two alkaline salt step gradients) are described for obtaining high concentrations of purified Mab in a lab-scale CHT chromatography column.

Key words Ceramic hydroxyapatite, Chromatography, Phosphate elution, Co-buffer, Surface neutralization, In situ regeneration, Alkaline salt elution, Adsorption, Desorption, Clearance

1 Introduction

Protein adsorption and desorption to and from CHT surfaces has recently been reviewed [1]. Acidic proteins bind through C (calcium)-sites, while basic proteins bind through P (phosphate)-sites [2]. Yet Mab in general binds to P-sites, although less easily so for acidic Mab than basic Mab. The adsorption–desorption mechanisms of CHT-protein interactions can be summarized as follows: amino groups adsorb to P-sites but are repelled by C-sites. The situation is opposite although more complex for carboxyl groups. While amine binding to P-sites and the initial attraction of carboxyl groups to C-sites are electrostatic, binding of carboxyl groups to C-sites involves formation of chelation coordination bonds which are stronger than anion exchange interactions. Phosphoryl groups on proteins and other solutes interact more strongly with C-sites than do carboxyl groups. Mobile phase conditions used to purify Mab with CHT chromatography fall into two categories [3].

Vincent Ossipow and Nicolas Fischer (eds.), *Monoclonal Antibodies: Methods and Protocols*, Methods in Molecular Biology, vol. 1131, DOI 10.1007/978-1-62703-992-5_15,

Buffered salt solutions are used to elute basic Mab, while phosphate solutions are used to elute acidic Mab. Linear gradient elution of Mab with alkaline salt or phosphate mobile phases is used as a screening method. The results from the screening method are then optimized to obtain highly purified Mab [4].

Mobile phase conditions that resolve Mab from cell culture impurities (e.g., DNA, host cell protein, endotoxin, adventitious virus, and Mab polymers) can affect the physical properties of the CHT when packed columns are used multiple times. If the intent of the lab scale purification is scale up to larger diameter columns, with the intent of numerous process cycles, the following discussion is important. The adsorption and desorption of hydroxonium ion (H^+) has recently been reviewed [4–6] and is the dominant factor in the irreversible damage to CHT. CHT adsorbs H^+ from buffer solutions and protein load solutions especially below pH 7. Protein loading is often conducted at pH 6.5 with very low buffer content, 2–5 mM phosphate, containing or lacking minor amounts of alkaline salts such as NaCl or KCl. These conditions enhance the adsorption of Mab to the surface of the CHT. Equilibration of the CHT column precedes protein loading usually at the same pH and buffer content as the load buffer. The tandem sequence of equilibration and load buffers saturates the surface of the CHT with H^+. In addition, the acidic properties of these buffers partially dissolve the CHT. During elution, H^+ desorbs from the surface of the CHT and is observed as an acidic pH transition in the column effluent. Concurrently, the effluent is also enriched with calcium ion displaced from the hydroxyapatite surface. The phosphate moiety remains in the hydroxyapatite matrix during the elution step but quickly displaces when the alkaline salt concentration declines. The physical changes of the matrix associated with these pH transitions generally do not affect the column performance or product purity for laboratory scale columns used up to 65 cycles, but process scale columns may be limited to much fewer cycles [7]. The latter has been addressed by a several alternative methods which can improve column lifetime at process scale: surface neutralization solutions preceding Mab elution, highly concentrated co-buffered elution, calcium phosphate fortified elution, and in situ surface regeneration, post elution (*see* **Note 1**).

2 Materials

Prepare all solutions using ultrapure water (prepared by purifying deionized water to attain a resistivity of 18 MΩ cm at 25 °C) and USP (United States Pharmacopoeia) grade reagents. Prepare and filter all solutions with 0.2 μm polyethersulfone (PES) or similar filters. Store filtered solutions at room temperature (unless indicated otherwise). Autoclave designated solutions at 121 °C for 20 min. Comply with all waste disposal regulations when disposing of waste materials.

2.1 Phosphate Linear Gradient Mobile Phases

1. Sanitization mobile phase (A): 1.0 N NaOH. Add 900 mL of water to a 2-L glass beaker. Add 40.0 g (*see* **Note 2**) NaOH pellets to the beaker and mix to dissolve. Allow to cool to room temperature and then quantitatively transfer the solution to a 1-L glass graduated cylinder. Add quantity sufficient water (Q.S.) to bring the volume to 1 L. Transfer the solution to a 1-L heavy wall bottle. Add a Teflon-coated stirring bar and vacuum-aspirate for 20 min (*see* **Note 3**).
2. Regeneration mobile phase (B): 0.40 M sodium phosphate, pH 7.0. Weigh 15.18 g of sodium phosphate monobasic monohydrate (*see* **Note 4**) and 77.74 g of sodium phosphate dibasic heptahydrate (*see* **Notes 4** and **5**) into a 2-L glass beaker. Add approximately 500 mL of hot water (*see* **Note 6**). Mix to dissolve the phosphate salts. Dilute with 400 mL of water, mix, and then allow the solution to cool to room temperature. Quantitatively transfer the solution to a 1-L glass graduated cylinder and then Q.S. to 1 L with water. Filter the solution and then aspirate to remove dissolved gas.
3. Equilibration and Wash mobile phase (C): 10 mM sodium phosphate, pH 6.8. Dilute 25.0 mL of Regeneration mobile phase to 900 mL with water. Adjust the pH to 6.8 with 1.0 N phosphoric acid and then Q.S. to 1 L with water. Filter the solution and then aspirate to remove dissolved gas.

2.2 Sodium Chloride Linear Gradient Mobile Phases

1. Prepare Sanitization and Regeneration mobile phases described in Subheading 2.1.
2. High salt mobile phase (D): 5 mM sodium phosphate, 1 M NaCl, pH 6.5. Dilute 12.5 mL of Regeneration mobile phase to 900 mL with water. Dissolve 58.45 g of sodium chloride. Adjust the pH to 6.5 with 1.0 N NaOH or 1.0 N phosphoric acid and then Q.S. to 1 L with water. Filter the solution as above. Autoclave and then recheck the pH when the solution has cooled to room temperature.
3. Gradient equilibration mobile phase (E): 5 mM sodium phosphate, pH 6.5. Dilute 12.5 mL of Regeneration mobile phase to 900 mL with water. Adjust the pH to 6.5 with 1.0 N HCl and then Q.S. to 1 L with water. Filter and then aspirate to remove dissolved gas.

2.3 Sodium Chloride Step Gradient Mobile Phases

1. Prepare Sanitization, Regeneration High salt, and Gradient equilibration mobile phases described in Subheading 2.2.

2.4 Buffered Sodium Chloride Step Gradient Mobile Phases

1. Prepare Sanitization, Regeneration, and Gradient equilibration mobile phases described in Subheading 2.3.
2. Buffered high salt mobile phase (F): 5 mM sodium phosphate, 75 mM MES, 1 M NaCl, pH 6.5. Dilute 12.5 mL of Regeneration mobile phase to 900 mL with water. Dissolve 58.45 g of sodium chloride and 15.97 g of MES monohydrate. Adjust the pH to 6.5 with 1.0 N NaOH and Q.S. to 1 L with water. Filter the solution. Autoclave and then recheck the pH when the solution has cooled to room temperature.

3 Methods

Conduct all purification protocols at room temperature. Store applied and collected Mab samples in an ice bucket. Set up liquid chromatography systems according to the manufacturer's instructions. Use a Bio-Scale MT2 (Bio-Rad Laboratories), Tricorn™ 5/20 column with adjustable end piece (GE Healthcare Life Sciences), or an Omnifit® 6.6/100 column with adjustable end piece (Diba Industries, Inc.).

3.1 Column Packing, Dry Method

1. Clean and then thoroughly dry the column parts. Assemble the dry parts except for the adjustable end piece.
2. Weigh 1.30 g of CHT Type I, 40 μm ceramic hydroxyapatite onto a 3″ × 3″ sheet of single-creased glassine weighing paper (*see* **Note 7**).
3. Transfer the CHT powder into the open end of the column. Tap the base of the column against a rubber cork until a constant bed height is achieved (*see* **Note 8**).
4. Insert the adjustable end piece such that it just touches the top of the tap-settled bed.

3.2 Liquid Chromatography System Setup and Column Conditioning (See Note 9)

1. Pump A mobile phase selection valve: Connect mobile phase A, C, and E to ports A1, A2, and A3.
2. Pump B mobile phase selection valve: Connect mobile phase B, D, and F to ports B1, B2, and B3.
3. Connect a 500 μL fixed-volume sample loop to the sample injector sample ports and then connect the column inlet line of the sample injector to the column outlet line which in turn is connected to the system's monitor line (UV, conductivity, and pH).
4. Prime the mobile phase lines, buffer selection valves, sample injector, and monitors. Calibrate the system if required (*see* **Note 10**).
5. Program the system pressure to 100 psi (*see* **Note 11**). Attach the dry-packed column outlet and inlet lines using the appropriate

Table 1
Phosphate linear gradient

Step	Step time (min)	Description	Mobile phase	Flow rate (mL/min)	Duration (min)
1	0.0	Isocratic	A	1.57	3.9
2	3.9	Isocratic	B	1.57	5.2
3	9.1	Isocratic	C	1.57	19.4
4	28.5	Load/inject	Sample	3.14	0.1
5	28.6	Zero UV	C	1.57	
6	28.6	Isocratic	C	1.57	1.3
7	29.9	Linear gradient	0 % C to 75 % B	1.57	19.4
8	49.3	Isocratic	B	1.57	3.9
9	53.2	Isocratic	A	1.57	3.9
	57.1	End of protocol			

threaded connectors. Position the column vertically on the system.

6. Set the flow rate to 3 mL/min. Condition the column for 15 min with mobile phase A. Readjust the adjustable end piece to the top of the packed bed if necessary.

3.3 Conducting Purification Methods with 3.4 mg of Mab Obtained from a Protein A Purification Step Using Acetate Elution at pH 2.5

The Mab eluate is neutralized with Tris base to pH 7 to 15 to 30 mM Tris acetate prior to applying it to the hydroxyapatite column.

1. Conduct the linear gradient method described in Table 1: Phosphate linear gradient. Archive the results.
2. Conduct the linear gradient method described in Table 2: NaCl linear gradient. Archive the results. Combine the results for the methods in Tables 1 and 2 to show the shape and position of the eluted Mab from the two linear gradient methods, Fig. 1.
3. Conduct the step gradient method described in Table 3: NaCl step gradient in which the concentration of the NaCl is 0.7 M. Archive the results.
4. Conduct the step gradient method described in Table 4: Buffered NaCl step gradient in which the concentration of NaCl is 0.7 M and MES is 52.5 mM. Combine the results for the methods in Tables 3 and 4 to show the shape and position of the eluted Mab, Fig. 2.

Table 2
NaCl linear gradient

Step	Step time (min)	Description	Mobile phase	Flow rate (mL/min)	Duration (min)
1	0.0	Isocratic	A	1.57	3.9
2	3.9	Isocratic	B	1.57	5.2
3	9.1	Isocratic	E	1.57	19.4
4	28.5	Load/inject	Sample	3.14	0.1
5	28.6	Zero UV	E	1.57	
6	28.6	Isocratic	E	1.57	1.3
7	29.9	Linear gradient	0 % E to 75 % B	1.57	8.0
8	49.3	Isocratic	B	1.57	3.9
9	53.2	Isocratic	A	1.57	3.9
	57.1	End of protocol			

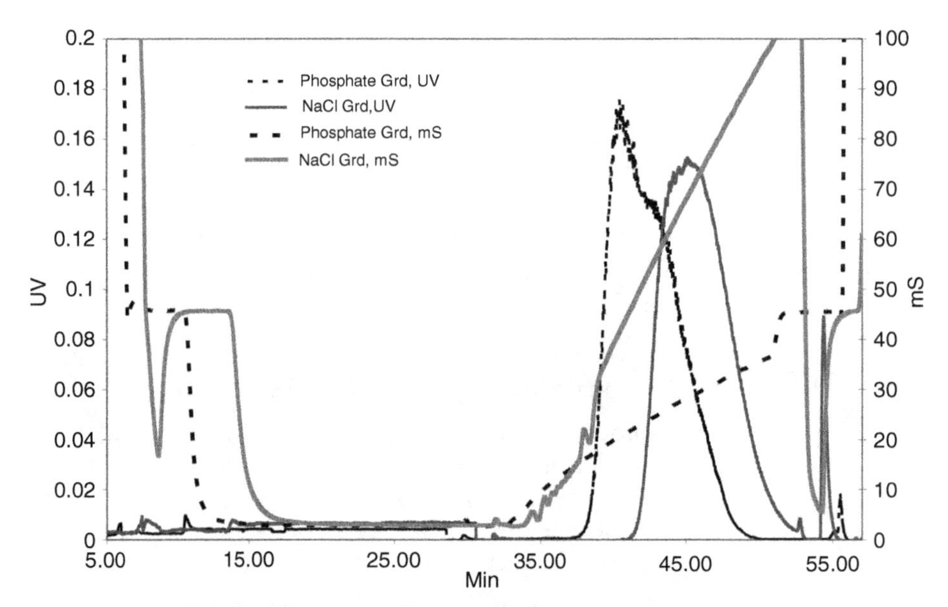

Fig. 1 Chromatogram of 3.4 mg of applied Mab using linear phosphate gradient or linear sodium chloride gradient elution. In both cases the Mab eluted over a broad range, 12 and 11 min. However, the chromatograms also indicate clearance from more strongly adsorbed constituents as indicated by UV peaks between 54 min and 57 min, respectively. Tests indicate that the Mab eluted by these linear gradient methods have reduced amounts of fermentation impurities and partial reduction in Mab polymer. The peak maximum for the phosphate gradient is 47 % mobile phase B. The peak maximum for the sodium chloride gradient is 70 % mobile phase D

Table 3
NaCl step gradient, 3.4 mg

Step	Step time (min)	Description	Mobile phase	Flow rate (mL/min)	Duration (min)
1	0.0	Isocratic	A	1.57	3.9
2	3.9	Isocratic	D	1.57	2.0
3	5.9	Isocratic	B	1.57	5.2
4	11.1	Isocratic	E	1.57	19.4
5	30.5	Load/inject	Sample	3.14	0.1
6	30.6	Zero UV	E	1.57	
7	30.6	Isocratic	E	1.57	1.3
8	31.9	Isocratic	30 % E, 70 % D	1.57	8.0
9	39.9	Isocratic	B	1.57	3.9
10	43.8	Isocratic	A	1.57	3.9
	47.7	End of protocol			

Table 4
Buffered NaCl step gradient, 3.4 mg

Step	Step time (min)	Description	Mobile phase	Flow rate (mL/min)	Duration (min)
1	0.0	Isocratic	A	1.57	3.9
2	3.9	Isocratic	F	1.57	2.0
3	5.9	Isocratic	B	1.57	5.2
4	11.1	Isocratic	E	1.57	19.4
5	30.5	Load/inject	Sample	3.14	0.1
6	30.6	Zero UV	E	1.57	
7	30.6	Isocratic	E	1.57	1.3
8	31.9	Isocratic	30 % E, 70 % F	1.57	8.0
9	39.9	Isocratic	B	1.57	3.9
10	43.8	Isocratic	A	1.57	3.9
	47.7	End of protocol			

3.4 Conducting Optimized Purification Method with 34 mg Applied Mab

The buffered NaCl step gradient is chosen in this case based on the results obtained above. The concentration of the NaCl in this example is necessary due to the basic *p*I of the Mab. Lower or higher concentrations of NaCl could be used depending on the *p*I of the Mab. The buffer type, MES is optimal for CHT lifetime as

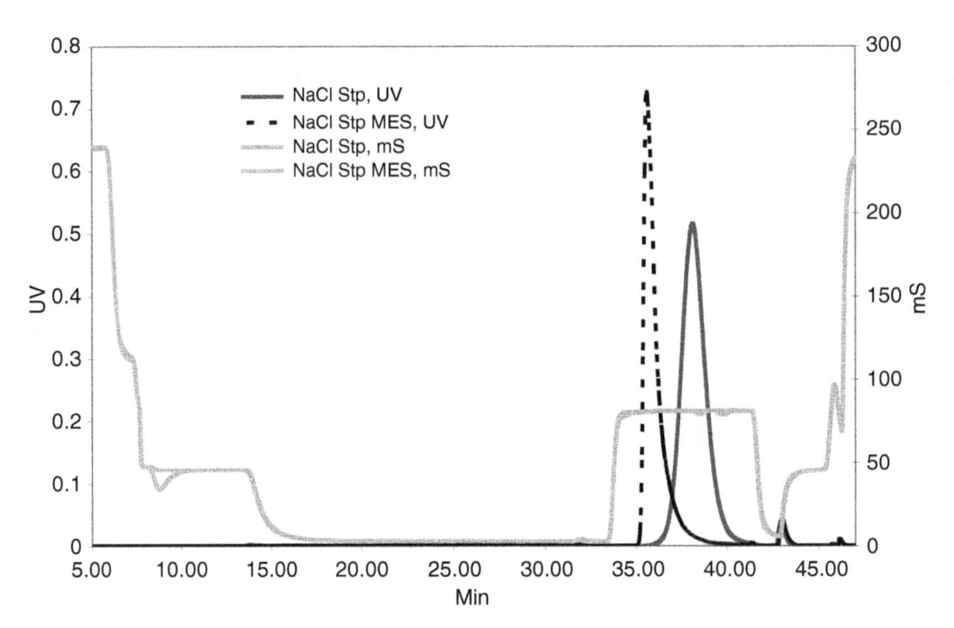

Fig. 2 Chromatogram of 3.4 mg of applied Mab using NaCl and MES buffered NaCl step gradients. In both cases the applied Mab eluted over a narrow range, 5 min and 3.5 min, respectively. The chromatograms indicate excellent clearance from more strongly adsorbed constituents as indicated by UV peaks between 42.8 and 47 min. Endotoxin, DNA and HCP were below the level of detection. Mab polymer was less than 0.1 %. Mobile phase F elutes the applied Mab earlier than mobile phase D due to the buffering capacity of MES for H^+ as shown in Fig. 3

it adsorbs hydrogen ion displaced from the CHT surface by the NaCl. In addition, MES is one of the few Good's buffers available in USP grade.

1. Conduct the step gradient method described in Table 5: Buffered NaCl step gradient in which the concentration of NaCl is 0.7 M and MES is 52.5 mM. Archive the results. Figure 3 compares the retention times of a basic *p*I Mab relative to 70 % buffered high salt mobile phase F and 70 % high salt mobile phase D. The retention times differ due to the H^+ concentration in the mobile phase during elution. Under acidic conditions proteins adsorb more strongly to the phosphate moiety of the CHT. Protein adsorption weakens as the H^+ concentration declines. The MES buffer component in mobile phase F neutralizes H^+ ion released from surface of hydroxyapatite by NaCl and hence decreasing Mab adsorption, allowing the Mab to be displaced by the NaCl earlier than with 70 % mobile phase D. Figure 4 demonstrates that purification over the loading range of 3.4–34 mg/mL is unaltered by mass load provided it does not exceed protein saturation. Not shown, this basic Mab purification was achieved at 50 mg/mL load on CHT Type I or 80 % of its maximum load.

Table 5
Buffered NaCl step gradient, 34 mg

Step	Step time (min)	Description	Mobile phase	Flow rate (mL/min)	Duration (min)
1	0.0	Isocratic	A	1.57	3.9
2	3.9	Isocratic	F	1.57	2.0
3	5.9	Isocratic	B	1.57	5.2
4	11.1	Isocratic	E	1.57	19.4
5	30.5	Load/inject	Sample	3.14	1.3
6	31.8	Zero UV	E	1.57	
7	31.8	Isocratic	E	1.57	1.3
8	33.1	Isocratic	30 % E, 70 % F	1.57	8.0
9	41.1	Isocratic	B	1.57	3.9
10	45.0	Isocratic	A	1.57	3.9
	48.9	End of protocol			

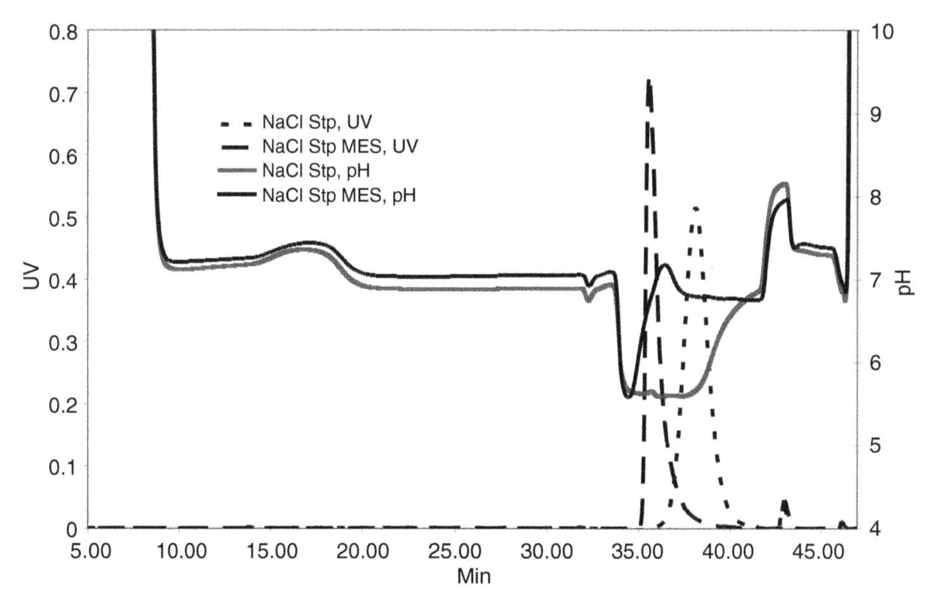

Fig. 3 The same chromatogram as shown in Fig. 2, but shows the pH trace for the 3.4 mg of applied Mab eluted using NaCl and MES buffered NaCl step gradients. The pH trace shows the difference between strongly buffered NaCl (mobile phase F) and weakly buffered NaCl (mobile phase D) and its effect on the elution time of the Mab

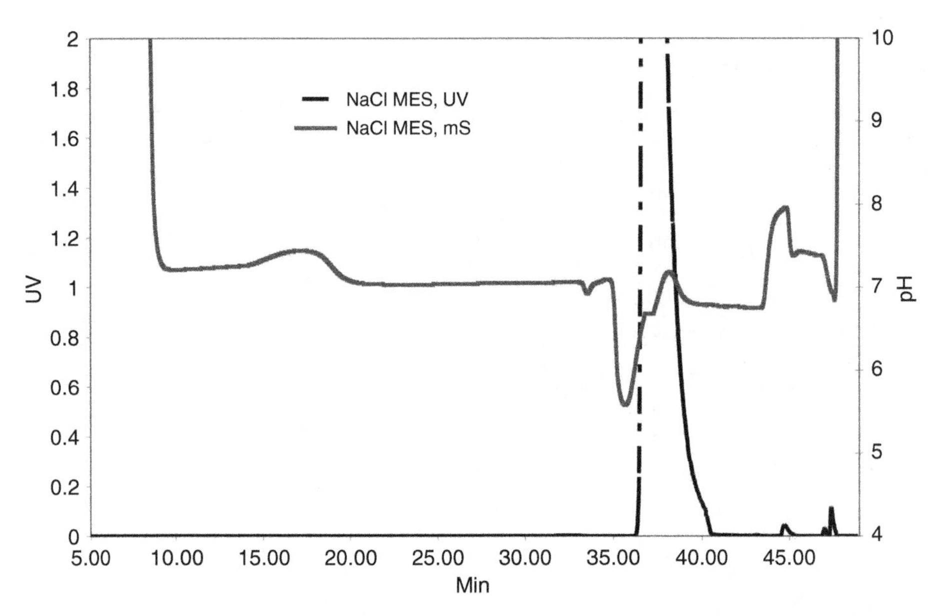

Fig. 4 Chromatogram of 34 mg of applied Mab using buffered NaCl step gradient. The capacity of CHT Type I for Mab approaches 50 mg/mL. This example illustrates that 2.0 mL of packed CHT Type I adsorbed 17 mg/mL and eluted in approximately 4.0 mL or about the same volume as the applied volume. This indicates that CHT Type I not only further purifies Mab but also has the ability to increase its concentration

4 Notes

1. For more information regarding the alternative methods for addressing process scale applications mentioned in the Introduction, consult Bio-Rad Laboratories.
2. USP grade NaOH pellets are free of insoluble matter and generally do not require sterile filtration.
3. Solutions of NaOH tend to contain large amounts of dissolved gasses. Degassing the solution will eliminate mobile phase flow rate failure and aberrations in detector responses (conductivity, UV, and pH) on liquid chromatographs.
4. Hydrated phosphate salts are pyrophosphate free. Avoid replacing hydrated phosphate salts with anhydrous phosphate salt as pyrophosphate amounts vary broadly. Pyrophosphate adsorbs to C-sites on the surface of CHT and alters its selectivity for Mab. Avoid the use of dodecahydrates as these spontaneously decompose to heptahydrates.
5. Suppliers of sodium phosphate dibasic heptahydrate list the excess water content of the salt. The 93.24 g weight should be adjusted accordingly.
6. Solutions of sodium phosphate salts especially above pH 6.8 can be prepared much faster when the water is warmed to about 70 °C.

7. Before opening the container of CHT, shake it to loosen the powder. Crease the glassine weighing paper and open it. The crease provides a channel in the paper which is used to transfer the weighed powder into the open end of the column.
8. The packing density of CHT is 0.63 g/mL. Accordingly, 1.3 g of resin should settle to between 53 and 54 mm in the Bio-Scale column, 100 and 110 mm in the Tricorn column, or 60–70 mm in the Omnifit column.
9. Use a liquid chromatography system that has a linear gradient option, mobile phase selection valves, fraction collection interface, and monitoring options that include UV, conductivity, and pH (e.g., Bio-Rad NGC™ Medium-Pressure Chromatography System).
10. Calibrate the flow rate for each pump according to the manufacturer's instructions. Calibrate the conductivity monitor and pH monitors before use. Use pH 7.00 and 4.00 standard buffers to calibrate the pH monitor to assure the best relative pH reading of the NaCl column effluent.
11. CHT 40 µm ceramic hydroxyapatite can be used at high linear velocities and high pressures. Expected pressure at the flow rate selected for this application (245 cm/h for the Bio-Scale column, 275 cm/h for the Omnifit column, and 480 cm/h for the Tricorn column) should be below 60 psi. Consult http://www.sensorsone.co.uk/pressure-units-conversion.html to convert from psi to other pressure units.

References

1. Cummings LJ, Snyder MA, Brisack K (2009) Protein chromatography on hydroxyapatite columns. In: Burgess RR, Deutscher MP (eds) Methods in enzymology, vol 463. Academic, Burlington, pp 387–404
2. Kandori K, Ishikawa T, Miyagawa K (2004) Adsorption of immunogamma globulin onto various synthetic calcium hydroxyapatite particles. J Colloid Interface Sci 273:406–413
3. Nakagawa T, Ishihara T, Yoshida H, Yoneya T, Wakamatsu K, Kadoya T (2010) Relationship between human IgG structure and retention time in hydroxyapatite chromatography with sodium chloride gradient elution. J Sep Sci 33: 37–45
4. Gagnon P (1996) Purification tools for monoclonal antibodies. Validated Biosystems, Tucson, AZ, p 87
5. Kwon KY, Wang E, Chung A, Chang N, Lee SW (2009) Effect of salinity on hydroxyapatite dissolution studied by atomic force microscopy. J Phys Chem C 113:3369–3372
6. Skartsila K, Spanos N (2007) Surface characterization of hydroxyapatite: potentiometric titrations coupled with solubility measurements. J Colloid Interface Sci 308:405–412
7. McCue JT, Cecchini D, Hawkins K, Dolinski E (2007) Use of an alternative scale-down approach to predict and extend hydroxyapatite column lifetimes. J Chromatogr A 1165:78–85

Chapter 16

Rapid Purification of Monoclonal Antibodies Using Magnetic Microspheres

Pauline Malinge and Giovanni Magistrelli

Abstract

Magnetic microspheres represent an interesting alternative to conventional chromatography resins in automated high-throughput protocols replacing centrifugation and filtration by magnetic separation. Some magnetic microspheres have unique features like high magnetite content and non-porous surface which allows them to migrate very fast in magnetic fields while binding target molecules with a low unspecific adsorption. Here, we describe the use of protein A or protein G-coated magnetic microspheres to purify quickly monoclonal antibodies from crude serum-free supernatants without the need of preliminary clarification or purification step. Using this method, multiple samples can be processed in parallel, a high level of purity can be obtained, and the purified IgG maintain their biological activity.

Key words Magnetic microspheres, Affinity purification, Desalting

1 Introduction

Magnetic microspheres are a versatile tool for a fast protein separation [1, 2]. Magnetic microspheres were optimized in terms of size, surface properties, and magnetic behavior (ferrite content). Smaller microspheres display higher surface area for a maximal binding capacity without sedimentation but with a lower speed of diffusion in complex media. On the opposite, large microspheres have a lower binding capacity but diffused rapidly in viscous liquids [3]. Thus, high recovery in complex matrices can be achieved using larger microspheres [4].

In addition, magnetic microspheres can be modified and coated with different components which allow bioseparation of cells [5, 6], setup of immunoassays [7], purification of IgG or recombinant proteins [8]. For example, biotinylated proteins can be captured on streptavidin-coated microspheres and the coated microspheres can be used for phage display selections [9] or IgG capture.

Vincent Ossipow and Nicolas Fischer (eds.), *Monoclonal Antibodies: Methods and Protocols*, Methods in Molecular Biology, vol. 1131, DOI 10.1007/978-1-62703-992-5_16, © Springer Science+Business Media New York 2014

Table 1
Characteristics of magnetic microspheres

Microsphere identification	Designation	Ligand	Size	Mean diameter (μm)	Magnetic pigment	Commercialized	Catalog number
A		Protein G	Small	0.36	37 % (v/v)	No	
B	BE-M07/1 Hydro	Protein G	Medium	0.99	47.5 % (v/v)	Yes	80 380 155
C	BE-M07/1 Lipo	Protein G	Medium	0.95	52 % (v/v)	Yes	80 380 156
D		Protein A	Small	0.36	37 % (v/v)	No	
E	BE-M06/1 Hydro	Protein A	Medium	0.99	47.5 % (v/v)	Yes	80 380 152
F	BE-M06/1 Lipo	Protein A	Medium	0.95	52 % (v/v)	Yes	80 380 153

Lipo = lipophilic and refers to microspheres with hydrophobic functional groups
Hydro = hydrophilic and refers to microspheres with carboxyl functional groups

This chapter describes the materials and methods to perform rapid purification of IgG using protein A or protein G-coated magnetic microspheres. IgG produced by transient transfection of CHO and HEK293 cell lines cultured in serum-free medium can be purified directly from the crude medium and multiple samples can be processed in parallel. Magnetic microspheres that differ in magnetic pigment content and in size are available (Table 1). For the application described here, smaller microspheres with lower magnetic pigment content (microspheres A and D) were found to be the most appropriate. Their slower diffusion in the crude medium was not found to be a limitation in the purification process. The purification efficiency was evaluated by analyzing the purified IgG on gel and testing their binding capacity in ELISA in comparison to IgG purified using classical Protein A sepharose.

2 Materials

All reagents and kits must be used and stored as recommended by the manufacturer.

2.1 Purification

1. Protein A or protein G magnetic microspheres (Merck Millipore, Table 1). Microspheres B, C, E, and F are commercially available, as mentioned in Table 1; microspheres A and D are available upon request to Merck Millipore.
2. Magnetic separation rack (Invitrogen or equivalent).

3. Protein A/G binding buffer (Thermo Scientific).
4. Rotator Stuart (Dynalab).
5. Phosphate Buffered Saline (PBS; Sigma-Aldrich).
6. Sodium Chloride (NaCl; Sigma-Aldrich).
7. Wash buffer: PBS buffer containing 1 M NaCl.
8. IgG elution buffer (Thermo Scientific).
9. Magnetic microspheres (Merck Millipore), stored at +4 °C.
10. Trizma® hydrochloride buffer solution at pH 9.0 (Sigma-Aldrich), stored at room temperature.
11. PD-10, NAP-10, or NAP-5 gel filtration desalting columns (GE Healthcare).
12. Amicon Ultra-4 PLTK Ultracel-PL membrane (Millipore).
13. Eppendorf Protein LoBind tubes (Eppendorf).

2.2 IgG Analysis by Gel Electrophoresis

1. NanoDrop® ND-1000 spectrophotometer (Witec AG) or equivalent.
2. NuPAGE® LDS Sample Buffer (Invitrogen).
3. β-mercaptoethanol (Sigma Aldrich).
4. NuPAGE® 4–12 % Bis-Tris Mini Gel (Invitrogen).
5. SeeBlue® Plus2 Pre-Stained Standard (Invitrogen).
6. NuPAGE® MES SDS Running Buffer (Invitrogen).
7. Electrophoresis Novex mini-cell (Invitrogen).
8. Power supply (Fischer Scientific).
9. SimplyBlue™ SafeStain (Invitrogen).

3 Methods

3.1 Purification

All buffers are stored at +4 °C. Before use, equilibrate buffers at room temperature.

3.1.1 Magnetic Microsphere Preparation

1. Gently mix the magnetic microspheres before use to obtain a homogeneous suspension (*see* **Note 1**). Transfer the magnetic microspheres in appropriate tubes for the magnet and add protein A/G binding buffer (*see* **Note 2**). The volume of microspheres necessary to purify the IgG is determined by the expected amount of product in the supernatant (*see* **Note 3**). A semi-quantitative ELISA can be performed before purification to evaluate IgG concentration in the supernatant and adapt magnetic microsphere quantity (*see* **Note 4**). Typically, we use 100 μL of microsphere suspension for 1.5 mL of culture supernatant and resuspend microspheres in 500 μL of binding buffer.

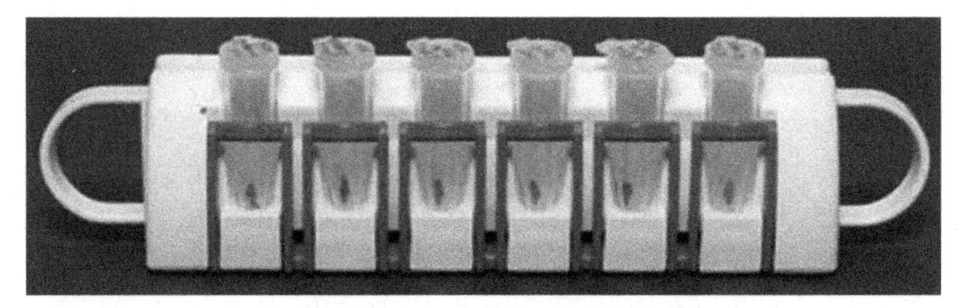

Fig. 1 Magnetic separation rack used for capturing magnetic microspheres

2. Separate the magnetic microspheres from buffer using a magnetic rack, as shown in Fig. 1 (*see* **Note 5**).
3. Remove supernatant by pipetting, remove tubes containing microspheres from the rack, add binding buffer (500 μL in this example), and resuspend manually the microspheres.
4. Repeat this wash step three times to equilibrate magnetic microspheres in the binding buffer and to discard components of the storage buffer.

3.1.2 Supernatant Incubation and Magnetic Microsphere Wash

1. Incubate the supernatant containing antibodies with the conditioned magnetic microspheres under rotation at 20 rpm for 10–15 min at room temperature (*see* **Notes 6** and 7).
2. Separate microspheres from the supernatant using the magnetic rack. Microspheres and bound antibodies are separated from unbound proteins present in the supernatant.
3. Remove the flow-through by aspiration (*see* **Note 8**).
4. Remove tubes from the magnetic rack.
5. Resuspend microspheres in wash buffer (500 μL in our case) (*see* **Note 9**).
6. Separate microspheres from wash buffer with the magnetic rack, aspirate the liquid, remove tubes from the magnetic rack, and resuspend microspheres in the same wash buffer, 500 μL.
7. Repeat this step three times to remove unspecific proteins bound to microspheres and proteins with low affinity for protein A or protein G.

3.1.3 IgG Elution

1. Incubate microspheres with an appropriate volume of IgG elution buffer, under agitation for 2 min at room temperature. For the described example, a volume of 50 μL of IgG elution buffer was used.
2. Separate microspheres on the magnetic rack and recover the elution buffer containing purified antibodies (*see* **Note 10**).
3. Instantly neutralize elution samples by adding 1/10 volume (5 μL) of Trizma® hydrochloride buffer solution at pH 9.0 (*see* **Note 11**).

4. As an alternative to neutralization and for unstable IgG in the neutralized elution buffer, a desalting step can be performed to exchange buffer directly after purification, as described below (*see* **Note 12**).

3.1.4 Desalting Using Gel Filtration Columns

According to technical specifications, the choice of the column is based on the sample volume.

1. Remove top and bottom caps from gel filtration desalting columns and allow excess of liquid drain by gravity flow.
2. Equilibrate the column with PBS and let the liquid flow by gravity.
3. Apply sample on the column.
4. Place a collection tube under the column.
5. Add PBS for elution.
6. Collect the eluate by gravity flow.
7. Aliquot the final product in Protein LoBind tubes to avoid loss of product on the tube inner wall.

3.1.5 Desalting Using Concentration Membranes

According to technical specifications, the choice of the membrane is based on sample volume and sample molecular weight.

1. Equilibrate the Amicon membrane with PBS by centrifugation.
2. Add the sample on the membrane.
3. Centrifuge and add PBS for desalting.
4. Repeat the addition of PBS and the centrifugation step to recover the purified and desalted sample in an appropriate volume of PBS.
5. Aliquot the final product in LoBind tubes to avoid loss of product on the tube inner wall.

3.2 IgG Analysis by Gel Electrophoresis

1. Quantify IgG with a NanoDrop® or equivalent equipment, using an appropriate extinction coefficient (we use 13.7 for IgG) and the appropriate blank solution (formulation buffer, PBS in this example).
2. Aliquot 1–2 μg of protein in 20 μL of PBS (*see* **Note 13**).
3. Add 5 μL of LDS loading buffer containing or not β-mercaptoethanol for analysis under reducing or non-reducing conditions, respectively (*see* **Note 14**). A 1 mL reducing solution contains 960 μL of LDS loading buffer and 40 μL of β-mercaptoethanol.
4. Heat samples at 95 °C for 5 min under reducing conditions or 2 min under non-reducing conditions.
5. Load samples and standard on the gel.
6. Start migration in MES running buffer at 180 V for 35 min (*see* **Note 15**).

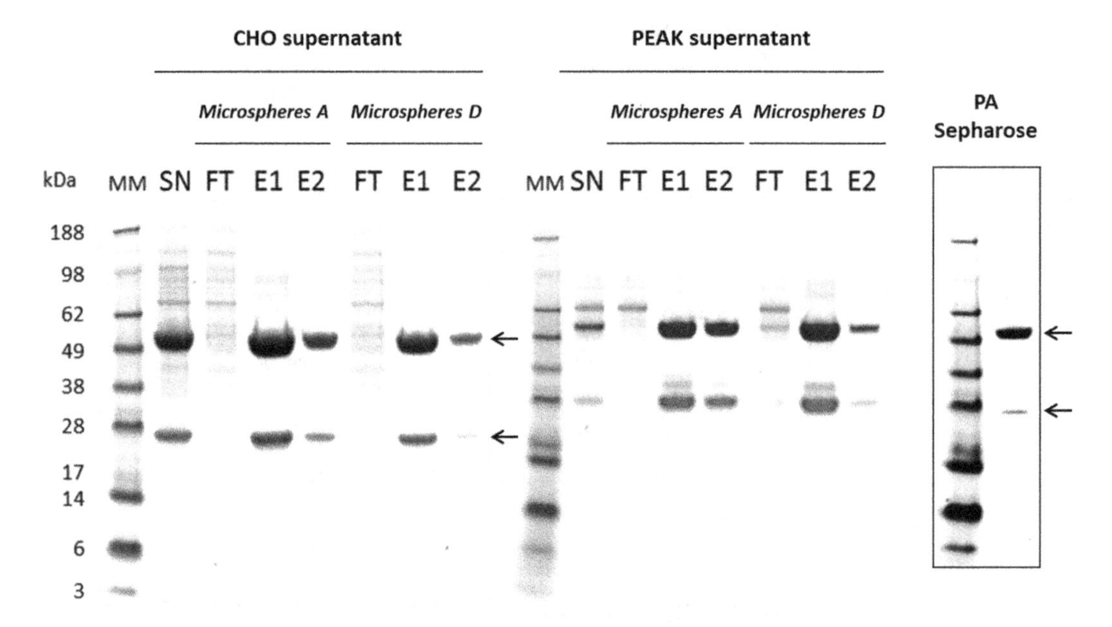

Fig. 2 Purification efficiency evaluated by SDS PAGE electrophoresis in denaturing and reducing conditions: supernatant harvested from mammalian cell cultures (SN), microsphere flow-through (FT), and elution fractions from Protein A or G purification containing the antibody of interest (E1 and E2). Under reducing conditions, two bands are seen in the elution fractions, corresponding to the heavy and the light chain of purified antibody. The same high purity level was obtained with this purification approach when compared to purification using Protein A (PA) Sepharose resin

7. After migration, recover the gel and wash with distilled water.
8. Add SimplyBlue™ SafeStain and stain the gel overnight.
9. Discard the SimplyBlue™ SafeStain, wash with distilled water, and de-stain gel in water under gentle agitation until the bands appear clearly.

The profiles on gel as well as the activity in ELISA of the IgG purified either with magnetic microspheres or sepharose were comparable (Figs. 2 and 3). Thus, protein A and protein G-coated magnetic microspheres are well suited for IgG purification from different cell culture supernatants.

4 Notes

1. Magnetic microspheres should not be resuspended by vortexing, but only by gentle manual agitation.
2. Use appropriate tips for microsphere sampling. If the volume to pipet is lower than 200 μL, use the self-sealing barrier pipette tips with cut extremity (such as Art 200G) or cut the extremity of a tip. If the volume is superior to 200 μL, classical

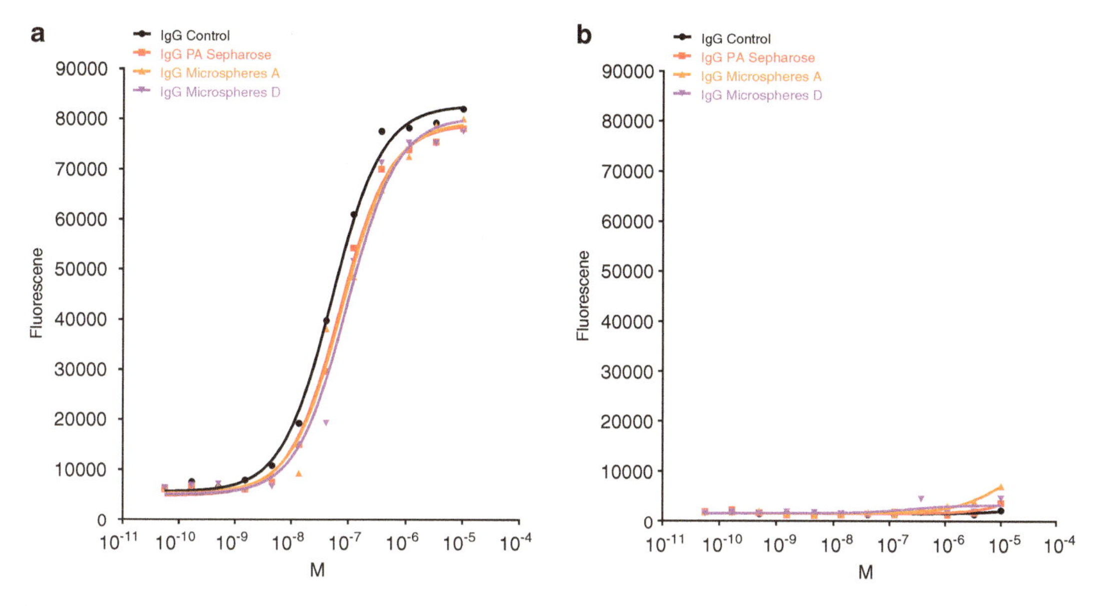

Fig. 3 Binding activity evaluated by ELISA for purified IgG from HEK293 cell culture supernatant with magnetic microspheres A and D and comparison with the same antibody purified using a PA Sepharose. (**a**). ELISA assay on a specific target: same activity is observed for all IgG. (**b**). Control ELISA on an irrelevant target to test the specificity of purified IgG (*see* **Note 16**)

1,000 μL tips can be used. Tube size, wash buffer volume, and type of magnetic rack must be adapted according to the volume of microspheres used.

3. Adapt the volume of microsphere suspension to be used to the amount of IgG estimated in the supernatant. The binding capacity is expressed as microgram of IgG per milligram of microspheres and the concentration of microspheres in the suspension is specified on the product data sheet. As an example, microspheres D have a binding capacity ≥50 μg of human IgG per mg of microsphere and the suspension contains 10 mg of microspheres per mL. It is recommended to work with an excess of microspheres and to consider the binding capacity a little lower than specified by the manufacturer.
4. The FastELYSA® (RD-Biotech) is a ready-to-use kit for human IgG quantification from culture medium. Only a small sample volume (maximum 20 μL) is required to obtain the concentration.
5. The separation time depends on the microsphere size and the magnetic pigment content and must be adapted to each experiment. By visual inspection, the buffer must be clear when all magnetic microspheres are captured by the magnet. Here, we let tubes sit for 5 min on the magnetic rack before pipetting the supernatant.

6. To control unspecific binding of medium components to magnetic microspheres, a small amount of microspheres can be incubated with medium from untransfected mammalian cells and treated as the other samples. Unspecific binding of contaminants can be evaluated on SDS PAGE.
7. Check that tubes are correctly closed and add Parafilm® around the caps as a security to avoid loss of product during rotation.
8. It is recommended to store the flow-through and all following wash fractions at +4 °C until the end of the purification and analysis process. If for any reason IgG did not bind to microspheres, or if microspheres were saturated, the flow-through can be used for further purification rounds.
9. The addition of NaCl in the PBS for wash steps allow for a more stringent wash with improved elution fractions purity. The unspecific binding of contaminants to microspheres is limited. This effect was tested by purifying in parallel two samples using wash solutions with and without NaCl (data not shown) and was previously described [10].
10. The elution step can be repeated to obtain several elution fractions. If higher concentrations are required, the elution buffer volume can be lowered and several elution rounds can be performed. After IgG quantification, fractions can be pooled in different batches according to concentration requirements.
11. The addition of Tris–HCl solution to the elution fractions allows reaching a physiological pH. If a different elution buffer is used, the volume of neutralization solution must be adapted. Some samples can precipitate or aggregates can be formed in the elution buffer; therefore, optimization of the elution buffer composition and pH is sometimes required.
12. Desalting of the elution fractions can be an alternative to neutralization for some IgG or if an appropriate formulation buffer was optimized. According to the elution sample volume, different materials can be used such as desalting columns from GE Healthcare: PD-10 columns for loading of 2.5 mL of sample, NAP-10 columns for loading of 1 mL of sample and NAP-5 columns for loading of 500 μL of sample. Another way to desalt samples is to use the Amicon Ultra-4 centrifugal filter unit with Ultracel membrane. By successive centrifugation and addition of formulation buffer, the IgG is conditioned in the new buffer. The appropriate cut-off must be chosen for the Amicon membrane.
13. A maximum of 5 μg of IgG must be loaded per well on a NuPAGE® gel. To obtain well-resolved bands for both heavy and light chains in denaturing and reducing conditions, 2 μg is sufficient.
14. β-mercaptoethanol must be manipulated under a fume hood.

15. If two gels are run in the same mini-cell, the voltage must be decreased to 150 V and the migration time increased to 45 min to avoid over heating of the gels during migration.
16. In the ELISA, appropriate controls must be included such as (1) an irrelevant target to evaluate binding specificity, (2) an irrelevant IgG to evaluate background signals, (3) a target specific IgG as positive control (if available).

Acknowledgments

This work was supported by Fabrice Sultan from Merck Estapor® who provided us with all magnetic microspheres.

References

1. Holschuh K, Schwämmle A (2005) Preparative purification of antibodies with protein A—an alternative to conventional chromatography. J Magn Magn Mater 293:345–348
2. Gao J, Li Z, Russell T, Li Z (2012) Antibody affinity purification using metallic nickel particles. J Chromatogr B 895–896:89–93
3. Hafeli UO (2003) Characterization of magnetic particles and microspheres and their magnetophoretic mobility using a digital microscopy method. Eur Cell Mater 3(2): 24–27
4. Merck Estapor Microspheres (2011) A versatile biotool for bioseparation, biopurification and biodetection, Bio-Estapor® microspheres, Brochure
5. Carriere D (1999) Whole blood Capcellia CD4/CD8 immunoassay for enumeration of CD4+ and CD8+ peripheral T lymphocytes. Clin Chem 45:92–97
6. Francois P (2003) Rapid detection of methicillin-resistant *Staphylococcus aureus* directly from sterile or nonsterile clinical samples by a new molecular assay. J Clin Microbiol 41(1):254–260
7. Stege PW, Raba J, Messina GA (2010) Online immunoaffinity assay-CE using magnetic nanobeads for the determination of anti-Helicobacter pylori IgG in human serum. Electrophoresis 31(20):3475–3481
8. Safarik I, Safarikova M (2004) Magnetic techniques for the isolation and purification of proteins and peptides. Biomagn Res Technol 2:7
9. Pande J, Szewczyk MM, Grover AK (2010) Phage display: concept, innovations, applications and future. Biotechnol Adv 26(6):849–858
10. Josic D, Lim Y-P (2001) Methods for purification of antibodies: analytical and preparative methods for purification for antibodies. Food Technol Biotechnol 39(3):215–226

Chapter 17

Generation of Cell Lines for Monoclonal Antibody Production

Krista Alvin and Jianxin Ye

Abstract

Monoclonal antibodies (mAbs) represent the largest group of therapeutic proteins with 30 products approved in the USA and hundreds of therapies currently undergoing clinical trials. The complex nature of mAbs makes their development as therapeutic agents constrained by numerous criteria such as quality, safety, regulation, and quantity. Identification of a clonal cell line expressing high levels of mAb with adequate quality attributes and generated in compliance with regulatory standards is a necessary step prior to a program moving to large-scale production for clinical material. This chapter outlines the stable transfection technology that generates clonal cell lines for commercial manufacturing processes.

Key words CHO cells, Monoclonal antibody production, Stable transfection, Single cell cloning, Cell line development

1 Introduction

Monoclonal antibodies (mAbs) and other therapeutic proteins have been the fastest-growing new therapies in the past decade [1, 2]. The general process for protein production in mammalian cells includes stable cell line generation. Choice of host cell expression system is a major determinant of a mAb drug candidate's ability to satisfy the critical attributes necessary to bring a drug to market. Mammalian host organisms are preferential to other expression host systems (e.g., bacteria, yeast, insect cells, transgenic plants [3–5]) because of their ability to properly fold and assemble complex molecules as well as to perform posttranslational modifications yielding antibodies that are similar to those naturally produced in humans [6–8]. The two predominant cell types used for production of recombinant mAb constructs, Chinese hamster ovary (CHO) and mouse myeloma (NS0), require stable integration of the gene of interest. Identification of a clonal cell line expressing high levels of mAb with adequate quality attributes and generated

Vincent Ossipow and Nicolas Fischer (eds.), *Monoclonal Antibodies: Methods and Protocols*, Methods in Molecular Biology, vol. 1131, DOI 10.1007/978-1-62703-992-5_17,

in compliance with regulatory standards is a necessary step prior to a program moving to large-scale production of clinical material.

The process of mammalian cell line development begins with stable transfection of the gene of interest using a selective plasmid vector. For CHO cells, two vector systems are most frequently employed to yield high producing transfectants: the dihydrofolate reductase (DHFR) system [9, 10] and glutamine synthetase (GS) system [11]. CHO cell lines such as CHO-DXB11 [12] and CHO-DG44 [13] are deficient in DHFR, so plasmids carrying the DHFR gene and the gene of interest that incorporate into the host cell genome allow transfectants to grow in media lacking glycine, hypoxanthine, and thymidine (GHT) [8]. DHFR transfected CHO cells can undergo gene amplification through iterative rounds of increased methotrexate (MTX) exposure, thus yielding higher protein expression [14]. Similarly, CHOK1 cells that incorporate the GS gene and the gene of interest are capable of growing in glutamine-free media [15]. GS transfected CHO cells can have increased selective pressure through introduction of the GS inhibitor methionine sulfoximine (MSX). Compared to the DHFR expression system, there are some advantages to using the GS system. GS avoids high ammonia levels found when utilizing the DHFR selection system and eliminates the use of an unstable amino acid—glutamine. The GS system also does not require multiple rounds of amplification to identify high producing cell lines thereby shortening timelines needed for cell line selection [16].

This chapter focuses on stable transfection technology for clonal cell line generation. Figure 1 summarizes the overall steps

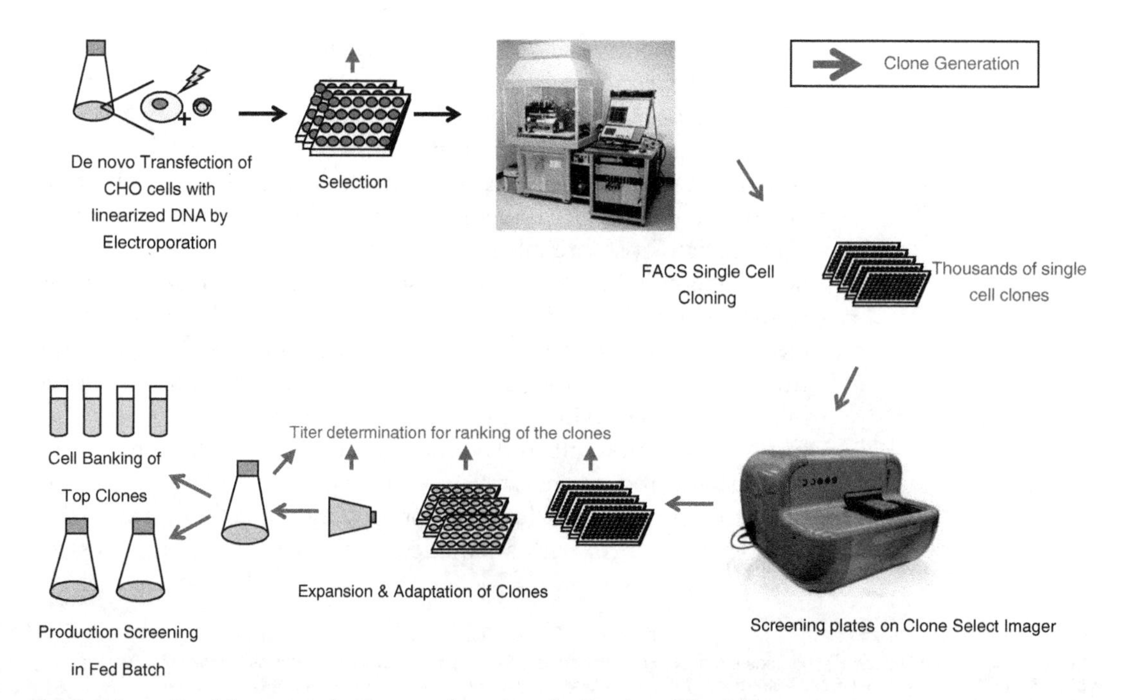

Fig. 1 Schematic of the expansion process from transfection to cell banking

involved in generating a desirable clonal cell line for commercial manufacturing processes, including transfection, selection of production pools, subcloning, screening of clonal cell lines, expansion, bioreactor evaluation, and cell banking.

2 Materials

1. Linearized plasmid containing selection marker and gene of interest.
2. Electroporation equipment, such as a Bio-Rad Gene Pulser Xcell electroporator (Bio-Rad Laboratories) or other transfection reagents.
3. Suspension adapted CHO cells or equivalent. For example, CHOK1 from ATCC adapted in serum-free chemically defined medium.
4. CD-CHO medium (Invitrogen), DMEM medium (SAFC), alphaMEM medium (Sigma), or other appropriate cell culture medium and cell culture supplements, such as glutamine.
5. Production medium and feeds for fed-batch process.
6. Appropriate selection agent (drug). For example, if glutamine synthetase is used as the selective marker, MSX is used as the selection reagent. Appropriate MSX concentration has to be determined by titration.
7. Appropriate tissue culture vessels, such as 96-well plates, 24-well plates, T25 flasks, and 125–500 mL vented shake flasks.
8. Standard cell culture equipment, such as CO_2 incubator, orbital shaker, centrifuge, and cell counter.
9. Equipment for metabolic measurements, such as BioProfile Analyzers (Nova Biomedical).
10. Equipment and materials for titer determination, such as ELISA reagents or Protein A HPLC columns and buffers.
11. A cell sorter, such as the FACS Vantage SE (Becton Dickinson) and related stains, such as 7-Amino-actinomycin D (7-AAD, Invitrogen).
12. Microscopic instruments, such as a standard inverted microscope with a 20× objective or the CloneSelect Imager (CSI, Molecular Devices) system.
13. Anti-human light chain antibody (Zymed Technologies, Carlsbad, CA).
14. HRP-anti human IgG pan specific antibody (BD Pharmingen, San Diego, CA).

15. TMB (3,3′,5,5′-tetramethylbenzidine) (Pierce, Rockford, IL).
16. Poros A column (Applied Biosystems, Foster City, CA).

3 Methods

3.1 Transfection

Mammalian cell transfection can be accomplished through various approaches, such as electroporation, calcium phosphate coprecipitation, Lipofectamine, etc. [17]. In our experiments, we use a Bio-Rad Gene Pulser Xcell electroporator (Bio-Rad Laboratories). The expression vector containing the gene of interest is typically linearized to enhance the integration events. The linearization site should be somewhere in the plasmid backbone to avoid impact on the expression of the selectable marker or gene of interest.

1. CHO cells are maintained in 125 mL shake flasks at 37 °C in a humidified 5 % CO_2 incubator on an orbital shaker. Shaker speed is 100 rpm. The culture medium is CD-CHO supplemented with 2 mM glutamine. Subculture the cells every 3–4 days, seeding at a concentration of 2×10^5 cells/mL.
2. Passage the cells 2 days prior to transfection at a cell density of 2×10^5 cells/mL.
3. Prior to transfection, measure the cell density and viability using a cell counter. The cell culture viability should be above 90 %.
4. For each transfection by electroporation, 40 μg of linearized DNA and 1×10^7 viable cells are used (*see* **Note 1**).
5. Pellet 1×10^7 viable cells by centrifugation (5 min at $200 \times g$).
6. Wash the pellet with 20 mL of DMEM medium. Pellet the cells by centrifugation.
7. Resuspend the cell pellet in appropriate amount of DMEM medium so that the total volume of cell suspension and DNA is equal to 800 μL.
8. Mix the cell suspension and DNA, and transfer the mixture into to a sterile 4 mm gap cuvette.
9. Load cuvette into electroporation carriage and pulse. The optimal electroporation parameters could be cell line dependent (*see* **Note 2**).
10. Immediately following electroporation, transfer the entire contents from the cuvette into 24 mL of culture medium. Transfer this cell suspension into five 24-well plates (0.2 mL cell suspension/well). Incubate at 37 °C in a humidified 5 % CO_2 incubator overnight.

3.2 Selection

The selection phase depends on the expression plasmid and cell line. Kill curves can be used to determine the appropriate selection pressure. During expansion, the selection pressure is maintained to avoid significant loss of productivity.

1. Approximately 24 h post-transfection, add 1 mL of culture medium with selection drug to each well in the 24-well plates. The total culture volume in each well will be about 1.2 mL. The appropriate concentration of selectable drug will be dependent on the drug and host cell line used. For example, we use 25 μM MSX for CHO cells (*see* **Note 3**).
2. Ten to 20 days post-transfection, sample each well for antibody titer determination using methods such as an ELISA assay previously described by de la Cruz Edmonds et al. [18]. Briefly, 96-well plates are coated with monoclonal anti-human light chain antibody as the capture reagent. The culture supernatant is then added followed by HRP-anti human IgG pan specific antibody. TMB is next added for enzymatic oxidation of HRP, resulting in a colorimetric change in intensity that is measured at 450 nm on a spectrophotometer.

3.3 FACS Analysis and Single Cell Cloning

Based on the titer results in the 24-well plates, the top cell lines are then single cell cloned. The top cell lines can be pooled together before single cell cloning, or the cells from each chosen well can be single cell cloned into its own 96-well plates.

1. Conditioned (spent) media are collected for use during single cell cloning. First, wild-type cultures seeded at 2×10^5 vc/mL are grown to mid log phase and harvested through centrifugation. Culture supernatants are collected and stored at 4 °C for no more than 7 days prior to use.
2. Transfected CHO cells are analyzed and sorted using a cell sorter, such as the FACS Vantage SE (Becton Dickinson). In combination with forward light scatter, 7-Amino-actinomycin D (7-AAD, Invitrogen) staining is used to eliminate dead cells and debris using a 670/30 nm band pass filter.
3. Single cell cloning in 96-well plates is verified using one of two microscopic methods. Manual microscopic screening of the inner surface area of a given well of a 96-well plate can be performed using a standard inverted microscope with a 20× objective. Automated microscopic screening can also be performed using instruments, such as the CloneSelect Imager System. Proper documentation of this step is required for a manufacturing cell line. Typically, 2–3 weeks post-sorting, isolated cell colonies will appear in 20–50 % of the wells depending on the medium and cells used. With optimized sorting settings, majority of the colonies originated from a single cell.

3.4 Expansion

The expansion phase is performed to increase the cell mass of the clones from a single cell in 96-well plates to shake flasks.

Selection pressure is maintained during the whole expansion process. At each expansion step, the productivity of each clone is measured, and the number of clones expanded to each stage is decreased based on the productivity results.

1. Fourteen to 21 days post-single cell sorting, transfer the cells from 96-well plates into 24-well plates followed by T-flasks. At each of these static stages, the productivity of each clone can be determined using an ELISA assay.
2. From the T-flask, the cultures are expanded into 125 mL vented shake flasks and incubated at 37 °C in a humidified 5 % CO_2 incubator on an orbital shaker set to 100 rpm.
3. Once in shake flasks, measure the cell viability and cell density, and passage the cells at 2×10^5 cells/mL every 3–4 days (*see* **Note 4**).
4. After 4–5 passages in shake flasks, the clones should be readapted to the suspension culture, and the cell viability should reach above 90 %. At this point, the clones are ready for more detailed analysis to find the final production cell line.
5. During the suspension growth, the productivity of each clone can be determined using protein A affinity capture HPLC. The supernatant is loaded onto a Poros A column and eluted through reduction of pH in the mobile phase. Detection is measured at 280 nm.

3.5 Fed Batch Productivity Evaluation and Final Clone Selection

The productivity evaluation phase is performed to mimic large-scale production conditions in order to find the final production cell line. Initially, shake flasks can be a good scale-down model for this evaluation. After a small subset of clones are identified as the final candidates, a final evaluation in bioreactors is necessary to select the final production clone with desirable productivity, process performance, and product quality attributes. The evaluation procedure in a 250 mL shake flask using a fed-batch process is discussed here.

1. Seed the cells in the production medium at an inoculation density of 2×10^5 cells/mL. To mimic a large-scale manufacturing process, selection agent usually is not included in this step. For a 250 mL shake flask, the initial working volume is 50 mL.
2. Four days post-inoculation, sample the culture for cell viability, cell density, glucose level, lactate level, ammonium level, pH, etc.
3. Based on the glucose measurements, add glucose to the culture to achieve 40 mM glucose on each feed day.
4. Feed the culture with medium feeds for fed-batch process as appropriate.

5. Repeat **steps 2–4** on days 6, 8, and 11 post-inoculation or as appropriate.
6. Harvest on day 14 or before the cell viability drops below 50 % (*see* **Note 5**).
7. The cell supernatant is purified through Protein A chromatography and the mAb quality attributes (such as glycans, aggregation, fragmentation, acidic variance, etc.) can be determined using different analytical methods.
8. Evaluation in bioreactors can be performed following similar procedures described in **steps 1–7**. Process optimization can also be performed during this step if needed.

3.6 Cell Line Stability Studies

After transfection, the gene of interest gets integrated in the host genome. However, the expression levels will not always be stable. Due to various factors, the gene expression levels are subject to change. For example, the gene of interest could be deleted due to genomic rearrangements, or the gene could be silenced due to epigenetic modifications. For a final manufacturing cell line, an evaluation is necessary to ensure that the selected clone has long-term stability.

1. Starting from the master cell bank, continue to passage the cells in the appropriate growth medium for 60–80 generations.
2. At a suitable interval such as every 20 generations, assess the productivity of the clones using the final production medium and feed (e.g., fed-batch process in shake flasks). In general, the cell lines are considered stable if the productivities decline less than 30 % after 60–80 generations, which will cover the actual manufacturing window.
3. Collect the cell culture broth and isolate the products from those cultures at different ages from **step 2**. The isolated product is subject to appropriate analytic assessments to ensure the product characteristics are consistent.

3.7 Cell Banking

Cell banking is a critical step in order to preserve the cells for future experiments and it is also the source of material for GMP cell banking, which will then be used to produce the clinical material. This can be done at any time during the cell expansion stage when there are enough cells to make the required number of vials.

1. Three or four days after inoculation, cells are harvested by centrifuging at $200 \times g$ for 5 min at room temperature. Pellet enough cells for the required number of vials, 1.0×10^7 cells/vial.
2. After centrifugation, the supernatant is removed and the cell pellet is resuspended in freezing medium at 1.0×10^7 cells/mL. The freezing medium is prepared by adding 10 % DMSO (Sigma) to the growth medium without the selection agent.

3. 1 mL aliquots of the concentrated cell suspension are prepared per vial and stored in a controlled rate freezing container (e.g., Nalgene Mr. Frosty containers in −70 °C freezers provide a −1 °C/min cooling rate). The following day, the vials are transferred to a liquid nitrogen freezer for long-term storage.

4 Notes

1. Circular plasmid DNA can be prepared using a variety of DNA preparation methods, such as Qiafilter kits (Qiagen) or CsCl gradient. Since the circular DNA is subjected to digestion, for preparation of linearized DNA, EDTA should be avoided in the buffer. After digestion, linearized DNA should be purified using the standard phenol–chloroform extraction method.
2. The optimal parameters for electroporation could be cell line dependent. A typical capacitance value is around 1,000 μF. The appropriate voltage ranges from 200 to 350 V.
3. The selective conditions may vary based on cell lines and selectable markers used. When hygromycin is used as a selectable marker, 150–300 μg/mL of hygromycin can be used for CHO cells. For Puromycin, 7–20 μg/mL can be used for CHO cells. When using glutamine synthetase for selection, 25–50 μM of MSX can be used for CHO cells.
4. At the first passage in shake flasks, the cells need to readapt to the suspension culture, and typically there is a short lag phase. After 4–7 days in suspension culture, the culture is ready to be passaged when the cell density reaches 5×10^5 cells/mL. If the cell density is too low, continue the culture for several more days before passaging. The cell viability will increase as passaging continues. Typical doubling time for CHO cells in an appropriate medium should be around 24 h.
5. The actual feeding schedule varies depending on the feeds and the cell lines. To align the principle of quick production, it is recommended to have a platform process established beforehand. However, if it is desired, process optimization can be performed, such as temperature, pH, feed composition, feed schedule, etc. The yield of mAbs from a fed-batch process depends on many factors, such as the antibody itself, expression system, the production cell line, medium, feeds, feeding schedule, culture duration, process parameters, etc. A wide range has been reported from hundreds of mg/L up to 8–10 g/L.

References

1. Li J, Zhu Z (2010) Research and development of next generation of antibody-based therapeutics. Acta Pharmacol Sin 31:1198–1207
2. Reichert JM (2008) Monoclonal antibodies as innovative therapeutics. Curr Pharm Biotechnol 9:423–430
3. Schmidt FR (2004) Recombinant expression systems in the pharmaceutical industry. Appl Microbiol Biotechnol 65(4):363–372
4. Schirrmann T et al (2008) Production systems for recombinant antibodies. Front Biosci 13: 4576–4594
5. Andersen DC, Krummen L (2002) Recombinant protein expression for therapeutic applications. Curr Opin Biotechnol 13(2): 117–123
6. Andersen DC, Reilly DE (2004) Production technologies for monoclonal antibodies and their fragments. Curr Opin Biotechnol 15(5): 456–462
7. Sethuraman N, Stadheim TA (2006) Challenges in therapeutic glycoprotein production. Curr Opin Biotechnol 17(4):341–346
8. Trill JJ, Shatzman AR, Ganguly S (1995) Production of monoclonal-antibodies in Cos and Cho cells. Curr Opin Biotechnol 6(5): 553–560
9. Kaufman RJ (2000) Overview of vector design for mammalian gene expression. Mol Biotechnol 16(2):151–160
10. Wurm FM (2004) Production of recombinant protein therapeutics in cultivated mammalian cells. Nat Biotechnol 22(11): 1393–1398
11. Birch JR, Mainwaring DO, Racher AJ (2008) Use of the Glutamine Synthetase (GS) expression system for the rapid development of highly productive mammalian cell processes. In: Knäblein DJ (ed) Modern biopharmaceuticals: design, development and optimization. Wiley-VCH Verlag GmbH, Weinheim, Germany
12. Graf LH, Chasin LA (1982) Direct demonstration of genetic alterations at the dihydrofolate-reductase locus after gamma-irradiation. Mol Cell Biol 2(1):93–96
13. Urlaub G et al (1986) Effect of gamma-rays at the dihydrofolate-reductase locus—deletions and inversions. Somat Cell Mol Genet 12(6): 555–566
14. Jun SC et al (2005) Selection strategies for the establishment of recombinant Chinese hamster ovary cell line with dihydrofolate reductase-mediated gene amplification. Appl Microbiol Biotechnol 69(2):162–169
15. Bebbington CR et al (1992) High-level expression of a recombinant antibody from myeloma cells using a glutamine synthetase gene as an amplifiable selectable marker. Biotechnology (N Y) 10(2):169–175
16. Kingston RE et al (2002) Amplification using CHO cell expression vectors. Curr Protoc Mol Biol chapter 16: Unit 16 23
17. Birch JR, Racher AJ (2006) Antibody production. Adv Drug Deliv Rev 58:671–685
18. de la Cruz Edmonds MC et al (2006) Development of transfection and high-producer screening protocols for the CHOK1SV cell system. Mol Biotechnol 34(2):179–190

Chapter 18

Expression and Purification of Recombinant Antibody Formats and Antibody Fusion Proteins

Martin Siegemund, Fabian Richter, Oliver Seifert, Felix Unverdorben, and Roland E. Kontermann

Abstract

In the laboratory-scale production of antibody fragments or antibody fusion proteins, it is often difficult to keep track on the most suitable affinity tags for protein purification from either prokaryotic or eukaryotic host systems. Here, we describe how such recombinant proteins derived from *Escherichia coli* lysates as well as HEK293 cell culture supernatants are purified by IMAC and by different affinity chromatography methods based on fusions to FLAG-tag, Strep-tag, and Fc domains.

Key words Antibody, scFv, Fusion protein, *E. coli*, HEK293, IMAC, FLAG, Strep, Fc fusion

1 Introduction

Recombinant antibody fragments are emerging components in the development of protein therapeutics for the treatment of cancer or inflammatory diseases. The rise of these smaller, genetically engineered antibody formats is based on several benefits compared with classical, whole IgG molecules. The compact size of, for example, single-chain variable fragments (scFv) facilitates a better tissue penetration and predestines such formats as building blocks for bivalent molecules using dimerization domains like the Fc moiety of IgG [1, 2] or the heavy chain domain 2 of IgM [3]. Such bivalent antibody formats are often preferred due to the increased avidity to their target antigen. Further, the genetic fusion of a recombinant miniantibody to an effector molecule was shown to mediate targeted cytotoxicity to tumor cells. Regarding such fusion proteins, preferred antibody moieties itself may comprise a dual functionality, which means (1) to bind a tumor antigen in support of a focused enrichment of the therapeutic protein at the tumor site and (2) to deliver a cytotoxic effector moiety for inhibition

Vincent Ossipow and Nicolas Fischer (eds.), *Monoclonal Antibodies: Methods and Protocols*, Methods in Molecular Biology, vol. 1131, DOI 10.1007/978-1-62703-992-5_18, © Springer Science+Business Media New York 2014

of tumor cell growth. Examples for effector molecules used in immunotherapeutics include cytokines [4–6], toxic proteins [7, 8], and RNases [9].

Derived from single-chain Fv molecules by reducing the linker length between V_H and V_L, diabodies and single-chain diabodies provide another compact format to achieve a dimerization and also a bispecificity of antigen binding. Bispecific antibody formats are widely used in experimental cancer therapy to direct effector cells of the immune system, such as cytotoxic T-cells and NK cells, to tumor cells, whereby respective antigens on both cells types are recognized by the bispecific molecule [10]. Another mode of action for bispecific antibodies is the simultaneous blocking of two different target molecules in order to gain an enhanced therapeutic effect, e.g., in the potential treatment of complex inflammatory diseases. A recent development in this field was the invention of dual-variable-domain immunoglobulins (DVD-Igs), a new bispecific and tetravalent antibody format with beneficial pharmacokinetics and industry-compatible manufacturing properties [11, 12].

To overcome the major disadvantage of small antibody formats, the rapid elimination from the blood stream due to renal filtration, the fusion of a small therapeutic drug to the Fc part of an antibody can extend the serum half-life via interaction with the neonatal Fc receptor. Further, by interaction with other Fc receptors, Fc fusion proteins can also mediate effector functions like ADCC (antibody-dependent cellular cytotoxicity) and CMC (complement-mediated cytotoxicity).

The growing variety of recombinant antibody formats requires in each specific case a decision for a suitable expression system and a practicable purification strategy. While most of the smaller formats without glycosylation sites like single-chain Fv, Fab fragments or (single-chain) diabodies can be also produced periplasmically in *Escherichia coli* with yields of ~1 mg/l in standard batch cultures, more complex antibody fusion proteins or molecules involving C_H2 or Fc domains demand an expression in CHO or HEK293 cells. In addition, the weak solubility of some scFv or diabody molecules in the periplasmic *E. coli* expression often also requires a eukaryotic expression system. The choice of an appropriate purification tag is another crucial parameter for protein purification in the laboratory scale. The relatively weak selectivity of the Ni-NTA matrix for the common 6×histidine tag facilitates one-step IMAC purification with a satisfying protein quality only from pure tissue culture supernatants. In contrast, the protein purity gained after IMAC of *E. coli* periplasmic or whole cell lysates is relatively crude and requires for sensitive applications often a time-consuming second purification step, e.g., ion exchange chromatography. As a critical issue, the presence of a 6×histidine tag in recombinant proteins was reported to have an influence on activity, biophysical properties, or immunogenicity in some cases [13, 14]. Alternatively to the 6×histidine tag, the invention of today well-established affinity tags including GST, MBP, CBP, myc, Strep, and FLAG enables a more specific isolation of recombinant proteins [15]. Among the affinity tags using a protein–peptide interaction as a strategy for purification, the FLAG tag [16, 17] and the Strep tag II [18, 19] are most commonly used due to the negligible effects on the recombinant protein, the mild and native purification conditions, and their suitability for one-step protein purifications even from crude lysates.

In this chapter, we show how recombinant antibody formats and antibody–cytokine fusion proteins are best expressed and purified in a laboratory scale. We cover the expression of recombinant proteins in *E. coli* and mammalian cells (HEK293) and the purification by fusion to 6×histidine, FLAG, and Strep tags as well as Fc moieties. Further, we describe for example Diabody-single-chain TRAIL fusion proteins and how the specific nature of these and similar complex fusion proteins facilitates a beneficial combination of two purification principles.

2 Materials

2.1 Expression and Purification of 6×His-Tagged Proteins (E. coli)

1. *E. coli* TG1 (chemically competent).
2. Plasmid DNA of a prokaryotic expression construct.
3. 2× TY medium: 16 g tryptone, 10 g yeast extract, 5 g NaCl in 1 l dH_2O.
4. LB plates (glucose, ampicillin): 10 g tryptone, 5 g yeast extract, 5 g NaCl, 15 g agar, 10 g glucose, 100 mg ampicillin in 1 l dH_2O (*see* **Note 1**).
5. Isopropyl thiogalactoside (IPTG): 1 M stock solution in sterile dH_2O (*see* **Note 2**).
6. Periplasmic preparation buffer (PPB): 30 mM Tris–HCl, 1 mM EDTA, 20 % (w/w) sucrose, pH 8.0.
7. Lysozyme solution: 10 mg/ml in dH_2O (*see* **Note 3**).
8. $MgSO_4$ solution: 1 M in dH_2O.
9. Dialysis tubing and dialysis tubing clips.
10. IMAC loading buffer: 50 mM Na_2HPO_4/NaH_2PO_4, 250 mM NaCl, 20 mM imidazole, pH 7.5 (*see* **Note 4**).
11. Magnetic stirrer, stir bar.
12. Ni-NTA agarose (Qiagen).
13. Roller mixer.
14. Poly-Prep® chromatography columns (Bio-Rad).
15. IMAC washing buffer: 50 mM Na_2HPO_4/NaH_2PO_4, 250 mM NaCl, 30 mM imidazole, pH 7.5 (*see* **Note 4**).
16. IMAC elution buffer: 50 mM Na_2HPO_4/NaH_2PO_4, 250 mM NaCl, 250 mM imidazole, pH 7.5 (*see* **Note 4**).

17. Bradford protein assay (5× stock solution, Bio-Rad).
18. Phosphate buffered saline (PBS, 10×): 0.1 M Na_2HPO_4, 20 mM KH_2PO_4, 27 mM KCl, 1.37 M NaCl, pH 7.4.

2.2 Generation of Stably Transfected HEK293 Cells and Expression of Recombinant Protein

1. General cell culture equipment (6-well plates, 250 and 500 ml tissue culture flasks, pipettes, Pasteur pipettes, suction system, clean bench, CO_2 incubator, centrifuge with capacity for 15 and 50 ml conical bluecaps, microscope, water bath, hemocytometer, vortex mixer).
2. HEK293 cell line.
3. RPMI 1640 medium (Life Technologies) with 5 % fetal calf serum (*see* **Note 5**).
4. 0.05 % trypsin-EDTA (Life Technologies).
5. Plasmid DNA of a eukaryotic expression construct (from a Midiprep, ultrapure, in ddH_2O).
6. Lipofectamine® 2000 (Life Technologies).
7. OptiMEM® I medium (Life Technologies).

2.3 Purification of 6×His-Tagged Proteins (Mammalian Cells)

1. Ammonium sulfate.
2. Poly-Prep® chromatography columns.
3. Ni-NTA agarose.
4. PBS.
5. IMAC wash buffer: 50 mM Na_2HPO_4/NaH_2PO_4, 250 mM NaCl, 30 mM imidazole, pH 7.5 (*see* **Note 4**).
6. Bradford protein assay.
7. IMAC elution buffer: 50 mM Na_2HPO_4/NaH_2PO_4, 250 mM NaCl, 250 mM imidazole, pH 7.5 (*see* **Note 4**).
8. Dialysis tubing and dialysis tubing clips.
9. Magnetic stirrer, stir bar.

2.4 Purification of FLAG-Tagged Proteins (Mammalian Cells)

1. Liquid chromatography columns, Luer Lock, bed volume 8 ml (Sigma).
2. Laboratory stand with clamp.
3. ANTI-FLAG® M2 affinity gel (Sigma).
4. Tris buffered saline (TBS, 10×): 0.5 M Tris–HCl, pH 7.4, 1.5 M NaCl.
5. Silicone tubing, 2 mm inner diameter.
6. Peristaltic pump.
7. 0.1 M glycine HCl, pH 3.5 (*see* **Note 3**).
8. 500 ml conical centrifuge bottles (Corning) with adaptors or alternatively glass flasks.

9. FLAG peptide (peptides & elephants or Sigma).
10. Stock solution of 20 % sodium azide in dH_2O.
11. Vivaspin® 20 PES centrifugal concentrators (Sartorius Stedim, *see* **Note 6**).
12. Dialysis tubing and dialysis tubing clips (*see* **Note 7**).
13. PBS.
14. Magnetic stirrer, stir bar.
15. Acrodisc® 13 mm syringe filter, 0.2 μm (Pall).
16. 5 ml single use syringe, syringe needles.

2.5 Two-Step Purification of Antibody Fusion Proteins (IMAC/Anti-FLAG)

1. Ni-NTA Agarose.
2. IMAC buffer A (10×): 0.5 M Na_2HPO_4/NaH_2PO_4, 3 M NaCl, pH 8.0.
3. IMAC buffer B: 0.05 M Na_2HPO_4/NaH_2PO_4, 0.3 M NaCl, 0.1 M imidazole, pH 8.0.

2.6 Strep-Tag Affinity Purification (Mammalian Cells)

1. 1 M Tris–HCl, pH 8.0.
2. Avidin (IBA Life Sciences).
3. Liquid chromatography columns, Luer Lock, bed volume 8 ml (Sigma).
4. Strep-Tactin® Sepharose (IBA Life Sciences).
5. Silicone tubing, 2 mm inner diameter.
6. Peristaltic pump.
7. Buffer W (washing buffer): 100 mM Tris–HCl, 150 mM NaCl, 1 mM EDTA, pH 8.0.
8. Buffer E (elution buffer): 100 mM Tris–HCl, 150 mM NaCl, 1 mM EDTA, 2.5 mM desthiobiotin, pH 8.0 (*see* **Note 8**).
9. Buffer R (regeneration buffer): 100 mM Tris–HCl, 150 mM NaCl, 1 mM EDTA, 1 mM HABA (hydroxy-azophenylbenzoicacid), pH 8.0.
10. Dialysis tubing and dialysis tubing clips.
11. PBS or TBS.
12. Magnetic stirrer, stir bar.
13. Vivaspin® 20 or Vivaspin® 2 PES centrifugal concentrators with MW cutoff suitable for the protein of interest.
14. 5 ml single use syringe, syringe needles.
15. Acrodisc® 13 mm syringe filter, 0.2 μm.

2.7 Purification of Fc Fusion Proteins by Protein A Affinity Chromatography (Mammalian Cells)

1. Ammonium sulfate.
2. Magnetic stirrer, stir bar.
3. TOYOPEARL® AF-rProtein A-650 F (Tosoh).
4. Poly-Prep® chromatography columns.

5. PBS.
6. Bradford protein assay.
7. Elution buffer: 100 mM glycine-HCl pH 2–4.
8. Neutralization buffer: 1 M Tris–HCl pH 7.5.
9. Column regeneration buffer: 0.1–0.5 M NaOH.
10. Dialysis tubing (cutoff 11 kDa, Sigma) and dialysis tubing clips.

3 Methods

3.1 Expression and Purification of 6×His-Tagged Proteins (E. coli)

Single-chain variable fragments (scFv) are monovalent and monospecific molecules (MW ~25 kDa) consisting of the two variable domains of the heavy (V_H) and light (V_L) chain of an antibody, which are connected via a flexible peptide linker of 15 amino acid residues (3× GGGGS). When expressing scFv in *E. coli*, an export of the polypeptide chain to the oxidative milieu of the periplasm is mandatory to ensure the formation of structurally relevant intramolecular disulfide bonds. To this end, the gene of the scFv fragment should be cloned into a vector (e.g., pAB1, ref. 20) that harbors a correspondent signal sequence (e.g., pelB). The affinity tag for purification and detection of the scFv fragments, for example 6×histidine tag, is commonly located at the C-terminus of the molecule (Fig. 1a). A short peptide spacer of at least five small neutral amino acid residues between the protein of interest and the affinity tag provides for accessibility concerning purification and detection.

1. Two days before cultivating and induction of the protein synthesis (day 0) transform the scFv encoding plasmid into a chemically competent *E. coli* strain. Therefore mix gently competent cells and 1 μg of plasmid DNA and incubate for 10 min on ice. For efficient transformation rates, incubate the mixture for 1 min at 42 °C (heat shock). Afterwards, add 1 ml of 2× TY (not containing any antibiotics) and incubate the cells for 30 min at 37 °C (shaking at 170 rpm). Then, harvest cells (centrifuge 6,000 ×*g*, 3 min; resuspend pellet in 100 μl 2× TY) and streak the suspension on a LB plate (glucose, ampicillin). Incubate the plate at 37 °C overnight (*see* **Note 9**).
2. (Day 1, afternoon) Inoculate 30 ml of 2× TY (1 % glucose, 100 μg/ml ampicillin) with one colony from the plate and culture cells at 37 °C overnight (shaking at 170 rpm).
3. (Day 2) Inoculate 1 l of 2× TY (0.1 % glucose, 100 μg ampicillin/ml) with 10 ml (1:100 dilution) of the overnight culture and incubate at 37 °C (shaking at 170 rpm) to an OD_{600} of 0.8–1.0 (*see* **Note 10**).

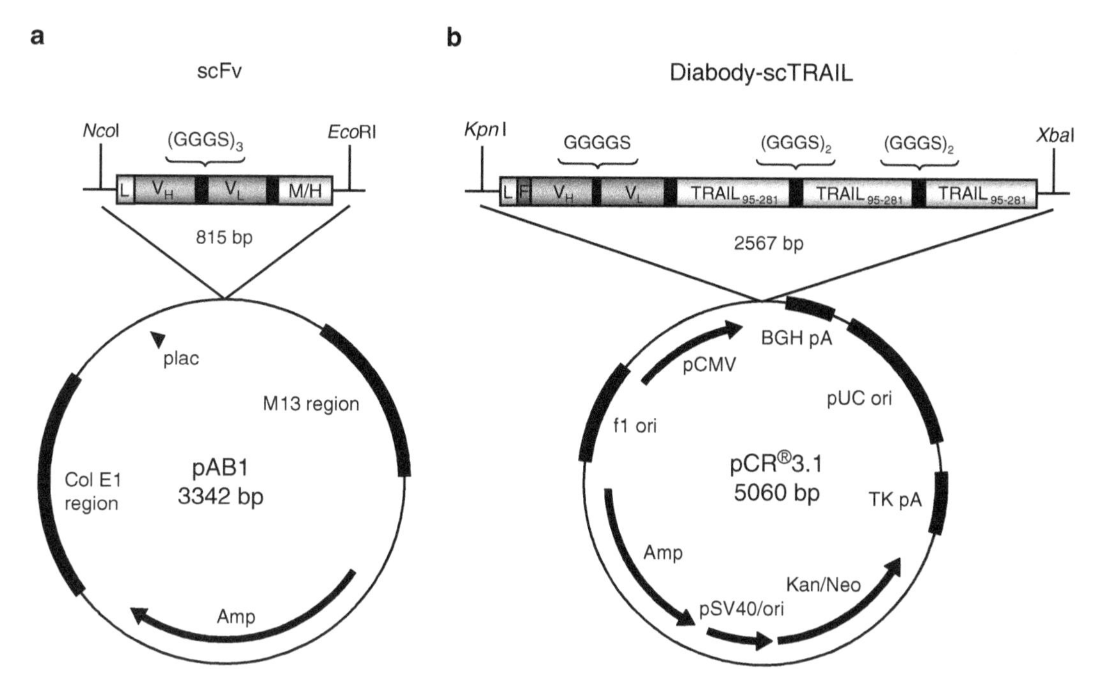

Fig. 1 (**a**) Representative DNA construct for periplasmic expression of a scFv in *E. coli* under the control of the lac promotor/operator. *L* pelB signal sequence, *M* Myc tag, *H* 6×histidine tag. (**b**) Schematic representation of a DNA construct for constitutive mammalian expression of a Diabody-single-chain TRAIL fusion protein under the control of the cytomegalovirus promoter. Polyadenylation of the transgenic mRNA is promoted by a site from the bovine growth hormone gene. Propagation and cloning of the plasmid in *E. coli* is facilitated by the pUC origin of replication and the ampicillin resistance-mediating beta-lactamase gene. The neomycin marker gene allows for selection of stably transfected mammalian cells using G418/geneticin as antibiotic. *L* leader peptide, *F* FLAG tag

4. Add IPTG (final concentration: 1 mM) and incubate the culture at room temperature (22–25 °C) for 3 h (shaking with 170 rpm).
5. Harvest the cells (6,000 × *g*, 10 min, 4 °C).
6. Resuspend the pellet in 50 ml PPB per liter bacterial culture (*see* **Note 11**).
7. Add 250 μl lysozyme solution (final concentration: 2.5 mg lysozyme/50 ml PPB), mix by inverting and incubate for 20 min on ice.
8. Add 0.5 ml $MgSO_4$ solution (final concentration: 10 mM) to stabilize the spheroblasts and mix by inverting.
9. Centrifuge (10,000 × *g*, 10 min, 4 °C) and collect the periplasmic preparation (supernatant).
10. Dialyze the periplasmic preparation against IMAC loading buffer at 4 °C overnight using a magnetic stirrer (*see* **Note 12**).

11. (Day 3) Collect periplasmic preparation and centrifuge (10,000 × *g*, 10 min, 4 °C) to pellet aggregates and residual cell debris (*see* **Note 13**).
12. For equilibrating Ni-NTA beads, transfer an appropriate amount of Ni-NTA beads into a 15 ml tube (if you are using more than 1 ml Ni-NTA beads, take a 50 ml tube) and centrifuge (2,000 × *g*, 5 min, 4 °C). Discard the supernatant and resuspend the beads in IMAC loading buffer (tenfold of the Ni-NTA bed volume). Repeat this step once (*see* **Note 14**).
13. Transfer the Ni-NTA beads to the periplasmic preparation and incubate rolling at 4 °C for at least 2 h (*see* **Note 15**).
14. Load the Ni-NTA bead suspension on a chromatography column and drain by gravity flow. Save the flow-through for analysis on SDS-PAGE (*see* **Note 16**).
15. For washing, load 16 column bed volumes (CV) of IMAC washing buffer on the column. Let the IMAC washing buffer drain completely by gravity flow before adding fresh buffer. Repeat this step until the wash fractions do not contain any protein (*see* **Note 17**). Save the wash fractions separately for analysis on SDS-PAGE.
16. For elution, load 1 CV IMAC elution buffer on the column. Again, let the IMAC elution buffer drain completely before adding fresh buffer. Repeat this step until all protein is eluted from the Ni-NTA beads. Check the protein content of the elution fractions by protein assay. Pool the fractions that contain sufficient amounts of protein to a main fraction, the other protein containing fractions can be pooled to a side fraction.
17. Dialyze main and side fraction against PBS at 4 °C overnight using a magnetic stirrer (at least a dilution of 1:1,000).
18. (Day 4) Centrifuge the dialyzed protein solution to pellet aggregates (8,000 × *g*, 1 min, 4 °C). Collect the clear supernatant and store the protein either in the fridge (stable for some weeks) or at −20 °C (stable for several months) (*see* **Note 18**).
19. Perform a SDS-PAGE analysis to check the quality of the purification. Therefore, load 5 μl of main and side fraction and 20 μl of the flow-through and the wash fractions on the gel (Fig. 2) (*see* **Note 19**).

3.2 Generation of Stably Transfected HEK293 Cells and Expression of Recombinant Protein

The first step is to clone the antibody construct into a mammalian expression vector, for example pCR3.1 (Life Technologies; G418 resistance) or pIRESpuro3 (Clontech; puromycin resistance). To allow for the secretion of the recombinant protein into the culture medium, the gene of interest must be fused to a sequence coding for an N-terminal export signal, commonly the leader peptide from Igκ light chain, which will be cleaved off during export. The affinity tag for detection and purification, e.g., FLAG tag or Strep-tag II, is

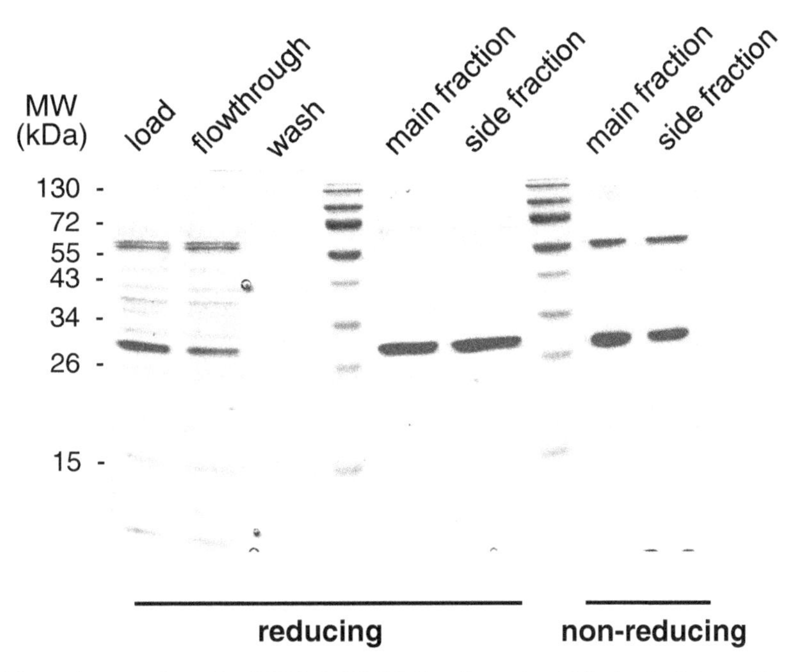

Fig. 2 Coomassie-stained SDS-PAGE (15 % polyacrylamide) of a Cysteine-scFv-6×His fusion protein purified from an *E. coli* periplasmic preparation. 20 μl (load, flow-through, wash) or 5 μl (main and side eluate fractions) was analyzed under reducing or non-reducing conditions

best placed either between the leader peptide and the recombinant protein or at the C-terminus of the construct to ensure best possible accessibility. Regarding this, a short peptide spacer comprising at least two small neutral amino acid residues (Gly, Ser or Ala) should be introduced between the affinity tag and adjacent protein sequences (Fig. 1b).

Before starting with the transfection, thaw a stock of HEK293 cells from an early passage and passage the cells two times in RPMI medium with 5 % FCS. Isolate Midiprep DNA of your construct(s). Dissolve the DNA in ddH_2O with a concentration of 0.5–2 μg/μl (*see* **Note 20**).

1. The day before transfection (day 0), trypsinize HEK293 cells and seed out $1.5–2 \times 10^6$ cells/well in a 6-well cell culture plate. Cultivate the cells overnight in 3 ml/well RPMI/5 % FCS at 37 °C, 5 % CO_2 (*see* **Note 21**).
2. (Day 1, afternoon) Prior to use, mix the Lipofectamine 2000 by vortexing or by pipetting up and down. For two 6-wells per construct, add 8 μg DNA and 16 μl Lipofectamine 2000 to two separate tubes containing 500 μl OptiMEM I each. Vortex both tubes shortly. Let the solutions incubate at room temperature for 5 min.
3. Combine both solutions and vortex shortly. Let the solution incubate at room temperature for 20 min.

4. Add 500 μl of the DNA-Lipofectamine 2000 complexes to each 6-well with HEK293 cells (*see* **Note 22**).
5. (Day 2) Trypsinize and expand transfected HEK293 cells to a 250 ml tissue culture flask (use one flask for each 6-well). Use RPMI/5 % FCS for cultivation.
6. (Day 3) Add 450 μg/ml G418 to the medium. Mix by rocking the flask gently (*see* **Note 23**).
7. Replace culture medium supplemented with 450 μg/ml G418 every second day (*see* **Note 24**).
8. A minimum of 2 weeks after starting antibiotic selection, preserve an aliquot of the stably transfected cells by cryoconservation. Expand the cells to 500 ml tissue culture flasks. You may now reduce the G418 concentration to 250 μg/ml (2.5–5 μg/ml when using a puromycin system). Let the cells grow to 90 % confluency. From one confluent 500 ml flask (the "master flask," which is always used for seeding out), split cells to all in all four flasks. When cells are at least 70 % confluent, exchange medium by 27 ml OptiMEM I (with exception of the master flask) and cultivate for 3 days (*see* **Note 25**).
9. After 3 days, pipette off the OptiMEM I medium and centrifuge it at 2,000 × *g* for 20 min at 4 °C. Collect the supernatant in a sterile bottle and store it at 4 °C. (Optional) Add again 25 ml OptiMEM I to the cells or let the cells in the flask revive for 1 day in RPMI/5 % FCS. Switch to OptiMEM I medium the next day and cultivate for another 3 days (*see* **Note 26**).

3.3 Purification of 6×His-Tagged Proteins Expressed in Mammalian Cells

To reduce degradation of the proteins carry out all steps at 4 °C and use cold buffers.

1. For handling smaller volumes of protein supernatant it is possible to concentrate the protein from the cell free OptiMEM I supernatant by ammonium sulfate precipitation. Therefore, add $(NH_4)_2SO_4$ stepwise to a final concentration of 390 g/l with continuous stirring. Upon complete resolution of the ammonium sulfate continue stirring for at least 1 h.
2. To pellet the precipitated protein, centrifuge the solution at 12,000 × *g* for 30 min at 4 °C and carefully discard the supernatant to not disrupt the very unstable pellet. Resuspend this pellet in 10 ml PBS.
3. To capture the His-tagged proteins, load 1 ml of Ni-NTA bead slurry onto the chromatography column (*see* **Note 27**). Remove the bead storage buffer by gravity flow and wash the beads in the column with 5 ml of PBS for equilibration (*see* **Note 28**).
4. Apply the protein solution slowly to the column to not disrupt the gel bed and drain by gravity flow (*see* **Note 29**). Save the flow-through for further analysis of the purification steps.

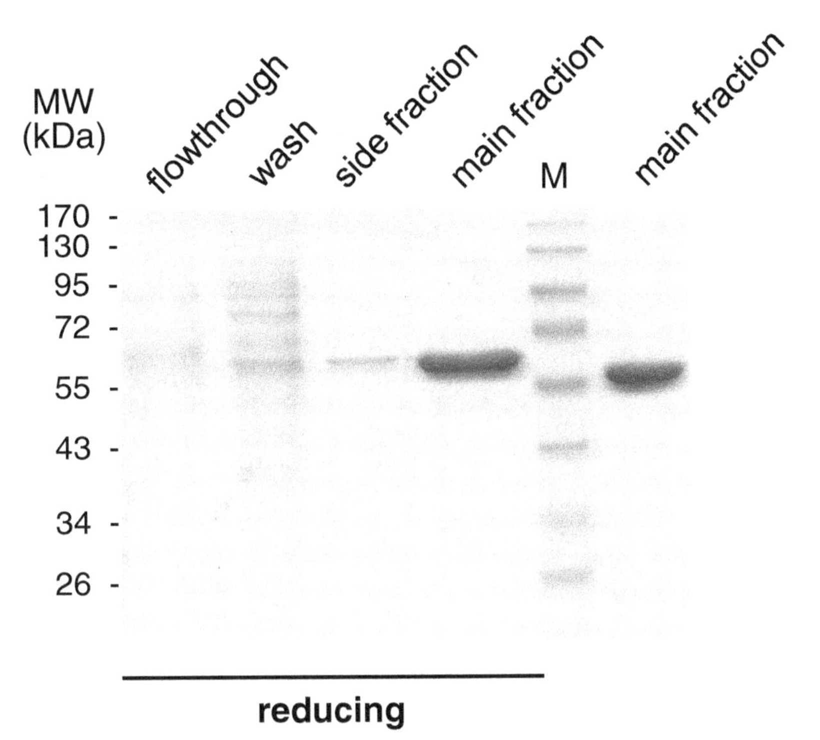

Fig. 3 Coomassie-stained SDS-PAGE (12 % polyacrylamide) of a scDb-6×His fusion protein purified from HEK293 tissue culture supernatant. 25 μl (flow-through, wash) or 5 μl (main and side eluate fractions) was analyzed under reducing or non-reducing conditions

5. For the wash step, run at least 20 ml (40 CV) of wash buffer containing 30 mM imidazole through the column and verify the protein content of the wash fractions via a Bio-Rad protein assay (*see* **Notes 17** and **30**). Collect the wash fractions separately.
6. When the wash step is completed, the bound protein can be eluted from the Ni-NTA beads by increasing the imidazole concentration to 250 mM. Subsequent to applying the elution buffer, collect 0.5 ml fractions and test them for their protein content using the Bio-Rad protein assay.
7. Pool the fractions with the highest protein content and dialyze overnight against 4 l of PBS (*see* **Note 31**).
8. To ascertain the quality of the purification, perform a SDS-PAGE analysis and load 5 μl of the dialyzed main and side fraction and 25 μl of the wash fraction and flow-through (Fig. 3).

3.4 Purification of FLAG-Tagged Proteins Expressed in Mammalian Cells

This protocol describes an efficient method for purification of up to 10 mg of FLAG-tagged recombinant protein from ~200 to ~900 ml tissue culture supernatant using a semi-batch procedure, which leads to higher yields than a column purification procedure. The method is suitable for the purification of antibody fragments as well as larger antibody fusion proteins. Unless indicated otherwise, the purification steps should be carried out at room temperature.

1. Mount the empty, clean chromatography column on a stand. Rinse the column twice with TBS, close the outlet of the column, and leave 2 ml TBS in the column.
2. Pipette an appropriate volume of M2 agarose in the column and allow the resin to settle for 5 min. Open the column outlet and allow the buffer to flow out (*see* **Note 32**).
3. Use 5 ml of TBS to wash adherent M2 agarose from the pipette and transfer the TBS to the column.
4. Connect the column lid with the silicone tubing and use the peristaltic pump to speed up draining by pressing air on the column. Drain the column (*see* **Note 33**).
5. Load 5–10 ml TBS on the column without disturbing the gel bed and let it drain completely.
6. For washing, apply three sequential column bed volumes (CV) of 0.1 M glycine HCl, pH 3.5 without disturbing the gel bed while loading. Let each aliquot drain completely before adding the next (*see* **Note 34**).
7. Repeat **step 5**. Then, fill the column with TBS for a second time and let it drain until 1-2 ml TBS remains above the gel bed. For equilibration, pump 10 CV of TBS through the material. After that, leave ~1–2 ml of TBS above the gel bed and close the column outlet.
8. Distribute the tissue culture supernatant to one or more 500 ml centrifuge bottles. Resuspend the M2 agarose in the overlaying TBS by pipetting up and down with a 5 ml serological pipette and distribute the suspension equally to the centrifuge bottles. Add TBS to completely recover the M2 agarose from the column. Fill the empty column with TBS (*see* **Note 35**).
9. Incubate the centrifuge bottles on a roller mixer at 4 °C for ~2 h.
10. Centrifuge bottles at 1,000–1,500 × *g* for 20 min (*see* **Note 36**).
11. Carefully remove the supernatant from the centrifuge bottles without disturbing the resin pellet and keep it in a separate bottle. Leave a few milliliters of supernatant to facilitate resuspension of the resin. Place the centrifuge bottles containing the resin pellets on ice.
12. (Optional) To maximize the resin recovery from the batch incubation, load the supernatant on the empty column by pumping at a high flow rate (8–10 ml/min). Reduce the flow rate, if column backpressure rises (*see* **Note 37**).
13. Use a 1 ml pipette to resuspend the resin and load the material on the empty column. Wash the 1 ml pipette and the wall of the centrifuge bottle with ~3 ml of TBS and load the wash on

the column. Drain column completely to form a compact column bed, but do not let the material run dry.

14. For washing, load five sequential CV of TBS and let each volume drain completely before adding the next.
15. For elution, load six sequential CV of TBS with 100 μg/ml of FLAG peptide and let each volume drain completely before adding the next.
16. Combine fractions containing protein and dialyze the eluate in at least 5 l of PBS at 4 °C overnight (*see* **Note 38**).
17. Repeat **steps 6** and 7, but add 0.02 % sodium azide to the TBS. Store the M2 agarose column at 4 °C (*see* **Note 39**).
18. The next day, recover the dialyzed eluate and centrifuge at 5,000 × *g* for 10 min at 4 °C. Transfer supernatant into a fresh tube.
19. Wash a Vivaspin 20 device twice with PBS according to the manufacturer's instruction and remove remaining PBS from the concentrate chamber after the second wash. Fill the concentrate chamber with the eluate and spin the device at the recommended speed (3,000–4,000 × *g* for 10,000 or 50,000 Da MW cutoff membranes) at 4 °C until the eluate is concentrated to a volume of ~4 ml (*see* **Note 40**).
20. Add ~15 ml PBS (when using a fixed angle rotor, reduce the volume) to the concentrate chamber and mix gently by pipetting. Spin the device until the eluate is concentrated to a volume of ~4 ml (*see* **Note 41**).
21. Repeat **step 20**, but spin the device until the eluate is concentrated to a volume of ~1–3 ml (*see* **Note 42**).
22. Mix the concentrated eluate in the retentate chamber by pipetting it up and down with a 200 μl pipette tip. Use a syringe with a needle to recover the concentrated protein (*see* **Note 43**).
23. Filter the concentrate with the 13 mm syringe filter into a sterile tube. Vortex the sample shortly and measure the protein concentration. Aliquot the purified protein and store it frozen or at 4 °C (*see* **Note 18**).

3.5 Two-Step Purification of Antibody Fusion Proteins (IMAC/Anti-FLAG)

This method can be used to purify FLAG-tagged antibody fusion proteins comprising an antibody fragment and a non-antibody moiety, for example a cytokine like TRAIL. As several other proteins, TRAIL has a natural affinity to Ni-NTA agarose even without fusion to a 6×histidine tag. Therefore, the following procedure is especially useful to get rid of protein fragments occurring during expression of larger antibody fusion proteins or to capture recombinant proteins from a large volume of supernatant in order to facilitate a subsequent anti-FLAG purification. When using IMAC for proteins without 6×histidine tags, the equilibration and wash buffers should contain no imidazole. In advance, you should check that your specific antibody fusion protein binds to Ni-NTA agarose in a downscaled pilot experiment.

For purification of 6×His-tagged proteins, please refer to Subheadings 3.1 and 3.3.

1. Follow the **steps 1–5** described in Subheading 3.4 using Ni-NTA agarose instead of anti-FLAG M2 affinity gel. Use ddH_2O instead of TBS in all steps (*see* **Note 44**).
2. For equilibration, run 10 CV of IMAC buffer A through the resin. After that, leave ~1–2 ml buffer A above the gel bed and close the column outlet.
3. Follow the **steps 8–14** described in Subheading 3.4 using Ni-NTA agarose instead of anti-FLAG M2 affinity gel. Use IMAC buffer A instead of TBS in all steps (*see* **Note 45**).
4. For elution, load six sequential CV of IMAC buffer B and let each volume drain completely before adding the next.
5. Combine fractions containing the protein.
6. (Optional) Dialyze the eluate in 3 l of PBS or TBS at 4 °C overnight (*see* **Note 46**).
7. Using as starting material the IMAC eluate in a 50 ml tube instead of tissue culture supernatant in large bottles, perform the anti-FLAG affinity purification described in Subheading 3.4.

3.6 Affinity Purification of Strep-Tagged Proteins Expressed in Mammalian Cells)

Due to the special binding kinetics of fusion proteins with Strep-tag II [18, 19] to the Strep-Tactin affinity matrix, a column purification instead of a batch purification is highly recommended. In general, Strep-tag purification works best, when the volume of protein extract for loading is not greater than 10 CV. To ensure higher yields from larger volumes of tissue culture supernatants containing low amounts of recombinant protein, a slow flow rate during protein loading is crucial. However, in this case, a fusion of the antibody format to the Twin-Strep-tag [21] may be more beneficial. Alternatively, if compatible with your specific antibody format, the volume of the supernatant can be reduced by ammonium sulfate precipitation (*see* Subheadings 3.3 and 3.7).

1. Take the volume of tissue culture supernatant intended for subsequent purification and check the pH. If necessary, adjust the pH to a value >7.5 by adding incremental amounts of 1 M Tris–HCl, pH 8.0.
2. Add 1.2 mg avidin to 100 ml of OptiMEM I tissue culture supernatant (*see* Subheading 3.2) and mix by inverting the bottle (*see* **Note 47**).
3. Load a chromatography column with the required amount of Strep-Tactin. Let the column drain by gravity flow to remove

the storage buffer. Equilibrate the material with 2 CV of Buffer W and let the column drain again (*see* **Note 48**).

4. Load ~5 ml of tissue culture supernatant on the equilibrated Strep-Tactin column and let some supernatant flow through the column by gravity. Using a peristaltic pump, load the residual tissue culture supernatant on the column. When the volume of supernatant exceeds 10 CV, it is important to maintain a flow rate of ~1 ml/min (*see* **Note 49**).
5. Wash the column five times with 1 CV of Buffer W, after tissue culture supernatant has completely entered the column (*see* **Note 50**).
6. For elution, add eight times 0.5 CV of Buffer E and collect the eluate in 0.5 CV fractions. Measure the protein concentration in the fractions and/or analyze fractions by SDS-PAGE.
7. If desired, combine fractions containing protein and dialyze the eluate in at least 3 l of PBS or TBS at 4 °C overnight. (optional) Concentrate and filter sterilize the eluate as described in Subheading 3.4 prior to storage (The PBS "wash" steps after initial concentration may be omitted in this case).
8. For regeneration of the Strep-Tactin matrix, wash the column three times with 5 CV of Buffer R. The color of the material will change from yellow to red. Remove Buffer R by adding two times 4 CV of Buffer W. Store the column at 4 °C overlaid with 2 ml of Buffer W.

3.7 Purification of Fc Fusion Proteins Expressed in Mammalian Cells by Protein A Affinity Chromatography

1. Place a glass bottle with a stir bar on a magnetic stirrer at 4 °C (cold room). Add ammonium sulfate stepwise (three to four steps) to the supernatant to a final concentration of 390 g per liter of OptiMEM I tissue culture supernatant to precipitate the protein. Always wait until the salt has dissolved before adding the next portion. After the addition of the last portion, let the supernatant stir (still at 4 °C) for at least 1 h. The supernatant should become turbid.
2. In the meantime, equilibrate the protein A resin. Take 0.5 ml of the slurry (beads in 20 % Ethanol in a 1:1 v/v mix) per liter of tissue culture supernatant, load it on a closed plastic chromatography column and let the resin settle. Open the column and let most of the 20 % EtOH drop into the waste. Apply 10–20 CV of PBS to the column to wash the beads (*see* **Note 28**).
3. Centrifuge the supernatant at 12,000 × *g* for 30 min at 4 °C. Discard the supernatant carefully and resuspend the pellet, containing the protein, in 20 ml of PBS per liter of tissue culture supernatant.
4. Apply the crude protein solution to the column and let it flow through it by the force of gravity (*see* **Notes 28** and **51**).

5. Apply at least 10 CV of PBS to the column, to wash away unspecifically bound protein. Test the protein content of the wash fractions by a Bradford protein assay or a similar quick assay (*see* **Note 17**).
6. Aliquot 50 μl of 1 M Tris–HCl (pH 7.5) into 10–20 1.5 ml reaction tubes. Elute the protein by adding 10–20 CV of glycine HCl, pH 2–4 in single aliquots of 500 μl to the column. Collect each eluted fraction in the prepared reaction tubes to neutralize the eluate instantly (*see* **Notes 52** and **53**).
7. Test protein content of the eluate fractions by the protein assay (*see* **Note 17**).
8. Regenerate the protein A resin by the addition of 10 CV of 0.1–0.5 M NaOH. When the entire regeneration buffer has passed through the column, add 5 ml 20 % ethanol for storage. Let half of the ethanol solution run through the column, seal it on bottom and top and keep it at 4 °C. Protein A resin can be used several times.
9. Keep samples of supernatant, crude protein solution, flow-through, wash (20 μl each), and eluted fractions (10 μl) for SDS-PAGE analysis (Fig. 4).
10. Wash the dialysis tubing both inside and outside with PBS and seal one side with a clip before filling the tube with the eluate.

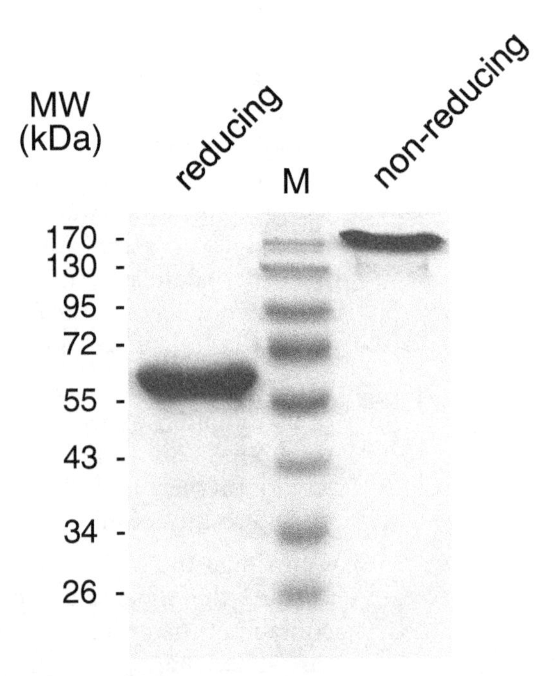

Fig. 4 Coomassie-stained SDS-PAGE (12 % polyacrylamide) of a scFv-Fc-6×His fusion protein purified from HEK293 tissue culture supernatant. 5 μl of eluate was analyzed under reducing or non-reducing conditions

Close the other end of the tube with another clip and avoid including air in the closed tube (*see* **Note 54**). Place the tube, containing the protein in at least a 1,000-fold excess of PBS and incubate stirring at 4 °C overnight.

11. The next day, take the dialysis tubes out of the buffer and carefully open the clip on the top (*see* **Note 55**). Transfer the sample to a 1.5 ml reaction tube and centrifuge at 16,100 × *g* for 1 min to remove precipitated protein. Transfer the supernatant to a new 1.5 ml reaction tube and determine the protein concentration.
12. Aliquot the protein samples in reasonable small volumes and store them at 4 °C (for short term) or −20 °C (*see* **Note 18**).

4 Notes

1. After autoclaving the solution, cool to a temperature of 60 °C and add glucose and ampicillin before pouring the plates.
2. Store in 1 ml aliquots at −20 °C.
3. Prepare fresh before use.
4. A 5× sodium phosphate buffer (250 mM Na_2HPO_4/NaH_2PO_4, 1.25 M NaCl) can be prepared as a stock solution. At the day of purification, add imidazole fresh to the 1× IMAC buffer and adjust pH to 7.5.
5. Heat-inactivate the fetal calf serum at 56 °C for 1 h before use.
6. For smaller molecules like scFvs or diabodies, concentrators with 10 kDa MW cutoff should work well. Preparations of larger antibody fusion proteins may be polished with 50 kDa cutoff concentrators, provided that the molecular weight of the protein is at least 1.5-fold greater than the cutoff. Where applicable, concentrators with 50 kDa MW cutoff offer an efficient way to remove smaller contaminating proteins or protein fragments.
7. The MW cutoff should be at least 5~10 kDa to ensure proper dialysis of the FLAG peptide.
8. Use buffers E and W without EDTA when purifying metalloproteins.
9. For the prokaryotic production of scFv molecules we use the *E. coli* strain TG1, which is an amber mutant and translates the nucleotide triplet TAG in glutamine instead of terminating the translation. The transformation is commonly performed with calcium chloride competent cells.
10. For the cultivation of the cells we recommend using flasks with baffles, which should be filled up to 50 % of the maximal volume to afford a sufficient air sparging. Flasks without baffles

should only be filled up to 25 % of the maximal volume. The reduced amount of glucose provides repression of the lac promoter until an OD_{600} of 0.8–1.0 is reached.

11. You can resuspend the pellet efficiently by firstly detach the pellet with the help of a Pasteur pipette and secondly dissolve the cells by swinging the beaker around. Finally, resuspend remaining cell aggregates by pipetting up and down.
12. We recommend using dialysis tubes with a cutoff size of 50 % of the protein of interest or smaller.
13. Due to the high osmotic properties of the PPB, the dialysis tube may be under pressure. Prior opening the dialysis tube, place the tube into a beaker to ensure collecting the whole preparation.
14. The Ni-NTA beads are stored in a 50 % aqueous suspension containing 30 % ethanol. Thus, 1 ml of Ni-NTA beads slurry contains 500 μl beads (column bed volume). We commonly use 500 μl Ni-NTA bead slurry per 1 l bacterial culture. 1 ml bed volume of Ni-NTA beads can bind up to 10 mg protein.
15. We recommend performing a batch purification due to higher purification yields compared to a column purification.
16. Let the periplasmic preparation drain completely but avoid the running dry of the beads bed.
17. It could prove beneficial to collect the first milliliters of the first wash fraction separately, if the IMAC washing buffer is too stringent and elutes the protein of interest. We test the protein content of the wash fractions using a Bio-Rad protein assay. Therefore, mix 10 μl of flow-through with 90 μl of 1× protein assay in one well of a 96-well plate and place the plate on a white surface (e.g., a sheet of paper). The washing of the Ni-NTA beads is sufficient when the protein assay does not stain blue.
18. When freezing the protein at −20 °C we recommend aliquoting the protein in small fractions to avoid freeze and thaw cycles. Before freezing your protein, test with a small aliquot, if freezing and thawing affects the solubility.
19. Note: For SDS-PAGE analysis of scFv molecules we use a gel containing 15 % polyacrylamide.
20. To further purify the DNA from contaminants and impurities, we recommend an *n*-butanol precipitation. To this, mix 50 μl of DNA with 500 μl *n*-butanol (unsaturated) by vigorous vortexing for 30 s in a 1.5 ml reaction tube. Centrifuge at 16,000×*g* at 16 °C for 20 min with the tube hinge facing outwards. A small DNA pellet is visible in most cases (however, when you do not see a pellet, don't worry). Pipette off the supernatant by placing the pipette tip on the tube wall opposite to the hinge (at this point, it is not necessary to remove the supernatant completely).

Add 500 μl of 70 % ethanol (pre-cooled at −20 °C) and mix shortly. Centrifuge at 16,000×*g* at 4 °C for 10 min with the tube hinge facing outwards. Pipette off the supernatant as in the former step (this time as much as possible). Let the DNA air dry under a clean bench until no liquid is remaining in the tube (approx. 15–30 min). Add 50 μl of sterile ddH_2O to the DNA and let it stand at 4 °C for ~1 h (or overnight). Mix by vortexing and flicking the tube and measure DNA concentration.

21. For one DNA construct to be transfected, seed out two wells of a 6-well plate.
22. Distribute the complexes dropwise over the whole area of the well. Mix by rocking the plate gently.
23. Do not replace the medium at this point. For vectors with puromycin resistance use 8 μg/ml puromycin for selection. Basically, the antibiotic concentration depends on the specific activity of the reagent and the cell line. To determine your individual conditions for selection, we recommend generating a "kill curve." To this, titrate the antibiotic and use for selection the lowest concentration, which is necessary to kill all cells during 4–6 days of incubation.
24. After 3 days of antibiotic incubation, non-transfected cells begin to die rapidly. A frequent medium exchange (every second day) is necessary at this stage to remove dead cells. After approximately 1 week, when only the usually small fraction of recombinant cells remains, a medium exchange every third or fourth day is sufficient.
25. It is sufficient to add the antibiotic only to the "master flask" used for seeding out. When HEK293 cells are split 1:4, it takes approximately 3–4 days until they reach the required confluency for a switch to OptiMEM I. We generally do not add FCS to the OptiMEM I to facilitate protein purification. We recommend to replace the RPMI/5 % FCS medium after 2 (3) days for boosting the cell growth.
26. Repeat the **steps 8** and **9** as required to collect enough tissue culture supernatant for purification. We found that stably transfected HEK293 cells tolerate at least two "sessions" with OptiMEM I (3 days each) before they detach for the most part. Check the supernatants by western blotting for expression and integrity of the recombinant protein.
27. 1 ml of bead slurry contains 500 μl of beads, which is enough to purify 6 mg of a 6×His-tagged scDb.
28. It is important that the liquid level never reaches the settled beads to avoid the running dry of the gel bed. At each time point of the purification procedure the column can be capped to avoid further draining.

29. Alternatively to loading the dissolved protein onto the column, the equilibrated beads in 1 ml PBS are transferred to the protein solution and incubated overnight on a roller mixer. The next day, the protein-bead suspension is added to an empty column and the supernatant is removed by gravity flow. This method is preferable, because it ensures a more homogenous and quantitative binding of the proteins to the column material.
30. Best washing results can be obtained when four times 5 ml of washing buffer are loaded and drained completely, before adding the next 5 ml.
31. Usually, fraction 2 and 3 contain most of the eluted protein. When the purified protein tends to precipitate after dialysis add, an additional elution fraction. By doing so, the concentration might decrease, but the yield can be increased.
32. M2 agarose is a 50 % slurry in glycerol containing buffer and is usually very viscous when taken directly from the −20 °C freezer. For easier pipetting we recommend to bring the M2 agarose to room temperature by gentle warming by hand. Rolling and inverting the bottle is important to avoid settling of the resin before pipetting. As a guideline for the required amount of M2 agarose, we found that 1 ml bed volume M2 agarose (=2 ml of the 50 % slurry) binds at least 1 mg of recombinant protein.
33. We recommend a flow rate of 2 ml/min. *Important!* Stop draining immediately by twisting up the column cap before the liquid level reaches the settled material to avoid running dry of the gel bed. This technique should be used in all steps where single aliquots of buffer or resin suspension are loaded on the column.
34. In order to protect the binding capacity of the M2 agarose, it is very important to leave the material in glycine HCl for not longer than 15 min.
35. If your centrifuge model is incompatible with Corning bottles, glass bottles can be used for batch incubation of tissue culture supernatant and M2 agarose. Resuspend the M2 agarose in small volumes of TBS to minimize the absorption of the material to the pipette. Do not overfill the bottles to ensure proper mixing during the following incubation step.
36. When using glass bottles instead of centrifugation bottles, let them stand for ~15 min to allow the resin to settle. Use a 10 ml serological pipette to load the settled resin on the empty column and let it drain. We recommend applying air pressure by the peristaltic pump to speed up draining. The supernatant still contains resin and should be centrifuged in 50 ml tubes at 1,000 × *g* for 10 min. Carefully pipette off the supernatant and

collect the resin in the chromatography column. Due to the adsorption of the resin to the bluecap tubes, it is better to minimize the number of tubes and to repeat the centrifugation step, if necessary.

37. M2 agarose is sometimes difficult to separate by centrifugation. This step will catch any unsettled resin in the supernatant. However, one should schedule ~2 h for pumping 1 l supernatant through the column. Keep the flow-through at 4 °C.
38. Because of the selectivity of M2 agarose and as fractions are usually not very concentrated, it is in principle not necessary to determine the protein concentration in each fraction. Simply combine the fractions in one collection tube prior to dialysis. Pre-cool the PBS at 4 °C.
39. You may stop neutralization with TBS as soon as the column flow-through has pH ~7.5 (can be checked using a pH indicator stick).
40. Depending on the protein concentration, this step will take approximately 45–60 min. Interrupt the centrifugation every 15–20 min to check the progress and to mix the sample by pipetting up and down using a 1 ml pipette. This is important to avoid the formation of concentration gradients at the membrane during centrifugation. Although the filtrate should be free of target protein, you should preserve it at 4 °C.
41. The “wash” with PBS removes a lot more FLAG peptide from the sample than the dialysis step alone.
42. In general, also depending on the expected protein yield, we recommend concentrating the eluate tenfold.
43. For maximum protein recovery, add 200 μl PBS to the empty concentrate chamber and flush the walls comprising the membranes with the buffer repeatedly. Combine the wash with the concentrate in the syringe.
44. Note: 1 ml bed volume of Ni-NTA agarose can bind 5–10 mg protein.
45. We prefer a batch purification protocol instead of a column purification due to more quantitative binding of antibody fusion proteins to the IMAC matrix.
46. As a possibility to speed up the procedure, we found that IMAC buffer B has no significant influence on the performance of the anti-FLAG affinity gel.
47. Avidin will capture free biotin in the medium which otherwise would disturb the purification by binding to Strep-Tactin. Adjust the amount of avidin for greater volumes of supernatant.
48. 1 ml bed volume of Strep-Tactin will bind 50–100 nmol Strep-tag fusion protein.

49. Depending on the volume of supernatant, this step may take several hours and is conveniently done overnight. When loading overnight, carefully adjust the flow rate to avoid running dry of the column. It is recommended to perform this step in a cold room at 4 °C.
50. From this point on, work once again with gravity flow.
51. Alternatively, the binding of the protein can be performed in a batch process to allow for a longer contact time of protein and beads. Equilibrate the beads by centrifugation (3,000 × *g*, e.g., in a 50 ml reaction tube), discard the supernatant and resuspend the beads in 10 ml PBS. Centrifuge the beads again, discard the supernatant and resuspend the beads in a 5 ml aliquot of the sterile filtered tissue culture supernatant or crude protein extract in PBS after ammonium sulfate precipitation. Combine this suspension with the main amount of the starting material and stir at 4 °C overnight. Centrifuge the suspension at 3,000 × *g*, discard most of the supernatant, resuspend the beads in 10 ml of the residual supernatant and apply to the column. Proceed with the protocol as described.
52. The suitable pH of the elution buffer depends on the specific interaction of the produced protein (e.g., ref. 22).
53. It can be helpful to let the first two to three drops run into the waste, to get rid of the residual wash buffer. Most of the protein usually elutes in the fractions 1–4.
54. Use clips that enclose some air to ensure that the dialysis tubing floats closely to surface. Alternatively, fix the tube at the clips to the beaker, to avoid contact of the tubing and the stir bar.
55. The tubing might have a high internal pressure. To avoid loss of protein, place a beaker underneath the tubing when you open the clip.

References

1. de Lorenzo C, Tedesco A, Terrazzano G et al (2004) A human, compact, fully functional anti-ErbB2 antibody as a novel antitumour agent. Br J Cancer 91:1200–1204
2. Jazayeri JA, Carroll GJ (2008) Fc-based cytokines. Prospects for engineering superior therapeutics. BioDrugs 22:11–26
3. Seifert O, Plappert A, Heidel N et al (2012) The IgM CH2 domain as covalently linked homodimerization module for the generation of fusion proteins with dual specificity. Protein Eng Des Sel 25:603–612
4. Müller D, Frey K, Kontermann RE (2008) A novel antibody-4-1BBL fusion protein for targeted costimulation in cancer immunotherapy. J Immunother 31:714–722
5. Ortiz-Sánchez E, Helguera G, Daniels TR et al (2008) Antibody-cytokine fusion proteins: applications in cancer therapy. Expert Opin Biol Ther 8:609–632
6. Siegemund M, Pollak N, Seifert O et al (2012) Superior antitumoral activity of dimerized targeted single-chain TRAIL fusion proteins under retention of tumor selectivity. Cell Death Dis 3:e295
7. Wolf P, Elsässer-Beile U (2009) Pseudomonas exotoxin A: from virulence factor to anti-cancer agent. Int J Med Microbiol 299:161–176
8. Potala S, Sahoo SK, Verma RS (2008) Targeted therapy of cancer using diphtheria toxin-derived immunotoxins. Drug Discov Today 13: 807–815

9. Schirrmann T, Krauss J, Arndt MAE et al (2009) Targeted therapeutic RNases (ImmunoRNases). Expert Opin Biol Ther 9:79–95
10. Müller D, Kontermann RE (2010) Bispecific antibodies for cancer immunotherapy: current perspectives. BioDrugs 24:89–98
11. Wu C, Ying H, Grinnell C et al (2007) Simultaneous targeting of multiple disease mediators by a dual-variable-domain immunoglobulin. Nat Biotechnol 25:1290–1297
12. Gu J, Ghayur T (2012) Generation of dual-variable-domain immunoglobulin molecules for dual-specific targeting. Meth Enzymol 502: 25–41
13. Khan F, Legler PM, Mease RM et al (2012) Histidine affinity tags affect MSP1(42) structural stability and immunodominance in mice. Biotechnol J 7:133–147
14. Wu L, Su S, Liu F et al (2012) Removal of the tag from His-tagged ILYd4, a human CD59 inhibitor, significantly improves its physical properties and its activity. Curr Pharm Des 18: 4187–4196
15. Waugh DS (2005) Making the most of affinity tags. Trends Biotechnol 23:316–320
16. Hopp TP, Prickett KS, Price VL et al (1988) A short polypeptide marker sequence useful for recombinant protein identification and purification. Nat Biotechnol 6:1204–1210
17. Einhauer A, Jungbauer A (2001) The FLAG peptide, a versatile fusion tag for the purification of recombinant proteins. J Biochem Biophys Methods 49:455–465
18. Schmidt TG, Skerra A (1993) The random peptide library-assisted engineering of a C-terminal affinity peptide, useful for the detection and purification of a functional Ig Fv fragment. Protein Eng 6:109–122
19. Skerra A, Schmidt TG (1999) Applications of a peptide ligand for streptavidin: the Strep-tag. Biomol Eng 16:79–86
20. Kontermann RE, Martineau P, Cummings CE et al (1997) Enzyme immunoassays using bispecific diabodies. Immunotechnology 3: 137–144
21. Dammeyer T, Timmis KN, Tinnefeld P (2013) Broad host range vectors for expression of proteins with (Twin-) Strep-tag, His-tag and engineered, export optimized yellow fluorescent protein. Microb Cell Fact 12:49
22. Duhamel RC, Schur PH, Brendel K et al (1979) pH gradient elution of human IgG1, IgG2 and IgG4 from protein A-sepharose. J Immunol Methods 31:211–217

Chapter 19

Purification of Antibodies and Antibody Fragments Using CaptureSelect™ Affinity Resins

Pim Hermans, Hendrik Adams, and Frank Detmers

Abstract

Ever since the introduction of bacterial derived surface proteins like protein A that demonstrate a natural binding reactivity towards antibodies, affinity chromatography has evolved into a well-established technology for the purification of antibodies and antibody fragments. Although high selectivity is provided by these types of affinity ligands, not all antibodies or antibody fragments are covered, which then forces the use of non-affinity-based processes that are less selective and often result in lower one-step purity and yield. To fill these gaps, we here describe a novel range of CaptureSelect™ affinity resins that enables immunoaffinity chromatography for a much broader range of antibody targets.

Key words Antibody purification, Affinity chromatography, Immunoaffinity chromatography, Antibody fragments, CaptureSelect™ affinity resins

1 Introduction

Affinity chromatography (AC) is defined as a liquid chromatographic technique that makes use of a reversible biological interaction for the separation and analysis of specific targets within a sample [1, 2]. The binding agent, or affinity ligand, that selectively interacts with the desired target is immobilized onto a solid support in order to create an affinity matrix that can be used in a column format. Due to its selectivity, which generates high target purity in a single step, the basic concept of AC is straightforward and generally applicable by following the steps of:

- Binding of the target under conditions that favor specific binding to the affinity matrix.
- Washing out of contaminants in the non-bound fraction.
- Recovering the bound target under conditions that favor desorption (elution).

Vincent Ossipow and Nicolas Fischer (eds.), *Monoclonal Antibodies: Methods and Protocols*, Methods in Molecular Biology, vol. 1131, DOI 10.1007/978-1-62703-992-5_19,

Since the introduction of *Staphylococcus aureus* protein A in the early 1970s as a bacterial derived affinity ligand that interacts with the Fc domain of many IgG antibodies [3], protein A-based affinity chromatography has matured as the most common technique for antibody purification and has been applied at large scale for the downstream processing of human IgG-based therapeutics [4, 5]. Other bacterial proteins like protein G and L also show high selectivity in antibody binding through recognition of either the IgG Fc domain (protein G) [6] or antibody light chains that belong to certain kappa subclasses (protein L) [7]. However, since affinity binding is restricted to antibodies or antibody fragments that display an epitope recognized by either of these affinity ligands, it is clear that antibody molecules lacking such an epitope are excluded from AC applications. Alternative chromatographic principles like those based on ion exchange (IEC) or hydrophobic interactions (HIC) are less selective for the target antibody and hamper high purity in a one-step process as typically obtained with AC. We here describe a set of antibody-based affinity resins (CaptureSelect™) that targets a broader variety of antibody epitopes and facilitates immunoaffinity chromatography (IAC) for a more complete range of antibody targets.

CaptureSelect™ affinity ligands are derived from the unique heavy chain antibodies found in *Camelidae* [8, 9]. These heavy chain antibodies are devoid of the entire light chain and C_H1 domain found in conventional antibodies. As a result, antigens are bound by the variable domain of the heavy chain only (V_HH). Because this variable domain exists as a single polypeptide chain, it is an extremely stable and easy to produce antibody fragment that retains the full antigen binding activity of the parent heavy chain antibody [10, 11]. These features are fully utilized when employed as binding agents in IAC applications since it combines high target selectivity by its antibody origin with the robustness of being a small sized affinity ligand that can withstand multiple cycles of binding, desorption, and regeneration conditions [12–18].

We have developed a range of CaptureSelect™ affinity resins directed against a unique panel of antibody sub-domains that provides a so-called Antibody Toolbox® for researchers and manufacturers to facilitate IAC-based purification of virtually any antibody format. An overview of the different antibody sub-domains covered by this Antibody Toolbox is schematically represented in Fig. 1. More detailed information on binding characteristics and applications for each CaptureSelect™ affinity resin is summarized in Table 1 and can be found in Subheading 4.

Notable differences of CaptureSelect™ affinity resins with protein A, G, or L-based products are further highlighted. For instance, the ability to distinguish between antibodies from different species, like CaptureSelect™ affinity resins that specifically bind to human antibodies without cross-reactivity towards mouse or bovine antibodies.

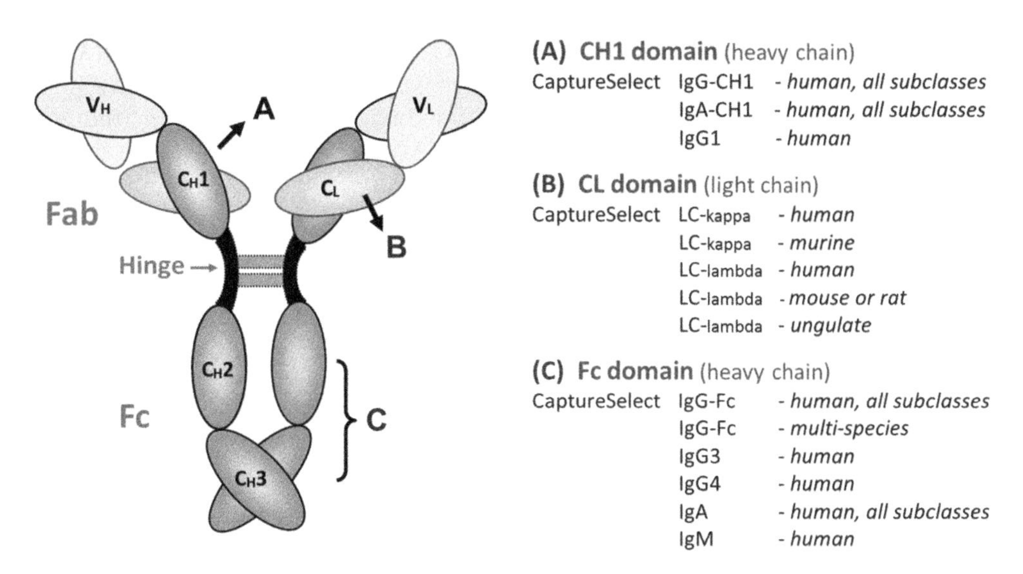

Fig. 1 Schematic representation of an antibody and its sub-domains recognized by the different CaptureSelect™ affinity resins. (A) The heavy chain CH1 domain present in the Fab portion (B) the constant domain of an antibody light chain (CL) of either the kappa—or lambda subclass present in the Fab portion of all antibodies (C) the Fc portion of an antibody consisting of two identical heavy chain regions

This feature is exploited by Kuroiwa et al. [12] who included aforementioned affinity resins in a process to specifically purify human polyclonal antibodies from transgenic cattle that co-expressed autologous bovine IgG antibodies. The range of affinity resins also includes tools for the purification of non-IgG isotypes, like IgM and IgA. For the purification of recombinant human IgA, Reinhart et al. [13] reported >95 % recovery from concentrated culture supernatants and exceptionally high IgA purity levels in just a single step using CaptureSelect™ IgA. Besides antibody purification, the selectivity provided by IAC can also be applied for the scavenging of antibody-related contaminants. Sun et al. [14] showed efficient removal of porcine IgA impurities using CaptureSelect™ IgA in a process to purify human Factor IX from transgenic sow milk. Fine selectivity is further illustrated by CaptureSelect™ affinity resins capable of separating different human IgG subclasses like IgG1, 3 and 4. Klooster et al. [15] describes the successful use of the CaptureSelect™ IgG4 affinity resin in the isolation of polyclonal IgG4 antibodies from crude human plasma. For antibody Fab fragments several options have now become available that, unlike protein A, G, or L, enable a generic approach for Fab purification by targeting the constant domains of a Fab fragment present on its heavy chain (C_H1) or light chain (C_L).

For smaller fragments that only consist of the variable domain(s) of an antibody, like single chain Fv fragments (scFv), Nanobodies® or other single domain antibodies, no CaptureSelect™

affinity resin is readily available that can cover this broad variety of variable domains. For these types of antibody constructs, however, purification is often facilitated by introduction of an affinity tag at the N- or C-terminus like the commonly used hexa-histidine or FLAG tag. We here describe the use of a novel affinity tag system that makes use of a CaptureSelect™ affinity ligand exhibiting high selectivity for a C-terminal tag (C-tag) that consists of only four residues (i.e., E-P-E-A). When fusing this short tag at the C-terminus of for example a single domain antibody, IAC can be applied for the purification of such domains from any type of feedstock [16].

All CaptureSelect™ affinity ligands are covalently linked to agarose-based solid supports that are capable of withstanding the flow rates and pressures that are typical for fast protein liquid chromatography (FPLC) applications. Besides for FPLC these resins can also be used for bench-top antibody purification using disposable spin- or gravity-flow columns. Due to the high selectivity provided by this type of affinity resins, the purification of antibodies and antibody fragments can be achieved in any type of feedstock ranging from mammalian, plant, or microbial based expression systems to crude body fluids like plasma, serum, or ascites. In general, the optimal binding condition for samples is around physiological pH and ionic strength. A good general binding buffer is Phosphate-buffered saline (PBS, pH 7.2–7.4). Recovery of the bound antibody target can be achieved by standard acidic elution buffers such as 100 mM Glycine pH 3, followed by neutralization with a Tris–HCl buffer pH 8. For whole IgG antibodies the dynamic binding capacity (DBC) of the affinity resins that can be achieved ranges between 10 and 25 mg IgG/ml resin using a linear flow of 150 cm/h. More detailed information on methods and applications is described in the sections below.

2 Materials

2.1 Packing of CaptureSelect Affinity Resins

1. The current range of CaptureSelect™ affinity resins is listed in Table 1. The affinity resins are named after the binding region of the immunoglobulin recognized and the species specificity.
2. GEHC Tricorn 5/20–10/600 or XK16/20–50/100 columns or equivalent.
3. Solutions for column packing: 150 mM NaCl or PBS (pH 7.4).

2.2 Packed Bed Format or FPLC Purification

1. GEHC AKTA Explorer or equivalent, including a fraction collector.
2. Polyclonal antibody mixtures, such as crude plasma, serum or intravenous immunoglobulin (IVIG) from plasma, or

Table 1
List of CaptureSelect™ affinity resins

CaptureSelect[a]™ product code	Sub-domain	Antibody formats	Main species reactivity	Negative	DBC[b] (mg/ml)	Elution	*See* Notes
IgG all subclasses							
IgG-CH1	CH1	Whole IgG Fab fragments	**Human** Pr, Hr, Dog, Cat, Gp, Do	**Bovine** Mo, Rat, Rb	>20	pH 3–2	**1** and **2**
IgG-Fc (Hu)	Fc	Whole IgG Fc fragments	**Human** Pr	**Bovine** Mo, Rat, Rb	>20	pH 3–2	**3**, **4**, **8**, and **9**
IgG-Fc (ms)	Fc	Whole IgG Fc fragments	**Human, Rat, Mouse** Pr, Gp, Bo, Hr, Gt, Sh	Chicken	>15	pH 3–2	**5**
Light chains							
LC-kappa (Hu)	CL-kappa	IgG, A, M, D, E Fab fragments	**Human** Pr	**Bovine** Mo, Rat, Rb	>20	pH 2	**6–9**
LC-kappa (mur)	CL-kappa	IgG, A, M, D, E Fab fragments	**Mouse, Rat** Gp	**Bovine, Horse** Hu, Rb	>10	pH 3	**10**
LC-lambda (Hu)	CL-lambda	IgG, A, M, D, E Fab fragments	**Human** Pr	**Bovine** Mo, Rat, Rb	>10	pH 3–2	**6–9**
LC-lambda (mouse)	CL-lambda	IgG, A, M, D, E Fab fragments	**Mouse**	**Bovine** Rat	>10	pH 3	**10**
LC-lambda (rat)	CL-lambda	IgG, A, M, D, E Fab fragments	**Rat**	**Bovine** Mouse	>10	pH 3	**10**
LC-lambda (ung)	CL-lambda	IgG, A, M, D, E Fab fragments	**Bovine, Horse** Do, Gt, Sh, Pig	**Mouse, Rat** Hu, Rb	10–15	pH 3	**11**

(continued)

Table 1 (continued)

CaptureSelect[a]™ product code	Sub-domain	Antibody formats	Main species reactivity	Negative	DBC[b] (mg/ml)	Elution	*See* Notes
IgG subclasses							
IgG1 (Hu)	CH1	Whole IgG1 Fab fragments	**Human** Pr	**Bovine** Mo, Rat, Rb	>8	pH 3	**12**
IgG3 (Hu)	Fc (hinge)	whole IgG3	**Human**	**Bovine** Mo, Rat, Rb	>6	pH 3	**12**
IgG4 (Hu)	Fc	whole IgG4 Fc fragments	**Human** Pr	**Bovine** Mo, Rat, Rb	>6	pH 3	**12**
Other isotypes							
IgM	Fc	IgM Fc fragments	**Human, Mouse, Rat** Pr	**Bovine** Rb	>2.5	pH 3–2	**13**
IgA (Hu)	Fc	IgA, sIgA Fc fragments	**Human** Pr	**Bovine** Mo, Rat, Rb	>8	pH 3	**9, 14–17**
IgA-CH1 (Hu)	CH1	IgA, sIgA Fab fragments	**Human** Pr	**Bovine** Mo, Rat, Rb	>6	pH 3 pH 7 (2 M $MgCl_2$)	**14–17**
Affinity TAG							
C-Tag	C-term peptide: -E-P-E-A	C-tagged Ab domains	**n.a.**	**n.a.**	>5	pH 3 mild neutral	**18**

Hu human, *Pr* primates, *Bo* bovine, *Hr* horse, *Gt* goat, *Sh* sheep, *Do* donkey, *Gp* guinea pig, *Mo* mouse, *Rb* rabbit, *Ch* chicken, *ms* multispecies, *mur* murine, *ung* ungulate (hoofed animals like horse, sheep, cow)

[a]For large-scale manufacturing processes that require cGMP standards several CaptureSelect based affinity resins are made available by GE Health Care (GEHC) (*see* **Note 22**)

[b]DBC determined for whole antibodies at a flow rate of 150 cm/h. For more detailed information on the Antibody Toolbox affinity resins, please refer to http://www.lifetechnologies.com/captureselect

recombinant monoclonal antibodies, Fc-fusions or Fab fragments from any type of feedstock ranging from mammalian, plant, or microbial based expression systems.

3. PBS (pH 7.2–7.4), which can be used as equilibration and as binding/running buffer.
4. 100 mM Glycine (pH 3), which can be used as elution buffer (*see* **Note 19**).
5. 1 M Tris–HCl (pH 8), which can be used as neutralization buffer.
6. 100 mM Glycine buffer (pH 2), which can be used as strip buffer.

2.3 Spin Column Format Purification

1. Appropriate CaptureSelect™ affinity resin.
2. Mobicol "F" spin columns (Mobitec) or equivalent (*see* **Note 20**).
3. Feedstock containing antibodies as described in Subheading 2.2, **item 2**.
4. Binding buffer: PBS (pH 7.4).
5. Elution buffer: 100 mM Glycine (pH 3).
6. Neutralization buffer: 1 M Tris–HCl (pH 8).
7. Strip buffer: 100 mM Glycine buffer (pH 2).
8. Rotating wheel or equivalent apparatus for the incubation and mixing of the spin columns.
9. Eppendorf centrifuge with variable speed or equivalent device.

2.4 Sodium Dodecyl Sulfate Polyacrylamide Gel Electrophoresis (SDS-PAGE)

Protein samples (start, flow-through, and elution fractions) can be analyzed by SDS-PAGE, followed by Coomassie Blue staining of the gels.

1. Novex Tris–Glycine SDS Sample Buffer (Life Technologies).
2. Novex 4–20 % Tris–Glycine gel (12 or 15 wells; Life Technologies).
3. XCell SureLock Mini-Cell (Life Technologies).
4. Novex 10× Tricine SDS Running Buffer (Life Technologies).
5. Coomassie brilliant blue (CBB) such as PhastGel Blue R (GEHC).
6. CBB destaining solution (40 % ethanol, 10 % HAc).
7. Heating block or water bath.
8. Powerpack 300 power supply (Bio-Rad laboratories) or equivalent.

2.5 Single-Step Purification of Polyclonal IgM from Human Serum by FPLC

1. Crude human serum.
2. CaptureSelect™ IgM affinity resin (*see* Table 1).
3. GEHC Tricorn 10/100 column or equivalent.
4. Binding buffer: PBS (pH 7.4).
5. Elution buffer: 100 mM Glycine (pH 2.0).
6. Neutralization buffer: 1 M Tris–HCl (pH 8).

2.6 Fab Isolation from Papain Digested Monoclonal Antibody Preps

1. Papain-immobilized agarose resin (Pierce).
2. Monoclonal antibody preparation (concentration >1 mg/ml).
3. CaptureSelect™ IgG-CH1 and CaptureSelect™ IgG-Fc (Hu) resin (*see* Table 1).
4. GEHC Tricorn 5/20 column or equivalent.
5. GEHC AKTA Explorer or equivalent, including a fraction collector.
6. Binding buffer: PBS (pH 7.4).
7. Elution buffer: 100 mM Glycine (pH 3).
8. Neutralization buffer: 1 M Tris–HCl (pH 8).
9. Dialysis tubing such as Spectra/Por 6 dialysis tubing (3.5 K MWCO) (Spectrumlabs).

2.7 Purification of a Single Domain Antibody Equipped with a C-Terminal E-P-E-A Tag by FPLC

1. *E. coli* feedstock containing a recombinant E-P-E-A tagged single domain antibody fragment.
2. CaptureSelect™ C-tag resin (*see* Table 1).
3. GEHC Tricorn 5/20–10/600 or XK16/20–50/100 columns or equivalent.
4. GEHC AKTA Explorer or equivalent, including a fraction collector.
5. Running buffer: PBS (pH 7.4).
6. Elution buffer: 100 mM Glycine (pH 2.0).
7. Neutralization buffer: 1 M Tris–HCl (pH 8).

3 Methods

3.1 Packing of CaptureSelect Affinity Resins

All CaptureSelect™ affinity resins are supplied in 20 % v/v ethanol and should be stored at +4 °C (39 °F). The resins can be packed much like any other chromatography adsorbent and depending on scale and equipment availability are suitable for slurry packing procedures.

1. It is recommended that a homogenous 1:1 distribution of resin and storage solution (20 % v/v ethanol) is established by careful mixing of the resin container, by inversion of the container or

by use of a paddle stirrer. The use of mechanical stirrers, homogenizers or similar devices is not recommended.

2. Once a homogenous slurry (1:1) has been achieved, the slurry should be poured down the side of the column into approximately 0.1 column volume (CV) of 150 mM NaCl or PBS. Alternatively, the slurry can be pumped into the column using a peristaltic pump.
3. The volume of slurry required depends on the final bed height, the preferred packing flow rate and bed compression. CaptureSelect™ affinity resins compresses under flow rate of 600 cm/h by 15–20 % and this may increase to 20–25 % compression at higher packing flow rates.
4. Once the required volume has been added to the column, the top plunger should be lowered to 2–3 cm above the slurry (or less in case small columns are used) and then packed at flow rate of 1.5× to 2× the operational flow rate (i.e., 225–300 cm/h for an operational flow rate of 150 cm/h). Recommended packing solutions are either 150 mM NaCl or PBS. Monitor pressure and bed compression, and when bed compression is complete stop the flow. Note that when buffer flow is stopped, the bed will expand.
5. Adjust the top plunger to ~0.5 cm below the packed bed height observed when the adsorbent was being packed under flow. Avoid the ingress of air.
6. Continue packing at the applied packing flow rate for 2 column volumes (CV) to confirm no further bed compression.
7. If bed compression does occur repeat **step 4**.
8. Once packing is complete, adjust flow to operational flow rate (e.g., 150 cm/h) and continue to pump through packing solution for 2 CVs.

3.2 Packed Bed Format or FPLC Purification

The antibody of interest can be isolated at preparative scale by applying FPLC. For most cases 1–5 ml columns will be sufficient (equivalent to at least ~5–50 mg of purified IgG per run, respectively), but the method is also applicable for larger columns.

1. Carefully pack the CaptureSelect™ affinity resin as described in Subheading 3.1.
2. Sample can be applied on the column using PBS (or equivalent) as loading buffer. The amount of sample that can be loaded is depending on the concentration of the target molecule in your sample and the DBC of the affinity resin (*see* Table 1). Typical flow rate for sample loading is around 150 cm/h.
3. After sample application, the column should be washed with running buffer (e.g., PBS, pH 7.4) until the baseline has been re-established. A typical wash is 5–10 CV's.

4. Elution of the antibodies from the affinity matrix is achieved by using 5 CV's of a suitable elution buffer (e.g., 100 mM Glycine pH 3) using a linear flow up to 300 cm/h. For direct neutralization, add 1 M Tris–HCl (pH 8) to the collection tube using 1/10 volume of the elution volume applied.
5. The CaptureSelect™ affinity resin can be stripped with for example 100 mM Glycine (pH 2) followed by re-equilibration in running buffer to allow a second affinity purification run.
6. If the column will not be used immediately, the matrix should be washed in water and stored in 20 % ethanol at 4 °C (39 °F).

3.3 Spin Column Format Purification

The following method is used for small purification or analytical purposes using disposable columns and common laboratory equipment (*see* Fig. 2).

1. It is recommended that a homogenous 1:1 distribution of resin and storage solution (20 % v/v ethanol) is established by careful mixing of the resin container, by inversion of the container or by use of a paddle stirrer. The use of mechanical stirrers, homogenizers or similar devices is not recommended.
2. For 100 μl of resin carefully apply 200 μl of the CaptureSelect™ resin slurry from **step 1** to the spin column and spin for 1 min at ~700 × *g* (~3,000 rpm in an Eppendorf centrifuge).
3. Wash the resin by adding 500 μl PBS (pH 7.4) or equivalent buffer, and subsequently spin for 1 min at ~700 × *g*. Repeat this step two times (three washes in total).
4. Load 100–700 μl of the antibody-containing sample on the affinity resin in the spin column. Close the spin column and incubate for 30–60 min on a rotating wheel.

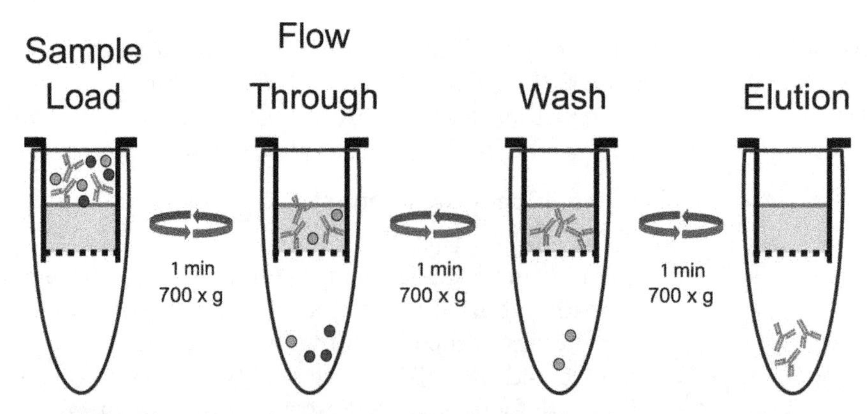

Fig. 2 Scheme for antibody isolation using spin-column format. 100–700 μl CaptureSelect™ affinity resin can be used in a Mobitec M1002S or M105035F spin column

5. Collect the non-bound proteins (flow-through) in a 2 ml collection tube by centrifugation of the spin column for 1 min at ~700 × *g*.
6. Wash the resin two to four times as described above.
7. Elute the captured antibody target of interest by adding 200–400 μl elution buffer (e.g., 100 mM Glycine pH 3) and collect the elution fraction by centrifugation at 700 × *g*. For direct neutralization, add 1 M Tris–HCl (pH 8) to the collection tube using 1/10 volume of the elution volume applied. For optimal recovery, additional elution steps may be required.
8. The CaptureSelect™ affinity resin can be stripped with for example 100 mM Glycine (pH 2) followed by re-equilibration in running buffer to allow a second spin column purification run.
9. If the spin column will not be used immediately, the resin should be stored in 20 % ethanol at 4 °C (39 °F).

3.4 SDS-PAGE

1. Mix 10 μl of antibody target-containing sample with 10 μl of SDS sample buffer and incubate the mixture for 5 min at 95 °C.
2. Load the sample aliquots into the wells of the gel. According to the gel, up to 11 or 14 samples (excluding a molecular weight marker) can be run simultaneously.
3. Run gels for 70 min at 125 V.
4. Remove gels from the cartridge and stain for 30 min using Coomassie staining solution.
5. De-stain the gel in CBB de-staining solution, and scan the gels for documentation.

3.5 Single-Step Purification of Polyclonal IgM from Human Serum by FPLC

The selectivity of CaptureSelect™ affinity resins allows one-step purification of antibodies from complex mixtures like crude serum. Here we demonstrate high selectivity in the isolation of polyclonal IgM from human serum using the CaptureSelect™ IgM affinity resin (*see* Fig. 3).

1. 7 ml of CaptureSelect™ IgM affinity resin (*see* Table 1) is packed in a Tricorn 10/100 column as described in Subheading 3.1.
2. Human serum is diluted tenfold in PBS (pH 7.4), and 100 ml of the diluted serum sample is loaded onto the column at a flow rate of 2 ml/min (~175 cm/h) using FPLC as described in Subheading 3.2.
3. After a washing step with PBS of 5 CV at 2 ml/min, the captured human IgM antibodies are eluted with 100 mM Glycine (pH 2), following neutralization by adding 1 M Tris–HCl (pH 8) to the collection tube using 1/10 volume of the elution volume applied.
4. Collected fractions are analyzed by SDS-PAGE as described in Subheading 3.4.

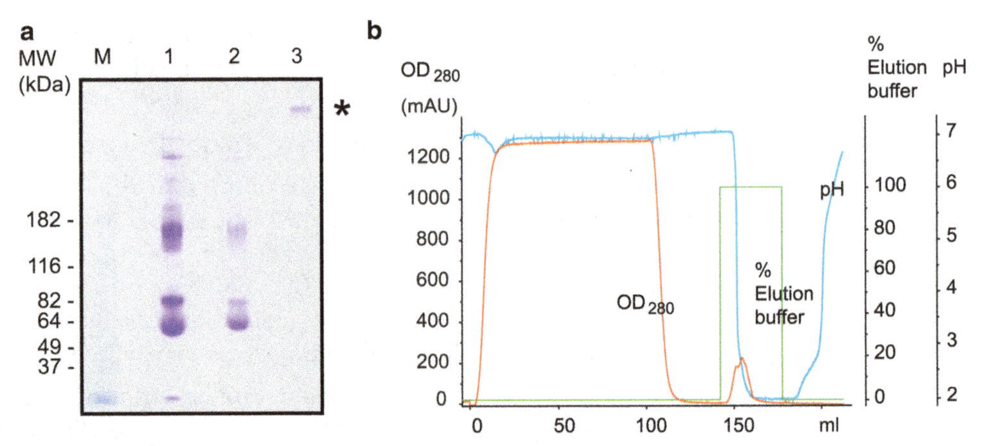

Fig. 3 IgM isolation by FPLC. Hundred milliliter of 10× in PBS buffer diluted human serum was loaded (at 2 ml/min) on a 7 ml CaptureSelect IgM column (10.5 cm bed height). As running buffer PBS was used, and for the elution 100 mM Glycine (pH 2.0) was applied. Samples taken during the purification were analyzed by non-reducing SDS-PAGE. (**a**) SDS-PAGE gel containing: *M* molecular weight marker, *1* 10× diluted human serum, *2* flow-through IgM column, *3* purified IgM fraction. The *asterisk* denotes the height of the IgM molecule on the SDS-PAGE gel. (**b**) Typical FPLC chromatogram obtained after IgM purification

3.6 Fab Isolation from Papain Digested Monoclonal Antibody Preps

Fab isolation kits available on the market are mostly based on the IgG digestion with immobilized papain. After digestion, the papain beads are removed and undigested IgG and Fc moieties are separated from the Fab fragments by Protein A affinity chromatography. At least three disadvantages to this method exist: (1) Release of protease of the papain beads will contaminate the final Fab isolate. (2) Protein A does not recognize Fc moieties from the IgG3 subclass. (3) Protein A has an intrinsic affinity for Fab fragments of the VH3 class. These latter two disadvantages will result in contamination of the Fab isolate with undigested IgG3 or Fc (subclass 3), and in a lower Fab yield when working with IgG monoclonals from subclass 3, respectively. Here we present an alternative method using the immobilized papain beads in combination with CaptureSelect™ IgG-CH1 and CaptureSelect™ IgG-Fc (Hu) resin. Application of this method will result in highly pure human IgG derived Fab isolates (*see* Fig. 4).

1. Approximately 5 mg of a purified human IgG monoclonal antibody (MAb) is digested according to the manufacturer's instructions supplied with the immobilized papain (*see* **Note 21**).
2. After removal of the papain beads, the supernatant, containing Fab, Fc, and undigested MAb, is loaded onto a 1 ml CaptureSelect™ IgG-Fc (Hu) column (Tricorn 5/20) using FPLC as described in Subheading 3.2.
3. The flow-through of this column is collected and subsequently loaded onto a 1 ml CaptureSelect™ IgG-CH1 column (Tricorn 5/20) using FPLC as described in Subheading 3.2.

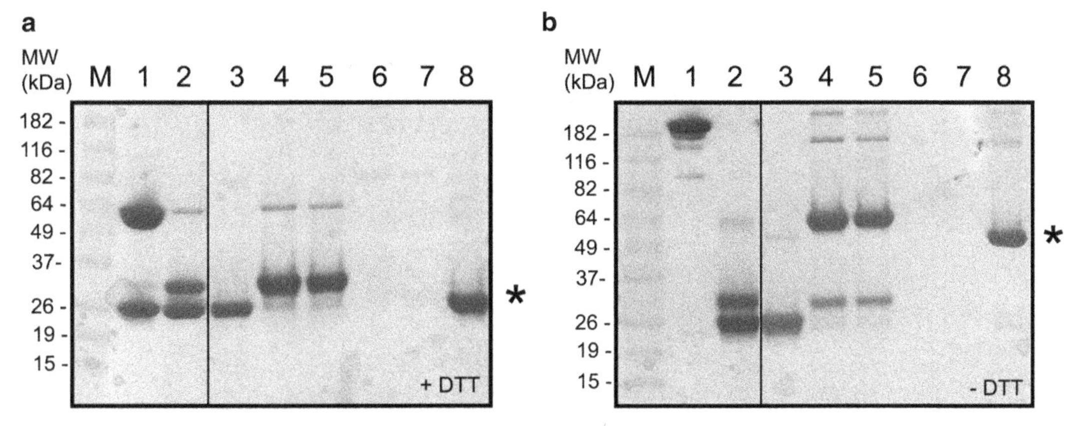

Fig. 4 Fab isolation of a monoclonal antibody. Samples obtained after different stages during the Fab isolation were analyzed by SDS-PAGE under (**a**) reducing and (**b**) non-reducing conditions. *M* molecular weight marker, *1* undigested MAb, *2* papain treated MAb, *3* flow-through CaptureSelect™ IgG-Fc (Hu) column, *4+5* elution IgG-Fc (Hu) column, *6+7* flow-through CaptureSelect™ IgG-CH1 column, *8* elution IgG-CH1 column. The *asterisk* denotes the height of the Fab fragment on the SDS-PAGE gel. Note that due to the high concentration of Cystein in the papain digestion mixture, IgG fragments are reduced even under non-reducing SDS-PAGE conditions ((**b**), *lanes 2–5*)

4. After washing of the IgG-CH1 column, the Fab fragments are eluted using 100 mM Glycine (pH 3), following neutralization by adding 1 M Tris–HCl (pH 8) to the collection tube using 1/10 volume of the elution volume applied.
5. The Fab fragments are dialyzed against PBS overnight, analyzed by SDS-PAGE as described in Subheading 3.4.

3.7 Purification of a Single Domain Antibody Equipped with a C-Terminal E-P-E-A Tag by FPLC

For antibody targets that solely consist of one or more variable domains such as VL, VH, or (sc)Fv, no affinity resins are currently available that can cover the broad variety that exists within this antibody format. In order to facilitate the purification, affinity tags are commonly introduced at the N- or C-terminus of the variable domain (e.g., hexa-histidine or FLAG tag). We here describe the use of a novel C-terminal affinity tag that consists of only four residues (i.e., E-P-E-A) and that is specifically captured by the CaptureSelect™ C-tag affinity resin (*see* **Note 18**). High selectivity is exemplified by the single step purification of a camelid single domain antibody fragment (V_HH) "C-tagged" with E-P-E-A from an *E. coli* periplasmic cell fraction (*see* Fig. 5).

1. 400 μl of CaptureSelect™ C-tag affinity resin (*see* Table 1) is packed in a Tricorn 10/100 column as described in Subheading 3.1.
2. An EPEA C-tagged camelid single domain antibody fragment (V_HH-EPEA) is spiked into a crude *E. coli* periplasmic cell fraction at a concentration of 1 mg/ml of which 2 ml are loaded

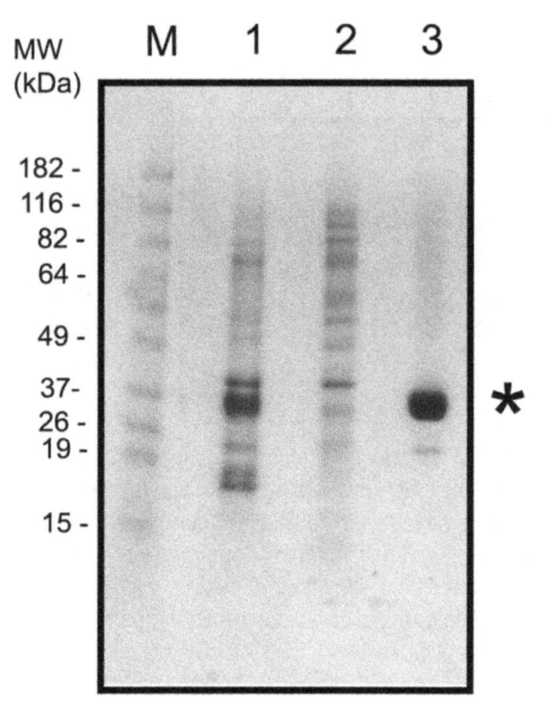

Fig. 5 C-tag purification of a single-domain antibody by FPLC. Two milliliters of *E. coli* periplasmic fraction containing the C-tagged V_HH-EPEA fragment was loaded at 150 cm/h on a 400 μl CaptureSelect C-tag column (2 cm bed height). Samples taken during the purification were analyzed by non-reducing SDS-PAGE. *M* molecular weight marker, *1* periplasmic fraction containing C-tagged V_HH-EPEA, *2* flow-through C-tag column, *3* purified C-tagged V_HH-EPEA fraction. The *asterisk* denotes the height of the V_HH-EPEA protein on the SDS-PAGE gel

onto the column at a flow rate of 150 cm/h using FPLC as described in Subheading 3.2.

3. After a washing step with PBS (pH 7.4) of 5 CV at 150 cm/h, the captured V_HH-EPEA construct is eluted with 100 mM Glycine (pH 2), following neutralization by adding 1 M Tris–HCl (pH 8) to the collection tube using 1/10 volume of the elution volume applied.
4. Collected fractions are analyzed by SDS-PAGE as described in Subheading 3.4.

4 Notes

1. The CaptureSelect™ IgG-CH1 affinity resin is unique in its coverage of binding to any human IgG derived Fab fragment independently of the type of light chain (kappa or lambda). So when dealing with feedstock materials that contain significant amounts of over-expressed free light chains, this resin will not cross-bind to these product-related contaminants.

2. Chimeric antibodies or Fab fragments thereof that are based on a human IgG format and display an Fv domain of nonhuman origin, such as mouse, can also be purified by the IgG-CH1 affinity resin.

3. The CaptureSelect™ IgG-Fc (Hu) affinity resin binds the Fc domain of all four human IgG subclasses. So, unlike protein A, this affinity resin also binds to human IgG3 antibodies and can be used to for example selectively purify (or deplete) the complete IgG repertoire from human serum or plasma samples (*see* ref. 17).

4. The CaptureSelect™ IgG-Fc (Hu) affinity resin binds to an epitope on the C_H3 domain within the human IgG Fc domain. Therefore, its binding reactivity to the Fc portion will not be affected by mutations or modifications of the C_H2 domain that are being introduced to for example alter Fc effector functions. Furthermore, binding to Fc will be retained even when the whole C_H2 domain is deleted and can be of use for C_H3 based fusion proteins. This will not be the case for protein A and G, since both these affinity ligands bind to the C_H2–C_H3 interface.

5. Like protein G, the CaptureSelect™ IgG-Fc multi-species (ms) affinity resin can be used for the purification of IgG antibodies from many different species (except chicken IgY). However, since the Fc interaction is less tight compared to protein G, less harsh elution conditions can be applied for antibody recovery. Efficient elution is observed using for example 0.1 M Glycine at pH 3.

6. The CaptureSelect™ resins LC-kappa (Hu) and LC-lambda (Hu) target the constant domain of human kappa—and lambda light chains ($C_L\kappa$ and $C_L\lambda$), respectively. Unlike protein L, which has affinity for some variable domains (V_L) of the kappa subclass (*see* ref. 7), these affinity ligands provide a more consistent coverage since both recognize a conserved epitope on the constant domain of the light chain (C_L) that is expressed within all human kappa—or lambda light chains. Furthermore, this binding is also irrespective of the isotype or subclass of the antibody heavy chain and thereby facilitates purification of any isotype variant within human IgG, IgM, IgA (*see* ref. 18), IgD, IgE, and Fab fragments thereof.

7. By combining LC-kappa™ (Hu)—and LC-lambda (Hu) affinity resins the whole human immunoglobulin repertoire can selectively be purified (or depleted) from human serum or plasma samples.

8. For polyclonal human IgG preparations, like intravenous immunoglobulin (IVIG) or obtained by IAC from human serum or plasma using CaptureSelect™ IgG-Fc (Hu), the kappa—and lambda light chain bearing antibodies (~60 % kappa and ~40 % lambda) can be separated by using LC-kappa (Hu) and/or LC-lambda (Hu) affinity resins.

9. The human species selectivity provided by for example CaptureSelect™ IgG-Fc (Hu), LC-kappa (Hu), LC-lambda

(Hu), and IgA (Hu) can be used when human antibodies are to be purified from feedstock samples that are contaminated with nonhuman antibodies, like those sourced from transgenic animals (*see* ref. 12) or from culture media supplemented with fetal calf serum (FCS) used for in vitro cell expression systems.

10. The set of CaptureSelect™ LC-kappa murine (mur), LC-lambda (mouse), and LC-lambda (rat) affinity resins can be used for the purification of any type of mouse—or rat monoclonal antibody (e.g., IgG, IgM, IgA) and Fab fragments thereof since it involves recognition of the antibody light chain (same principle as for LC-kappa (Hu) and LC-Lambda (Hu), *see* **Note 6**). Since no cross-binding is observed for bovine IgG with any of the above affinity ligands, FCS containing culture feedstocks can be applied for purification.
11. The CaptureSelect™ LC-lambda ungulate (ung) affinity resin binds to the lambda light chain of antibodies from "hoofed animals" like cattle (bovine), horse, donkey, goat, and sheep. For the bovine species, ~95 % of all antibodies expressed contain a light chain of the lambda subclass (*see* ref. 19). Therefore, the LC-lambda (ung) affinity resin already allows purification of the majority of bovine immunoglobulins. For example, since no cross-binding is observed for mouse and rat antibodies, it can also be used to remove any bovine IgG contaminants that might be present in for example protein A or G purified monoclonal antibody preps derived from FCS-containing culture supernatants.
12. The CaptureSelect™ affinity resins IgG1 (Hu), IgG3 (Hu), and IgG4 (Hu) can be used for purification (or depletion) of human IgG subclass antibodies from human serum—or plasma samples.
13. Besides purification of human IgM from serum or cell culture media, the CaptureSelect™ IgM resin can also be used for mouse and rat IgM monoclonals. Since no cross-binding is observed for bovine antibodies, FCS-containing culture feedstocks can be applied for purification.
14. For purification of human IgA antibodies CaptureSelect™ IgA (Hu) and IgA-CH1 (Hu) can be used. Both resins bind to human IgA-1 and 2 subclass as well as monomeric—and dimeric IgA and to secretory IgA (sIgA).
15. The CaptureSelect™ IgA (Hu) resin binds human IgA through recognition of an epitope on the IgA Fc domain.
16. The IgA-CH1 (Hu) affinity ligand binds to the C_H1 domain of the IgA heavy chain and can also be used for the purification of IgA Fab fragments. In addition to elution under acidic conditions (e.g., 0.1 M Glycine pH 3), efficient release of IgA molecules is also obtained under neutral pH conditions using 20 mM Tris–HCl, 2.0 M $MgCl_2$ at pH 7.0 as elution buffer.

17. For purification of IgA Fab or Fab2 fragments from papain or pepsin digests, the CaptureSelect™ IgA (Hu) resin can be used to remove IgA Fc fragments and partly or non-digested IgA molecules since it binds to the IgA Fc domain. Further purification of the Fab or Fab2 fragments can then be achieved using CaptureSelect™ IgA (Hu) in a second step (similar to IgG derived Fab fragments as shown in Fig. 3).
18. The CaptureSelect™ C-tag affinity matrix purifies C-terminal E-P-E-A tagged proteins with high affinity and selectivity from complex mixtures like cytoplasm or periplasmic fractions in a one step process (*see* ref. 16). Target binding is achieved at a pH range between 6 and 8 and NaCl up to 150 mM, like PBS (pH 7.4) and under denaturing conditions in the presence of urea (up to 8 M) or guanidine HCl (up to 1 M). Elution is performed at acidic conditions using for example 100 mM Glycine or 20 mM citric acid (pH 3) or under mild elution conditions at neutral pH like 20 mM Tris–HCl, 2 M $MgCl_2$ (pH 7) or 20 mM Tris–HCl, 1 M NaCl, 50 % (v/v) propylene glycol (pH 7), when target proteins are purified that cannot withstand acidic treatment.
19. As alternative buffer 20 mM citric acid (pH 3) can be used. Take care that the Ig-containing samples are neutralized after the elution. This can be done with 1/10 volume of 1 M Tris–HCl (pH 8).
20. Spin columns should be equipped with filters having a pore size of less than 100 μm. The MoBiTec spin columns can be used for sample volumes up to 700 μl in combination with a 2 ml collection tube.
21. The manufacturer gives a general protocol for the digestion of (human) IgG. When optimization is required, the following parameters can be varied: IgG concentration and digestion time. Instead of using immobilized papain, free papain can be also used with the method. The latter one is a more cost-effective method when making Fab fragments on preparative scale.
22. For large-scale manufacturing processes that require cGMP standards the following CaptureSelect™ based affinity resins are made available by GE Health Care (GEHC) in their product range for antibody affinity chromatography:
    - IgSelect: affinity resin targeting the Fc domain of human IgG antibodies (*see* refs. 20, 21).
    - KappaSelect: affinity resin binding to the constant domain of human kappa light chains (*see* **Notes 6–9**).
    - LambdaFabSelect: affinity resin binding to the constant domain of human lambda light chains (*see* **Notes 6–9**).
    - More information on bioprocess scale CaptureSelect™ affinity resins can be found on www.lifetechnologies.com/captureselect.

References

1. Ettre L (1993) Nomenclature for chromatography (IUPAC recommendations 1993). Pure Appl Chem 65(4):819–872
2. Hage D (1999) Affinity chromatography: a review of clinical applications. Clin Chem 45(5):593–615
3. Hjelm H, Hjelm K, Sjöquist J (1972) Protein A from *Staphylococcus aureus*: its isolation by affinity chromatography and its use as an immunosorbent for isolation of immunoglobulins. FEBS Lett 28(1):73–76
4. Shukla A et al (2007) Review, downstream processing of monoclonal antibodies: application of platform approaches. J Chromatogr B 848(1):28–39
5. Hober S, Nord K, Linhult M (2007) Protein A chromatography for antibody purification. J Chromatogr B 848(1):40–47
6. Grodzki A, Berenstein E (2010) Affinity chromatography: protein A and protein G sepharose. Methods Mol Biol 588:33–41
7. Nilson B et al (1992) Protein L from Peptostreptococcus magnus binds to the kappa light chain variable domain. J Biol Chem 267(4):2234–2239
8. Hamers-Casterman C et al (1993) Naturally occurring antibodies devoid of light chains. Nature 363:446–448
9. Muyldermans S (2001) Single domain antibodies: current status. J Biotechnol 74(4):277–302
10. van der Linden R et al (1999) Comparison of physical properties of llama VHH antibody fragments and mouse monoclonal antibodies. Biochim Biophys Acta 1431(1):37–46
11. Frenken L et al (2000) Isolation of antigen specific Llama V_HH antibody fragments and their high level secretion by *Saccharomyces cerevisiae*. J Biotechnol 78(1):11–21
12. Kuroiwa Y et al (2009) Antigen-specific human polyclonal antibodies from hyperimmunized cattle. Nat Biotechnol 27(2):173–181
13. Reinhart D, Weik R, Kunert R (2012) Recombinant IgA production: single step affinity purification using camelid ligands and product characterization. J Immunol Methods 378(1–2):95–101
14. Sun Y et al (2012) Pilot production of recombinant human clotting factor IX from transgenic sow milk. J Chromatogr B 898:78–89
15. Klooster R et al (2012) Muscle-specific kinase myasthenia gravis IgG4 autoantibodies cause severe neuromuscular junction dysfunction in mice. Brain 135(4):1081–1101
16. Hermans P et al (2012) Reinventing affinity tags: innovative technology designed for routine purification of C-terminal EPEA-tagged recombinant proteins. Genet Eng Biotechnol News 32(17):48–49
17. Klooster R et al (2007) Improved anti-IgG and HSA affinity ligands: clinical application of VHH antibody technology. J Immunol Methods 324(1–2):1–12
18. Beyer T et al (2009) Serum-free production and purification of chimeric IgA antibodies. J Immunol Methods 346:26–27
19. Aitken R, Hosseini A, MacDuff R (1999) Structure and diversification of the bovine immunoglobulin repertoire. Vet Immunol Immunopathol 72(1–2):21–29
20. Low D, O'Leary R, Pujar N (2007) Future of antibody purification. J Chromatogr B 848(1): 48–63
21. Liu J et al (2009) Comparison of camelid antibody ligand to protein A for monoclonal antibody purification. Biopharm Int 22(9): 35–43

Chapter 20

Reformatting of scFv Antibodies into the scFv-Fc Format and Their Downstream Purification

Emil Bujak, Mattia Matasci, Dario Neri, and Sarah Wulhfard

Abstract

The scFv-Fc format allows for rapid characterization of candidate scFvs isolated from phage display libraries before conversion into a full-length IgG. This format offers several advantages over the phage display-derived scFv, including bivalent binding, longer half-life, and Fc-mediated effector functions. Here, a detailed method is presented, which describes the cloning, expression, and purification of an scFv-Fc fragment, starting from scFv fragments obtained from a phage display library. This method facilitates the rapid screening of candidate antibodies, prior to a more time-consuming conversion into a full IgG format. Alternatively, the scFv-Fc format may be used in the clinic for therapeutic applications.

Key words scFv-Fc format, scFv reformatting, Phage display, Monoclonal antibodies

1 Introduction

Many therapeutic antibodies that are currently on the market are used in full IgG format [1]. This antibody format offers several advantages including long half-life, bivalent (or bispecific) binding as well as the ability to induce antibody-dependent cell-mediated cytotoxicity (ADCC) and complement dependent cytotoxicity (CDC). A phage display library, where a large and diverse repertoire of antibody fragments is displayed on the surface of a phage particle, is an important source of fully human high-affinity antibody fragments. The single chain variable fragment (scFv), with its size of just 27 kDa, represents the part of an antibody directly responsible for binding to an antigen and as such lends itself to display on the surface of a filamentous phage. The process of screening large phage display libraries yields high-affinity scFv binders against virtually any antigen [2, 3]. The small size and monovalent nature of scFv fragments results in pharmacokinetic (PK) properties, which are desirable for some clinical applications. The high degree of tumor penetration is countered by fast off-rates and poor retention at the tumor site [4]. ScFv fragments are rapidly

Vincent Ossipow and Nicolas Fischer (eds.), *Monoclonal Antibodies: Methods and Protocols*, Methods in Molecular Biology, vol. 1131, DOI 10.1007/978-1-62703-992-5_20,

cleared from the blood as their size is below the renal filtration threshold. In fact, over 90 % of the injected dose may be cleared from the blood within 1 h [5]. This PK profile is ideally suited for radio-immunoimaging applications.

For many applications, it is convenient to reformat an scFv into a different format. The modular nature of an antibody molecule allows for the design of a myriad of different antibody formats, e.g., diabody, Fab, small immunoprotein, scFv-Fc, IgG, or fused to other proteins [6].

Engineering an scFv into a full-length IgG format represents one useful avenue towards understanding the clinical potential of a candidate scFv, obtained from a phage display library, prior to the more lengthy conversion into the full IgG format. The human IgG antibody, with its long half-life, high degree of retention at the tumor site, as well as effector functions mediated by its Fc domain, represents a highly desirable format for some clinical applications. However, engineering of IgG requires separate cloning steps for variable heavy (V_H) and light (V_L) chains, making the preparation of a stable cell line a tedious and time-consuming process [7]. Additionally, the large size of IgG also means that it has a relatively poor ability to extravasate from the blood vessels and penetrate into tumor tissue. Nevertheless, many marketed monoclonal antibodies come in this format, demonstrating its usefulness in the clinic.

The scFv-Fc represents an alternative to the IgG format, which retains many of the desirable characteristics of IgG (bivalency, half-life, solubility, ADCC, and CDC). This format consists of an scFv fused in one polypeptide chain to the hinge region, CH2 and CH3 domains of human IgG1 (Fig. 1). The scFv-Fc format retains the affinity and specificity of the parental scFv, while offering several advantages over the phage display derived scFv. Multivalent antibodies usually have a higher apparent affinity constant compared to their monovalent counterparts. This behavior is attributed to their higher avidity, as they are more likely to rebind to the antigen, once the dissociation from the antigen at one or both binding sites occurs [8]. In fact, the superior tumor retention of scFv homodimers over scFv monomers is predominantly attributed to their bivalent binding nature than to their increased molecular weight [9].

The size strongly affects the rate of clearance of an antibody from the blood and normal tissue, as well as its ability to extravasate from blood vessels and penetrate into tumor tissue. Full IgG antibodies owe their long half-life to their high molecular weight (150 kDa) as well as to recycling by neonatal Fc receptors (FcRn) present on endothelial cells and circulating monocytes. These cells internalize serum IgGs, which bind to FcRn inside acidic endosomal vesicles. This IgG–FcRn complex is then transported back to the cell membrane from which the IgG is released into the circulation, thus effectively extending its half-life [10].

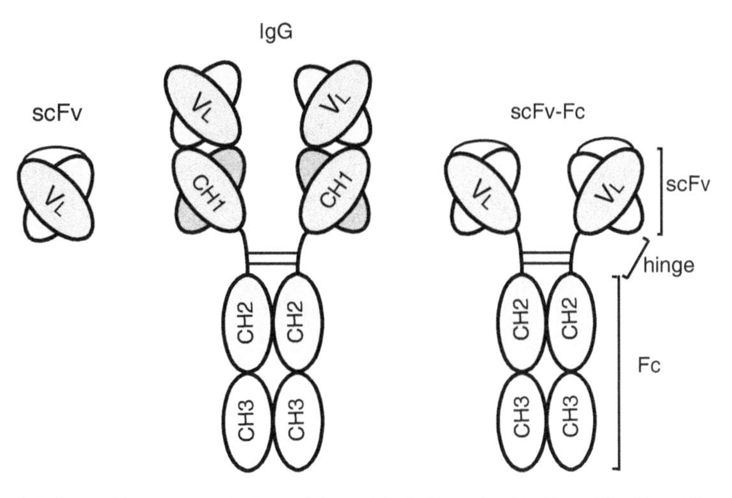

Fig. 1 Schematic representation of the scFv, IgG, and scFv-Fv antibodies. *VL* variable light chain, *CH1* human IgG1 heavy chain constant domain 1, *CH2* human IgG1 heavy chain constant domain 2, *CH3* human IgG1 heavy chain constant domain 3. Two polypeptide chains of the scFv-Fc form a dimer through two disulfide bonds in the hinge region

The scFv-Fc-coated tumor cells can bind Fc receptors expressed on phagocytic cells as well as natural killer (NK) cells, facilitating ADCC that in turn leads to tumor cell killing. Additionally, the complement system undergoes activation by Fc domains of antibodies, eliciting the complement dependent cytotoxicity. These Fc receptors trigger the activation of cellular immunity, which effectively makes them the key link between a monoclonal antibody therapeutic and the adaptive immune response. In fact, it is through a successful activation and mobilization of the innate immune system by ADCC and CDC that the peptides derived from the lysosomal degradation of tumor cells are presented on dendritic cells. Exposed in a complex with MHC class I and II molecules, they induce the activation of CD8+ T cells and CD4+ helper T cells respectively [11].

Many clinically used therapeutic antibodies, such as the anti-CD20 IgG antibody, rituximab, and anti-EGFR IgG antibody, cetuximab, have ADCC and CDC at the core of their tumor-killing mechanism of action. This effectively demonstrates the significance of an Fc domain for the activation of the immune system in combating cancer. In fact, the antitumor action of trastuzumab and rituximab is lower in Fc-receptor-deficient mice than in wild-type mice [12]. Additionally, the importance of CDC is best demonstrated by the fact that antitumor activity of rituximab is markedly reduced in C1q-deficient mice [13].

The scFv-Fc format, with its straightforward cloning and expression procedure, allows for rapid characterization of candidate scFvs isolated from phage display libraries with functional ADCC and CDC capabilities before the more complicated conversion into a full-length IgG. Here, a detailed method describes the cloning, expression, and purification of scFv-Fc fragments from the corresponding scFv fragment, obtained from a phage display library. This method encompasses the cloning of an scFv fragment as well as of the hinge region, CH2 and CH3 domains of human IgG1 followed by their PCR assembly and insertion into a mammalian expression vector, pCEP4, by means of a HindIII–NotI double digestion. Human embryo kidney cells (HEK293-EBNA), which have previously been adapted to grow in suspension, are transiently transfected, followed by subsequent protein A affinity purification of the scFv-Fc fusion protein of interest.

Currently, an scFv-Fc antibody, SBI-087 (Pfizer), which targets CD-20 on the surface of B cells, is in Phase II clinical trials for rheumatoid arthritis and systemic lupus erythematosus [14, 15], demonstrating the potential and usefulness of this format for clinical applications.

2 Materials

2.1 Equipment

The experiments described here can be performed using standard molecular biology laboratory equipment including autoclave, refrigerator, freezer, Cary 300 UV–Visible Spectrophotometer, microcentrifuge (Eppendorf 5417R), thermocycler (Biometra T3000), Shaking Benchtop Incubator (New Brunswick Scientific Innova 4000), Incubator (Binder WTC), Incubation Shaker (Infors HT Minitron), Tabletop centrifuge (Thermo Fisher scientific, Heraeus megafuge 1.0R), Superspeed centrifuge (Sorvall RC5c plus), Tubing pump (Ismatec), and an electroporator (Electro Cell Manipulator BTX 600R). General laboratory consumables used include MilliQ water (Millipore), 15 and 50 mL centrifuge tubes (BD Biosciences), 1.5 and 2 mL micro tubes (Sarstedt), filtered pipette tips, disposable pipettes, and cell culture flasks.

2.2 Cloning of the scFv-Fc Fragment and Insertion into the pCEP4 Vector Using Restriction Enzymes

1. Glycerol stock of TG1 *E. coli* bacteria that harbors the scFv of interest within a pHEN [16] vector is obtained after the screening of a phage display library [2, 3].
2. Hinge, CH2 and CH3 domains of human IgG1 are amplified from the pcDNA3-L19-IgG1 vector [17].
3. 2× YT-broth.
4. 10 cm agar plates, 2× YT/Amp/Glu: 2× TY broth, 100 μg/mL Ampicillin, and 1 % Glucose. Autoclave 2× YT medium together with 15 g/L of agar and let it cool down to 50 °C in a water bath before adding ampicillin and the 20 % glucose solution in MilliQ water.

5. Agar.
6. 14 mL round bottom tube with snap cap (BD Biosciences).
7. QIAprep Spin Miniprep Kit (Qiagen).
8. PCR tubes (Sarstedt).
9. Primers are obtained from Operon. The stock concentration of primers is 10 μM, while the concentration used in PCR reactions is 0.5 μM.

   Hind III Leader Seq Fo
   5′CCCAAGCTTGTCGACCATGGGCTGGAGCC TGATCCTCCTGTTCCTCGTCGCTGTGGC3′

   Leader Seq DP47 Fo
   5′TCCTCCTGTTCCTCGTCGCTGTGGCTAC AGGTGTGCACTCGGAGGTGCAGCTGTTGG AGTCTGGG3′

   DPL16 Hinge Ba
   5′GTTTTGTCACAAGATTTGGGCTCGCCTAGG ACGGTCAGCTTGGTCCC3′

   DPK22 Hinge Ba
   5′GTTTTGTCACAAGATTTGGGCTCTTTG ATTTCCACCTTGGTCCCTTG3′

   Hinge CH2 Fo
   5′GAGCCCAAATCTTGTGACAAAACTCACA CATGCCCACCGTGCCCAGCACC3′

   CH3 NotI Stop Ba
   5′TTTTCCTTTTGCGGCCGCTTATTAC CCGGAGACAGGGAGAGGC3′

   pCEPfo
   5′AGCAGAGCTCGTTTAGTGAACCG3′

   pCEPba
   5′GTGGTTTGTCCAAACTCATC3′

   Fdseq long
   5′GACGTTAGTAAATGAATTTTCTGTATGAGG3′

   LMB3long
   5′CAGGAAACAGCTATGACCATGATTAC3′

10. Expand High Fidelity PCR system (Roche).
11. PCR nucleotide mix (dNTP, Roche). The stock concentration of the nucleotide mix is 10 mM, while the concentration used in PCR reactions is 0.5 mM.
12. Agarose DNA gel 1 %. Mix the agarose powder with the TBE buffer to 1 % and heat in a microwave until completely melted. Add ethidium bromide to a final concentration of 0.1 μg/mL and pour into a casting tray containing a sample comb and allow to solidify at room temperature.

13. TBE buffer: 89 mM Tris base, 89 mM boric acid, 2 mM EDTA pH = 8.
14. Agarose.
15. Ethidium bromide.
16. GelPilot DNA loading dye, 5× (Qiagen).
17. Smart Ladder MW-1700-10 (Eurogentec).
18. Electrophoresis power supply (EPS 301, GE Healthcare).
19. Gel Extraction Kit (Qiagen).
20. pCEP4 vector (Invitrogen).
21. Hind III and NotI HF (New England Biolabs).
22. QIAquick PCR Purification Kit (Qiagen).
23. T4 DNA ligase (New England Biolabs).
24. Qiagen Mini Prep kit (Qiagen).
25. Electroporation cuvette 0.2 cm (Gene Pulser Bio-Rad).
26. Red Taq DNA polymerase ready mix (Sigma-Aldrich).
27. Electrocompetent TG1 *E. Coli* (Stratagene).
28. Ampicillin.
29. Glucose.

2.3 Transient Gene Expression in HEK293-EBNA Cells and Protein A Purification

1. LB medium.
2. HEK293-EBNA (Invitrogen).
3. Freestyle 293 medium (Invitrogen).
4. RPMI medium (Gibco).
5. Polyethylenimine (PEI).
6. Pluronic F68 (Invitrogen).
7. Sodium Butyrate.
8. Antibiotic-antimycotic (Gibco).
9. 500 mL centrifuge bottles (Sorvall Dry-Spin Polypropylene Bottles).
10. Ultra-linked Protein A resin (Sino Biological).
11. Liquid chromatography columns (Sigma-Aldrich).
12. Buffer A: 100 mM NaCl, 0.1 % Tween20, 500 μM EDTA pH = 8, in PBS.
13. Buffer B: 500 mM NaCl, 500 μM EDTA, in PBS.
14. Triethylamine (TEA, Fluka).

2.4 Amber Stop Codon Mutation Procedure

1. Primers are obtained from Operon. The stock concentration of primers is 10 μM, while the concentration used in PCR reactions is 0.5 μM.

DP47 mtoq Ba
5′CTGGAGCCTGGCGGACCCAGCTCAT CTCAGCCTGGCTAAAGGTGAATCCA3′

DP47 mtoq Fo
5′GATGAGCTGGGTCCGCCAGGC TCCAGGGAAGGGGCTGGAGTGGGTCTCAGCTATT3′

3 Methods

3.1 Cloning of the scFv-Fc Fragment and Insertion into the pCEP4 Vector

1. Use the glycerol stock of TG1 *E. coli* bacteria that harbors the scFv of interest within a pHEN vector. Streak bacteria, which are stored at −80 °C, on a small 2× YT/Amp/Glu plate and incubate overnight at 37 °C.
2. Pick one colony to inoculate 2 mL of 2× YT medium supplemented with ampicillin (100 μg/mL) and glucose (1 %) in a 14 mL round bottom tube with snap cap. Shake for 6–8 h in a shaking incubator at 37 °C and 180 RPM.
3. Transfer the bacterial culture into a 2 mL micro tube, using a sterile 2 mL pipette and spin down in a microcentrifuge at 20,800 × *g* and 4 °C. Decant the supernatant and subject the pellet to DNA extraction.
4. Perform DNA extraction using the QIAprep Spin Miniprep Kit, adhering to the manufacturer's protocol. Use 50 μL of MilliQ water for elution of DNA from the QIAprep spin column. Quantify the DNA content of the eluate by measuring its absorbance at 260 nm in a quartz cuvette using a UV spectrophotometer. The absorbance value of 1 corresponds to a DNA concentration of 50 ng/μL (*see* **Note 1**).
5. Obtain the amino-acid sequence from the nucleotide sequence of the scFv using the Web tool at www.expasy.org/translate and rule out the presence of stop codons. Use N-glyc (www.cbs.dtu.dk/services/NetNGlyc) and O-glyc (www.cbs.dtu.dk/services/NetOGlyc) Web tools to identify any potential glycosylation sites. If the absence of stop codons and glycosylation sites in any of the complementarity determining regions (CDRs) has been confirmed, proceed to **step 6**. Otherwise go to Subheading 3.3 (*see* **Note 2**).
6. Perform the PCR amplification of the scFv using Leader Seq DP47 Fo forward primer and either DPK22 Hinge Ba or DPL16 Hinge Ba backward primer [2, 3] (*see* **Note 3**). Amplify the hinge region, CH2 and CH3 domains of human IgG1 from the pcDNA3-L19-IgG1 vector [17] using Hinge CH2 Fo forward primer and CH3 NotI Stop Ba backward primer. In order to amplify the scFv and Fc DNA, prepare PCR mixes in two PCR tubes as follows: 0.5 μL of mini prep template

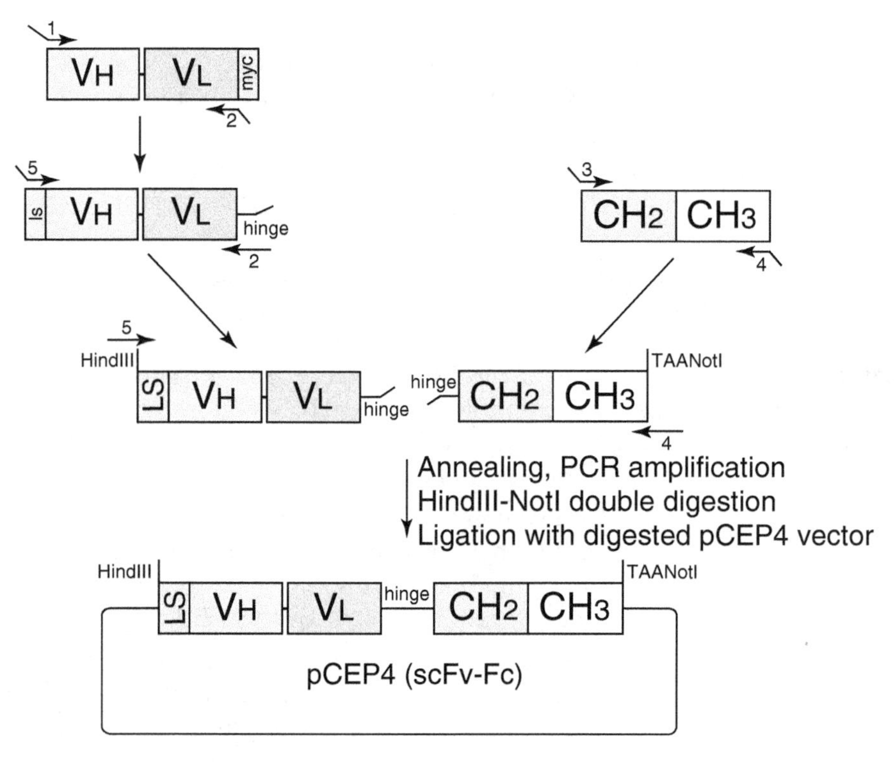

Fig. 2 Schematic representation of the assembly of scFv and Fc into an scFv-Fc fragment and the subsequent cloning into the pCEP4 vector. *1* Leader Seq DP47 Fo; *2* DPL16 Hinge Ba or DPK22 Hinge Ba; *3* Hinge CH2 Fo; *4* CH3 NotI Stop Ba; *5* Hind III Leader Seq Fo; *myc* polypeptide protein tag; *ls* incomplete secretion sequence, appended by primer 1; *LS* complete secretion sequence, appended by primer 5

DNA (~100 ng), 1 μL of forward primer, 1 μL of backward primer, 1 μL of dNTP, 0.5 μL of HiFi polymerase, 2 μL of HiFi polymerase system buffer, and 14 μL of MilliQ water (*see* **Note 4**). In parallel, prepare the negative control PCR mix as follows: 1 μL of Fdseq primer, 1 μL of LMB3long primer, 1 μL of dNTP, 2 μL of Expand High Fidelity buffer, 0.5 μL of Expand High Fidelity enzyme mix, and 14.5 μL of MilliQ water (*see* **Note 5**). Use the following thermocycler program: lid temperature 99 °C, 94 °C 2 min (94 °C 20 s, 60 °C 30 s, 72 °C 1 min), repeat 25 times, 72 °C 10 min, and 4 °C pause (Fig. 2).

7. Spike the PCR mix with 5 μL of Pilot Gel loading dye and load onto a 1 % agarose gel in TBE buffer using an 8-well electrophoresis comb and Smart ladder as a reference. Run the gel using the following settings: 100 V, 60 mA for 30 min. Visualize the bands under UV light. Both lanes should contain a band of 900 base pairs (bp), which correspond to the amplified scFv as well as the amplified hinge region, CH2 and CH3 domains. Separately cut out both 900 bp bands with a razor blade and transfer into two individual 1.5 mL micro tubes.

Extract DNA from the gel using Qiagen Gel Extraction Kit according to the manufacturer's protocol. Perform the final elution step with 30 μL of MilliQ water. The negative control should not yield any bands (*see* **Note 6**).

8. Perform the PCR assembly of the two DNA fragments using the following PCR mix: 1 μL of scFv DNA (amplified in **step 6**), 1 μL of Fc DNA (amplified in **step 6**), 1 μL of Hind III Leader Seq Fo primer, 1 μL of CH3 NotI Stop ba primer, 1 μL of dNTP, 2 μL of Expand High Fidelity buffer, 0.5 μL of Expand High Fidelity enzyme mix, and 12.5 μL of MilliQ water. Use the following thermocycler program: lid temperature 99 °C, 94 °C 2 min (94 °C 20 s, 60 °C 30 s, 72 °C 1.5 min), repeat 25 times, 72 °C 10 min, and 4 °C pause (*see* **Note 7**).
9. Spike the PCR mix with 5 μL of Gel Pilot DNA loading dye and load onto a 1 % agarose gel in TBE buffer using an 8-well electrophoresis comb and Smart ladder as a reference. Run the gel using the following settings: 100 V, 60 mA for 30 min. Visualize and cut out the 1,500 bp band under UV light. This band corresponds to the amplified scFv-Fc. Transfer into a 1.5 mL micro tube and extract DNA from the gel using Qiagen Gel Extraction Kit according to the manufacturers protocol. Perform the final elution step with 30 μL of MilliQ water. Determine DNA concentration by measuring the absorbance at 260 nm.
10. Perform double digestions of both the scFv-Fc DNA fragment and pCEP4 plasmid in parallel with HindIII and NotI digestion enzymes. Typically, one should use the entire quantity of the assembled PCR DNA. Transfer 25 μL of assembled PCR DNA (usually contains 0.8–2 μg of DNA) into a 1.5 mL micro tube, add 1 μL of HindIII (20 U/μL), 1 μL of NotI HF (20 U/μL), 5 μL of NEB buffer 2 (10×), 0.5 μL of NEB BSA (100×), and 17.5 μL of MilliQ water. In parallel, transfer 5 μg of pCEP4 plasmid DNA (volume depends on the concentration) into a new 1.5 mL micro tube, add 1.5 μL of HindIII (20 U/μL), 1.5 μL of NotI HF (20 U/μL), 5 μL of NEB buffer 2 (10×), 0.5 μL of NEB BSA (100×), and MilliQ water up to a final volume of 50 μL (*see* **Note 8**). Mix thoroughly and incubate at 37 °C for 20 h (*see* **Note 9**). Following this, add 12.5 μL of Pilot loading dye into each of the digestion reactions and separate by electrophoresis on a 1 % agarose gel in TBE buffer using an 8-well electrophoresis comb and Smart ladder as a reference. In parallel, load 5 μg of the non-digested pCEP4 plasmid that was previously spiked with 5× Pilot loading dye and load in a separate well (*see* **Note 10**). The double-digested pCEP4 plasmid and scFv-Fc DNA will be visible as bright bands of ~10.5 kb and ~1.5 kb, respectively. Afterwards perform gel extraction, using the QIAquick Gel Extraction

Kit, according to the manufacturer's protocol. Elute with 30 μL of MilliQ water. Quantify the concentration of DNA in each of the reactions by measuring absorbance at 260 nm.

11. Perform ligation by mixing double digested scFv-Fc DNA with a double digested pCEP4 plasmid DNA inside a 1.5 mL micro tube. Typically, mix 6 μL of scFv DNA (30 ng/μL) with 6 μL of pCEP4 plasmid DNA (60 ng/μL), add 2 μL of buffer for T4 DNA ligase, 5 μL of MilliQ water and 1 μL of T4 DNA ligase (*see* **Note 11**). Mix well and incubate at room temperature for 30 min. In parallel, perform a negative control ligation reaction, which is essentially identical to the one above, but contains MilliQ water instead of scFv-Fc DNA. Afterwards perform PCR clean up, using the QIAquick PCR Purification Kit according to the manufacturer's protocol. Elute with 30 μL of MilliQ water (*see* **Note 12**).
12. Transform two separate 50 μL aliquots of electrocompetent TG1 *E. coli* by adding 5 μL of the purified ligation mix to a 0.5 mL micro tube containing bacteria. Transfer the mixture into a pre-chilled 2 mm electroporation cuvette and perform electroporation using the following parameters: capacitance of 25 μF, voltage of 2.5 kV and resistance of 175 Ω. Immediately after the electrical impulse, add 70 μL of 2× YT medium and streak the entire quantity of electroporated bacteria onto a 10 cm 2× YT/Amp/glu agar plate (pre-warmed at 37 °C) using a sterile Pasteur pipette. Incubate overnight at 37 °C (*see* **Note 13**).
13. Perform colony screening using Red Taq polymerase mix. Transfer 12 μL of RedTaq polymerase mix into each of the seven separate PCR tubes and add 10 μL of RedTaq polymerase water, 1 μL of pCEP4fo primer, and 1 μL of pCEP4ba primer. In parallel, prepare 2 mL of 2× YT medium supplemented with 1 % glucose and 100 μg/mL of ampicillin and transfer into six separate 14-mL round bottom tubes with snap caps. In total pick six individual colonies using a pipette tip and dip the tip first into a PCR tube followed by inoculation of 2× YT medium. Incubate media at 37 °C, 180 RPM for 6–8 h. Afterwards, transfer the bacterial culture into a 2 mL micro tube, spin down in a microcentrifuge at 20,800 ×*g* for 2 min, discard the supernatant, and store the pellets at −20 °C. The seventh PCR reaction should not be spiked with a colony and serves as a negative control (*see* **Note 14**). Thermocycler should be run as follows: lid temperature 99 °C, 94 °C 5 min (94 °C 20 s, 53 °C 30 s, 72 °C 1.5 min), repeat 25 times, 72 °C 10 min, and 4 °C pause. Afterwards, separate the seven PCR reactions by electrophoresis on a 1 % agarose gel in TBE buffer using an 8-well electrophoresis comb and Smart ladder as a reference. Run the gel at 100 V, 60 mA for 30 min. Visualize the bands under UV light and using a razor blade,

cut out those corresponding to 1,500 bp. Extract DNA from the gel using Qiagen Gel Extraction Kit. Perform DNA sequencing using pCEPfo primer in order to obtain the N-terminal sequence of DNA insert and pCEPba primer for the C-terminal part (*see* **Note 15**).

14. When the colony containing the right scFv-Fc sequence has been identified, extract the plasmid DNA from the corresponding bacterial pellets using Qiagen Mini Prep kit. Following the manufacturer's protocol and eluting with 50 μl of MilliQ water, one should in the end obtain 10–30 μg of plasmid.
15. Store the obtained mini-prep plasmid DNA that harbors the scFv-Fc of verified sequence at −20 °C.

3.2 Transient Gene Expression in HEK293-EBNA Cells and Protein A Purification

1. Preparation of Maxi-prep plasmid DNA. Transform a 50 μL aliquot of electrocompetent TG1 *E. coli* (other strains can be used, e.g., DH5α) by adding 0.5 μL of the mini-prep DNA (~100 ng) to a micro tube containing bacteria. Transfer the mixture into a pre-chilled 2 mm electroporation cuvette and electroporate using the following parameters: capacitance of 25 μF, voltage of 2.5 kV, resistance of 175 Ω. Immediately after the impulse add 70 μL of 2× YT medium and streak bacteria onto a 10 cm 2× YT/Amp/glu agar plate (pre-warmed at 37 °C) using a sterile inoculation loop. Incubate overnight at 37 °C (*see* **Note 16**).
2. Pick a single colony from the agar plate and inoculate 400 mL of LB medium supplemented with 100 μg/mL of ampicillin in a 2 L Erlenmeyer flask. Incubate overnight (16–20 h) in a shaking incubator at 37 °C and 180 RPM. Transfer the bacterial culture into a centrifuge bottle and spin down for 5 min at 10,800 × *g* and 4 °C. Decant the supernatant and remove any residual liquid by keeping the tube inverted for 5 min on a paper towel. Resuspend bacterial pellet with 10 mL of Qiagen buffer P1, which has previously been supplemented with RNase, and transfer into a 50 mL centrifuge tube. Proceed with the plasmid DNA extraction using the Qiagen Maxi Prep kit adhering to manufacturer's instructions. Perform the final elution step with 1 mL of Qiagen TE buffer. Quantify the DNA concentration by measuring absorbance at 260 nm. Typically, the plasmid concentration is ~400–600 μg/mL. Use the obtained DNA for transient gene expression of HEK cells with the aim of producing the scFv-Fc fragments.
3. Preparation of HEK293-EBNA cells for transfection (*see* **Note 17**). Take out a 1 mL cryovial containing five to ten million cells from liquid nitrogen and thaw quickly in a 37 °C water bath. When the cells are almost completely thawed, decontaminate the outer side of the vial with 70 % (vol/vol) ethanol. Inside

the laminar flow chamber, transfer the entire content of the vial into a 15 mL centrifuge tube containing 10 mL of Freestyle 293 pre-warmed to 37 °C. Centrifuge at 115 ×*g* for 5 min at room temperature in a refrigerated tabletop centrifuge with swinging bucket rotor. Inside the laminar flow hood, aspirate and discard the supernatant, gently resuspend the pellet with a 5 mL pipette in 10 mL fresh FreeStyle 293 medium pre-warmed to 37 °C, and transfer the cell suspension into a 250 mL bottle. Shake the bottle gently to mix the cell suspension and incubate for 1 day in a shaking cell culture incubator with an atmosphere containing 5 % CO_2 at 37 °C.

The viability of cells after this procedure should be >90 %, as determined by Trypan Blue staining using a hemocytometer. If viability is above 90 % then dilute cells to 0.5 million per mL and allow to grow for 2–3 days in the same conditions. Split cells every 3–4 days down to 0.5 million cells per mL, with the aim of expanding them (*see* **Note 18**).

4. The day before transfection dilute cells down to one million per mL using a fresh Freestyle 293 medium. Immediately prior to transfection count the cells and adjust their concentration to 20 million per mL in the RPMI medium that has previously been warmed up to 37 °C. This is to be done inside a laminar flow chamber by transferring the adequate volume of cells into a 50 mL centrifuge tube, which is then spun down at 115 ×*g* for 5 min at room temperature in a tabletop centrifuge with swinging bucket rotor. Following this, inside the laminar-flow chamber aspirate the supernatant and resuspend cells with the required volume of RPMI medium, so that the final cell concentration is 20 million per mL. Transfer the cells into a sterile bottle, the volume of which should be at least five times bigger than the final volume of the production batch. Add 1.25 μg of DNA per one million of cells, followed by addition of 3.75 μg of PEI per one million of cells. Incubate for 3 h in a shaking cell culture incubator with an atmosphere containing 5 % CO_2 at 37 °C and the shaking speed of 100 RPM. After 3 h have passed, add Freestyle293 medium, supplemented with 0.1 % Pluronic F68, 3 mM Sodium butyrate and Antibiotic-antimycotic, so that the final cell concentration in the medium is one million per mL. Place the bottle back in the shaking cell culture incubator with an atmosphere containing 5 % CO2 at 37 °C and 100 RPM.
5. On day 6 after transfection, transfer the cell culture into a 500 mL centrifuge bottle and spin down in a superspeed centrifuge at 13,680 ×*g* and 4 °C for 30 min. Filter the supernatant through a 0.22 μm filter using a vacuum pump.
6. Rinse one liquid chromatography column with deionized water. Load the column with 2 mL of 50 % Protein A-Sepharose slurry (*see* **Note 19**). Wash the resin with two column volumes

of MilliQ water in order to remove the storage buffer. Following this, wash the resin with two column volumes of PBS. Close the exit nozzle with a cap and load the filtered supernatant (*see* **Note 20**). Connect the column with tubing and attach to a tubing pump. The pump flow should be set at 1 mL/min (*see* **Note 21**).

7. After the entire quantity of the supernatant has passed through the column, commence the washing procedure with Buffer A and Buffer B consecutively. Wash only the column containing protein A, while ensuring that the buffers are carefully applied onto the column using a pipette. The washing should be allowed to proceed until the OD of the flow-through at 280 nm goes below 0.01 (*see* **Note 22**).
8. At this point the column can be detached from the tubing pump and elution can be performed using 15 mL of 100 mM TEA, pH = 11. Always use freshly prepared elution buffer. Before the elution step, set up enough 1.5 mL micro tubes so that all elution fractions can be collected. Apply the elution buffer with utmost care, by applying it against the column wall, as it is essential not to disturb the resin bed (*see* **Note 23**). Collect 1 mL fractions of the eluate in 1.5 mL micro tubes containing 200 μL of Tris–HCl (pH = 8). Quantify the protein concentration by measuring absorbance at 280 nm. Pool together those fractions that have $OD_{280\ nm} > 0.1$ (*see* **Note 24**). Transfer pooled protein fractions into a dialysis bag with 12–14 kDa cut-off that had previously been washed with MilliQ water and closed on one side with a plastic clamp. Close the bag from the other side with a plastic clamp and place it in a beaker, which contains 200 times higher volume of PBS as compared to the volume of eluted protein. Leave over-night at 4 °C under constant stirring (*see* **Note 25**).
9. The day after take the protein out of the dialysis bag, filter through a 0.22 μm filter and determine the concentration once again. Typical yields are 5–10 mg/L of culture. Take a small quantity of protein for further characterization while snap-freezing the remainder in liquid nitrogen and storing it at −80 °C.
10. The characterization of the produced protein is to be performed using SDS-PAGE gel electrophoresis, gel filtration and BIAcore analysis methods.

3.3 Mutation of the Amber Stop Codon

Here, a procedure is presented that explains how to mutate an amber stop codon (TAG), present in the CDR1 of a heavy chain, into a codon for glutamine (CAG) (*see* **Note 26**). As the scFv-Fc fragment will be expressed in mammalian cells, presence of the amber stop codon will result in truncation of the protein (*see* **Note 2**). This procedure is therefore essential in order to successfully reformat an scFv into an scFv-Fc format (Fig. 3).

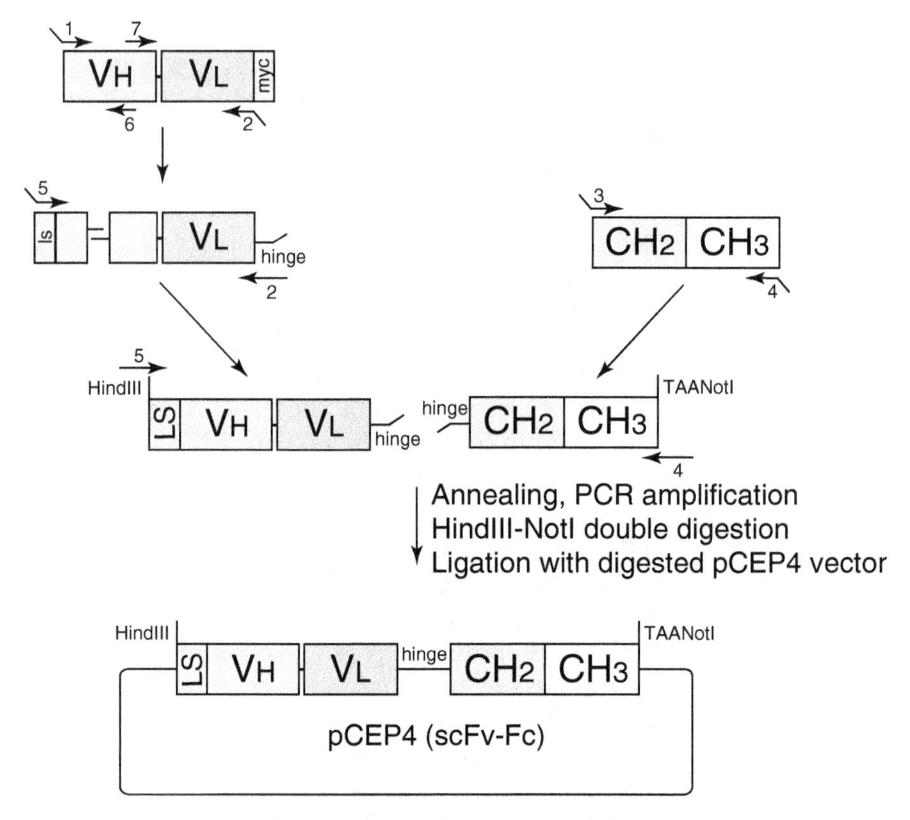

Fig. 3 Schematic representation of the procedure for mutation of the amber stop codon into glutamine. *1* Leader Seq DP47 Fo; *2* DPL16 Hinge Ba or DPK22 Hinge Ba; *3* Hinge CH2 Fo; *4* CH3 NotI Stop Ba; *5* Hind III Leader Seq Fo; *6* DP47 mtoq Ba; *7* DP47 mtoq Fo; *myc* polypeptide protein tag; *ls* incomplete secretion sequence, appended by primer 1; *LS* complete secretion sequence, appended by primer 5

1. Three PCR reactions should be prepared and run in parallel. Prepare a common PCR mix for all three as follows: 3 μL of dNTP, 1.5 μL of HiFi polymerase, 6 μL of HiFi polymerase system buffer, and 42 μL of MilliQ water. Transfer 17.5 μL of the PCR mix into each of the three PCR tubes. Add into the first tube 0.5 μL of scFv mini prep DNA (~100 ng), 1 μL of Leader Seq DP47 Fo, and 1 μL of DP47 mtoq Ba primer. Add into the second tube 0.5 μL of scFv mini prep DNA, 1 μL of DP47 mtoq Fo primer, and 1 μL of either DPL16 Hinge ba or DPK22 Hinge Ba primer (depending on the nature of the light chain). Transfer into the third PCR tube 0.5 μL of Fc DNA template, 1 μL of Hinge CH2 Fo forward primer, and 1 μL of CH3 Stop NotI Ba backward primer. The thermocycler program should be set as follows: lid temperature 99 °C, 94 °C 2 min (94 °C 20 s, 60 °C 30 s, 72 °C 1 min), repeat 25 times, 72 °C 10 min, and 4 °C pause.
2. Spike the PCR mix with 5 μL of Pilot Gel loading dye and separate by electrophoresis on a 1 % agarose gel in TBE buffer

using an 8-well electrophoresis comb and Smart ladder as a reference. Run the gel using the following settings: 100 V, 60 mA for 30 min. The bands can be visualized under UV light and they should be ~200 bp, ~700 bp, and ~900 bp in length corresponding to reactions 1, 2, and 3 respectively. Excise the bands of correct size using a razor blade and transfer gel pieces into three separate 1.5 mL micro tubes. Extract DNA from the gel using Qiagen Gel Extraction Kit according to the manufacturer's protocol. Perform the final elution step with 30 μL of MilliQ water.

3. Use the DNA obtained by gel extraction from reactions one and two to perform the assembly. Transfer 1 μL of reaction 1 DNA (amplified in **step 1**), 1 μL of reaction 2 DNA (amplified in **step 1**), 1 μL of HindIII Leader Seq Fo forward primer, 1 μL of either DPL16 Hinge Ba or DPK22 Hinge Ba primer (depending on the nature of the light chain), 1 μL of dNTP, 2 μl of HiFi polymerase system buffer, 0.5 μL of HiFi polymerase, and 11.5 μL of MilliQ water. The following thermocycler program should be used: lid temperature 99 °C, 94 °C 2 min (94 °C 20 s, 60 °C 30 s, 72 °C 1.5 min), repeat 25 times, 72 °C 10 min, and 4 °C pause.
4. Spike the PCR mix with 5 μL of Pilot Gel loading dye and separate by electrophoresis on a 1 % agarose gel in TBE buffer using an 8-well electrophoresis comb and Smart ladder as a reference. Run the gel using the following settings: 100 V, 60 mA for 30 min. Visualize the amplified DNA under UV light and excise those bands, which correspond to ~900 bp using a razor blade and transfer them into a 1.5 mL micro tube. Extract DNA from the gel using Qiagen Gel Extraction Kit according to the manufacturer's protocol. Perform the final elution step with 30 μL of MilliQ water. Proceed to **step 8** in Subheading 3.1.

4 Notes

1. The absorbance value should lie between 0.1 and 1 for a reliable DNA quantification. In order to achieve this, one should typically prepare either a 1:50 or 1:20 dilution of DNA in MilliQ water.
2. Single chain variable fragments originate from phage display libraries where their complementarity determining regions (CDR) are randomized. Depending on the library design, CDR3 regions of both heavy and light chains can be randomized or often all six CDR regions are randomized. While the library design precludes appearance of stop codons in CDRs, randomization may result in the occurrence of the amber stop

codon sequence within a CDR. Stop codons prevent translation of the remaining sequence, which results in truncation of the protein. Nevertheless, functional scFv with an amber stop codon in the sequence does get expressed, due to the fact that TG1 suppressor strain of *E. coli* is used in phage display. This strain of *E. coli* reads the amber stop codon as glutamine, thus allowing for the expression of a full-length scFv antibody. This ability of the TG1 suppressor strain is essential for phage display as it allows for scFv to be expressed both freely in solution and in a complex with pIII minor phage coat protein on the phage surface. However, as here the scFv-Fc will be expressed in mammalian cells, the presence of the amber stop codon will result in truncation of the antibody. Consequently, its functionality will be lost.

Another potential issue when using mammalian expression systems, which is absent in the case of bacterial expression systems, is the glycosylation of amino acids within the CDR regions.

The glycosylation of amino acid residues or the presence of an amber stop codon within the CDR regions either absolutely abrogates or significantly hampers the binding of an antibody to the antigen. It is therefore essential to verify the absence of amber stop codons as well as glycosylation sites in CDRs. Should any of these unfavorable codons be identified they should be mutated using PCR.

3. Two light chain germ line genes are used in the construction of the ETH2-GOLD, PHILO-1, and PHILO-2. Namely, these are DPK22 and DPL16.
4. The Leader Seq DP47 Fo primer anneals to the N-terminus of the heavy chain of scFv and appends a part of the leader sequence to it. The DPL16 Hinge ba and DPK22 Hinge ba primers anneal to the C-terminus of the scFv, adding a part of the hinge region to it. The Hind III Leader Seq Fo primer appends the remainder of the leader sequence together with the HindIII restriction enzyme digestion sequence to the N-terminus. In parallel, amplification of the CH2 and CH3 domains is performed using the Hinge CH2 Fo forward primer and the CH3 NotI Stop Ba backward primer. These primers append the N-terminus with a part of the hinge region and the C-terminus with a stop codon followed by the NotI restriction enzyme digestion sequence.
5. Primers that are used for the reformatting procedure anneal to the framework region of an scFv, which is shared by all scFvs within one library, and is not specific for a particular scFv. It is therefore essential to eliminate any possibility of contamination with a random scFv. The negative control reaction is used to identify any potential contamination from a random scFv that might stem from a phage display library used in house.

The pair of primers that are used in the negative control amplification, Fdseq and LMB3 long, allows for the amplification of any scFv within the cloning site of the pHEN vector, which is used in library construction.

6. The presence of a 900 bp band in the negative control reaction suggests that the material used for the amplification is contaminated with an scFv. Proceeding with reformatting beyond this point bears great risk of amplifying an scFv different to the one intended. Thus, it is essential to stop at this point, prepare fresh aliquots of primers, dNTP, buffers, and polymerase and start the entire process from the very beginning.
7. The assembly is performed through the overlapping hinge sequences of both scFv and Fc. Concomitantly, the Hind III Leader Seq Fo primer appends the remainder of the leader sequence to the N-terminus.
8. Typically, 5–10 U of digestion enzyme should be used per 1 μg of DNA. The reaction should ideally be performed in the final volume of 50 μL.
9. The reaction can also be performed in a thermocycler which should be programmed as follows: 37 °C for 20 h, 65 °C for 20 min, and 4 °C pause. As both of the restriction enzymes are heat sensitive they can be inactivated at 65 °C for 20 min, resulting in cessation of their activity.
10. The non-digested vector will show two bands under UV light: one at ~10.5 kilobase pairs (kb), corresponding to the circular two-stranded plasmid DNA and another one above it, which corresponds to nicked DNA where only one DNA strand has been cut.
11. The molar concentration of the scFv DNA should be at least three times higher than of the plasmid DNA.
12. The aim of the negative control ligation is to verify that the plasmid was digested by both restriction enzymes and not only by one. A plasmid, which has been cut with only one digestion enzyme, can undergo re-ligation without incorporation of the insert. Ideally, the agar plate with the negative control reaction should contain at least tenfold fewer colonies when compared to the agar plate with the test ligation reaction.
13. The time constant should be above 5 ms for the electroporation to be successful. If the constant is below 5 ms, the electroporation should be repeated with another aliquot of TG1 *E. coli*. When electroporating a plasmid that originates from a ligation mix, low DNA concentration may negatively affect the success of an electroporation reaction. Thus, should problems occur regarding the number of colonies on an agar plate, one could consider increasing the quantity of the ligation mix used in electroporation.

14. The negative control reaction should not yield any bands. The presence of a ~1,500 bp band under UV light suggests contamination with an irrelevant scFv-Fc. Should this occur, prepare fresh aliquots of primers, dNTP, polymerase buffer, polymerase enzyme and repeat the PCR screening.
15. Sequencing may either be outsourced to an external company or performed according to established protocols.
16. The quantity of mini-prep DNA may be increased up to 2 μL should the electroporation fail to yield any colonies on an agar plate. Additionally, prior to streaking out, the electroporation mixture may be subjected to shaking in an incubator at 37 °C and 180 RPM up to 60 min. This should increase the electroporation efficiency. Alternatively, the entire quantity of the electroporation mix could be streaked out, which would result in a higher number of colonies on an agar plate.
17. HEK293-EBNA cell line is suited for expression of fragments cloned inside the pCEP4 plasmid as this vector can be maintained episomally inside the cell. The Epstein-Barr Virus replication origin (oriP) and nuclear antigen (encoded by the EBNA-1 gene) is carried by this plasmid to permit extrachromosomal replication in HEK cells.
18. Alternatively, the CHO-S cell line can be used for the expression of scFv-Fc constructs. In order to do so, the entire scFv-Fc DNA fragment should be inserted in the pcDNA3.1 vector using the identical procedure as presented in Subheading 3.1, **step 10**. However, transfection of the HEK293-EBNA cell line usually results in higher expression yields.
19. The protein A binds specifically to the Fc domain of an scFv-Fc. Additionally, protein A binds to the heavy variable domain, which originates from the DP47 germ line gene.
20. This procedure prevents accumulation of air bubbles in the system. The protein A resin must always be kept in solution.
21. Setting the flow at speed higher than 1 mL/min might preclude the capture of scFv-Fc antibodies by the protein A resin, which results in a lower purification efficiency.
22. Typically, ten column volumes of Buffer A and five column volumes of Buffer B are sufficient for a successful washing procedure.
23. It is of great importance not to perturb the resin beads while applying the elution buffer, as this might result in a lower concentration of the eluted protein.
24. The protein A resin can be regenerated by washing with two column volumes of TEA, followed by two column volumes of PBS. Afterwards, add two column volumes of MilliQ water

followed two by column volumes of 20 % (vol/vol) ethanol. Store the resin in 20 % (vol/vol) ethanol at 4 °C indefinitely.

25. The purification procedure should ideally be performed inside a cold room. Alternatively, keeping the supernatant on ice throughout the process allows for the purification to be performed at room temperature.

26. The Leader Seq DP47 Fo primer anneals to the N-terminus of the heavy chain of scFv and appends a part of the leader sequence to it. The DP47 mtoq Ba primer anneals to the CDR1 region and mutates an amber stop codon (TAG), present in CDR1 of the heavy chain, into a glutamine codon (CAG). The DP47 mtoq Fo primer anneals to the framework region downstream of the CDR1 and together with either DPL16 Hinge ba or DPK22 Hinge ba primers amplifies the remainder of the scFv DNA. The Hind III Leader Seq Fo primer appends the remainder of the leader sequence together with the HindIII restriction enzyme digestion sequence to the N-terminus. The DPL16 Hinge ba and the DPK22 Hinge ba primers anneal to the C-terminus of the scFv, adding a part of the hinge region to it. In parallel, amplification of the CH2 and CH3 domains is performed using Hinge CH2 Fo forward primer and CH3 NotI Stop Ba backward primer. These primers append the N-terminus with a part of the hinge region and the C-terminus with a stop codon followed by the NotI restriction enzyme sequence. Following this, the assembly of the scFv-Fc DNA by overlap extensions is performed using Hind III Leader Seq Fo and CH3 NotI Stop Ba primers.

References

1. Walsh G (2010) Biopharmaceutical benchmarks 2010. Nat Biotechnol 28:917–924
2. Villa A, Lovato V, Bujak E et al (2011) A novel synthetic naïve human antibody library allows the isolation of antibodies against a new epitope of oncofetal fibronectin. MAbs 3:264–272
3. Silacci M, Brack S, Schirru G et al (2005) Design, construction, and characterization of a large synthetic human antibody phage display library. Proteomics 5:2340–2350
4. Cuesta ÁM, Sainz-Pastor N, Bonet J et al (2010) Multivalent antibodies: when design surpasses evolution. Trends Biotechnol 28:355–362
5. Tarli L, Balza E, Viti F et al (1999) A high-affinity human antibody that targets tumoral blood vessels. Blood 94:192–198
6. Holliger P, Hudson PJ (2005) Engineered antibody fragments and the rise of single domains. Nat Biotechnol 23:1126–1136
7. Zuberbühler K, Palumbo A, Bacci C et al (2009) A general method for the selection of high-level scFv and IgG antibody expression by stably transfected mammalian cells. Protein Eng Des Sel 22:169–174
8. Crothers D, Metzger H (1972) The influence of polyvalency on the binding properties of antibodies. Immunochemistry 9:341–357
9. Adams GP, Tai M-S, Mccartney JE et al (2006) Avidity-mediated enhancement of in vivo tumor targeting by single-chain Fv dimers. Clin Cancer Res 12:1599–1605
10. Roopenian DC, Akilesh S (2007) FcRn: the neonatal Fc receptor comes of age. Nat Rev Immunol 7:715–725
11. Weiner LM, Surana R, Wang S (2010) Monoclonal antibodies: versatile platforms for cancer immunotherapy. Nat Rev Immunol 10:317–327

12. Clynes RA, Towers TL, Presta LG et al (2000) Inhibitory Fc receptors modulate in vivo cytotoxicity against tumor targets. Nat Med 6:443–446
13. Cragg MS, Glennie MJ (2004) Antibody specificity controls in vivo effector mechanisms of anti-CD20 reagents. Blood 103:2738–2743
14. (2012) Study evaluating the SBI-087 in subjects with systemic lupus erythematosus, NCT00714116. http://clinicaltrials.gov/ct2/show/NCT00714116?term=sle&rank=58
15. (2012) Study evaluating the efficacy and safety of SBI-087 in seropositive subjects with active rheumatoid arthritis, NCT01008852. http://clinicaltrials.gov/ct2/show/NCT01008852?term=SBI-087&rank=4
16. Hoogenboom HR, Griffiths AD, Johnson KS et al (1991) Multi-subunit proteins on the surface of filamentous phage: methodologies for displaying antibody (Fab) heavy and light chains. Nucleic Acids Res 19:4133–4137
17. Borsi L, Balza E, Bestagno M et al (2002) Selective targeting of tumoral vasculature: comparison of different formats of an antibody (L19) to the ED-B domain of fibronectin. Int J Cancer 102:75–85

Part III

Anitbody Characterization and Modification

Chapter 21

Antibody V and C Domain Sequence, Structure, and Interaction Analysis with Special Reference to IMGT®

Eltaf Alamyar, Véronique Giudicelli, Patrice Duroux, and Marie-Paule Lefranc

Abstract

IMGT®, the international ImMunoGeneTics information system® (http://www.imgt.org), created in 1989 (Centre National de la Recherche Scientifique, Montpellier University), is acknowledged as the global reference in immunogenetics and immunoinformatics. The accuracy and the consistency of the IMGT® data are based on IMGT-ONTOLOGY which bridges the gap between genes, sequences, and three-dimensional (3D) structures. Thus, receptors, chains, and domains are characterized with the same IMGT® rules and standards (IMGT standardized labels, IMGT gene and allele nomenclature, IMGT unique numbering, IMGT Collier de Perles), independently from the molecule type (genomic DNA, complementary DNA, transcript, or protein) or from the species. More particularly, IMGT® tools and databases provide a highly standardized analysis of the immunoglobulin (IG) or antibody and T cell receptor (TR) V and C domains. IMGT/V-QUEST analyzes the V domains of IG or TR rearranged nucleotide sequences, integrates the IMGT/JunctionAnalysis and IMGT/Automat tools, and provides IMGT Collier de Perles. IMGT/HighV-QUEST analyzes sequences from high-throughput sequencing (HTS) (up to 150,000 sequences per batch) and performs statistical analysis on up to 450,000 results, with the same resolution and high quality as IMGT/V-QUEST online. IMGT/DomainGapAlign analyzes amino acid sequences of V and C domains and IMGT/3Dstructure-DB and associated tools provide information on 3D structures, contact analysis, and paratope/epitope interactions. These IMGT® tools and databases, and the IMGT/mAb-DB interface with access to therapeutical antibody data, provide an invaluable help for antibody engineering and antibody humanization.

Key words IMGT, Immunoglobulin, Antibody, Antibody humanization, High-throughput sequencing, Immunogenetics, Immunoinformatics, IMGT-ONTOLOGY, IMGT Collier de Perles, IMGT unique numbering

1 Introduction

Standardized sequence, structure, and interaction analysis of immunoglobulins (IG) or antibodies, with reference to IMGT®, is crucial for a better understanding and a better comparison of the monoclonal antibody (mAb) specificity, affinity, half-life, and effector properties [1–8]. IMGT®, the international ImMunoGeneTics

Vincent Ossipow and Nicolas Fischer (eds.), *Monoclonal Antibodies: Methods and Protocols*, Methods in Molecular Biology, vol. 1131, DOI 10.1007/978-1-62703-992-5_21, © Springer Science+Business Media New York 2014

information system® (http://www.imgt.org) [9], was created in 1989 by Marie-Paule Lefranc at Montpellier, France (Centre National de la Recherche Scientifique (CNRS) and Université Montpellier 2), in order to standardize the immunogenetics data and to manage the huge diversity of the antigen receptors, IG or antibodies, and T cell receptors (TR) [10–13]. IMGT® is the global reference in immunogenetics and immunoinformatics [14, 15], and its standards have been endorsed by the World Health Organization–International Union of Immunological Societies (WHO–IUIS) Nomenclature Committee [16, 17] and the WHO International Nonproprietary Name (INN) Programme [18]. IMGT® provides a common access to standardized and integrated data from genome, proteome, genetics, and three-dimensional (3D) structures and comprises 7 databases (for sequences, genes, and 3D structures), 17 online tools, and more than 15,000 pages of Web resources [9]. The accuracy and the consistency of the IMGT® data are based on IMGT-ONTOLOGY, the first ontology for immunogenetics and immunoinformatics [19–26]. The concepts of identification, classification, description, and numerotation (and corresponding rules and standards) have been generated, respectively, from the axioms of IDENTIFICATION (standardized keywords) [27], DESCRIPTION (standardized labels) [28], CLASSIFICATION (gene and allele nomenclature) [29], and NUMEROTATION (IMGT unique numbering and IMGT Collier de Perles) [30, 31].

This chapter comprises basic information on V and C domains with reference to IMGT® (Subheading 2) and the protocols of the most popular IMGT® tools and databases (Subheadings 3–6) which are widely used for a standardized analysis of the antibody V and C domains. IMGT/V-QUEST [32–37] (Subheading 3) is the IMGT® online tool for the analysis of IG and TR nucleotide sequences. IMGT/V-QUEST identifies the variable (V), diversity (D), and junction (J) genes in rearranged IG and TR sequences and, for the IG, the nucleotide (nt) mutations and amino acid (AA) changes resulting from somatic hypermutations. The tool integrates IMGT/JunctionAnalysis [38, 39] for the detailed characterization of the V-D-J or V-J junctions and IMGT/Automat [40, 41] for a complete sequence annotation. IMGT/HighV-QUEST [37, 42–44] (Subheading 4) is the high-throughput version of IMGT/V-QUEST and was implemented to answer the needs of data analysis from the next-generation sequencing (NGS). This tool analyzes up to 150,000 sequences per run and provides statistical analysis on 450,000 results, with the same degree of resolution and high-quality results as IMGT/V-QUEST.

IMGT/DomainGapAlign [45–47] (Subheading 5) is the IMGT® online tool for the analysis of amino acid sequences and two-dimensional (2D) structures of domains. IMGT/3Dstructure-DB [45, 48, 49] (Subheading 6) and associated tools provide information on 3D structures, contact analysis, and paratope/epitope interactions. IMGT® tools and databases run against IMGT reference directories built from IMGT/LIGM-DB, the IMGT® nucleotide database [50], and from IMGT/GENE-DB, the IMGT® gene database [51].

2 Basic Information for V and C Domains with Special Reference to IMGT®

IMGT-ONTOLOGY concepts allow bridging the gap between sequences and 3D structures by characterizing receptors, chains, and domains, with the same IMGT rules and standards (IMGT standardized labels, IMGT nomenclature, IMGT unique numbering, IMGT Collier de Perles), independently from the molecule type (genomic DNA (gDNA), complementary DNA (cDNA), transcript, or protein) or from the species [25–31]. Interestingly, they also allow bridging the gap between genomic data and 3D structures, the exon delimitations being used to delimit domains in IMGT® [52–56]. In this section, the focus is on V and C domains [30, 31] which are characteristics of the immunoglobulin superfamily (IgSF) to which belong the IG and TR, antigen receptors of the gnasthostomata (vertebrates with jaws), and the IgSF other than IG and TR of vertebrates and invertebrates. Basic information for V and C domains with reference to IMGT® is necessary for an effective use of IMGT® databases and tools and comprises the IMGT unique numbering [30, 52–56] and IMGT Collier de Perles [31, 57–60].

2.1 Domain Types

2.1.1 V Domain

A V domain (Fig. 1) comprises about 100 amino acids and is made of nine antiparallel beta strands (A, B, C, C′, C″, D, E, F, and G) linked by beta turns (AB, CC′, C″D, DE, and EF) or loops (BC, C′C″, and FG) and forming a sandwich of two sheets [30, 52–54] (Table 1). The sheets are closely packed against each other through hydrophobic interactions giving a hydrophobic core and joined together by a disulfide bridge between first-CYS at position 23 in the B-STRAND in the first sheet and the second-CYS 104 in the F-STRAND in the second sheet. The V domain includes the V-DOMAIN of the IG and TR that corresponds to the V-J-REGION or the V-D-J-REGION encoded by V-(D)-J rearrangements [10, 11] (*see* **Note 1**) and the V-LIKE-DOMAIN of the IgSF other than IG and TR [61–63].

2.1.2 C Domain

A C domain (Fig. 2) comprises about 100 amino acids and is made of seven antiparallel beta strands (A, B, C, D, E, F, and G) linked by beta turns, a transversal strand (CD) and loops (BC and FG), on two sheets [30, 55] (Table 2). A C domain has a topology and a three-dimensional structure similar to that of a V domain but without the C′ and C″ strands and the C′C″ loop. The C domain includes the C-DOMAIN of the IG and TR [10, 11] and the C-LIKE-DOMAIN of the IgSF other than IG and TR [61–65].

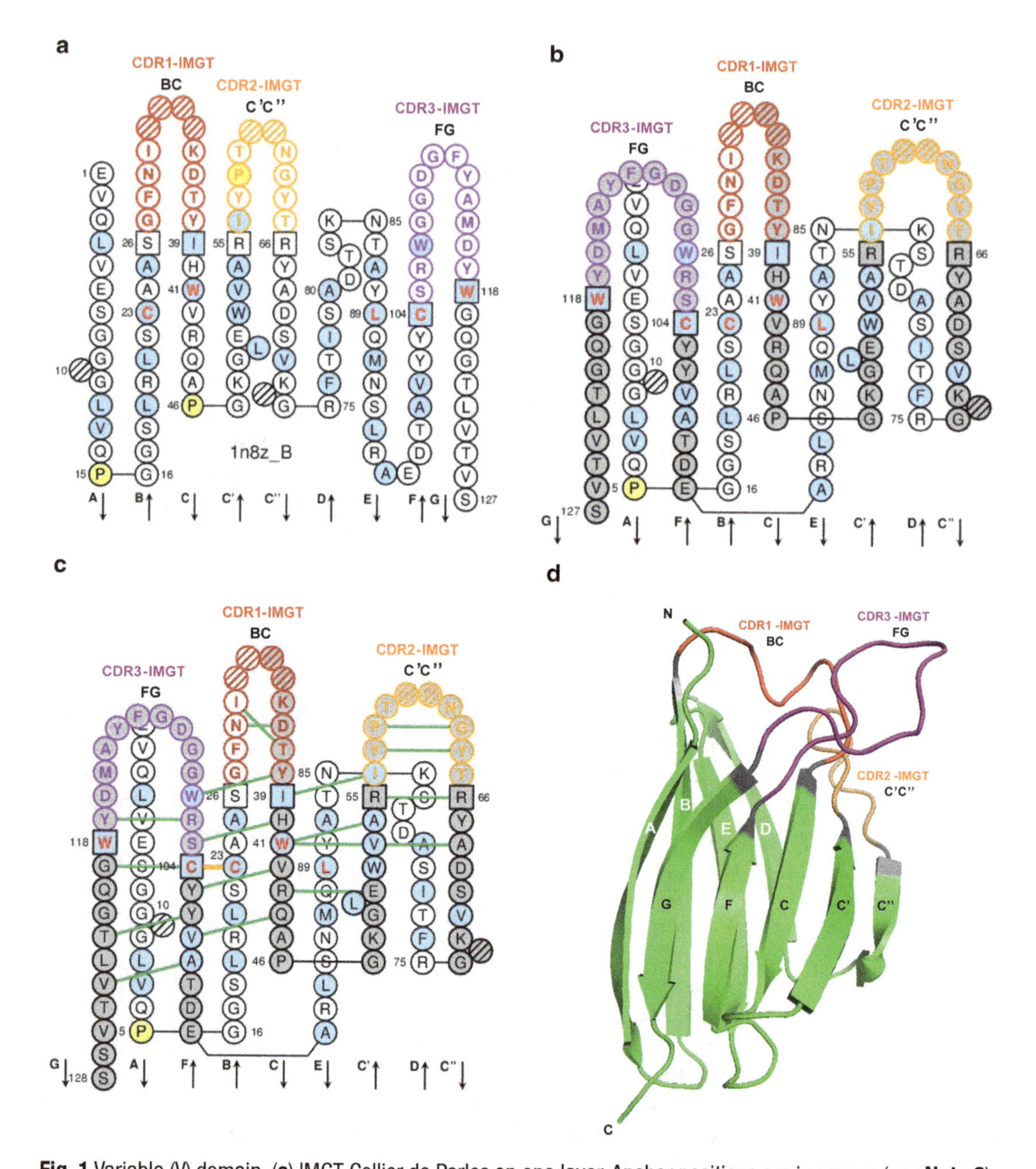

Fig. 1 Variable (V) domain. (**a**) IMGT Collier de Perles on one layer. Anchor positions are in square (*see* **Note 2**). Hatched positions correspond to gaps according to the IMGT unique numbering [30, 31, 53, 54]. Proline is shown in *yellow*. Positions with *bold* (online *red*) *letters* indicate the four conserved positions that are common to a V domain and to a C domain: first-CYS 23, CONSERVED-TRP 41, hydrophobic 89, second-CYS 104, and a fifth conserved position that is specific to the IG and TR V-DOMAIN: J-TRP or J-PHE 118 [30, 31, 53, 54] (Table 1). In an IG or a TR V-DOMAIN, the hydrophobic amino acids (hydropathy index with positive value) and tryptophan (W) found at a given position in more than 50 % of sequences are displayed (online with a *blue* background color). The G-STRAND (or FR4-IMGT) is the C-terminal part of the J-REGION. The FR4-IMGT is at least composed of nine or ten amino acids beyond the phenylalanine F (J-PHE 118) or tryptophan W (J-TRP 118) of the motif F/W-G-X-G that characterizes the J-REGION. The example is the VH of trastuzumab (IMGT/mAb-DB and IMGT/2Dstructure-DB, http://www.imgt.org). The CDR-IMGT lengths are [8.8.13]. (**b**) IMGT Collier de Perles on two layers. (**c**) IMGT Collier de Perles on two layers with hydrogen bonds (1n8z_B from IMGT/3Dstructure-DB, http://www.imgt.org). (**d**) Graphical 3D representation based on the FR-IMGT and CDR-IMGT delimitations (Table 1), obtained using Pymol (http://www.pymol.org) and "IMGT numbering comparison" (IMGT/Structure-DB, http://www.imgt.org)

Table 1
V domain strands and loops, IMGT positions and lengths, based on the IMGT unique numbering for V domain (V-DOMAIN and V-LIKE-DOMAIN) [30, 53, 54]

V domain strands and loops[a]	IMGT positions	Lengths[b]	Characteristic Residue@Position[c]	V-DOMAIN FR-IMGT and CDR-IMGT
A-STRAND	1–15	15 (14 if gap at 10)		FR1-IMGT
B-STRAND	16–26	11	First-CYS 23	
BC-LOOP	27–38	12 (or less)		CDR1-IMGT
C-STRAND	39–46	8	CONSERVED-TRP 41	FR2-IMGT
C′-STRAND	47–55	9		
C′C″-LOOP	56–65	10 (or less)		CDR2-IMGT
C″-STRAND	66–74	9 (or 8 if gap at 73)		FR3-IMGT
D-STRAND	75–84	10 (or 8 if gaps at 81, 82)		
E-STRAND	85–96	12	Hydrophobic 89	
F-STRAND	97–104	8	Second-CYS 104	
FG-LOOP	105–117	13 (or less, or more)		CDR3-IMGT
G-STRAND	118–128	11 (or 10)	V-DOMAIN J-PHE 118 or J-TRP 118[d]	FR4-IMGT

[a]IMGT® labels (concepts of description) are written in capital letters
[b]In number of amino acids (or codons)
[c]Residue@Position is an IMGT® concept of numerotation that numbers the position of a given residue (or that of a conserved property amino acid class) based on the IMGT unique numbering
[d]In the IG and TR V-DOMAIN, the G-STRAND (or FR4-IMGT) is the C-terminal part of the J-REGION, with J-PHE or J-TRP 118 and the canonical motif F/W-G-X-G at positions 118–121. The JUNCTION refers to the CDR3-IMGT plus the two anchors second-CYS 104 and J-PHE or J-TRP 118

2.2 IMGT Unique Numbering

2.2.1 IMGT Unique Numbering for V Domain

The V domain strands and loops and their delimitations and lengths are based on the IMGT unique numbering for V domain (V-DOMAIN and V-LIKE-DOMAIN) [30, 53, 54] (Table 1). The loop length (number of amino acids (or codons), that is, number of occupied positions) is a crucial and original concept of IMGT-ONTOLOGY. The lengths of the loops BC (or CDR1-IMGT for V-DOMAIN), C′C″ (or CDR2-IMGT for V-DOMAIN), and FG (or CDR3-IMGT for V-DOMAIN) characterize the V domain. The lengths of the three loops BC, C′C″, and FG, delimited by the anchors (*see* **Note 2**), are shown in number of amino acids (or codons), into brackets and separated by dots. For example [9.6.9] means that the BC, C′C″, and FG loops (or CDR1-IMGT, CDR2-IMGT, and CDR3-IMGT for a V-DOMAIN) have a length of 9, 6, and 9 amino acids (or codons), respectively.

In the IG and TR V-DOMAIN, position 104 (second-CYS) and position 118 (J-PHE or J-TRP) that belong to the FR3-IMGT and FR4-IMGT, respectively, are the anchors of the CDR3-IMGT. They correspond to the 5′ and 3′ ends of the JUNCTION (*see* **Note 2**). The G-STRAND is the C-terminal part of the

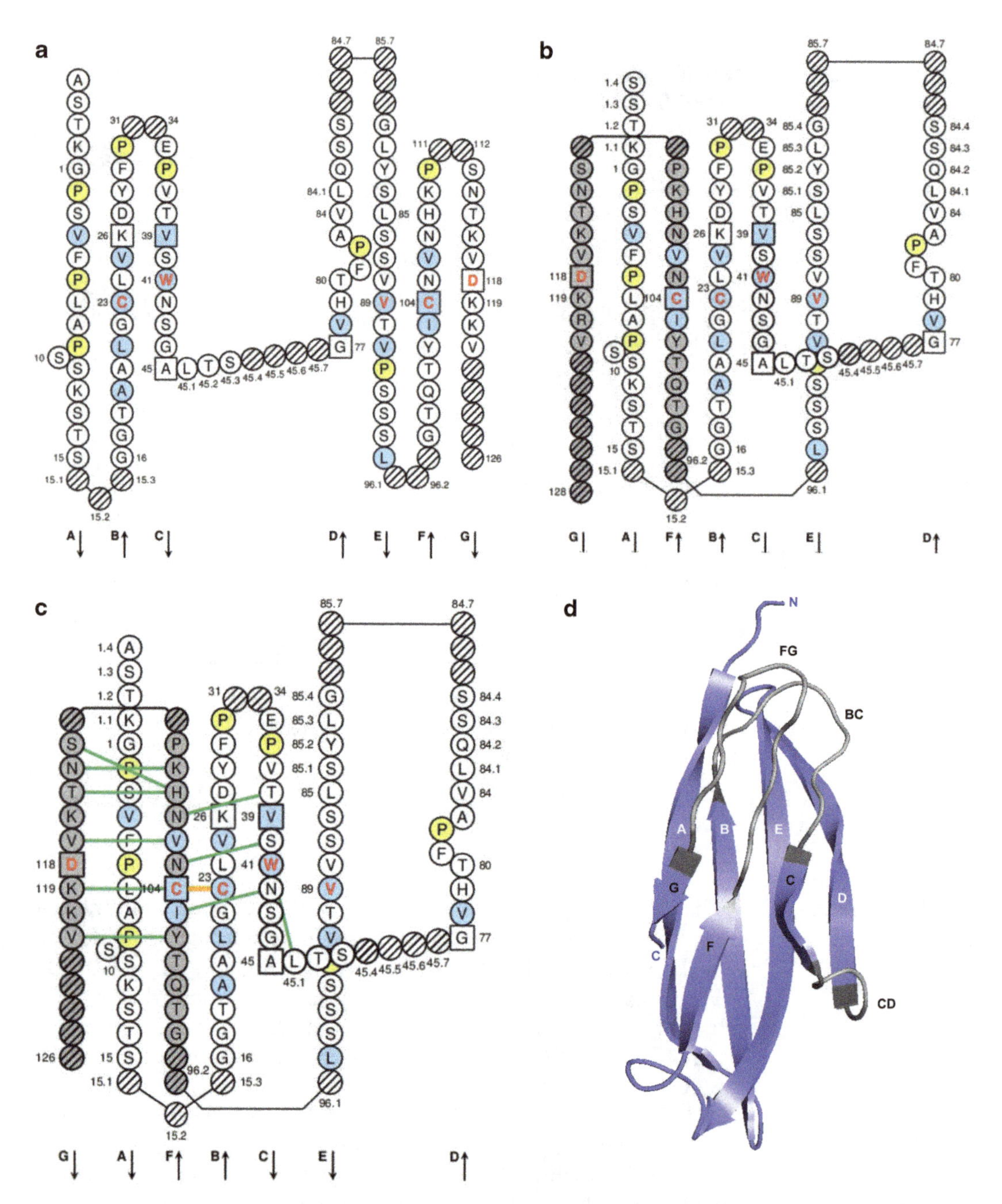

Fig. 2 Constant (C) domain. (**a**) IMGT Collier de Perles on one layer. Anchor positions are in square (*see* **Note 2**). Hatched positions correspond to gaps according to the IMGT unique numbering [30, 31, 55]. Proline is shown in *yellow*. Positions with *bold* (online *red*) *letters* indicate the four conserved positions that are common to a C domain and to a V domain: first-CYS 23, CONSERVED-TRP 41, hydrophobic 89, second-CYS 104 [30, 31, 55] (Table 2), and position 118 which is only conserved in V-DOMAIN. The example is the CH1 of trastuzumab (IMGT/mAb-DB and IMGT/2Dstructure-DB, http://www.imgt.org). (**b**) IMGT Collier de Perles on two layers. (**c**) IMGT Collier de Perles on two layers with hydrogen bonds (1n8z_B from IMGT/3Dstructure-DB, http://www.imgt.org). (**d**) Graphical 3D representation based on the IMGT strand and loop delimitations (Table 2), obtained using Pymol (http://www.pymol.org) and "IMGT numbering comparison" (IMGT/Structure-DB, http://www.imgt.org)

Table 2
C domain strands, turns, and loops and IMGT positions and lengths based on the IMGT unique numbering for C domain (C-DOMAIN and C-LIKE-DOMAIN) [30, 55]

C domain strands, turns, and loops[a]	IMGT positions	Lengths[b]	Characteristic Residue@ Position[c]
A-STRAND AB-TURN	1–15 15.1–15.3	15 (14 if gap at 10) 0–3	
B-STRAND BC-LOOP	16–26 27–31 34–38	11 10 (or less)	First-CYS 23
C-STRAND CD-STRAND D-STRAND DE-TURN	39–45 45.1–45.9 77–84 84.1–84.7 85.1–85.7	7 0–9 8 (or 7 if gap at 82) 0–14	CONSERVED-TRP 41
E-STRAND EF-TURN	85–96 96.1–96.2	12 0–2	Hydrophobic 89
F-STRAND FG-LOOP G-STRAND	97–104 105–117 118–128	8 13 (or less, or more) 11 (or less)	Second-CYS 104

[a]IMGT® labels (concepts of description) are written in capital letters
[b]In number of amino acids (or codons)
[c]Residue@Position is an IMGT® concept of numerotation that numbers the position of a given residue (or that of a conserved property amino acid class) based on the IMGT unique numbering

J-REGION, with J-PHE or J-TRP 118 and the canonical motif F/W-G-X-G at positions 118–121 [30, 53, 54].

2.2.2 IMGT Unique Numbering for C Domain

The C domain strands, turns, and loops and their delimitations and lengths are based on the IMGT unique numbering for C domain (C-DOMAIN and C-LIKE-DOMAIN) [30, 55] (Table 2). The loops BC and FG and the transverse strand CD are delimited by the anchors (*see* **Note 2**).

2.3 IMGT Collier de Perles

IMGT Collier de Perles [31, 57–59] are standardized IMGT 2D graphical representations of protein domains that can be obtained using the IMGT/Collier de Perles tool [60] available from the IMGT® home page at http://www.imgt.org (for sequences already gapped according to the IMGT unique numbering) or using the IMGT/Collier de Perles tool integrated in IMGT/V-QUEST [32–37] (starting from IG or TR V domain nt sequences) or integrated in IMGT/DomainGapAlign [45–47] (e.g., starting from IG or TR V or C domain AA sequences). IMGT Collier de Perles for V and C domains, on one or two layers, are also available in IMGT/2Dstructure-DB (for amino acid sequences) and on two layers with hydrogen bonds in IMGT/3Dstructure-DB (for 3D structures) [45, 48, 49].

3 IMGT/V-QUEST

The link to the IMGT/V-QUEST tool is available from the IMGT® home page, http://www.imgt.org, in the "IMGT tools" section. Clicking on that link gives access to the IMGT/V-QUEST welcome page.

3.1 IMGT/V-QUEST Welcome Page

1. In the IMGT/V-QUEST welcome page, locate the section "Analyse your immunoglobulin (IG) or antibody nucleotide sequences." The locus (IGH, IGK, or IGL for IG sequences) will be automatically identified by the tool.
2. Locate the species or the taxon that corresponds to the IMGT reference directory against which you want your sequences to be analyzed (*see* **Note 3**).
3. Click on the species (for instance "Human") or the taxon. This will give access to the IMGT/V-QUEST search page (Fig. 3).

3.2 IMGT/V-QUEST Search Page

3.2.1 Sequence Submission

IMGT/V-QUEST can analyze IG-rearranged nucleotide sequences from genomic DNA or cDNA, indifferently. Sequences must be submitted in FASTA format (up to 50 sequences per run) (Fig. 3a). The sequences must be copied and pasted in the text area "Type (or copy/paste) your nucleotide sequence(s) in FASTA format" or uploaded as a text file by selecting the option "Or give the path access to a local file containing your sequence(s) in FASTA format" and using the "Browse" (or "Parcourir" button) to select the file. Before launching the analysis, the user can select the result display (Subheading 3.2.2) and/or "Advanced parameters" (Subheading 3.2.3).

3.2.2 Selection of the Result Display

The user can choose between three types of display for the results, "Detailed view," "Synthesis view," or "Excel file" (Fig. 3b), and the parameters of the analysis can be customized with a large choice of options in "Advanced parameters" (Fig. 3c).

1. "Detailed view"
   "Detailed view" provides the results for each sequence, individually. They may be displayed in HTML (by default) or Text format, with a user-defined number (nb) of nucleotides per line in alignments (60 by default) and nb of aligned reference sequences (5 by default). There is a choice of 14 different result displays for the analysis of the V-DOMAIN (Fig. 3b). The results most commonly requested are selected by default. You can check or uncheck the result displays as needed.
2. "Synthesis view"
   The "Synthesis view" consists of an interesting alternative to the "Detailed view" if several sequences of the same run use the same V gene and allele. Indeed, sequences expressing the

a

WELCOME !
to IMGT/V-QUEST Search page
THE INTERNATIONAL IMMUNOGENETICS INFORMATION SYSTEM®

Im Muno Gene Tics Information system®
http://www.imgt.org

Citing IMGT/V-QUEST :
Brochet, X. et al., Nucl. Acids Res. 36, W503-508 (2008). PDF PMID: 18503082
Giudicelli, V., Brochet, X., Lefranc, M.-P.,Cold Spring Harb Protoc. 2011 Jun 1;2011 (6). pii: pdb.prot5633. doi: 10.1101/pdb.prot5633.
PMID: 21632778 Abstract also in *IMGT booklet with generous provision from Cold Spring Harbor (CSH) Protocols* PDF (high res) PDF (lower res)

IMGT/V-QUEST programme version: 3.2.30 (28 January 2013) - IMGT/V-QUEST reference directory release: 201237-3 (12 September 2012)

Analyse your immunoglobulin (IG) or antibody sequences

Your selection: Human

Your sequences are compared to the **Human** ***(Homo sapiens)*** **IG set** from the IMGT/V-QUEST reference directory sets
Sequence sets to test IMGT/V-QUEST are available here

Sequence submission

◉ **Type (or copy/paste) your sequence(s) in FASTA format**

```
>SEQ1
atggactggacctggagcatccttttcttggtggcagcagcaacaggtgcccactcccag
gttcagctggtgcagtctggagctgaggtgaagaagcctggggcctcagtgaaggtctcc
tgcaaggcttctggttacacctttaccagctatggtatcagctgggtgcgacaggcccct
ggacaagggcttgagtggatgggatggatcagcgcttacaatggtaacacaaactatgca
cagaagctccagggcagagtcaccatgaccacagacacatccacgagcacagcctacatg
gagctgaggagcctgagatctgacgacacggccgtgtattactgtgcgagagggggcagc
tcggaccccaggactttgatatctggggccaagggacaatggtcaccgtctcttcaggg
```

○ **Or give the path access to a local file containing your sequence(s) in FASTA format**

[ ] Browse

Start | Clear the form

b

Display results

◉ **A. Detailed view** ◉ HTML ○ Text Nb of nucleotides per line in alignments: 60 Nb of aligned reference sequences: 5

1. ☑ Alignment for V-GENE
2. ☐ Alignment for D-GENE
3. ☑ Alignment for J-GENE
4. ☑ Results of IMGT/JunctionAnalysis
   ○ with full list of eligible D-GENEs
   ◉ without list of eligible D-GENEs
5. ☐ Sequence of the JUNCTION ('nt' and 'AA')
6. ☑ V-REGION alignment
7. ☑ V-REGION translation
8. ☑ V-REGION protein display
9. ☐ V-REGION mutation and AA change table
10. ☐ V-REGION mutation and AA change statistics
11. ☐ V-REGION mutation hot spots
12. ☐ Sequences of V-, V-J- or V-D-J- REGION ('nt' and 'AA') with gaps in FASTA and access to IMGT/PhyloGene for V-REGION ('nt')
13. ☐ Annotation by IMGT/Automat
14. IMGT Collier de Perles
   ◉ link to IMGT/Collier-de-Perles tool
   ○ IMGT Collier de Perles (for a nb of sequences < 5)
   ○ no IMGT Collier de Perles

○ **B. Synthesis view** ○ HTML ○ Text Nb of nucleotides per line in alignments: Summary table sequence order

1. ☑ Alignment for V-GENE
2. ☑ V-REGION alignment
3. ☑ V-REGION translation
4. ☑ V-REGION protein display
5. ☑ V-REGION protein display (with AA class colors)
6. ☑ V-REGION protein display (only AA changes displayed)
7. ☑ V-REGION most frequently occurring AA
8. ☑ Results of IMGT/JunctionAnalysis

○ **C. Excel file**

1. ☑ Summary
2. ☑ IMGT-gapped-nt-sequences
3. ☑ nt-sequences
4. ☑ IMGT-gapped-AA-sequences
5. ☑ AA-sequences
6. ☑ Junction
7. ☑ V-REGION-mutation-and-AA-change-table
8. ☑ V-REGION-nt-mutation-statistics
9. ☑ V-REGION-AA-change-statistics
10. ☑ V-REGION-mutation-hot-spots
11. ☑ Parameters

c

Advanced parameters

Selection of IMGT reference directory set	F+ORF+in-frame P	◉ With all alleles ○ With allele *01 only
Search for insertions and deletions in V-REGION	○ Yes (slower, the nb of submitted sequences in a single run is limited to 10)	◉ No
Parameters for IMGT/JunctionAnalysis	Nb of accepted D-GENE in IGH JUNCTION (default is 1) default	Nb of accepted mutations: default in 3'V-REGION; default in D-REGION; default in 5'J-REGION
Parameters for "Detailed view"	Nb of nucleotides to exclude in 5' of the V-REGION for the evaluation of the nb of mutations (in results 9 and 10)	Nb of nucleotides to add (or exclude) in 3' of the V-REGION for the evaluation of the alignment score (in results 1)

Fig. 3 The IMGT/V-QUEST search page. This page comprises, from top to bottom, three sections: (**a**) Sequence submission, (**b**) "Display results," and (**c**) "Advanced parameters." The page shown is for IG ("Analyse your immunoglobulin (IG) or antibody sequences"). The species (or taxon) selection from the IMGT/V-QUEST welcome page is recalled at the top of the page (here "human")

same V gene and allele will be aligned together. The results may be displayed in HTML (by default) or Text format, with a user-defined nb of nucleotides per line in alignments (60 by default) and a "Summary table sequence order" ("V-GENE and allele name" or "input"). There is a choice of eight different result displays (six for the alignments of sequences, one for the most frequently occurring AA, and one for results of IMGT/JunctionAnalysis) (Fig. 3b).

3. "Excel file"

   "Excel file" allows the users to get the results of the analysis in a spreadsheet with the choice of 11 sheets (Fig. 3b). The "Summary" and "Parameters" sheets are always provided.

3.2.3 "Advanced Parameters"

Default values of "Advanced parameters" have been set for classical analysis (Fig. 3c). They should be modified for more sophisticated queries or for unusual sequences (e.g., for sequences with insertions or deletions). The options include:

1. "Selection of IMGT reference directory set" with a choice of four sets:
   (a) F + ORF,
   (b) F + ORF + in-frame P (by default),
   (c) F + ORF including orphons,
   (d) F + ORF + in-frame P including orphons,

   where F is functional, ORF is open reading frame (ORF), and P is pseudogene (*see* **Note 3**). This allows sequences to be compared with only relevant gene sequences (e.g., orphon sequences are relevant for genomic but not for expressed repertoire studies). The selected set can also be chosen either "With all alleles" or "With allele *01 only" (*see* **Note 4**).
2. "Search for insertions and deletions in V-REGION" ("Yes" or "No"): Somatic hypermutations by nucleotide insertions and deletions in the V-REGION are rare events that are known to occur in normal and malignant cells [66]. By default, IMGT/V-QUEST does not search for insertions and deletions. If "Yes" is selected, the analysis is slower and the number of submitted sequences in a single run is limited to 10 (*see* **Note 5**).
3. "Parameters for IMGT/JunctionAnalysis":
   (a) "Nb of accepted D-GENE": This selection is provided for the IGH junctions.
   (b) "Nb of accepted mutations" in 3′V-REGION, D-REGION, and 5′J-REGION (*see* **Note 6**).
4. "Parameters" for "Detailed view":
   (a) "Nb of nucleotides to exclude in 5′ of the V-REGION for the evaluation of the nb of mutations": in case of primer-specific nucleotides.

Sequence number 1: SEQ1

Sequence compared with the human IG set from the IMGT reference directory

```
>SEQ1
atggactggacctggagcatccttttcttggtggcagcagcaacaggtgcccactcccag
gttcagctggtgcagtctggagctgaggtgaagaagcctggggcctcagtgaaggtctcc
tgcaaggcttctggttacacctttaccagctatggtatcagctgggtgcgacaggcccct
ggacaagggcttgagtggatgggatggatcagcgcttacaatggtaacacaaactatgca
cagaagctccagggcagagtcaccatgaccacagacacatccacgagcacagcctacatg
gagctgaggagcctgagatctgacgacacggccgtgtattactgtgcgagagggggcagc
tcggaccccaggactttgatatctggggccaagggacaatggtcaccgtctcttcaggg
```

Result summary:	Productive IGH rearranged sequence (no stop codon and in-frame junction)		
V-GENE and allele	Homsap IGHV1-18*01 F	score = 1435	identity = 100.00% (288/288 nt)
J-GENE and allele	Homsap IGHJ3*02 F	score = 205	identity = 90.00% (45/50 nt)
D-GENE and allele by IMGT/JunctionAnalysis	Homsap IGHD6-13*01 F	D-REGION is in reading frame 1	
FR-IMGT lengths, CDR-IMGT lengths and AA JUNCTION	[25.17.38.11]	[8.8.13]	CARGGSSDPRTFDIW

Fig. 4 IMGT/V-QUEST "Detailed view" result page: Sequence and "Result summary." The SEQ1 is the human IGH rearranged sequence with the accession number AF062212 in the IMGT/LIGM-DB database [50]

(b) "Nb of nucleotides to add (or exclude) in 3′ of the V-REGION for the evaluation of the alignment score": in case of low (or high)-exonuclease activity.

3.3 IMGT/V-QUEST Output for Detailed View

The "Detailed view" results are displayed in "A. Detailed results for the IMGT/V-QUEST analysed sequences." The top of this page indicates the number of analyzed sequences with links to individual results. Use the link associated with a sequence name to go directly to its individual result that comprises the sequence and "Result summary" followed by the different displayed results as selected in the search page.

3.3.1 Sequence and "Result Summary"

The sequence and "Result summary" are shown at the top of each individual result (Fig. 4). The IMGT reference directory set against which the sequence was analyzed (e.g., human IG) is indicated. A sequence submitted in antisense orientation will be shown as complementary reverse sequence that is in V gene sense orientation. The "Result summary" provides a crucial feature that is the evaluation of the functionality of the V domain-rearranged sequence performed automatically by IMGT/V-QUEST: productive (if no stop codon in the V-(D)-J region, and in-frame junction) or unproductive (if stop codons in the V-(D)-J region and/or out-of-frame junction). It also summarizes the main characteristics of the analyzed V domain sequence which include (a) the names of the closest "V-GENE and allele" (e.g., IGHV1-18*01) and "J-GENE and allele" (e.g., IGHJ3*02) (*see* **Note 7**) with the alignment score (*see* **Note 8**), the percentage of identity, and the ratio of the number of identical nt/number of aligned nt; (b) the name of the closest "D-GENE and allele" (IGHD6-13*01) determined by

IMGT/JunctionAnalysis [38, 39] with the D-REGION reading frame; and (c) the FR-IMGT lengths (e.g., [25.17.38.11]), the CDR-IMGT lengths (e.g., [8.8.13]), and the AA JUNCTION sequence which characterizes a V domain. The information shown in Fig. 4 (online in orange), "Productive IGH rearranged sequence" and "100.00 %," is used by clinicians analyzing the level of mutations in chronic lymphocytic leukemia (CLL) IGHV-rearranged sequences as a prognostic criterion [75] (*see* **Note 9**). IMGT/V-QUEST provides warnings (not shown) that appear as notes in red to alert the user, if potential insertions or deletions are suspected in the V-REGION (*see* **Note 10**) or if other possibilities for the J-GENE and allele names are identified (*see* **Note 11**).

3.3.2 Sequence and Result Summary with the Option "Search for Insertions and Deletions"

If insertions and/or deletions are suspected, as mentioned above (Subheading 3.3.1), the user can go back to the IMGT/V-QUEST search page and select the option "Search for insertions and deletions" in "Advanced parameters." If indeed insertions and/or deletions are detected, they will be described in the "Result summary" row (Fig. 5) with their localization in FR-IMGT or CDR-IMGT, the nb of inserted or deleted nt, and, for insertions, the inserted nt, the presence or the absence of frameshift, the V-REGION codon from which the insertion or the deletion starts, and the nt position in the user-submitted sequence. The insertions are highlighted in capital letters in the user sequence (Fig. 5), and the tool runs a classical IMGT/V-QUEST search after having removed the insertion(s) from the user sequence. In case of deletions, the tool adds gaps to replace the identified deletions before running a classical IMGT/V-QUEST search. Users should be aware that an insertion or a deletion at the beginning of FR1-IMGT or at the end of the FR3-IMGT may not be detected.

3.3.3 Alignments for V, D, and J Genes and Alleles

The alignments for V, D, and J genes and alleles, if selected in the search page, display the alignments with the five closest V, D, and J gene alleles, respectively, with their alignment score and identity percentage. The alignment for D-GENE and allele should be considered with caution since it may show discrepancies with the results obtained by IMGT/JunctionAnalysis [38, 39] (*see* **Note 12**).

3.3.4 "Results of IMGT/JunctionAnalysis" and "Sequence of the JUNCTION ('nt' and 'AA')"

The results of IMGT/JunctionAnalysis comprise the following (Fig. 6):

1. "Analysis of the JUNCTION": It provides the name of the D gene and allele (*see* **Note 7**) for IGH sequences (*see* **Note 12**) and shows the details of the junction at the nucleotide level (nucleotides trimmed at the V, D, and J gene ends and N nucleotides added by the TdT) (*see* **Note 1**) with an accurate delimitation of the 3′V-REGION (from the second-CYS 104 to the 3′ end of the V-REGION), D-REGION, and 5′J-REGION (from the 5′ end of the J-REGION to J-TRP 118 or J-PHE 118)

Sequence number 2: seq2

Sequence compared with the human IG set from the IMGT reference directory

```
>SEQ2
atgtctgtctccttcctcatcttcctgcccgtgctgggcctcccatggggtgtcctgtca
caggtacagctgcagcagtcaggtccaggactggtgaagccctcgcagaccctctcactc
acctgtgccatctccggggacagtgtctctAGCagcaacggtgttgcttggaactgggtc
aggcagtccccatcgagaggccttgagtggctgggaaggacatactacaggtccaagtgg
tataatgattatgcagtatctgtgaaaagtcgaataaccatcaacccagacacatccaag
aaccagttctccctgcagctgaactctgtgactcccgaagacacggctgtgtattactgt
gcaagaggccgttggaccgcttttgatttctggggccaagggacaatggtcaccgtctct
tcc
```

Result summary:

Nucleotide insertions have been detected and automatically removed for this analysis: they are displayed as capital letters in the user submitted sequence above.

localization in V-REGION	nb of inserted nt	inserted nt	causing frameshift	from V-REGION codon	from nt position in user submitted sequence
CDR1-IMGT	3	AGC	no	34	151

IMGT/V-QUEST results after removal of the insertion(s):

Potentially productive IGH rearranged sequence: no stop codon and in-frame junction

(Check also your sequence with BLAST against IMGT/GENE-DB reference sequences to eventually identify out-of-frame pseudogenes)

V-GENE and allele	Homsap IGHV6-1*01 F	score = 1444	identity = 98.65% (293/297 nt) [98.32% (292/297 nt)]
J-GENE and allele	Homsap IGHJ3*01 F, or Homsap IGHJ3*02 F	score = 191	identity = 87.76% (43/49 nt)
D-GENE and allele by IMGT/JunctionAnalysis	Homsap IGHD3-16*02 F	D-REGION is in reading frame 2	
FR-IMGT lengths, CDR-IMGT lengths and AA JUNCTION	[25.17.38.11]	[10.9.10]	CARGRWTAFDFW

Fig. 5 IMGT/V-QUEST "Detailed view" result page: Sequence and "Result summary" with the option "Search for insertions and deletions." An insertion of 3 nt was detected by IMGT/V-QUEST. The analysis is then performed after having removed the insertion. The SEQ2 is the human IGH rearranged sequence with the accession number L21957 in the IMGT/LIGM-DB database [50]

(Subheading 2.2.1). Dots represent the nucleotides that have been trimmed by the exonuclease. The number of mutations in the 3′V-REGION, D-REGION, and 5′J-REGION is indicated under "Vmut," "Dmut," and "Jmut," respectively (Fig.6), and the corresponding mutated nucleotides are underlined in the sequence (*see* **Note 6**). Nucleotides of the N-REGION are displayed in N1 and N2. The ratio of the number of g+c nucleotides to the total number of N nucleotides is indicated under "Ngc" (Fig. 6).

2. "Translation of the JUNCTION": It displays the junction with amino acids colored according to the 11 IMGT "physicochemical" classes [76] (*see* **Note 13**), frame ("+" and "–" indicate in

4. Results of IMGT/JunctionAnalysis

Maximum number of accepted mutations in: 3'V-REGION = 2, D-REGION = 4, 5'J-REGION = 2
Maximum number of accepted D-GENE = 1

Analysis of the JUNCTION

D-REGION is in reading frame 1.

Click on mutated (underlined) nucleotide to see the original one:

Input	V name	3'V-REGION	N1	D-REGION	N2	5'J-REGION	J name	D name	Vmut	Dmut	Jmut	Ngc
SEQ1	Homsap IGHV1-18*01	tgtgcgagag.	ggg	.........gcagctcggac	cccagga	.....cttttgatatctgg	Homsap IGHJ3*02	Homsap IGHD6-13*01	0	2	0	8/10

Translation of the JUNCTION

Click on mutated (underlined) amino acid to see the original one:

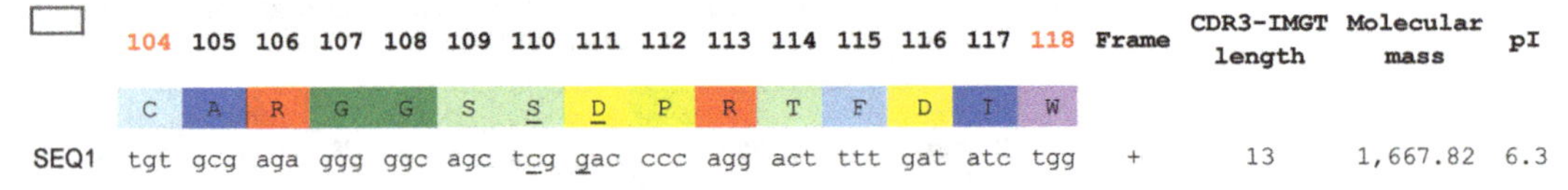

	104	105	106	107	108	109	110	111	112	113	114	115	116	117	118	Frame	CDR3-IMGT length	Molecular mass	pI
	C	A	R	G	G	S	S	D	P	R	T	F	D	I	W				
SEQ1	tgt	gcg	aga	ggg	ggc	agc	tcg	gac	ccc	agg	act	ttt	gat	atc	tgg	+	13	1,667.82	6.3

5. Sequence of the JUNCTION ('nt' and 'AA')

```
104 105 106 107 108 109 110 111 112 113 114 115 116 117 118
 C   A   R   G   G   S   S   D   P   R   T   F   D   I   W
tgt gcg aga ggg ggc agc tcg gac ccc agg act ttt gat atc tgg
```

Input for IMGT/JunctionAnalysis

```
>SEQ1,Homsap_IGHV1-18*01,Homsap_IGHJ3*02
tgtgcgagagggggcagctcggaccccaggacttttga-
tatctgg
```

Fig. 6 IMGT/V-QUEST "Detailed view" result page: "Results of IMGT/JunctionAnalysis" (with "Analysis of the JUNCTION" and "Translation of the JUNCTION") and "Sequence of the JUNCTION ('nt' and 'AA')." The SEQ1 is the human IGH rearranged sequence with the accession number AF062212 in the IMGT/LIGM-DB database [50]. "Ngc" is 8/10 as there are 8 "c" or "g" in N1 + N2 and the total of nt is 10 (ggg (N1), cccagga (N2))

frame and out of frame, respectively), CDR3-IMGT length, molecular mass, and isoelectric point (pI). In case of frameshifts, gaps (represented by one or two dots) are inserted to maintain the J-REGION frame. The corresponding codon which cannot be translated is represented by "#."

The "Sequence of the JUNCTION ('nt' and 'AA')" provides the JUNCTION in nt and AA according to the IMGT unique numbering and in the FASTA format with the correctly formatted header required as input by IMGT/JunctionAnalysis online [38, 39].

3.3.5 "V-REGION Alignment," "V-REGION Translation," and "V-REGION Protein Display"

The results provide three displays of the V-REGION [37]:

1. The "V-REGION alignment according to the IMGT unique numbering" displays the nt sequences with the FR-IMGT and CDR-IMGT delimitations.
2. The "V-REGION translation" displays the nt sequence and deduced AA translation of the input sequence, aligned with

the closest germline V-REGION and with the FR-IMGT and CDR-IMGT delimitations.

3. The "V-REGION protein display" displays the deduced AA translation of the input sequence, aligned with the V-REGION translation of the closest germline V-GENE, with the FR-IMGT and CDR-IMGT delimitations, and with, on the third line of the alignment and shown in bold, the AA of the input sequence which are different from the closest germline V-REGION.

3.3.6 Analysis of the Mutations: "V-REGION Mutation and AA Change Table," "V-REGION Mutation and AA Change Statistics," and "V-REGION Mutation Hot Spots"

The analysis of the mutations in the V-REGION is performed by comparison of the analyzed sequence with the closest germline V gene and allele.

1. The "V-REGION mutation and AA change table" lists the nt mutations and the corresponding AA changes if the mutations are not silent [37]. They are described for each FR-IMGT and CDR-IMGT, with their nt and codon position according to the IMGT unique numbering [30, 53, 54] and for the AA changes according to the IMGT AA classes [76]. For example c1 > g, Q1 > E (++−) means that the nt mutation (c > g) at nt 1 leads to an AA change (Q > E) at codon 1 (*see* **Note 13**).
2. The "V-REGION mutation and AA change statistics" comprises two tables, the first one for "Nucleotide (nt) mutations" described as silent/nonsilent, transitions/transversions, and the second one for "Amino acid (AA) changes" [37]. For both tables, results are given for the V-REGION and for FR-IMGT and CDR-IMGT. Statistics are calculated up to the 3′ end of the V-REGION identified in the input sequence (this includes the 3′ last two identical nucleotides with the closest germline V-REGION). The numbers in parentheses, in the V-REGION and CDR3-IMGT columns, correspond to the statistics calculated up to the 3′ end of the closest germline V-REGION and therefore may include nt and AA differences due to the junction diversity.

In order to avoid to count sequence differences due to the 5′ primer in "V-REGION mutation and AA change table" and in "V-REGION mutation and AA change statistics," before launching the analysis it is possible to exclude in 5′ of the V-REGION a given number of nucleotides, defined by the user, in "Advanced parameters" (parameters for "Detailed view" in the IMGT/V-QUEST search page).

The positions of the mutations have been correlated with the presence of specific patterns in the germline V gene and allele (*see* Somatic hypermutations, in IMGT Education: http://www.imgt.org/IMGTeducation/Tutorials/IGandBcells/_UK/SomaticHypermutations/). The "V-REGION mutation hot spots" table [37] shows the localization of the hot spot patterns (a/t)a (or wa) and (a/g)g(c/t)(a/t) (or rgyw) and their complementary reverse motifs t(a/t) (or tw) and (a/t)(a/g)c(c/t) (or wrcy) in the closest germline V gene and allele.

3.3.7 Sequences of V-, V-J-, or V-D-J-REGION ("nt" and "AA") with Gaps in FASTA and Access to IMGT/PhyloGene for V-REGION ("nt")

This result provides nt and AA sequences with gaps according to the IMGT unique numbering [30, 53, 54] and includes the "V-REGION nucleotide sequence in FASTA format" with access to the IMGT/PhyloGene tool [77], "V-REGION amino acid in FASTA format," "V-REGION amino acid on one line," V-J or V-D-J-REGION nt and AA sequences in FASTA format, and V-J or V-D-J-REGION AA sequence on one line [37]. In that option, the J reading frame of sequences with an out-of-frame junction is not restored (the sequence does not include "#"), and a note alerts the user.

3.3.8 Annotation by IMGT/Automat

The results of the analysis of the input sequence by IMGT/Automat [40, 41] provide a full automatic annotation for the V-J-REGION or the V-D-J-REGION using IMGT® standardized labels [29].

3.3.9 IMGT Collier de Perles

The "IMGT Collier de Perles" [31, 57–59] can be displayed either as a "link to IMGT/Collier de Perles tool" or as a direct "IMGT Collier de Perles (for a number of sequences >5)" representation integrated in IMGT/V-QUEST results, depending on the user selection in the IMGT/V-QUEST search page (*see* **Note 14**).

3.4 IMGT/V-QUEST Output for "Synthesis View"

The "Synthesis view" results are displayed in "B. Synthesis for the IMGT/V-QUEST analysed sequences." At the top of the page, the number of analyzed sequences is indicated.

3.4.1 Summary Table

The "Summary table" (Fig. 7) displays one row for each input sequence with the corresponding results, including (a) the name of the sequence (sequence ID), (b) the name of the closest V-GENE and allele (*see* **Note 7**), (c) the functionality of the V domain-rearranged sequence (productive or unproductive) (*see* **Note 3**), (d) the V-REGION score (*see* **Note 8**), (e) the V-REGION percentage of identity with, between parentheses, the ratio of number of identical nt/number of aligned nt (*see* **Note 15**), (f) the name of the closest J-GENE and allele (*see* **Note 16**), and (g) provided according to the IMGT/JunctionAnalysis results the D-GENE and allele name, D reading frame, CDR-IMGT lengths, and AA JUNCTION and the JUNCTION frame. In the absence of results of IMGT/JunctionAnalysis, only the AA JUNCTION defined by IMGT/V-QUEST is displayed.

3.4.2 Results of IMGT/JunctionAnalysis

Results are displayed per locus (e.g., IGH, IGK, or IGL for IG sequences). Below the "Summary table," use the links associated to the locus name(s) (e.g., IGH) to display "Results of

B. Synthesis for the IMGT/V-QUEST analysed sequences

Number of analysed sequences: 6

Sequences compared with the human IG set from the IMGT reference directory

- **Summary table:**

Sequence ID	V-GENE and allele	Functionality	V-REGION score	V-REGION identity % (nt)	J-GENE and allele	D-GENE and allele	D-REGION reading frame	CDR-IMGT lengths	AA JUNCTION	JUNCTION frame
SEQ1	Homsap IGHV1-18*01 F	Productive	1390	98,26% (283/288 nt)	Homsap IGHJ6*03 F	Homsap IGHD2-2*01 F	2	[8.8.22]	CARDSFGYCSSTSCPYYYYYMDVW	in-frame
SEQ2	Homsap IGHV1-18*01 F, or Homsap IGHV1-18*03 F	Productive	1264	93,40% (269/288 nt)	Homsap IGHJ4*01 F	Homsap IGHD2-2*01 F	1	[8.8.11]	CARDNYQLLWEYW	in-frame
SEQ3	Homsap IGHV1-18*01 F	Productive	1327	95,83% (276/288 nt)	Homsap IGHJ4*02 F	Homsap IGHD2-15*01 F	2	[8.8.18]	CARAPGYCSGGGCYRGDDYW	in-frame
SEQ4	Homsap IGHV1-18*01 F	Productive	1068	89,43% (237/265 nt)	Homsap IGHJ3*02 F	Homsap IGHD7-27*01 F	1	[8.8.12]	CARGQSEEAAFEIW	in-frame
SEQ5	Homsap IGHV1-18*01 F	Productive	1281	99,61% (258/259 nt)	Homsap IGHJ4*02 F	Homsap IGHD3-10*01 F	2	[8.8.13]	CARDRYGSGSNFDYW	in-frame
SEQ6	Homsap IGHV3-30*04 F	Productive	1435	100,00% (288/288 nt)	Homsap IGHJ4*02 F	Homsap IGHD6-19*01 F	2	[8.8.15]	CARDRSIAVAQYYFDYW	in-frame

- **Results of IMGT/JunctionAnalysis for : IGH junctions**
- **Alignment with the closest alleles:**

  The analysed sequences are aligned with the closest allele (with number of aligned sequences in parentheses):
  Homsap IGHV1-8*01 (5) Homsap IGHV3-30*04(1)

Fig. 7 IMGT/V-QUEST "Synthesis view" result page: Top of the page. The SEQ1, SEQ2, SEQ3, SEQ4, SEQ5, and SEQ6 correspond to the Z18851, Z92891, X15611, HM704452, AF103300, and AB021511 accession numbers, respectively, from the IMGT/LIGM-DB database [50]

IMGT/JunctionAnalysis" for sequences identified by the tool as belonging to the same locus (Fig. 8). Results are similar to those obtained for individual sequences detailed in Subheading 3.3.4.

3.4.3 Alignment with the Closest Alleles

The synthesis results provide six different displays (if all were selected): "Alignment for V-GENE," "V-REGION alignment according to the IMGT unique numbering," and "V-REGION translation" and "V-REGION" protein displays in three different formats [37].

3.4.4 "V-REGION Most Frequently Occurring AA per Position and per FR-IMGT and CDR-IMGT"

This section shows, for each FR-IMGT and CDR-IMGT, and for each position according to the IMGT unique numbering [30, 53, 54], the most frequently occurring AA [37].

3.5 IMGT/V-QUEST Output for Excel File

"Excel file" allows the users to open and save a spreadsheet including the results of the IMGT/V-QUEST analysis. The file contains 11 sheets (if all were selected in the IMGT/V-QUEST search page, *see* Subheading 3.2.2). The "Summary" and "Parameters" are always selected.

The content of the Excel file 11 sheets is equivalent to the content of the 11 result files (in CVS format) of IMGT/HighV-QUEST detailed in Table 3 (Subheading 4.4).

4 IMGT/HighV-QUEST

IMGT/HighV-QUEST [37, 42–44], the high-throughput version of IMGT/V-QUEST, can analyze large numbers of IG- and TR-rearranged sequences (up to 150,000) in a single run.

4.1 IMGT/HighV-QUEST Welcome Page

As for the other IMGT® databases and tools, IMGT/HighV-QUEST is freely available for academics. However, the IMGT/HighV-QUEST welcome page requires user identification and provides, for new users, a link to register. User identification has been set to avoid nonrelevant use and overload of the server and to contact the user if needed. The user identification gives access to the IMGT/HighV-QUEST search page.

4.2 IMGT/HighV-QUEST Search Page

4.2.1 Sequence Submission

The IMGT/HighV-QUEST search page [44] (Fig. 9) is very similar to the classical IMGT/V-QUEST search page with however, at the top (Fig. 9a), four additional fields as follows:

(a) The analysis title (which must be provided by the user).

(b) The selection of the species.

(c) The selection of the receptor type or locus: IMGT/HighV-QUEST allows the user to restrict the analysis to a particular

8. Results of IMGT/JunctionAnalysis

Results for the IGH junctions

Analysis of the JUNCTIONs

Click on mutated (underlined) nucleotide to see the original one:

Input	V name	3'V-REGION	P	N1	D-REGION	N2	P	5'J-REGION	J name	D name
SEQ1	Homsap IGHV1-18*01	tgtgcgagaga	t	tcgttc	.gggtattgtagtagtaccagctg.......	tcc		....ctactactactactacatggacgtctgg	Homsap IGHJ6*03	Homsap IGHD2-2*01
SEQ2	Homsap IGHV1-18*01	tgtgcgagaga		caat	...............tatcaactactgtg..	g		........gaatactgg	Homsap IGHJ4*01	Homsap IGHD2-2*01
SEQ3	Homsap IGHV1-18*01	tgtgcgagag.		caccg	.ggatattgtagtggggtgggtgctac...	agaggaga		.......tgactactgg	Homsap IGHJ4*02	Homsap IGHD2-15*01
SEQ4	Homsap IGHV1-18*01	tgtgcgagag.		gtcaat	....ctgagga	agcc		....gcttttgaaatctgg	Homsap IGHJ3*02	Homsap IGHD7-27*01
SEQ5	Homsap IGHV1-18*01	tgtgcgagaga	tc	g	......ctatggttcggggagt.........	a		...actttgactactgg	Homsap IGHJ4*02	Homsap IGHD3-10*01
SEQ6	Homsap IGHV3-30*04	tgtgcgagaga	tc	gga	..gtatagcagtggc......	cc	agt	actactttgactactgg	Homsap IGHJ4*02	Homsap IGHD6-19*01

Translation of the JUNCTIONs

Click on mutated (underlined) amino acid to see the original one:

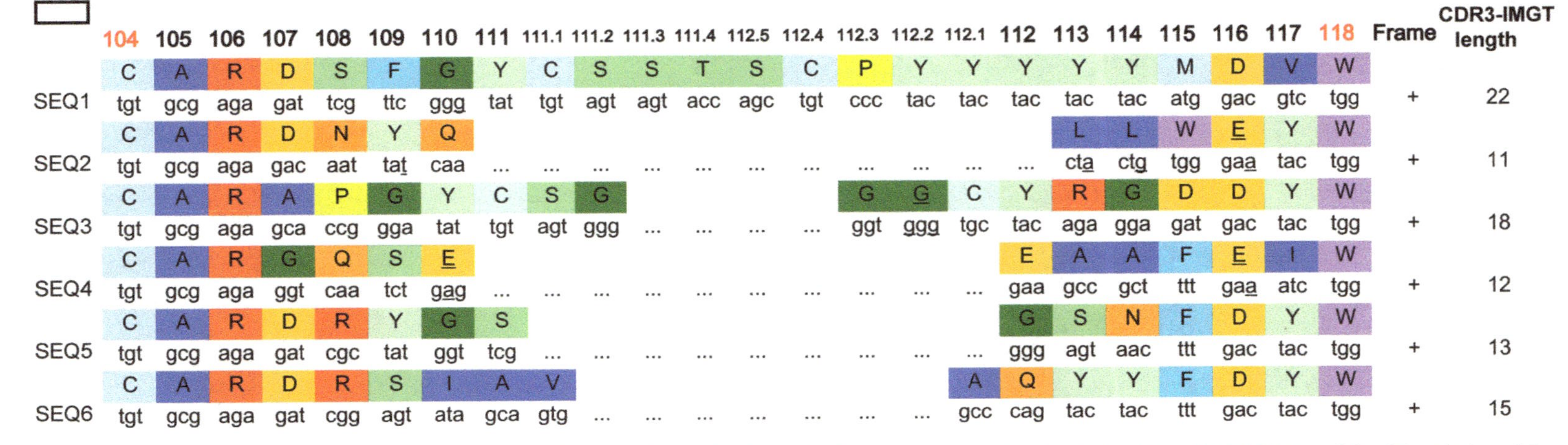

	104	105	106	107	108	109	110	111	111.1	111.2	111.3	111.4	112.5	112.4	112.3	112.2	112.1	112	113	114	115	116	117	118	Frame	CDR3-IMGT length
	C	A	R	D	S	F	G	Y	C	S	S	T	S	C	P	Y	Y	Y	Y	Y	M	D	V	W		
SEQ1	tgt	gcg	aga	gat	tcg	ttc	ggg	tat	tgt	agt	agt	acc	agc	tgt	ccc	tac	tac	tac	tac	tac	atg	gac	gtc	tgg	+	22
	C	A	R	D	N	Y	Q												L	L	W	E	Y	W		
SEQ2	tgt	gcg	aga	gac	aat	tat	caa	...	...	...	...	...	...	...	...	...	...	...	cta	ctg	tgg	gaa	tac	tgg	+	11
	C	A	R	A	P	G	Y	C	S	G					G	G	C	Y	R	G	D	D	Y	W		
SEQ3	tgt	gcg	aga	gca	ccg	gga	tat	tgt	agt	ggg	...	...	...	...	ggt	ggg	tgc	tac	aga	gga	gat	gac	tac	tgg	+	18
	C	A	R	G	Q	S	E											E	A	A	F	E	I	W		
SEQ4	tgt	gcg	aga	ggt	caa	tct	gag	...	...	...	...	...	...	...	...	...	...	gaa	gcc	gct	ttt	gaa	atc	tgg	+	12
	C	A	R	D	R	Y	G	S										G	S	N	F	D	Y	W		
SEQ5	tgt	gcg	aga	gat	cgc	tat	ggt	tcg	...	...	...	...	...	...	...	...	...	ggg	agt	aac	ttt	gac	tac	tgg	+	13
	C	A	R	D	R	S	I	A	V								A	Q	Y	Y	F	D	Y	W		
SEQ6	tgt	gcg	aga	gat	cgg	agt	ata	gca	gtg	...	...	...	...	...	...	...	gcc	cag	tac	tac	ttt	gac	tac	tgg	+	15

Fig. 8 IMGT/V-QUEST "Synthesis view" result page: Results of IMGT/JunctionAnalysis. The results show the "Analysis of the JUNCTIONs" and the "Translation of the JUNCTIONs" for the IGH sequences of the submitted analysis (same accession numbers as in Fig. 7 legend). Owing to the figure format, Vmut, Dmut, Jmut, Ngc, molecular mass, and p*I* are not shown

Table 3
Content of the 11 IMGT/HighV-QUEST result files in CVS format (results equivalent to those of the Excel file of IMGT/V-QUEST online)

File number	File name	Number of columns	Result content[a]
#1	"Summary"	25–29	– Identity percentage with the closest V, D, and J genes and alleles – FR-IMGT and CDR-IMGT lengths – Amino acid (AA) JUNCTION – Description of insertions and deletions if any
#2	"IMGT-gapped-nt-sequences"	18	– Nucleotide (nt) sequences gapped according to the IMGT unique numbering for the labels V-D-J-REGION, V-J-REGION, V-REGION, FR1-IMGT, CDR1-IMGT, FR2-IMGT, CDR2-IMGT, and FR3-IMGT – nt sequences of CDR3-IMGT, JUNCTION, J-REGION, and FR4-IMGT
#3	"Nt-sequences"	63–78	– nt sequences of all labels that can be automatically annotated by IMGT/Automat
#4	"IMGT-gapped-AA-sequences"	18	– AA sequences gapped according to the IMGT unique numbering for the labels V-D-J-REGION, V-J-REGION, V-REGION, FR1-IMGT, CDR1-IMGT, FR2-IMGT, CDR2-IMGT, and FR3-IMGT – AA sequences of CDR3-IMGT, JUNCTION, J-REGION, and FR4-IMGT
#5	"AA-sequences"	18	– Same columns as "IMGT-gapped-AA-sequences" (#4), but sequences of labels are without IMGT gaps
#6	"Junction"	33, 46, 66 or 77	– Results of IMGT/JunctionAnalysis (33 columns for IGL and IGK (also for TRA and TRG) sequences, 46 (if one D), 66 (if two D), or 77 (if 3 D) columns for IGH (also for TRB and TRD) sequences)
#7	"V-REGION-mutation-and-AA-change table"	11	– List of mutations (nt mutations, AA changes, AA class identity (+) or change (−)) for V-REGION, FR1-IMGT, CDR1-IMGT, FR2-IMGT, CDR2-IMGT, FR3-IMGT, and germline CDR3-IMGT
#8	"V-REGION-nt-mutation-statistics"	130	– Number (nb) of nt positions including IMGT gaps, nb of nt, nb of identical nt, total nb of mutations, nb of silent mutations, nb of nonsilent mutations, nb of transitions (a>g, g>a, c>t, t>c), and nb of transversions (a>c, c>a, a>t, t>a, g>c, c>g, g>t, t>g) for V-REGION, FR1-IMGT, CDR1-IMGT, FR2-IMGT, CDR2-IMGT, FR3-IMGT, and germline CDR3-IMGT

(continued)

Table 3 (continued)

File number	File name	Number of columns	Result content[a]
#9	"V-REGION-AA-change-statistics"	189	– nb of AA positions including IMGT gaps, nb of AA, nb of identical AA, total nb of AA changes, nb of AA changes according to AAclassChangeType (+++, ++–, +–+, +––, –+–, ––+, –––), and nb of AA class changes according to AAclassSimilarityDegree (nb of Very similar, nb of Similar, nb of Dissimilar, nb of Very dissimilar) for V-REGION, FR1-IMGT, CDR1-IMGT, FR2-IMGT, CDR2-IMGT, FR3-IMGT, and germline CDR3-IMGT
#10	"V-REGION-mutation-hotspots"	8	– Hot spots motifs ((a/t)a, t(a/t), (a/g)g(c/t)(a/t), (a/t)(a/g)c(c/t)) detected in the closest germline V-REGION with positions in FR-IMGT and CDR-IMGT
#11	"Parameters"		– Date of the analysis – IMGT/V-QUEST programme version, IMGT/V-QUEST reference directory release – Parameters used for the analysis: Species, receptor type or locus, IMGT reference directory set, and advanced parameters

[a] Files #1–10 comprise systematically sequence identification (sequence name, functionality, and names of the closest V, D, and J genes and alleles). The files #7–10 that report the analysis of mutations are used mostly for immunoglobulins (IG)

locus, for example "IGH." Values for receptor type or locus are displayed only if the corresponding IMGT reference directory is available.

(d) The choice of e-mail notifications: Because of their large number of sequences and use of the tool by multiple users at the same time, the analyses are first queued on the IMGT® server, and they are performed depending on the available resources [44] (in May 2013, 570 users from 38 countries have been registered and the number of submitted sequences was superior to 510 million with 58 % from the USA, 19 % from the EU, and 23 % from the remaining world). Therefore, the results are not displayed immediately, and users may choose to be notified by e-mail: when analysis is queued, when analysis is submitted, when analysis is completed (selected by default), and/or before the results are removed (always selected). These choices are not exclusive.

Sequences are submitted by giving a path access to a local file (the copy-and-paste option is not available). The file, in simple text

Fig. 9 IMGT/HighV-QUEST search page. This page comprises, from top to bottom, three sections: (**a**) Sequence submission, (**b**) "Display results," and (**c**) "Advanced parameters"

format, must contain the nucleotide sequences (from 1 to 150,000 sequences) in FASTA format. Before launching the analysis, the user can customize the result display (Subheading 4.2.2) and/or "Advanced parameters" (Subheading 4.2.3).

4.2.2 "Display Results"

IMGT/HighV-QUEST provides two types of results: "Detailed View" (individual result files) and "Files in CSV" (equivalent of the "Excel file" of IMGT/V-QUEST) (Fig. 9b). The customization is identical to that of IMGT/V-QUEST (except that IMGT Collier de Perles results and IMGT/Phylogene are not available in "Detailed View").

4.2.3 "Advanced Parameters"

The customization of "Advanced parameters" (Fig. 9c) is identical to that of IMGT/V-QUEST.

4.3 IMGT/HighV-QUEST "Analysis History" Table and Results Download

From their account, the user can check the status of his/her analyses at any time by displaying the "Analysis history" table (available from a link from the top of the search page) [44]. The table displays each analysis with its title, user name, status of the analysis (queued, running, completed, error), submission date (with a predicted completion time, if relevant), number of submitted sequences, IMGT/V-QUEST reference directory "species" and "receptor type or locus" (as selected by the user), and actions that can be performed by the user (download, remove).

When the analysis is completed, the user can "download" the results as a single file in ZIP format. The size of the ZIP file (in Mb) and the number of included files are indicated in the "Analysis history" table. For analyses having less than the expected number of files in the results file, a warning is shown. If the user removes a completed analysis, all related files (sequences and results) are definitively deleted from the server. A user may cancel a queued or a running analysis at any time.

4.4 IMGT/HighV-QUEST Downloaded ZIP File

The downloaded ZIP file (Fig. 10) contains a main folder with 11 files (equivalent to the 11 sheets of the Excel file provided by the classical IMGT/V-QUEST) in CSV format (Table 3) and one subfolder with individual files, in Text, for each sequence (providing "Detailed view" results) [44]. Text and CSV formats have been chosen in order to facilitate statistical studies for further interpretation and knowledge extraction.

4.5 IMGT/HighV-QUEST Statistical Analyses

Statistical analyses are performed on results of completed analyses (up to 450,000 sequences) [44]. The user chooses completed results (several results can be chosen), a title for the statistical analysis, e-mail notifications, and whether he/she wants a separate copy of graphical elements in the final results. The user can also

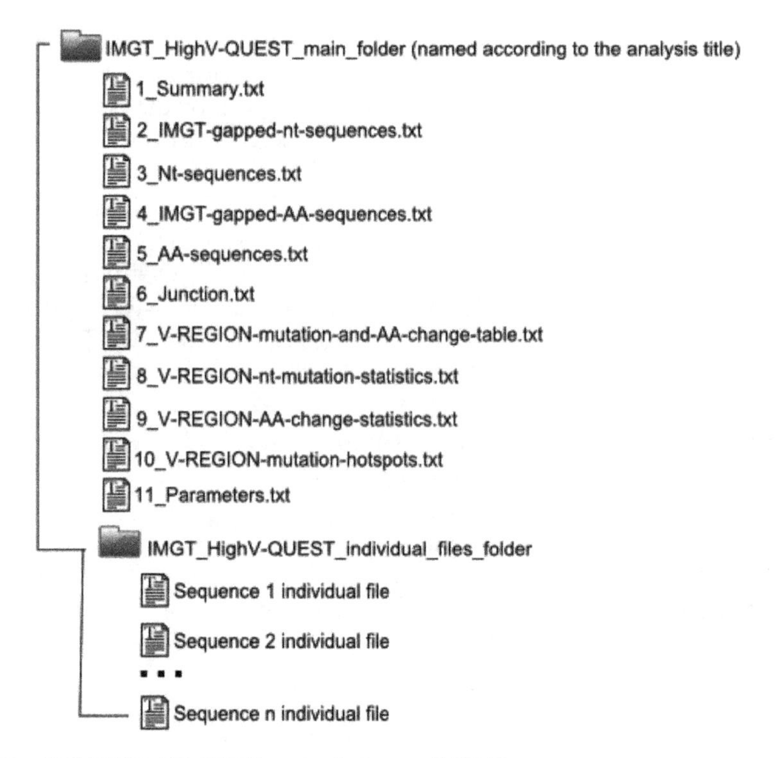

Fig. 10 IMGT/HighV-QUEST content of a result ZIP file

enter a comment of up to 500 characters and choose to include it in the final PDF reports. All chosen results for a statistical analysis must be compatible (same locus, same value for the options "Search for insertions/deletions" and "IMGT reference directory set") and without warnings. Statistics are performed on "1 copy." The status of all statistical analyses can be checked using the "Statistics history" table.

The downloaded result file is a single file in ZIP format, comprising five PDF reports:

1. Summary.
2. Number of "1 copy," "single allele," and "several alleles (or genes)" tables and histograms.
3. V, (D), and J gene and allele tables and histograms for "single allele" and "several alleles (or genes)."
4. CDR3-IMGT tables and histograms.
5. Filtered-out sequences and another general PDF report containing all these results together. If chosen by the user, the results contain also the graphical elements in a directory named "graphics."

5 IMGT/DomainGapAlign

5.1 IMGT/DomainGapAlign Welcome Page

1. The IMGT/DomainGapAlign welcome page [46] is accessed by clicking the link IMGT/DomainGapAlign ("IMGT tools" section) in the IMGT® home page at http://www.imgt.org.
2. In the IMGT/DomainGapAlign welcome page (Fig. 11), locate the text area and paste your amino acid sequences in FASTA format. Alternatively, you can upload a file. A precise

WELCOME !

to IMGT/DomainGapAlign

THE
INTERNATIONAL
IMMUNOGENETICS
INFORMATION SYSTEM®

http://www.imgt.org

Align and "IMGT-gap" your domain amino acid sequence

Citing IMGT/DomainGapAlign : Ehrenmann F., Kaas Q. and Lefranc M.-P. Nucleic Acids Res., 38, D301-307 (2010). PMID: 19900967 Abstract Full PDF

Paste your protein sequence(s) in FASTA format below

Sequence sets to test IMGT/DomainGapAlign are available here

Upload a file Parcourir... Reset

Select a domain type V

Select a species All species English name

Smith-Waterman score above 0

Displayed alignments 3 Display IMGT Colliers de Perles

Align and IMGT-gap my sequence(s) Clear the form

Advanced parameters

Alignment	E-value	200
Putting gaps in the sequence	Gap penalty for query	-5
	Gap penalty for reference	-20

Fig. 11 IMGT/DomainGapAlign welcome page

delimitation of the domain sequences is not required; however, if the sequence contains several domains, the sequence should be split between the different domains. Several domain amino acid sequences can be analyzed simultaneously (up to 50), but each sequence must have a distinct name and belong to the same domain type. If not, the query needs to be launched for each domain type, successively. If the limits and the numbers of domains of an amino acid sequence are unknown, you can progressively analyze the protein, shortening the sequence once a domain has been identified by the tool (it should be reminded that the first domain identified by the tool is not necessarily the first one in the protein). Sequences to test IMGT/DomainGapAlign are available by clicking on the link "here."

3. In the "Select a Domain type" drop-down list, select a domain type (e.g., for IG, V, or C) (detailed in Subheading 2.1).
4. In the "Select a Species" drop-down list, select a species. If the selection is "All species," the IMGT/DomainGapAlign tool will compare your sequence with all sequences available for the selected domain type in the IMGT domain reference directory (*see* **Note 17**).
5. In the "Smith–Waterman score above" drop-down list, you can modify the Smith–Waterman score for the alignments to display. Selecting a higher score corresponds to a higher selection of the results to display. For example, choosing a "Smith–Waterman score above 200" will only provide and display alignments for which the Smith–Waterman score is superior to 200 (*see* **Note 18**).
6. In the "Displayed alignments" drop-down list, select the number of alignments to display (by default 3) and tick off the checkbox if you want to display IMGT Collier de Perles.
7. Check the "Advanced parameters" section if you would like to modify parameters. For alignment, IMGT/DomainGapAlign use parameters by default: The *E*-value (*see* **Note 19**) is set to 200, the "Gap penalty for query" (relative to the user sequence) is −5, and the "Gap penalty for reference" (relative to the IMGT reference sequence) is −20 (*see* **Note 20**). These parameters can be modified for special queries.
8. Press the "Align and IMGT-gap my sequence(s)" button to launch the analysis. This will return the IMGT/DomainGap Align results page.

5.2 IMGT/DomainGapAlign Results Page

The IMGT/DomainGapAlign results page [46] comprises three parts: the results alignments, results summary and AA change tables, and IMGT Collier de Perles.

5.2.1 Results Alignments

The top of the results page for V domain (Fig. 12) and C domain (Fig. 13) displays the following:

1. The "Sequence name" (as provided by the user).
2. The "Closest reference gene and allele(s) from the IMGT domain directory" section: The domain type (for IG, V, or C) and the species as selected by the user are indicated in the section title (online in orange). The following results are displayed: "Gene and allele name" (*see* **Note 7**), "Species," "Domain number," "Smith–Waterman Score" (online in orange), and label of the domain (online in color) as identified in the closest reference gene and allele domain with its "percentage of identity" and "overlap" score. If several closest gene and alleles are displayed, the user can select "Align your sequence with" to display the corresponding alignment. For V-DOMAIN (Fig. 12), the closest reference gene and alleles section shows the results for the V-REGION and J-REGION of the V and J genes and alleles, respectively.
3. The alignment(s) with the domain of the closest gene and allele from the IMGT domain directory: The domain type (for IG, V, or C) and the species as selected by the user are indicated in the section title (online in orange). The alignments are shown, based on the IMGT unique numberings for V domain [30, 53, 54] (Fig. 12) and C domain [30, 55] (Fig. 13). Below the alignment, the label of the domain (online in color), as identified by the tool for the closest reference gene and allele, is indicated with a horizontal line. For the V-DOMAIN (Fig. 12), the V-REGION, N (for (N-D)-REGION), and J-REGION are delimited by IMGT/DomainGapAlign.
4. The region(s) and domain(s) identified in your sequence (by comparison with the closest genes and alleles): The species and gene and allele name identified by IMGT/DomainGapAlign are recalled (online in dark red). The domains (or regions for the V-DOMAIN) identified by the tool are colored according to the IMGT color menu.
5. The links to the sequence, with or without gaps, in FASTA format (HTML page or downloadable).

5.2.2 Results Summary and AA Change Tables

The results summary (by comparison with the closest gene and allele) and AA changes for V domain (Fig. 12) and C domain (Fig. 13) are shown as tables:

1. The results summary table has three columns that are common to the V and C domains and four additional columns for the V domain. The three common columns are:

a

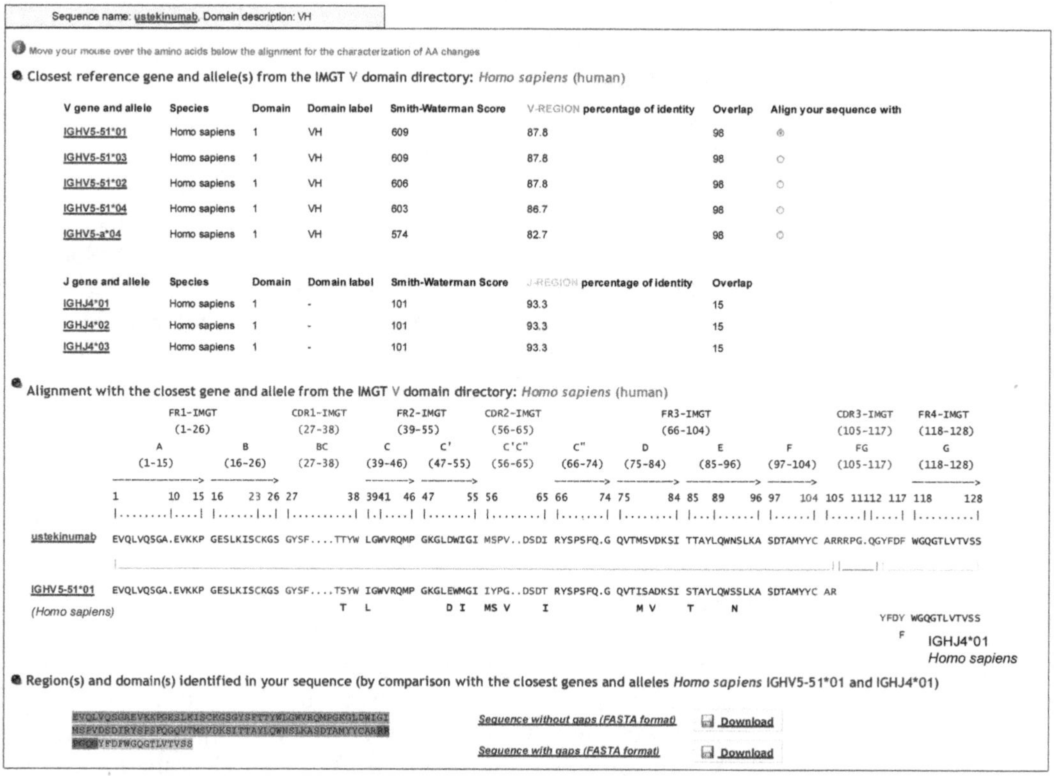

Sequence name: ustekinumab, Domain description: VH

Move your mouse over the amino acids below the alignment for the characterization of AA changes

Closest reference gene and allele(s) from the IMGT V domain directory: *Homo sapiens* (human)

V gene and allele	Species	Domain	Domain label	Smith-Waterman Score	V-REGION percentage of identity	Overlap	Align your sequence with
IGHV5-51*01	Homo sapiens	1	VH	609	87.8	98	◉
IGHV5-51*03	Homo sapiens	1	VH	609	87.8	98	○
IGHV5-51*02	Homo sapiens	1	VH	606	87.8	98	○
IGHV5-51*04	Homo sapiens	1	VH	603	86.7	98	○
IGHV5-a*04	Homo sapiens	1	VH	574	82.7	98	○

J gene and allele	Species	Domain	Domain label	Smith-Waterman Score	J-REGION percentage of identity	Overlap
IGHJ4*01	Homo sapiens	1	-	101	93.3	15
IGHJ4*02	Homo sapiens	1	-	101	93.3	15
IGHJ4*03	Homo sapiens	1	-	101	93.3	15

Alignment with the closest gene and allele from the IMGT V domain directory: *Homo sapiens* (human)

```
                  FR1-IMGT            CDR1-IMGT     FR2-IMGT        CDR2-IMGT                 FR3-IMGT                   CDR3-IMGT      FR4-IMGT
                   (1-26)              (27-38)      (39-55)         (56-65)                  (66-104)                    (105-117)      (118-128)
                A            B            BC          C        C'        C'C"       C"         D           E           F         FG            G
             (1-15)       (16-26)      (27-38)    (39-46)  (47-55)    (56-65)    (66-74)    (75-84)     (85-96)     (97-104)  (105-117)     (118-128)
             1        10  15 16     23 26 27       38 3941  46 47    55 56     65 66     74 75      84 85  89    96 97    104 105 11112 117 118      128
ustekinumab  EVQLVQSGA.EVKKP GESLKISCKGS GYSF....TTYW LGWVRQMP GKGLDWIGI MSPV..DSDI RYSPSFQ.G QVTMSVDKSI TTAYLQWNSLKA SDTAMYYC ARRRPG.QGYFDF WGQGTLVTVSS

IGHV5-51*01  EVQLVQSGA.EVKKP GESLKISCKGS GYSF....TSYW IGWVRQMP GKGLEWMGI IYPG..DSDT RYSPSFQ.G QVTISADKSI STAYLQWSSLKA SDTAMYYC AR
(Homo sapiens)                                  T    L                D I  MS V     I             M V        T     N
                                                                                                                                YFDY WGQGTLVTVSS
                                                                                                                                  F   IGHJ4*01
                                                                                                                                      Homo sapiens
```

Region(s) and domain(s) identified in your sequence (by comparison with the closest genes and alleles *Homo sapiens* IGHV5-51*01 and IGHJ4*01)

EVQLVQSGAEVKKPGESLKISCKGSGYSFTTYWLGWVRQMPGKGLDWIGI
MSPVDSDIRYSPSFQGQVTMSVDKSITTAYLQWNSLKASDTAMYYCARRR
PGQGYFDFWGQGTLVTVSS

Sequence without gaps (FASTA format) Download

Sequence with gaps (FASTA format) Download

Fig. 12 IMGT/DomainGapAlign results for a V-DOMAIN. (**a**) Result alignments. The domain amino acid (AA) sequence is aligned with the closest germline V-REGION and J-REGION, with IMGT gaps and delimitations of the FR-IMGT and CDR-IMGT according to the IMGT unique numbering [30, 53, 54]. In this example, the sequence is the VH of ustekinumab (IMGT/mAb-DB and IMGT/2Dstructure-DB, http://www.imgt.org). The V-REGION and J-REGION of the ustekinumab VH is identified as having 87.8 and 93.3 % identity with the *Homo sapiens* IGHV5-51*01 and IGHJ4*01, respectively. (**b**) Results summary and AA change tables. The results summary table provides IMGT molecule name, sequence name, V-REGION identity percentage, CDR-IMGT lengths, number of different AA in CDR1- and CDR2-IMGT, FR-IMGT lengths, number of different AA in FR-IMGT, and total number of AA changes. The AA change tables show AA changes for strands and loops and for FR-IMGT and CDR-IMGT [30, 53, 54]. (**c**) IMGT Collier de Perles. AA changes are highlighted (with online *pink* border)

b

Results summary (by comparison with the closest genes and alleles *Homo sapiens* IGHV5-51*01 and IGHJ4*01)

IMGT molecule name	Sequence name	V-REGION identity percentage	CDR-IMGT lengths	Number of different AA in CDR1- and CDR2-IMGT	FR-IMGT lengths	Number of different AA in FR-IMGT	Total number of AA changes in V-DOMAIN
ustekinumab	8954_H	87.8%	[8.8.12]	5	[25.17.38.11] = 91 AA	7	12

▸ AA changes in strands and loops

Strands	Number of different AA	AA changes
A strand (1-15)	0	-
B strand (16-26)	0	-
C strand (39-46)	1	I39>L (+ + +) very similar
C' strand (47-55)	2	E51>D (+ - +) similar M53>I (+ + -) similar
C" strand (66-74)	0	-
D strand (75-84)	2	I78>M (+ + -) similar A80>V (+ - +) similar
E strand (85-96)	2	S85>T (+ - +) similar S92>N (- - -) very dissimilar
F strand (97-104)	0	-
G strand (118-128)	0	-

Loops	Number of different AA	AA changes
BC loop (27-38)	1	S36>T (+ - +) similar
C'C" loop (56-65)	4	I56>M (+ + -) similar Y57>S (+ - -) dissimilar G59>V (- - -) very dissimilar T65>I (- - -) very dissimilar

▸ AA changes in FR-IMGT and CDR-IMGT

FR-IMGT	Number of different AA	AA changes
FR1-IMGT (1-26)	0	-
FR2-IMGT (39-55)	3	I39>L (+ + +) very similar E51>D (+ - +) similar M53>I (+ + -) similar
FR3-IMGT (66-104)	4	I78>M (+ + -) similar A80>V (+ - +) similar S85>T (+ - +) similar S92>N (- - -) very dissimilar
FR4-IMGT (118-129)	0	-

CDR-IMGT	Number of different AA	AA changes
CDR1-IMGT (27-38)	1	S36>T (+ - +) similar
CDR2-IMGT (56-65)	4	I56>M (+ + -) similar Y57>S (+ - -) dissimilar G59>V (- - -) very dissimilar T65>I (- - -) very dissimilar

c

IMGT Colliers de perles for ustekinumab

On one layer with AA changes

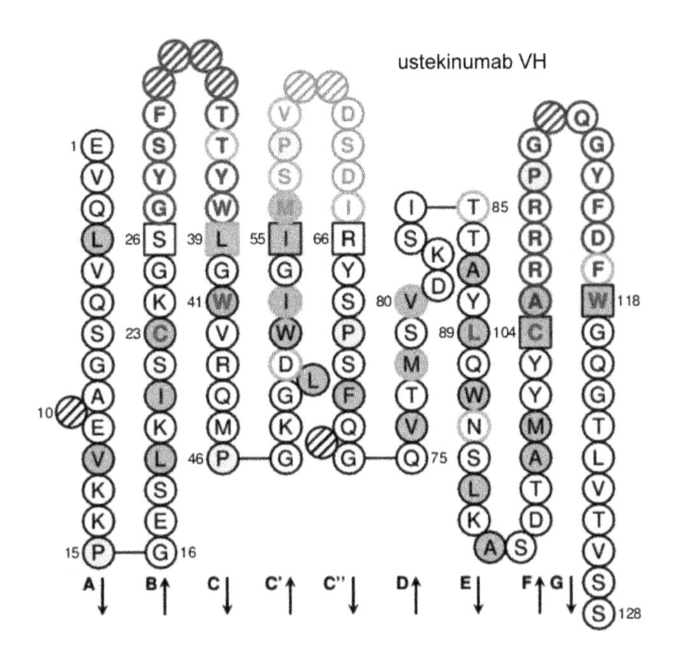

Fig. 12 (continued)

a

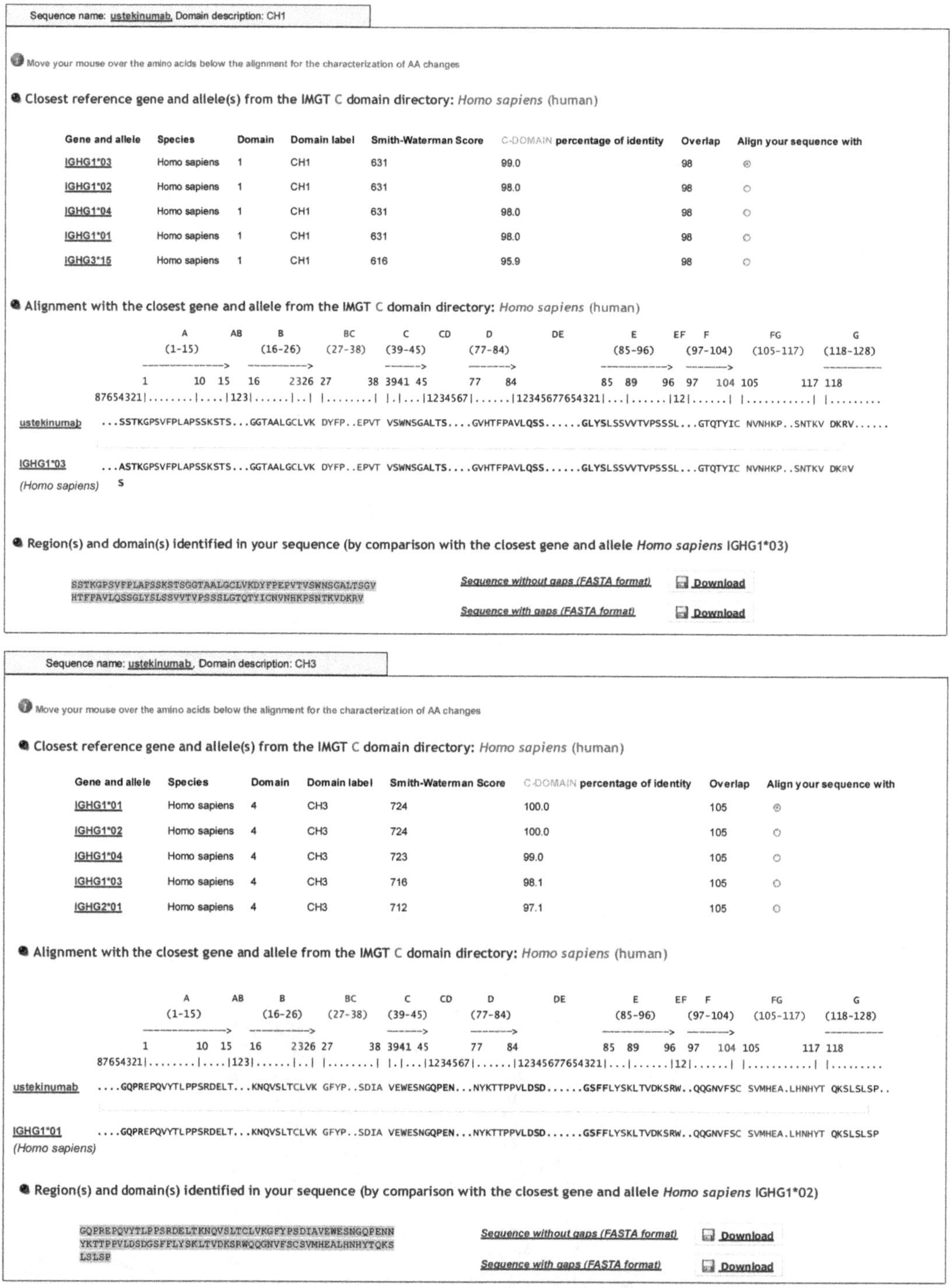

Sequence name: ustekinumab, Domain description: CH1

Move your mouse over the amino acids below the alignment for the characterization of AA changes

Closest reference gene and allele(s) from the IMGT C domain directory: *Homo sapiens* (human)

Gene and allele	Species	Domain	Domain label	Smith-Waterman Score	C-DOMAIN percentage of identity	Overlap	Align your sequence with
IGHG1*03	Homo sapiens	1	CH1	631	99.0	98	◉
IGHG1*02	Homo sapiens	1	CH1	631	98.0	98	○
IGHG1*04	Homo sapiens	1	CH1	631	98.0	98	○
IGHG1*01	Homo sapiens	1	CH1	631	98.0	98	○
IGHG3*15	Homo sapiens	1	CH1	616	95.9	98	○

Alignment with the closest gene and allele from the IMGT C domain directory: *Homo sapiens* (human)

```
                  A          AB      B          BC        C      CD     D              DE            E        EF   F           FG          G
                (1-15)            (16-26)     (27-38)  (39-45)       (77-84)                      (85-96)      (97-104)   (105-117)   (118-128)
           1        10  15  16       2326 27       38 3941 45       77     84               85  89      96  97    104 105         117 118
   87654321|........|....|123|......|..| |........| |.|...|1234567|......|12345677654321|...|......|12|......| |...........| |.........
ustekinumab  ...SSTKGPSVFPLAPSSKSTS...GGTAALGCLVK DYFP..EPVT VSWNSGALTS....GVHTFPAVLQSS......GLYSLSSVVTVPSSSL...GTQTYIC NVNHKP..SNTKV DKRV......

IGHG1*03     ...ASTKGPSVFPLAPSSKSTS...GGTAALGCLVK DYFP..EPVT VSWNSGALTS....GVHTFPAVLQSS......GLYSLSSVVTVPSSSL...GTQTYIC NVNHKP..SNTKV DKRV
(Homo sapiens) S
```

Region(s) and domain(s) identified in your sequence (by comparison with the closest gene and allele *Homo sapiens* IGHG1*03)

```
SSTKGPSVFPLAPSSKSTSGGTAALGCLVKDYFPEPVTVSWNSGALTSGV
HTFPAVLQSSGLYSLSSVVTVPSSSLGTQTYICNVNHKPSNTKVDKRV
```

Sequence without gaps (FASTA format) Download

Sequence with gaps (FASTA format) Download

Sequence name: ustekinumab, Domain description: CH3

Move your mouse over the amino acids below the alignment for the characterization of AA changes

Closest reference gene and allele(s) from the IMGT C domain directory: *Homo sapiens* (human)

Gene and allele	Species	Domain	Domain label	Smith-Waterman Score	C-DOMAIN percentage of identity	Overlap	Align your sequence with
IGHG1*01	Homo sapiens	4	CH3	724	100.0	105	◉
IGHG1*02	Homo sapiens	4	CH3	724	100.0	105	○
IGHG1*04	Homo sapiens	4	CH3	723	99.0	105	○
IGHG1*03	Homo sapiens	4	CH3	716	98.1	105	○
IGHG2*01	Homo sapiens	4	CH3	712	97.1	105	○

Alignment with the closest gene and allele from the IMGT C domain directory: *Homo sapiens* (human)

```
                  A          AB      B          BC        C      CD     D              DE            E        EF   F           FG          G
                (1-15)            (16-26)     (27-38)  (39-45)       (77-84)                      (85-96)      (97-104)   (105-117)   (118-128)
           1        10  15  16       2326 27       38 3941 45       77     84               85  89      96  97    104 105         117 118
   87654321|........|....|123|......|..| |........| |.|...|1234567|......|12345677654321|...|......|12|......| |...........| |.........
ustekinumab  ....GQPREPQVYTLPPSRDELT...KNQVSLTCLVK GFYP..SDIA VEWESNGQPEN...NYKTTPPVLDSD......GSFFLYSKLTVDKSRW..QQGNVFSC SVMHEA.LHNHYT QKSLSLSP..

IGHG1*01     ....GQPREPQVYTLPPSRDELT...KNQVSLTCLVK GFYP..SDIA VEWESNGQPEN...NYKTTPPVLDSD......GSFFLYSKLTVDKSRW..QQGNVFSC SVMHEA.LHNHYT QKSLSLSP
(Homo sapiens)
```

Region(s) and domain(s) identified in your sequence (by comparison with the closest gene and allele *Homo sapiens* IGHG1*02)

```
GQPREPQVYTLPPSRDELTKNQVSLTCLVKGFYPSDIAVEWESNGQPENN
YKTTPPVLDSDGSFFLYSKLTVDKSRWQQGNVFSCSVMHEALHNHYTQKS
LSLSP
```

Sequence without gaps (FASTA format) Download

Sequence with gaps (FASTA format) Download

Fig. 13 IMGT/DomainGapAlign results for a C-DOMAIN. (**a**) Result alignments. The CH1 of the gamma1 heavy chain (H-GAMMA-1) of ustekinumab is aligned with the CH1 of IGHG1*03 (with 1 AA change CH1 A1.4>S). The presence of CH1 R120 (with I103) corresponds to the G1m3 allotype [9]. The CH3 of the gamma1 chain of ustekinumab is aligned with the CH3 of IGHG1*01 with 100 % identity. The presence of CH3 D12 and L14 corresponds to the G1m1 allotype [9]. The origin of the simultaneous presence of G1m3 and G1m1 on the ustekinumab gamma1 chain (IMGT note in IMGT/2Dstructure-DB) is not known (this may result from antibody

b

Results summary (by comparison with the closest gene and allele *Homo sapiens* IGHG1*03)

IMGT molecule name	Sequence name	C-DOMAIN identity percentage	Total number of AA changes in C-DOMAIN
ustekinumab	8954_H	99.0%	1

Strands	Number of different AA	AA changes
A strand (1.9-1.1,1-15)	1	A1.4>S (- + -) dissimilar
B strand (16-26)	0	-
C strand (39-45)	0	-
CD strand (45.1-45.7)	0	-
D strand (77-84)	0	-
E strand (85-96)	0	-
F strand (97-104)	0	-
G strand (118-128)	0	-

Turns	Number of different AA	AA changes
AB turn (15.1-15.3)	0	-
DE turn (84.1-84.7, 85.7-85.1)	0	-
EF turn (96.1-96.2)	0	-

Loops	Number of different AA	AA changes
BC loop (27-36)	0	-
FG loop (105-117,111.1-111.6,112.1-112.6)	0	-

Results summary (by comparison with the closest gene and allele *Homo sapiens* IGHG1*02)

IMGT molecule name	Sequence name	C-DOMAIN identity percentage	Total number of AA changes in C-DOMAIN
ustekinumab	8954_H	100.0%	0

Strands	Number of different AA	AA changes
A strand (1.9-1.1,1-15)	0	-
B strand (16-26)	0	-
C strand (39-45)	0	-
CD strand (45.1-45.7)	0	-
D strand (77-84)	0	-
E strand (85-96)	0	-
F strand (97-104)	0	-
G strand (118-128)	0	-

Turns	Number of different AA	AA changes
AB turn (15.1-15.3)	0	-
DE turn (84.1-84.7, 85.7-85.1)	0	-
EF turn (96.1-96.2)	0	-

Loops	Number of different AA	AA changes
BC loop (27-36)	0	-
FG loop (105-117,111.1-111.6,112.1-112.6)	0	-

Fig. 13 (continued) engineering or from the expected—but not yet sequenced at the nucleotide level—IGHG1*08 allele (*see* table 4 in [9]). (**b**) Results summary and AA change tables for CH1 and CH3 based on the alignments. The results summary provides IMGT molecule name, sequence name, C-DOMAIN identity percentage, and total number of AA changes. AA changes are shown for strands (A, B, C, CD, D, E, F, and G), turns (AB, DE, and EF), and loops (BC and FG) [31, 55]. (**c**) IMGT Collier de Perles of the CH1 and CH3 domains of the H-GAMMA-1 chain of ustekinumab. AA changes are highlighted (online *pink* border)

- Sequence name.
- Domain identity percentage (for the V-DOMAIN, the percentage is calculated on the V-REGION).
- Total number of AA changes in the domain.

c IMGT Colliers de perles for ustekinumab

On one layer with AA changes

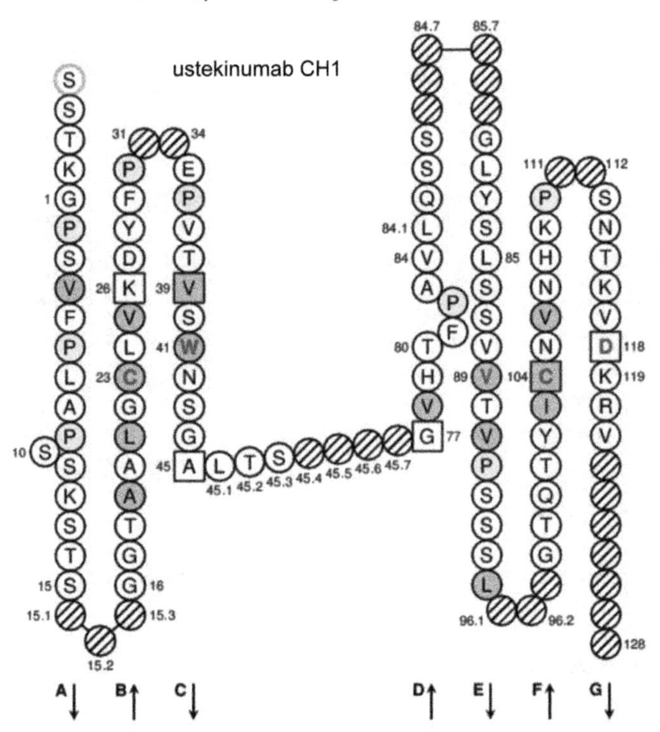

IMGT Colliers de perles for ustekinumab

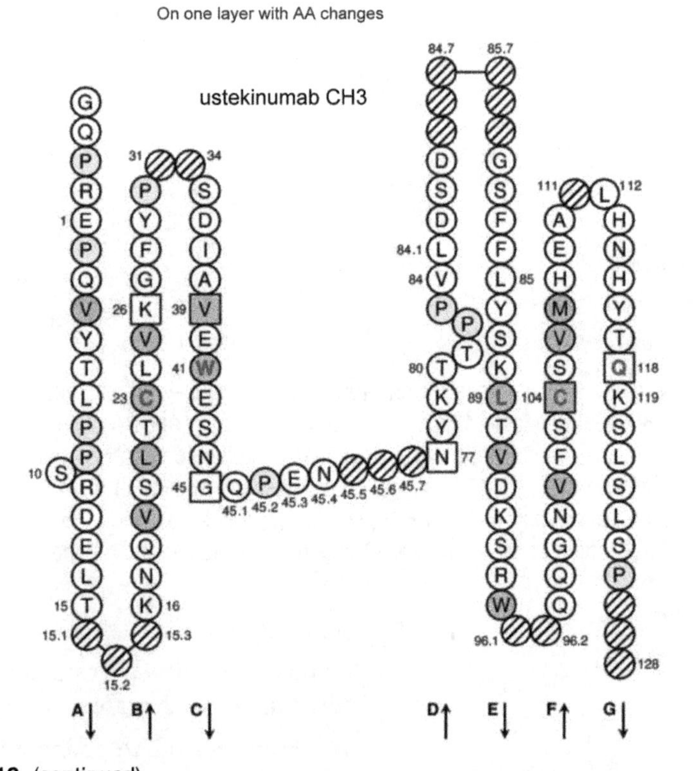

Fig. 13 (continued)

The four additional column characteristics of the V-DOMAIN (Fig. 12) are "CDR-IMGT lengths," "Number of different AA in CDR1-IMGT and CDR2-IMGT," "FR-IMGT lengths," and "Number of different AA in FR-IMGT" (V-DOMAIN).

2. The tables for AA changes are located below the results summary and provide AA changes in:
   - Strands and loops (for V domain) (Fig. 12).
   - Strands, turns, and loops (for C domain) (Fig. 13).
   - In these tables, IMGT AA changes are described according to the IMGT AA hydropathy, volume, and physicochemical classes [76] (*see* **Note 13**).

5.2.3 IMGT Collier de Perles

If selected in the IMGT/DomainGapAlign welcome page (**step 6** in Subheading 5.1), IMGT Collier de Perles on one and two layers and with or without AA changes are displayed. IMGT Collier de Perles for the V domain and C domain are shown in Figs. 12 and 13.

6 IMGT/3Dstructure-DB

For each IMGT/3Dstructure-DB entry [45, 48, 49], there is one IMGT/3Dstructure-DB card which provides access to all data related to that entry ("Chain details," "Contact analysis," "Paratope and epitope" (if relevant), "3D visualization Jmol or QuickPDB," "Renumbered IMGT file," "IMGT numbering comparison," "References and links," "Printable card"). This card has been used as a model for the IMGT/2Dstructure-DB card.

6.1 IMGT/3Dstructure-DB "Chain Details"

The IMGT/3Dstructure-DB "Chain details" section comprises information first on the chain itself and then per domain. Chains and domains are described using the IMGT-ONTOLOGY concepts (standardized labels, IMGT gene and allele names, IMGT unique numbering) [25–31].

6.1.1 Chain Characteristics

They comprise:

1. "Chain ID."
2. "Chain length" in amino acids.
3. "IMGT chain description" with the delimitations of the different domains.
4. "Chain sequence" with delimitations of the regions and domains, highlighting of AA (in orange color) that are different from the closest genes and alleles, and links to *Sequence in FASTA format* and to *Sequence in IMGT format.*

6.1.2 V Domain Characteristics

They comprise:

1. "IMGT domain description" (e.g., VH (1-120) [D1]).
2. "IMGT gene and allele name" (*see* **Note 7**) with the percentage of identity for the V (e.g., IGHV2-70*01 (86.90 %) (human)) and a link to *Alignment details.*
3. "IMGT gene and allele name" (*see* **Note 7**) with the percentage of identity for the J (e.g., IGHJ6*03 (72.20 %) (human) as well as other alleles giving the same percentage of identity) and a link to *Alignment details.*
4. "2D representation": Links to IMGT Collier de Perles on one layer or IMGT Collier de Perles on two layers.
5. "Contact analysis": A link to domain contacts (overview).
6. "CDR-IMGT lengths" (e.g., [10.7.12]).
7. "Sheet composition" (e.g., [A′BDE] [A″CC′C″FG]).
8. The V domain amino acid sequence with CDR-IMGT delimitations and highlighting of AA (in orange color) that are different from the closest V and J genes and alleles.
9. Link to *IMGT/DomainGapAlign results.*

6.1.3 C Domain Characteristics

They comprise:

1. "IMGT domain description" (e.g., CH1 (121-212) [D1]).
2. "IMGT gene and allele name" (*see* **Note 7**) with the percentage of identity for the corresponding C domain (here, CH1) of the closest C gene and allele (e.g., IGHG1*01 (100.00 %) (human), IGHG1*02 (100.00 %) (human), IGHG1*03 (99.00 %) (human)) and a link to *Alignment details.*
3. "2D representation": Links to IMGT Collier de Perles on one layer or IMGT Collier de Perles on two layers.
4. "Contact analysis": A link to domain contacts (overview).
5. "Sheet composition" (e.g., [ABDE] [CFG]).
6. The C domain amino acid sequence with CDR-IMGT delimitations and highlighting of AA (in orange color) that are different from the closest C gene and allele.
7. Link to *IMGT/DomainGapAlign results.*

6.2 IMGT/3Dstructure-DB Contact Analysis

The IMGT/3Dstructure-DB Contact analysis provides extensive information on the atom pair contacts between domains and/or chains and on the internal contacts in an IMGT/3Dstructure-DB entry. This information is displayed at different levels as follows:

1. Domain contacts (overview).
2. Domain pair contacts ("DomPair") that provides information on the contacts between two domains or between a domain and a ligand.

IMGT Residue@Position card

Residue@Position: 114 - TYR (Y) - VH - 3hmx_H

General information:	
PDB file numbering	105
IMGT file numbering	114
Residue full name	Tyrosine
Formula	C9 H11 N1 O3

IMGT LocalStructure@Position:	
Secondary structure	Coil
Phi (in degrees)	-85.55
Psi (in degrees)	179.80
ASA (in square angstrom)	1.6

Interactions with other IMGT Residue@Position

IMGT Num	Residue	Domain	Chain	Atom pair contacts	Polar	Hydrogen Bond	Non Polar
85.2	PHE F	C-LIKE	3hmx_A	6	0	0	6
	ALA A	C-LIKE	3hmx_A	16	3	1	13
26	ASN N	C-KAPPA	3hmx_L	1	1	0	0

Fig. 14 IMGT Residue@Position card. The IMGT Residue@Position card "114 – TYR(Y) – VH – 3hmx_H" (*see* **Note 21**) is from IMGT/3DstructureDB (http://www.imgt.org)

3. IMGT Residue@Position card ("R@P") that provides information on the contacts between "a given amino acid at a given position in a domain" (or IMGT Residue@Position) with amino acids of other domains (Fig. 14) (*see* **Note 21**): The IMGT Residue@Position cards can be accessed directly from the amino acid sequences of the IMGT/3Dstructure-DB card or from the IMGT Collier de Perles, by clicking on one AA.

Atom pair contacts identify interactions between atoms of two IMGT Residue@Position (*see* **Note 22**). They can be queried, at each level ("R@P," "DomPair," overview), by atom contact types (non-covalent, polar, hydrogen bond, etc.) and/or atom contact categories (backbone/backbone (BB), side chain/side chain (SS), etc.) [48, 49].

6.3 IMGT Paratope and Epitope

In an IG/antigen (Ag) complex, the amino acids in contact at the interface between the IG and the Ag constitute the paratope on the IG V-DOMAIN surface and the epitope on the Ag surface. Clicking on the "IMGT paratope and epitope" tag (displayed in the IMGT/Structure-DB card, only if relevant) gives access to "IMGT paratope and epitope details" (Fig. 15). IMGT paratope and epitope are determined automatically for the IG/Ag 3D structures in IMGT/3Dstructure-DB by combining contact analysis with an interaction scoring function [45]. Each AA which belongs

IMGT paratope and epitope details

For the **IMGT Residue@Position card** of a given residue, click on its letter in a sequence.
Differences with the closest IMGT allele sequence are in orange.

Epitope details of Lysozyme C [hen egg white] (HEL) EC:3.2.1.17 D18A [1a2y_C] and D1.3 mAb, anti-lysozyme C [Chicken] (FV-HEAVY_KAPPA) *Mus musculus* (house mouse) [1a2y_A,1a2y_B]

Lysozyme C [hen egg white] (HEL) EC:3.2.1.17 D18A Chain(s): **1a2y_C**

type	discontinuous
residues	N G Y S N G N K G T D V Q R Epitope IMGT Residue@Position cards
With positions	N(19_C)+GYS(22-24_C)+N(27_C)+GN(102-103_C)+KGTDVQ(116-121_C)+R(125_C)

Paratope details of D1.3 mAb, anti-lysozyme C [Chicken] [Chain(s): 1a2y_A,1a2y_B]

Paratope **D1.3 mAb, anti-lysozyme C [Chicken]** Chain(s): 1a2y_A,1a2y_B Domain(s): V-KAPPA 1a2y_A (V1_A), VH 1a2y_B (V1_B),

type	discontinuous
residues	Y Y Y T F W S T G Y W G D R D Y R
With positions	Y(38V1_A)+Y(55V1_A)+Y(56V1_A)+T(66V1_A)+F(107V1_A)+W(108V1_A)+S(109V1_A)+T(35V1_B)+G(36V1_B)+Y(37V1_B)+W(57V1_B)+G(58V1_B)+D(59V1_B)+R(108V1_B)+D(109V1_B)+Y(110V1_B)+R(111V1_B)

Fig. 15 IMGT paratope and epitope details. The IMGT paratope and epitope details are from IMGT/3DstructureDB (http://www.imgt.org)

to the paratope is defined by its position in an IG or a TR V-DOMAIN [54]. Each AA that belongs to the epitope is defined by its position in the chain in the 3D structure or, if the antigen belongs to an IgSF or an MhSF protein and if the epitope is part of a characterized V, C, or G domain, by its position in the domain according to the IMGT unique numbering [54–56].

6.4 IMGT/DomainSuperimpose and IMGT/StructuralQuery

IMGT/DomainSuperimpose allows to superimpose the 3D structures of two domains from IMGT/3Dstructure-DB. IMGT/StructuralQuery [48] allows to retrieve the IMGT/3Dstructure-DB entries based on specific structural characteristics of the intramolecular interactions: phi and psi angles, accessible surface area, type of atom contacts, distance in angstrom between amino acids, Residue@Position contacts, and, for V-DOMAIN, CDR-IMGT length or pattern.

6.4.1 Availability and Citation

Authors who use IMGT® databases and tools are encouraged to cite this chapter and to quote the IMGT® home page, http://www.imgt.org. Online access to IMGT® databases and tools is freely available for academics and under licences and contracts for companies.

7 Notes

1. The molecular synthesis of the IG and TR V-DOMAIN is particularly complex and unique (IMGT Education, http://www.imgt.org). It includes several mechanisms that occur at the DNA level: combinatorial rearrangements of the V, D, and J genes that code the V-DOMAIN; exonuclease trimming at the ends of the V, D, and J genes; and random addition of nucleotides by the terminal deoxynucleotidyl transferase (TdT)

that create the junction N-diversity and somatic hypermutations (for IG only) [10, 11].

2. Anchors are positions that belong to strands and represent anchors for the loops of the V and C domains (and by extension to the CD strand of the C domain that does not have the C′–C″ loop). Anchor positions are shown in square in IMGT Collier de Perles. Positions 26 and 39 are anchors of the BC-LOOP (CDR1-IMGT in V-DOMAIN). Positions 55 and 66 are anchors of the C′–C″ loop of the V domain (CDR2-IMGT in V-DOMAIN), whereas positions 45 and 77 are anchors of the CD-STRAND of the C domain. Positions 104 and 118 are anchors of the FG-LOOP (CDR3-IMGT in V-DOMAIN). The JUNCTION of a V-DOMAIN includes the anchors 104 and 118 and is therefore two amino acids longer than the corresponding CDR3-IMGT (positions 105–117).

3. IMGT reference directories have been set up for species which have been extensively studied, such as human and mouse. This also holds for the other species or taxons with incomplete IMGT reference directory sets. In those cases, results should be interpreted considering the status of the IMGT reference directory (information on the updates on the IMGT® Web site). Links to the IMGT/V-QUEST reference directory sets are available from the IMGT/V-QUEST welcome page. The IMGT/V-QUEST reference directory sets include IMGT reference sequences from all functional (F) genes and alleles, all ORF, and all in-frame pseudogene (P) alleles. Functionality is according to the IMGT Scientific chart rules (http://www.imgt.org/textes/IMGTScientificChart/SequenceDescription/IMGTfunctionality.html), based on the concepts of identification, generated from the IDENTIFICATION axiom [27]. The functionality of the user-analyzed sequences, which are rearranged, will be either productive (no stop codon in the V-(D)-J region and in-frame junction) or unproductive (stop codons in the V-(D)-J region and/or out-of-frame junction).

4. By definition, the IMGT reference directory sets contain one sequence for each allele. In IMGT®, an allele is a polymorphic variant of a gene, which is characterized by the mutations of its sequence at the nucleotide level, identified in its core coding sequence, and compared to the gene allele reference sequence, designated as allele *01. By default, the user sequences are compared with all genes and alleles. However, the option "With allele *01 only" is useful for (a) "Detailed view," if the user sequences need to be compared with different genes, and (b) "Synthesis view," if the user sequences which use the same gene need to be aligned together (independently of the allelic polymorphism).

5. This option is usually selected in a second step after a first analysis by IMGT/V-QUEST that provided warnings for potential insertions or deletions (Subheading 3.3.1).
6. By default, for the IGH locus, the maximum number of accepted mutations is 2, 4, and 2 mutations (parameters "2,4,2") for the 3′V-REGION, D-REGION, and 5′J-REGION, respectively. For the IGK and IGL loci, the maximum number of accepted mutations is 7 in the 3′V-REGION and 7 in the 5′J-REGION (parameters "7,7"). Moreover, in case of unmutated IGHV genes (no mutation from FR1-IMGT to FR3-IMGT), the tool does not allow any mutation in the 3′V-REGION and 5′J-REGION and automatically limits the number of allowed mutations in the D-REGION to 2 (parameters "0,2,0") in order to reflect the low probability of somatic hypermutations.
7. The IMGT® IG and TR gene names [10–13] were approved by the Human Genome Organisation (HUGO) Nomenclature Committee (HGNC) in 1999 [67, 68] and were endorsed by the WHO–IUIS Nomenclature Subcommittee for IG and TR [16, 17]. The IMGT® IG and TR gene names are the official reference for the genome projects and, as such, have been entered in IMGT/GENE-DB [51], in the Genome Database (GDB) [69], in LocusLink at the National Center for Biotechnology Information (NCBI) [70], in Entrez Gene (NCBI, USA) when this database (now designated as "Gene") superseded LocusLink [71], in Ensembl at the European Bioinformatics Institute (EBI) [72], and in the Vertebrate Genome Annotation (Vega) Browser [73] at the Wellcome Trust Sanger Institute (UK). HGNC, Gene (NCBI), Ensembl, and Vega have direct links to IMGT/GENE-DB [51]. Since 2008, IMGT gene and allele names have been used in the description of mAb (INN suffix—mab) and of fusion proteins for immunological applications (FPIA, INN suffix—cept) from the WHO/INN programme [18, 74].
8. The score of the alignment for two sequences is calculated by counting +5 for each position where nucleotides are identical (match) and −4 for each position with different nucleotides (mismatch).
9. In addition to the analysis of the somatic mutations, results of IMGT/V-QUEST, frequently used by clinicians, also include the sequences of the V-(D)-J junctions which are used in the synthesis of specific probes for the follow-up of residual diseases in leukemias and lymphomas.
10. Potential insertions or deletions are suspected by IMGT/V-QUEST when the V-REGION score is very low (less than 200), the percentage of identity is less than 85 %, and/or the input

sequence has different CDR1-IMGT and/or CDR2-IMGT lengths, compared to those of the closest germline V.

11. The note (in red online) indicates the additional J-GENE and allele names and the criterion used for their identification (usually the highest number of consecutive identical nucleotides).
12. The way to identify the closest germline D is different between IMGT/V-QUEST and IMGT/JunctionAnalysis since the evaluation of the alignment score is different. In case of discrepancy, the results of IMGT/JunctionAnalysis are the most accurate. If the option "with full list of eligible D-GENE" was selected in "Display view" (Subheading 3.2.2) (Fig. 3), its results allow comparing the IMGT/JunctionAnalysis D gene identification with all D genes which match the junction with their corresponding score. The alignment provided by IMGT/V-QUEST is still provided, although less accurate, as it is less stringent and displays several D genes and alleles, and therefore may help solving some ambiguous cases.
13. The 20 amino acids have been classified in 11 IMGT "physicochemical" classes which are based on "hydrophathy," "volume," and "chemical" characteristics (http://www.imgt.org/, in IMGT Education > Aide-mémoire > "Amino acids"). The AA changes are described according to the IMGT AA hydropathy, volume, and physicochemical classes [76]. For example Q1 > E (++−) means that in the AA change (Q>E), the two amino acids belong to the same hydropathy (+) and volume (+) classes but to different physicochemical property (−) classes. Four types of AA changes are identified in IMGT: very similar (+++), similar (++−, +−+), dissimilar (−−+, −+−, +−−), and very dissimilar (−−−).
14. IMGT/Collier de Perles tool can be customized to display the amino acids colored according to their hydropathy, volume, or IMGT physicochemical classes [76]. By default, the IMGT Collier de Perles are displayed on one layer. They can also be displayed on two layers in order to get a graphical representation closer to the 3D structure. In the case of sequences with an out-of-frame junction, the J reading frame is not restored (the sequence does not include "#").
15. A note (in red online) may appear with the V-GENE and allele name when potential insertions or deletions are suspected (criteria detailed in **Note 10**). In those cases, the alignment for this sequence has to be checked in "A. Detailed view," using the advanced parameters "Search for insertions and deletions in V-REGION" (Subheading 3.2.3).
16. A note (in red online) may appear to inform on other possibilities for the J-GENE and allele name as described in **Note 11**.

17. The IMGT domain reference directory is the IMGT reference directory for V, C, and G domains [45]. It is manually curated and contains the amino acid sequences of the domains delimited according to the IMGT rules (based on the exon delimitations) [30, 54–56]. Sequences are from the IMGT repertoire [9] or from IMGT/GENE-DB [51] (*see* **Note** 7). Owing to the particularities of the V-DOMAIN synthesis [10, 11] there is no V-DOMAIN in the IMGT reference directory. Instead, the directory comprises the translation of the IG and TR germline V and J genes (V-REGION and J-REGION, respectively). The IMGT domain reference directory provides the IMGT "gene" and "allele" names ("CLASSIFICATION" axiom) (*see* **Note** 7). Data are comprehensive for human and mouse IG and TR and human MH, whereas for other species and IgSF and MhSF they are added progressively. The IMGT domain reference directory comprises domain sequences of functional (F), ORF, and in-frame pseudogene (P) genes (*see* **Note 3**). As IMGT alleles are characterized at the nucleotide level (*see* **Note 4**), identical sequences at the amino acid level may therefore correspond to different alleles in the IMGT domain reference directory.
18. The Smith–Waterman algorithm is a well-known algorithm for performing local sequence alignment [78]. The algorithm determines identical regions between two nucleotide or protein sequences. Instead of looking at the complete sequence, the Smith–Waterman algorithm compares segments of all possible lengths and optimizes the similarity measure. The higher the score, the better the alignment in a series of results. Selecting a higher score in IMGT/DomainGapAlign corresponds to a display of higher sequence similarities.
19. The expect value (*E*-value) is a parameter that describes the number of hits one can "expect" to see by chance, for a given score of the match, when searching a database of a particular size. Each hit is associated to a score and an *E*-value. For example, an *E*-value of 1 assigned to a match "means" that in a database of a particular size one can expect one hit with a similar score simply by chance. The lower the *E*-value, or the closer it is to zero, the more "significant" the match is. A way of measuring the significance of a score of an alignment is to consider its *E*-value. Decreasing the *E*-value parameter for the alignment corresponds to a higher selection of the results to display.
20. The alignment is performed by a modified Smith–Waterman algorithm that considers the IMGT gap as a full amino acid and which discriminates the creation of gaps in the IMGT reference sequence. The asymmetry management of insertions between the user query and the reference sequences allows the user to modify one or the other gap penalty.

21. An "IMGT Residue@Position" is defined by the IMGT position numbering in a V, C, or G domain (or if not characterized, in the chain), the AA name (three-letter and between parentheses one-letter abbreviation), the IMGT domain description, and the IMGT chain ID, e.g., "114 – TYR(Y) – VH – 3hmx_H." Its characteristics are reported in an IMGT Residue@Position card (or "R@P") which includes (1) general information (PDB file numbering, IMGT file numbering, residue full name, and formula); (2) structural information "IMGT LocalStructure@Position" (secondary structure, Phi and Psi angles (in degrees), and accessible surface area (ASA) (in square angstrom)); and (3) detailed contact analysis.
22. Atom pair contacts are obtained in IMGT/3Dstructure-DB by a local program in which atoms are considered to be in contact when no water molecule can take place between them [48].

Acknowledgments

We are grateful to Gérard Lefranc for helpful discussion, to Souphatta Sasorith for help with the figures, and to the IMGT® team for its expertise and constant motivation. IMGT® is a registered trademark of CNRS. IMGT® is a member of the International Medical Informatics Association (IMIA). IMGT® is currently supported by CNRS, the Ministère de l'Enseignement Supérieur et de la Recherche (MESR), University Montpellier 2, the Agence Nationale de la Recherche (ANR) Labex MabImprove (ANR-10-LABX-53-01), and the Région Languedoc-Roussillon (Grand Plateau Technique pour la Recherche (GPTR)). This work was granted access to the HPC resources of CINES under the allocation 036029 (2010–2013) made by Grand Equipement National de Calcul Intensif (GENCI).

References

1. Lefranc MP (2004) IMGT, the International ImMunoGenetics information system®. In: Lo BKC (ed) Antibody engineering methods and protocols, vol 248, 2nd edn, Methods in molecular biology. Humana Press, Totowa, NJ, pp 27–49
2. Lefranc MP (2009) Antibody databases and tools: the IMGT® experience. In: An Z (ed) Therapeutic monoclonal antibodies: from bench to clinic, chap. 4. Wiley, New York, pp 91–114
3. Lefranc M-P (2009) Antibody databases: IMGT®, a French platform of world-wide interest [in French]. Bases de données anticorps: IMGT®, une plate-forme française d'intérêt mondial. Médecine/Sciences 25: 1020–1023
4. Ehrenmann F, Duroux P, Giudicelli V, Lefranc MP (2010) Standardized sequence and structure analysis of antibody using IMGT®. In: Kontermann R, Dübel S (eds) Antibody engineering, chap. 2, vol 2. Springer, Berlin, pp 11–31
5. Lefranc MP, Ehrenmann F, Ginestoux C, Duroux P, Giudicelli V (2012) Use of IMGT® databases and tools for antibody engineering and humanization. In: Chames P (ed) Antibody engineering. Methods in molecular biology, chap. 1, vol 907. Humana, New York, pp 3–37

6. Magdelaine-Beuzelin C, Kaas Q, Wehbi V, Ohresser M, Jefferis R, Lefranc M-P, Watier H (2007) Structure-function relationships of the variable domains of monoclonal antibodies approved for cancer treatment. Crit Rev Oncol Hematol 64:210–225
7. Jefferis R, Lefranc M-P (2009) Human immunoglobulin allotypes: possible implications for immunogenicity. MAbs 1(4):332–338
8. Lefranc MP, Lefranc G (2012) Human Gm, Km and Am allotypes and their molecular characterization: a remarkable demonstration of polymorphism. In: Christiansen F, Tait B (eds) Immunogenetics, vol 882, Methods in molecular biology. Humana, New York, pp 635–680
9. Lefranc M-P, Giudicelli V, Ginestoux C, Jabado-Michaloud J, Folch G, Bellahcene F, Wu Y, Gemrot E, Brochet X, Lane J, Regnier L, Ehrenmann F, Lefranc G, Duroux P (2009) IMGT®, the International ImMunoGeneTics information system®. Nucleic Acids Res 37:D1006–D1012
10. Lefranc M-P, Lefranc G (2001) The immunoglobulin FactsBook. Academic, London, UK, pp 1–458
11. Lefranc M-P, Lefranc G (2001) The T cell receptor FactsBook. Academic, London, pp 1–398
12. Lefranc MP (2000) Nomenclature of the human immunoglobulin genes. In: Coligan JE, Bierer BE, Margulies DE, Shevach EM, Strober W (eds) Current protocols in immunology. Wiley, Hoboken, NJ, pp 1–37
13. Lefranc MP (2000) Nomenclature of the human T cell receptor genes. In: Coligan JE, Bierer BE, Margulies DE, Shevach EM, Strober W (eds) Current protocols in immunology. Wiley, Hoboken, NJ, pp 1–23
14. Lefranc MP (2011) IMGT, the International ImMunoGeneTics information system. Cold Spring Harb Protoc 6:595–603. doi:10.1101/pdb.top115, pii: pdb.top115
15. Lefranc MP (2012) IMGT® information system. In: Dubitzky W, Wolkenhauer O, Cho K, Yokota H (eds) Encyclopedia of systems biology. Springer, LLC012 (in press). doi:10.1007/978-1-4419-9863-7
16. Lefranc M-P (2007) WHO-IUIS Nomenclature Subcommittee for immunoglobulins and T cell receptors report. Immunogenetics 59:899–902
17. Lefranc M-P (2008) WHO-IUIS Nomenclature Subcommittee for immunoglobulins and T cell receptors report August 2007, 13th International Congress of Immunology, Rio de Janeiro, Brazil. Dev Comp Immunol 32:461–463
18. Lefranc M-P (2011) Antibody nomenclature: from IMGT-ONTOLOGY to INN definition. MAbs 3(1):1–2
19. Giudicelli V, Lefranc M-P (1999) Ontology for immunogenetics: IMGT-ONTOLOGY. Bioinformatics 15:1047–1054
20. Lefranc M-P, Giudicelli V, Ginestoux C, Bosc N, Folch G, Guiraudou D, Jabado-Michaloud J, Magris S, Scaviner D, Thouvenin V, Combres K, Girod D, Jeanjean S, Protat C, Yousfi Monod M, Duprat E, Kaas Q, Pommié C, Chaume D, Lefranc G (2004) IMGT-ONTOLOGY for immunogenetics and immunoinformatics. In Silico Biol 4:17–29
21. Lefranc M-P (2004) IMGT-ONTOLOGY and IMGT databases, tools and web resources for immunogenetics and immunoinformatics. Mol Immunol 40:647–660
22. Lefranc M-P, Clément O, Kaas Q, Duprat E, Chastellan P, Coelho I, Combres K, Ginestoux C, Giudicelli V, Chaume D, Lefranc G (2005) IMGT-choreography for immunogenetics and immunoinformatics. In Silico Biol 5:45–60
23. Lefranc M-P, Giudicelli V, Regnier L, Duroux P (2008) IMGT®, a system and an ontology that bridge biological and computational spheres in bioinformatics. Brief Bioinform 9:263–275
24. Duroux P, Kaas Q, Brochet X, Lane J, Ginestoux C, Lefranc M-P, Giudicelli V (2008) IMGT-Kaleidoscope, the formal IMGT-ONTOLOGY paradigm. Biochimie 90: 570–583
25. Giudicelli V, Lefranc MP (2012) IMGT-ONTOLOGY 2012. Frontiers in bioinformatics and computational biology. Front Genet 3:79
26. Lefranc MP (2012) IMGT-ONTOLOGY. In: Dubitzky W, Wolkenhauer O, Cho K, Yokota H (eds) Encyclopedia of systems biology. Springer, LLC012 (in press). doi:10.1007/978-1-4419-9863-7
27. Lefranc MP (2011) From IMGT-ONTOLOGY IDENTIFICATION axiom to IMGT standardized keywords: for immunoglobulins (IG), T cell receptors (TR), and conventional genes. Cold Spring Harb Protoc 6:604–613. doi:10.1101/pdb.ip82, pii: pdb.ip82
28. Lefranc MP (2011) From IMGT-ONTOLOGY DESCRIPTION axiom to IMGT standardized labels: for immunoglobulin (IG) and T cell receptor (TR) sequences and structures. Cold Spring Harb Protoc 6:614–626. doi:10.1101/pdb.ip83, pii: pdb.ip83
29. Lefranc MP (2011) From IMGT-ONTOLOGY CLASSIFICATION axiom to IMGT standardized gene and allele nomenclature: for immunoglobulins (IG) and T cell receptors (TR).

Cold Spring Harb Protoc 6:627–632. doi:10.1101/pdb.ip84, pii: pdb.ip84
30. Lefranc M-P (2011) IMGT unique numbering for the variable (V), constant (C), and groove (G) domains of IG, TR, MH, IgSF, and MhSF. Cold Spring Harb Protoc 6:633–642. doi:10.1101/pdb.ip85, pii: pdb.ip85
31. Lefranc M-P (2011) IMGT Collier de Perles for the variable (V), constant (C), and groove (G) domains of IG, TR, MH, IgSF, and MhSF. Cold Spring Harb Protoc 6:643–651. doi:10.1101/pdb.ip86, pii: pdb.ip86
32. Giudicelli V, Chaume D, Lefranc M-P (2004) IMGT/V-QUEST, an integrated software for immunoglobulin and T cell receptor V-J and V-D-J rearrangement analysis. Nucleic Acids Res 32:W435–W440
33. Giudicelli V, Lefranc M-P (2005) Interactive IMGT on-line tools for the analysis of immunoglobulin and T cell receptor repertoires. In: Veskler BA (ed) New research on immunology. Nova Science Publishers Inc., New York, pp 77–105
34. Brochet X, Lefranc M-P, Giudicelli V (2008) IMGT/V-QUEST: the highly customized and integrated system for IG and TR standardized V-J and V-D-J sequence analysis. Nucleic Acids Res 36:W503–W508
35. Giudicelli V, Lefranc MP (2008) IMGT® standardized analysis of immunoglobulin rearranged sequences. In: Ghia P, Rosenquist R, Davi F (eds) Immunoglobulin gene analysis in chronic lymphocytic leukemia, chap. 2. Wolters Kluwer Health, Bologna, pp 33–52
36. Giudicelli V, Brochet X, Lefranc MP (2011) IMGT/V-QUEST: IMGT standardized analysis of the immunoglobulin (IG) and T cell receptor (TR) nucleotide sequences. Cold Spring Harb Protoc 6:695–715. doi:10.1101/pdb.prot5633, pii: pdb.prot5633
37. Alamyar E, Duroux P, Lefranc MP, Giudicelli V (2012) IMGT® tools for the nucleotide analysis of immunoglobulin (IG) and T cell receptor (TR) V-(D)-J repertoires, polymorphisms, and IG mutations: IMGT/V-QUEST and IMGT/HighV-QUEST for NGS. In: Christiansen F, Tait B (eds) Immunogenetics. Methods in molecular biology, vol 882. Humana, New York, pp 569–604
38. Yousfi Monod M, Giudicelli V, Chaume D, Lefranc M-P (2004) IMGT/JunctionAnalysis: the first tool for the analysis of the immunoglobulin and T cell receptor complex V-J and V-D-J JUNCTIONs. Bioinformatics 20:i379–i385
39. Giudicelli V, Lefranc M-P (2011) IMGT/JunctionAnalysis: IMGT standardized analysis of the V-J and V-D-J Junctions of the rearranged immunoglobulins (IG) and T cell receptors (TR). Cold Spring Harb Protoc 6:716–725. doi:10.1101/pdb.prot5634, pii: pdb.prot5634
40. Giudicelli V, Protat C, Lefranc MP (2003) The IMGT strategy for the automatic annotation of IG and TR cDNA sequences: IMGT/Automat. In: Proceedings of the European conference on computational biology (ECCB 2003), INRIA (DISC/Spid), Paris, DKB-31, p 103–104
41. Giudicelli V, Chaume D, Jabado-Michaloud J, Lefranc M-P (2005) Immunogenetics sequence annotation: the strategy of IMGT based on IMGT-ONTOLOGY. Stud Health Technol Inform 116:3–8
42. Alamyar E, Giudicelli V, Duroux P, Lefranc MP (2010) IMGT/HighV-QUEST: a high-throughput system and web portal for the analysis of rearranged nucleotide sequences of antigen receptors—high-throughput version of IMGT/V-QUEST. Poster, *11èmes Journées Ouvertes de Biologie, Informatique et Mathématiques (JOBIM)*, Montpellier, 7–9 Sept 2010, Poster #60. http://www.jobim2010.fr/?q=fr/node/55. Accessed 31 Jan 2013
43. Alamyar E, Giudicelli V, Duroux P, Lefranc MP (2011) IMGT/HighV-QUEST 2011. Poster, 12èmes Journées Ouvertes de Biologie, Informatique et Mathématiques (JOBIM), Paris, 28 June –1 July 2011
44. Alamyar E, Giudicelli V, Shuo L, Duroux P, Lefranc M-P (2012) IMGT/HighV-QUEST: the IMGT® web portal for immunoglobulin (IG) or antibody and T cell receptor (TR) analysis from NGS high throughput and deep sequencing. Immunome Res 8(1):26
45. Ehrenmann F, Kaas Q, Lefranc M-P (2010) IMGT/3Dstructure-DB and IMGT/DomainGapAlign: a database and a tool for immunoglobulins or antibodies, T cell receptors, MHC, IgSF and MhcSF. Nucleic Acids Res 38:D301–D307
46. Ehrenmann F, Lefranc MP (2011) IMGT/DomainGapAlign: IMGT standardized analysis of amino acid sequences of variable, constant, and groove domains (IG, TR, MH, IgSF, MhSF). Cold Spring Harb Protoc 6:737–749. doi:10.1101/pdb.prot5636, pii: pdb.prot5636
47. Ehrenmann F, Lefranc MP (2012) IMGT/DomainGapAlign: the IMGT® tool for the analysis of IG, TR, MHC, IgSF and MhcSF domain amino acid polymorphism. In: Tait B, Christiansen F (eds) Immunogenetics. Methods in molecular biology, chap. 33, vol 882. Humana, New York, pp 605–633

48. Kaas Q, Ruiz M, Lefranc M-P (2004) IMGT/3Dstructure-DB and IMGT/StructuralQuery, a database and a tool for immunoglobulin, T cell receptor and MHC structural data. Nucleic Acids Res 32:D208–D210
49. Ehrenmann F, Lefranc M-P (2011) IMGT/3Dstructure-DB: querying the IMGT database for 3D structures in immunology and immunoinformatics (IG or antibodies, TR, MH, RPI, and FPIA). Cold Spring Harb Protoc 6:750–761. doi:10.1101/pdb.prot5637, pii: pdb.prot5637
50. Giudicelli V, Duroux P, Ginestoux C, Folch G, Jabado-Michaloud J, Chaume D, Lefranc M-P (2006) IMGT/LIGM-DB, the IMGT® comprehensive database of immunoglobulin and T cell receptor nucleotide sequences. Nucleic Acids Res 34:D781–D784
51. Giudicelli V, Chaume D, Lefranc M-P (2005) IMGT/GENE-DB: a comprehensive database for human and mouse immunoglobulin and T cell receptor genes. Nucleic Acids Res 33:D256–D261
52. Lefranc M-P (1997) Unique database numbering system for immunogenetic analysis. Immunol Today 18:509
53. Lefranc M-P (1999) The IMGT unique numbering for immunoglobulins, T cell receptors and Ig-like domains. Immunologist 7:132–136
54. Lefranc M-P, Pommié C, Ruiz M, Giudicelli V, Foulquier E, Truong L, Thouvenin-Contet V, Lefranc G (2003) IMGT unique numbering for immunoglobulin and T cell receptor variable domains and Ig superfamily V-like domains. Dev Comp Immunol 27:55–77
55. Lefranc M-P, Pommié C, Kaas Q, Duprat E, Bosc N, Guiraudou D, Jean C, Ruiz M, Da Piedade I, Rouard M, Foulquier E, Thouvenin V, Lefranc G (2005) IMGT unique numbering for immunoglobulin and T cell receptor constant domains and Ig superfamily C-like domains. Dev Comp Immunol 29:185–203
56. Lefranc M-P, Duprat E, Kaas Q, Tranne M, Thiriot A, Lefranc G (2005) IMGT unique numbering for MHC groove G-DOMAIN and MHC superfamily (MhcSF) G-LIKE-DOMAIN. Dev Comp Immunol 29:917–938
57. Ruiz M, Lefranc M-P (2002) IMGT gene identification and Colliers de Perles of human immunoglobulins with known 3D structures. Immunogenetics 53:857–883
58. Kaas Q, Lefranc M-P (2007) IMGT Colliers de Perles: standardized sequence-structure representations of the IgSF and MhcSF superfamily domains. Curr Bioinforma 2:21–30
59. Kaas Q, Ehrenmann F, Lefranc M-P (2007) IG, TR and IgSf, MHC and MhcSF: what do we learn from the IMGT Colliers de Perles? Brief Funct Genomic Proteomic 6:253–264
60. Ehrenmann F, Giudicelli V, Duroux P, Lefranc M-P (2011) IMGT/Collier de Perles: imgt standardized representation of domains (IG, TR, and IgSF Variable and constant domains, MH and MhSF groove domains). Cold Spring Harb Protoc 6:726–736. doi:10.1101/pdb.prot5635, pii: pdb.prot5635
61. Duprat E, Kaas Q, Garelle V, Lefranc G, Lefranc M-P (2004) IMGT standardization for alleles and mutations of the V-LIKE-DOMAINs and C-LIKE-DOMAINs of the immunoglobulin superfamily. Recent Res Dev Hum Genet 2:111–136
62. Garapati VP, Lefranc M-P (2007) IMGT Colliers de Perles and IgSF domain standardization for T cell costimulatory activatory (CD28, ICOS) and inhibitory (CTLA4, PDCD1 and BTLA) receptors. Dev Comp Immunol 31:1050–1072
63. Bernard D, Hansen JD, du Pasquier L, Lefranc M-P, Benmansour A, Boudinot P (2005) Costimulatory receptors in jawed vertebrates: conserved CD28, odd CTLA4 and multiple BTLAs. Dev Comp Immunol 31:255–271
64. Bertrand G, Duprat E, Lefranc M-P, Marti J, Coste J (2004) Characterization of human FCGR3B*02 (HNA-1b, NA2) cDNAs and IMGT standardized description of FCGR3B alleles. Tissue Antigens 64:119–131
65. Duprat E, Lefranc M-P, Gascuel O (2006) A simple method to predict protein binding from aligned sequences - application to MHC superfamily and beta2-microglobulin. Bioinformatics 22:453–459
66. Belessi C, Davi F, Stamatopoulos K, Degano M, Andreou TM, Moreno C, Merle-Béral H, Crespo M, Laoutaris NP, Montserrat E, Caligaris-Cappio F, Anagnostopoulos AZ, Ghia P (2006) IGHV gene insertions and deletions in chronic lymphocytic leukemia: "CLL-biased" deletions in a subset of cases with stereotyped receptors. Eur J Immunol 36:1963–1974
67. Wain HM, Bruford EA, Lovering RC, Lush MJ, Wright MW, Povey S (2002) Guidelines for human gene nomenclature. Genomics 79:464–470
68. Bruford EA, Lush MJ, Wright MW, Sneddon TP, Povey S, Birney E (2008) The HGNC Database in 2008: a resource for the human genome. Nucleic Acids Res 36:D445–D448
69. Letovsky SI, Cottingham RW, Porter CJ, Li PW (1998) GDB: the Human Genome Database. Nucleic Acids Res 26(1):94–99

70. Maglott DR, Katz KS, Sicotte H, Pruitt KD (2000) NCBI's LocusLink and RefSeq. Nucleic Acids Res 28(1):126–128
71. Maglott D, Ostell J, Pruitt KD, Tatusova T (2007) Entrez Gene: gene-centered information at NCBI. Nucleic Acids Res 35:D26–D31
72. Stabenau A, McVicker G, Melsopp C, Proctor G, Clamp M, Birney E (2004) The Ensembl core software libraries. Genome Res 14:929–933
73. Wilming LG, Gilbert JG, Howe K, Trevanion S, Hubbard T, Harrow JL (2008) The vertebrate genome annotation (Vega) database. Nucleic Acids Res 36:D753–D760
74. World Health Organization (2009) General policies for monoclonal antibodies. INN Working Document 09.251. http://www.who.int/medicines/services/inn/en. Accessed 18 Nov 2009
75. Ghia P, Stamatopoulos K, Belessi C, Moreno C, Stilgenbauer S, Stevenson F, Davi F, Rosenquist R (2007) ERIC recommendations on IGHV gene mutational status analysis in chronic lymphocytic leukemia. Leukemia 21:1–3
76. Pommié C, Levadoux S, Sabatier R, Lefranc G, Lefranc M-P (2004) IMGT standardized criteria for statistical analysis of immunoglobulin V-REGION amino acid properties. J Mol Recognit 17:17–32
77. Elemento O, Lefranc M-P (2003) IMGT/PhyloGene: an on-line tool for comparative analysis of immunoglobulin and T cell receptor genes. Dev Comp Immunol 27:763–779
78. Smith TF, Waterman MS (1981) Identification of common molecular subsequences. J Mol Biol 147(1):195–197

Chapter 22

Measuring Antibody Affinities as Well as the Active Concentration of Antigens Present on a Cell Surface

Palaniswami Rathanaswami

Abstract

Measuring the affinity of a therapeutic antibody to its antigen, expressed in its native form on a cell surface, is an important aspect to understanding its in vivo potency. Measured affinities can also help in selecting the best antibody for therapy. The on-cell binding affinity of antibodies was determined in the past by labelling the antibody using radioactive, fluorescent, or other probes. Labelling the antibody could potentially modify the structure of the antibody and hence could alter its original affinity. Here, we describe a label-free method to measure the affinity of antibodies to their target antigens that are expressed on the surface of cells and the number of active antigen molecules present on a given cell. In addition, this method can also be used to measure the affinity of a ligand to its receptor on a cell.

Key words On-cell binding, Antibody affinity measurements, High affinity, Membrane-expressed proteins, Cell surface-expressed antigens, Transmembrane proteins, KinExA, Native binding, Equilibrium binding measurements, Label-free method

1 Introduction

Monoclonal antibodies (mAbs) have emerged as important therapeutic molecules for the treatment of human disease [1, 2]. When choosing an appropriate antibody for therapeutic purposes, in addition to its specificity to target antigen, the affinity of the antibody to its target is also important. A direct correlation exists between the affinity and potency of an antibody [3, 4]. High-affinity mAbs, in contrast to low-affinity mAbs, target tumors better [5] and also enhance the antibody-dependent cellular cytotoxicity (ADCC)-mediated killing of tumor cells even at low antigen expression levels [6]. Since the antigen is expressed on the cell membrane, if we measure affinity of the antibody to the soluble antigen counterpart, typically the extracellular domain, it may not correlate with antibody's binding affinity to native form of the antigen on cells. In addition, purification of membrane proteins is difficult and may result in a loss of stability or partially functional

Vincent Ossipow and Nicolas Fischer (eds.), *Monoclonal Antibodies: Methods and Protocols*, Methods in Molecular Biology, vol. 1131, DOI 10.1007/978-1-62703-992-5_22,

protein [7]. There are reports that describe the affinity measurements of antibodies to cell surface-expressed molecules. However they typically modify the antibodies with fluorescent or radioactive labels so that they can be detected [8, 9]. Moreover, chemical modification of an antibody for labelling can lead to binding constants that are 10–100-fold lower than the native antibodies [10, 11]. In addition, methods using radiolabeled antibodies require extensive cell washing after the equilibrium is reached [9] and that might lead to the dissociation of antibody from the cell membrane and disturb the equilibrium. Fluorescence-activated cell sorting (FACS) methods have been described to measure the affinity of cell membrane-expressed antigens [12, 13]. However these methods used a linear transformation of the data to calculate the affinity which can sometimes conceal deviations from simple single-site binding patterns that can be recognized by nonlinear analysis [14]. We [15] and others [16] used kinetic exclusion assay (KinExA) to measure the affinity of an antibody to cell surface receptors and overcome many of the above problems. Here I describe the method in which soluble antigen or labelling of an antibody is not required for measuring the affinity of a purified antibody to its cell membrane-expressed target. In addition this methodology also allows one to calculate the active concentration of the antigen present on the surface of the cell. This method involves the titration of cells expressing the target on the membrane with a known concentration of the specific antibody. After allowing sufficient time for the reaction to reach equilibrium, the amount of unbound antibody is measured in KinExA instrument. The K_d is calculated using the software provided by the vendor.

2 Materials

1. Use analytical grade reagents for preparing all buffers. Prepare all buffers using ultrapure water under sterile conditions. After preparation, filter the buffers using 4.5 μm filter (*see* **Note 1**).
2. Sample buffer: Dissolve 500 mg NaN_3 in 1 l of PBS, pH 7.4 and store at room temperature.
3. Blocking buffer: Dissolve 1 g BSA in 100 ml sample buffer and store at 4 °C.
4. Assay buffer: Prepared by dissolving 100 mg BSA in 1 l of sample buffer and stored at 4 °C.
5. Human mAbs were purified using Protein A-Sepharose affinity chromatography. The purity of the antibodies was confirmed by visualization of a single band on a nonreducing SDS-PAGE analysis and Coomassie blue staining. The antibody (Ab) concentration was determined by UV absorbance at A280 using the molar extinction coefficient.

6. Capturing beads: Add 100 μg of goat anti-human IgG (Immunodiagnostics, Bellingham, WA, USA) in 1 ml of PBS buffer pH 7.4 to 200 mg of PMMA beads (Sapidyne Instruments, Boise, ID, USA) in a 5 ml falcon tube and rotate overnight at room temperature (*see* **Note 2**). Centrifuge briefly, and remove the liquid from the beads. Add 1 ml of blocking buffer, and block the beads for 1 h by rotating at room temperature (*see* **Note 3**). Transfer the beads to KinExA bead vial, and adjust the volume to 35 ml with assay buffer.
7. Detection antibodies: Cy5-conjugated, affinity-purified goat anti-human Fcγ fragment-specific (minimum cross reactive to other species) IgG (Jackson ImmunoResearch Laboratories, Inc., West Grove, PA, USA) was diluted to 1 μg/ml using the assay buffer (*see* **Note 4**).
8. Cell culture: Cells are cultured in an appropriate cell culture medium, with 5 % CO_2 and at 37 °C. At the time of affinity measurements, the cells will be incubated with mAb in the same medium containing 0.05 % NaN_3 (*see* **Note 5**).
9. KinExA instrument: KinExA 3200 instrument (Sapidyne Instruments, Boise, ID, USA), that is equipped with pressure transducer, flow cell camera, and internal instrument controller, is used for the KinExA. KinExA Pro software comes with the instrument that is used for all calculations and analysis.

3 Methods

3.1 Preliminary Experiments

1. Signal test: Set the KinExA instrument to run different volumes (100 μl to 10 ml) of 10 and 100 pM binding site concentration (an IgG has two binding sites) of mAb, over PMMA beads coated with goat anti-Hu IgG, at a rate of 0.25 ml/min (*see* **Note 6**). Following each individual run the machine is set to run a sample buffer wash, 1 ml of Cy5-conjugated, goat anti-human Fcγ fragment-specific IgG (1 μg/ml) (0.25 ml/min) and a sample buffer wash. The signal over baseline for each concentration of mAb at each volume will be obtained by KinExA. Find out the volume that gives a usable signal above the baseline for 10 and 100 pM concentration of binding sites (*see* **Note 7**).
2. Cell range finding: Titrate cells expressing the target of interest. Take 200 million cells/ml and perform 8 serial 1 in 10 dilutions in cell culture media containing 0.05 % NaN_3, in 5 ml Falcon tubes. Include one tube that contains only media without cells. Add 1 ml of 200 pM binding sites diluted in cell culture medium to 1 ml of each dilution of cells (after the addition of mAb, the final cell concentration will start from 100 million/ml and the binding site concentration of mAb will be 100 pM).

Incubate the mixture overnight by rotating at 5 rpm, 37 °C. Centrifuge the cells at 2,400 rpm ($400 \times g$) for 4 min, take the supernatant that contains the unbound mAb, and run in KinExA instrument set to run: 500 μl of supernatant, a sample buffer wash, 1 ml of Cy5-conjugated goat anti-human Fcγ fragment-specific IgG (1 μg/ml), and a sample buffer wash. The signal over baseline for each concentration of cells will be obtained by KinExA. Find out the cell concentration that completely bound all mAb (i.e., no free mAb is left in the supernatant), and hence no signal over baseline is obtained (*see* **Note 8**).

3.2 Affinity Measurement

3.2.1 Experiment 1: Low mAb Concentration

1. Prepare 12 ml of cell culture media containing 0.05 % NaN_3 in a 15 ml Falcon tube (Tube 1) (*see* **Note 9**).
2. Prepare 12 ml of cells expressing the target at a concentration of 50 million cells/ml or twice the cell concentration that completely bound a mAb in the range finding experiment (*see* **Note 10**), in the cell culture media with 0.05 % NaN_3 (Tube 2).
3. Dispense 6 ml of cell culture medium with 0.05 % NaN_3 in tubes 3–13.
4. Mix tube 2 thoroughly, and transfer half of the volume (6 ml) to the next tube. Repeat this process through tube 12, and discard the final 6 ml taken from tube 12. Tube 13 contains no cells and serves as the 100 % signal sample.
5. Prepare 80 ml of cell culture media with 0.05 % NaN_3 and 20 pM binding sites of mAb (*see* **Note 11**).
6. Add 6 ml of the 20 pM binding site solution to each tube with the exception of tube 1 (*see* **Note 12**).
7. Allow the binding to reach equilibrium by rotating the tubes at 5 rpm for 36 h at 37 °C (*see* **Note 13**).
8. At the end of the incubation, centrifuge the cells at 2,400 rpm (or $400 \times g$) for 4 min.
9. Transfer the supernatant containing the unbound mAb into fresh tubes, being careful not to disturb the pelleted cells.
10. Place the 13 sample lines from the KinExA instrument in the 13 sample tubes containing the supernatants.
11. Connect the KinExA bead vial that contains the goat anti-human IgG-coated PMMA beads (used in the signal-test experiment) to the bead vial holder of the instrument.
12. Adequate sample buffer (PBS/NaN_3, 4 l) and secondary labelled Ab (Cy5-conjugated, goat anti-human Fcγ fragment-specific IgG, 45 ml) are taken and connected to their respective lines.
13. Using a KinExA 3200 instrument, open "Standard K_d program" and choose the "Affinity, whole cell" as the experiment type.

14. Under "Timing set up" tab, choose "Hard bead handling" from "Edit" menu (*see* **Note 14**). In the same tab, set the volume of lines 1–13 to run to 3,500 μl and the flow rate to 0.25 ml/min (*see* **Note 15**). In the cells/ml column, fill in the final cell concentration in each tube as millions/ml. Set the injection volume to run 1 ml at 0.25 ml/min. Add 10 pM in the "binding site concentration" window.
15. In the instrument tab of the program file, set the "Number of cycles" to run as 3 (*see* **Note 16**).
16. Start to run the samples (*see* **Note 17**).
17. After the run is over, analysis can be done as explained in K_d analysis.

3.2.2 Experiment 2: High mAb Concentration (See **Note 18***)*

1. Prepare 4 ml of cell culture media containing 0.05 % NaN_3 in a 5 ml Falcon tube (tube 1).
2. Prepare 4 ml of cells expressing the target at a concentration of 50 million cells/ml in the cell culture media with 0.05 % NaN_3 (Tube 2).
3. Dispense 2 ml of cell culture medium with 0.05 % NaN_3 in tubes 3–13.
4. Mix the tube 2 thoroughly, and transfer half of the volume (2 ml) to the next tube. Repeat this process through tube 12, and discard the final 2 ml taken from tube 12. Tube 13 contains no cells and serves as the 100 % signal sample.
5. Prepare 26 ml of cell culture media with 0.05 % NaN_3 and 200 pM binding sites of mAb.
6. Add 2 ml of the 200 pM binding site solution to each tube with the exception of tube 1.
7. Allow the binding to reach equilibrium by rotating the tubes at 5 rpm for 36 h at 37 °C.
8. At the end of the incubation, centrifuge the cells at 2,400 rpm (or 400 × *g*) for 4 min.
9. Transfer the supernatant containing the unbound mAb into fresh tubes, being careful not to disturb the pelleted cells.
10. Place the 13 sample lines from the KinExA instrument in the 13 sample tubes containing the supernatants.
11. Connect the KinExA bead vial that contains the goat anti-human IgG-coated PMMA beads (used in the signal test experiment) to the bead vial holder of the instrument.
12. Adequate sample buffer (PBS/NaN_3, 4 l) and secondary labelled Ab (Cy5-conjugated, goat anti-human Fcγ fragment-specific IgG, 45 ml) are taken and connected to their respective lines.

13. Using a KinExA 3200 instrument, open "Standard K_d program" and choose the "Affinity, whole cell" as the experiment type.
14. Under "Timing set up" tab, choose "Hard bead handling" from "Edit" menu. In the same tab, set the volume of lines 1–13 to run to 500 μl and the flow rate to 0.25 ml/min. In the cells/ml column, fill in the final cell concentration in each tube as millions/ml. Set the injection volume to run 1 ml at 0.25 ml/min. Add 100 pM in the "binding site concentration" window.
15. In the instrument tab of the program set the "Number of cycles" to 3.
16. Start to run the samples.
17. After the run is over, analysis can be done as explained in K_d analysis.

3.2.3 K_d Analysis

1. For each experiment, under "Binding Signals" tab, click "ignore" run 1, 14, and 27 which are the signals obtained when there is no cell or mAb in the sample (blank signal) (*see* **Note 19**).
2. Go to "Binding Curve" tab and click "Analyse" button in "Tools" menu.
3. The software will calculate the K_d, expression level (EL) of the target, ratio (binding site concentration of mAb over K_d), non-specific binding (NSB), and % error for the experiment that will be displayed in the "Binding Curve" tab.
4. Under "Error Curve(s)" tab, the calculated 95 % confidence interval for the K_d and EL will be displayed.

3.2.4 N-Curve Analysis

1. Open "N-curve analysis" under "Tools" menu in the KinExA Pro software.
2. From "Edit" menu, add Experiment 1 and Experiment 2 for analysis.
3. Choose "Affinity, Whole cell" in the "Analysis method" window.
4. Go to "Binding Curve" tab and click "Analyze n-curve" button in "Tools" menu.
5. The software will calculate the overall K_d, expression level (EL) of the target, and % error by fitting to data from both experiments, and these will be displayed in the "Binding Curve" tab (Fig. 1a).
6. Under "Error Curve(s)" tab, the calculated 95 % confidence interval from both experiments, for the K_d and EL and their "Best Fit" curves, will be displayed (Fig. 1b).

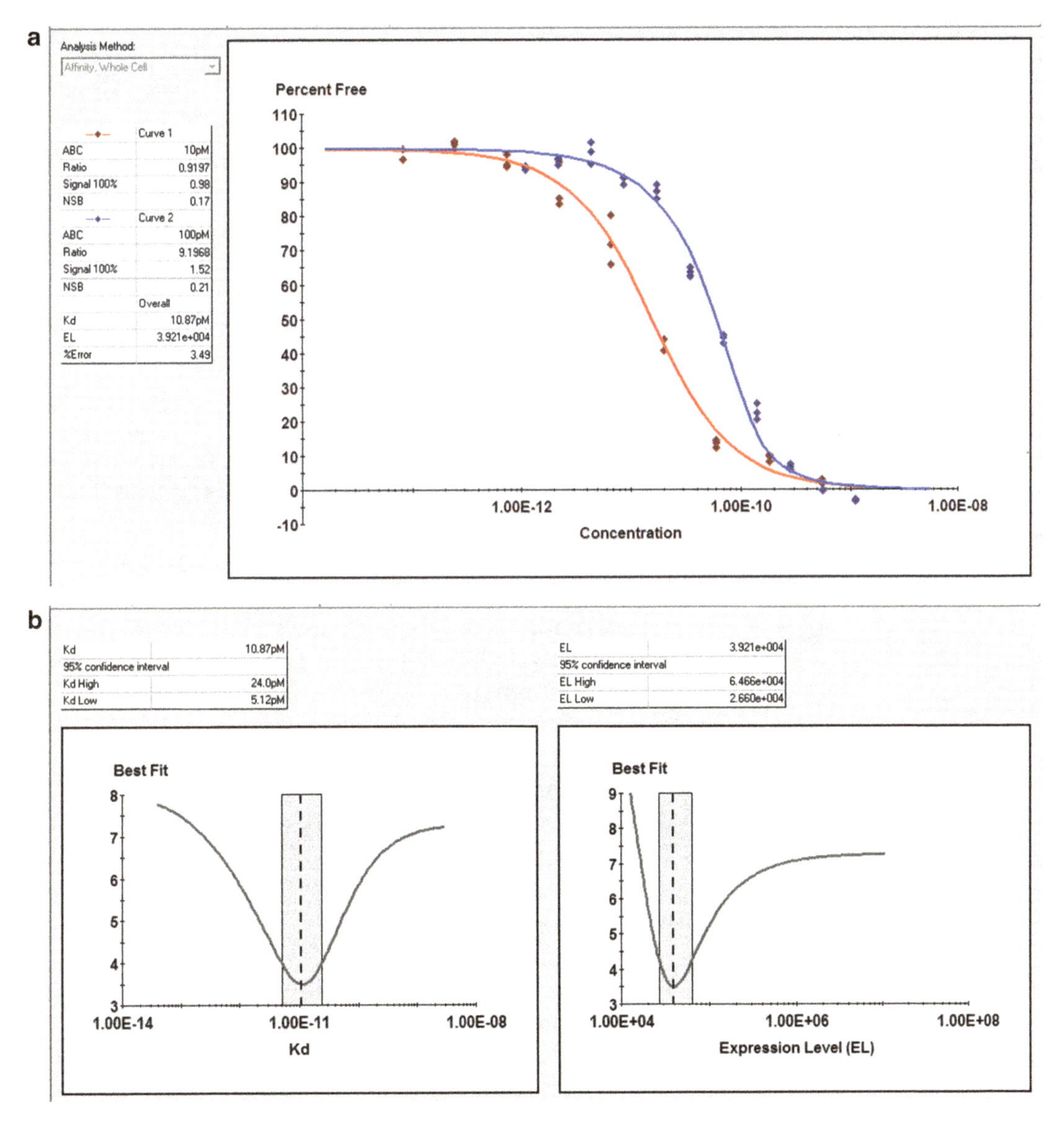

Fig. 1 Affinity (K_d) determination and antigen expression level measurement for cell membrane-expressed antigens and an antibody. A final concentration of 25×10^6/ml of antigen-expressing cells were serially diluted and incubated with 10 pM or 100 pM final active binding site concentration of mAb in the presence of 0.05 % NaN_3 and allowed to equilibrate. The unbound free mAb left in the supernatant was measured as explained in the text. (**a**) Plotting of the measured % free mAb (Left hand side curve for 10 pM, right hand side curve for 100 pM) against the calculated antigen concentration in N-curve analysis using "affinity, whole cell" method. (**b**) Plotting the residual fit error for non-optimal values for K_d (or the antigen expression level) with re-optimized values of the antigen expression level (or K_d value). The calculated K_d is equal to 10.87 pM with a 95 % confidence interval of 5.12 pM (K_d low) and 21.0 pM (K_d high). The expression level (EL) of antigen calculated to be 3.921e4 with a 95 % confidence interval of 2.660e4 (EL low) and 6.466e4 (EL high)

4 Notes

1. Since the buffers are being used for maintaining cells, it is important to keep them sterile.
2. We coat proteins via passive adsorption using PMMA beads. PMMA beads can also be coated at 4 °C. 300 mg of PMMA beads are sufficient for running two experiments. We also chemically couple proteins to Azlactone (Sapidyne Instruments, Boise, ID, USA), NHS-activated Sepharose 4 Fast Flow, and CNBr-activated Sepharose 4B beads (Amersham Biosciences, Inc., Quebec, QC, Canada) as per the manufacturer's recommended protocol.
3. The blocked beads can be stored at 4 °C until use. Adsorption-coated PMMA beads can be stored for a week. The chemically coupled beads can be stored longer without losing their binding activity.
4. We store the Cy5-labeled detection antibodies at 1 mg/ml in the presence of glycerol at −20 °C. Diluted working solutions of detection antibodies should be used immediately. We also use Cy5-conjugated, affinity-purified goat anti-human (H+L)-specific (minimum cross-reactive to other species) IgG (Jackson ImmunoResearch Laboratories, Inc., West Grove, PA, USA).

   Cy5 label has been discontinued recently. Alexa-labelled detection antibodies are now available and can be used for the same purpose at a working concentration of 250–500 ng/ml.
5. If the affinity is in single-digit pM range, cells expressing 50,000–100,000 receptors/cell can be effectively used in the experiment. NaN_3 is used to prevent internalization of the receptor during incubation with the antibody. We regularly check for receptor internalization, if any, before starting the full experiment by incubating the antibody with and without NaN_3 and measure the maximum binding intensity by FACS. If addition of NaN_3 is necessary to prevent internalization, we check the cell health by Trypan blue exclusion method.
6. We choose to run equilibrium experiments at two different concentrations of Abs, one equal to or below the anticipated K_d and another about tenfold higher than K_d. In this example, we set our equilibrium experiments at 10 and 100 pM binding site concentrations of mAb, and hence a signal test at these concentrations was determined.
7. In our hands, running about 5 ml of 10 pM mAb and about 1 ml of 100 pM mAb gives a signal >0.5 V for PMMA beads coated with goat anti-Hu IgG. Even though we aim at getting a signal of >0.5 V above the baseline, in some cases we have made successful measurements with smaller signals as low as 0.2 V above the baseline.

8. For cells having moderate expression of the receptor, about 10–20 million cells/ml generally bind completely the 100 pM of binding sites.
9. It is always useful to include the blank that contains only medium (no cells and no mAb). Even though this value will not be included in the K_d calculation later, it will give a good idea about the stickiness, if any, of the secondary labelled Ab to the flow cell and the beads.
10. The concentration of cells in tube 2 (50 million cells/ml) is determined from the cell range finding test. If a final cell concentration of 25 million cells/ml bound the Ab (final binding site concentration of 100 pM) completely, we take twice that cell concentration, since it will be diluted 2× after addition of the antibody solution. Here we performed 1:2 titrations of the cells. If a larger range is preferred, 1:3 dilutions can also be performed.
11. When added an equal volume of 2× binding site concentration of Ab to cells, the resulting final Ab concentration will be 1× (in this case 10 pM binding sites).
12. Tubes 2–12 will contain titrating concentrations of cells with a final Ab-binding site concentration of 10 pM. Tube 13 will have only Ab without any cells. The signal obtained for this will be used as 100 % free Ab in K_d calculation.
13. For high-affinity interactions (K_d < 100 pM), the experiment with the low Ab concentration, we should allow enough time for the reaction to reach equilibrium. We normally incubate for about 12 h for an expected K_d of 1 nM, 24 h for an expected K_d of 100 pM–1 nM, 36 h for an expected K_d range of 10–100 pM, and about 72 h for an expected K_d of <5 pM. These also depend on the association rate of the interaction. When the K_d and on-rate are unknown, then it is important to run a preliminary time course experiment to determine the time at which the equilibrium is reached.
14. For PMMA beads, "Hard bead handling" should be chosen. We also use proteins coupled to sepharose beads. For sepharose beads, "Soft bead handling" should be chosen.
15. The volume to run is determined from the signal test experiment. We normally run the sample and the labelled Ab at a flow rate of 0.25 ml/min. Depending on the experiment, these flow rates can be adjusted from 0.25 to 5 ml/min and independently from each other.
16. We normally run the samples and measure the signal in triplicates. The machine will run the samples 1–13 sequentially and then run two more cycles.
17. When we start the experiment, the program will prompt the user to save it in a folder. All data will be saved in this folder for subsequent K_d analysis.

18. For measuring antibody affinities to cell surface antigens, it is important to run two experiments by taking the Ab at two different concentrations (tenfold differences in concentration) and use it in n-curve analysis. In KinExA analysis, one concentration is used as the reference (either antigen or Ab), and the other is optimized in the analysis. Since the antigen concentration on the surface of the cell is unknown we will use Ab concentration as the known parameter. The best way to ensure getting both the antigen concentration and K_d is to measure at two different Ab concentrations (of the same lot) and use the n-curve analysis. It is also important to use the exact same batch of cells and to perform the two experiments at the same time since the expression level of antigen and hence its effective concentration can drift.
19. The blank signal value is not necessary for the software to calculate the K_d and hence should be ignored from the analysis.

Acknowledgements

The funding for this work was provided by the Amgen internal funding.

References

1. Veronese ML, O'Dwyer PJ (2004) Monoclonal antibodies in the treatment of colorectal cancer. Eur J Cancer 40:1292–1301
2. Adams GP, Weiner LM (2005) Monoclonal antibody therapy of cancer. Nat Biotechnol 23:1147–1157
3. Zuckier LS, Berkowitz EZ, Sattenberg RJ, Zhao QH, Deng HF, Scharff MD (2000) Influence of affinity and antigen density on antibody localization in a modifiable tumor targeting model. Cancer Res 60:7008–7013
4. Rathanaswami P, Roalstad S, Roskos L, Su QJ, Lackie S, Babcook J (2005) Demonstration of an in vivo generated sub-picomolar affinity fully human monoclonal antibody to interleukin-8. Biochem Biophys Res Commun 334:1004–1013
5. Andrew SM, Johnstone RW, Russell SM, McKenzie IF, Pietersz GA (1990) Comparison of in vitro cell binding characteristics of four monoclonal antibodies and their individual tumor localization properties in mice. Cancer Res 50:4423–4428
6. Velders MP, van Rhijn CM, Oskam E, Fleuren GJ, Warnaar SO, Litvinov SV (1998) The impact of antigen density and antibody affinity on antibody-dependent cellular cytotoxicity: relevance for immunotherapy of carcinomas. Br J Cancer 78:478–483
7. Booth PJ (2003) The trials and tribulations of membrane protein folding in vitro. Biochim Biophys Acta 1610:51–56
8. Tam SH, Sassoli PM, Jordan RE, Nakada MT (1998) Abciximab (ReoPro, chimeric 7E3 Fab) demonstrates equivalent affinity and functional blockade of glycoprotein IIb/IIIa and alpha(v) beta3 integrins. Circulation 98:1085–1091
9. Trikha M, Zhou Z, Nemeth JA, Chen Q, Sharp C, Emmell E, Giles-Komar J, Nakada MT (2004) CNTO 95, a fully human monoclonal antibody that inhibits alphav integrins, has antitumor and antiangiogenic activity in vivo. Int J Cancer 110:326–335
10. Debbia M, Lambin P (2004) Measurement of anti-D intrinsic affinity with unlabeled antibodies. Transfusion 44:399–406
11. Siiman O, Burshteyn A (2000) Cell surface receptor-antibody association constants and enumeration of receptor sites for monoclonal antibodies. Cytometry 40:316–326
12. Barr TA, Heath AW (2001) Functional activity of CD40 antibodies correlates to the position

of binding relative to CD154. Immunology 102:39–43
13. Benedict CA, MacKrell AJ, Anderson WF (1997) Determination of the binding affinity of an anti-CD34 single-chain antibody using a novel, flow cytometry based assay. J Immunol Methods 201:223–231
14. Klotz IM (1997) Ligand-Receptor Energetics: A guide for the perplexed. John Wiley and Sons, Inc., New York
15. Rathanaswami P, Babcook J, Gallo M (2008) High-affinity binding measurements of antibodies to cell-surface-expressed antigens. Anal Biochem 373:52–60
16. Xie L, Mark Jones R, Glass TR, Navoa R, Wang Y, Grace MJ (2005) Measurement of the functional affinity constant of a monoclonal antibody for cell surface receptors using kinetic exclusion fluorescence immunoassay. J Immunol Methods 304:1–14

Chapter 23

Determination of Antibody Structures

Robyn L. Stanfield

Abstract

While antibodies share a conserved structural framework, their complementarity-determining region loops are highly variable in size and sequence. Even more variable are the potential ways these loops can be used to interact with antigen. Thus, X-ray crystal structures of antibody Fab fragments and Fab–antigen complexes are critical for a detailed understanding of the antibody–antigen recognition process. This chapter describes the basic procedures necessary for the crystallization and structure determination of antibody Fab fragments by X-ray crystallography.

Key words Antibody, Antigen, Crystal, Structure, X-ray, Fab

1 Introduction

Antibodies and their fragments are becoming increasingly important as therapeutics for diseases such as cancer, Alzheimer's disease, and various immune disorders [1] and neutralizing antibodies targeting viruses such as HIV-1 and influenza are helping to define critical epitopes of concern for vaccine design efforts [2]. Less high profile but equally of interest are antibodies for use as purification agents and blotting reagents and to facilitate the crystallization of otherwise intractable proteins [3]. In order to better understand how antibodies interact with their target antigens, crystallization and high-resolution structure determination of the antibodies and their antigen complexes are necessary.

There are currently well over 1,000 structures of antibodies or their fragments in the Protein Data Bank (PDB) [4]. Structures have been determined for several intact antibodies and many Fabs and single-chain Fv (scFv) fragments, as well as light-chain dimers, heavy-chain-only antibodies from camelids and sharks, human V_HH domains, and Fc fragments. The abundance of models makes molecular replacement the phasing method of choice for crystal structure determination. While inter-domain flexibility needs to be taken into account during structure determination, the availability

Vincent Ossipow and Nicolas Fischer (eds.), *Monoclonal Antibodies: Methods and Protocols*, Methods in Molecular Biology, vol. 1131, DOI 10.1007/978-1-62703-992-5_23, © Springer Science+Business Media New York 2014

of fast computers and powerful software usually makes the process very straightforward. In this chapter we focus on the structure determination of Fab fragments; however, the techniques are equally applicable to other antibody fragments.

2 Materials

2.1 Crystallization and Sample Mounting

1. Pure Fab or Fab–antigen complex at a final concentration of 5–20 mg/ml (*see* **Note 1**) and approximately 0.5–1 μl of sample for each crystallization condition to be tested, i.e., 12–24 μl for one 24-well tray.
2. Sitting-drop crystallization plates and sealing tape (*see* **Note 2**).
3. Crystallization screens that can be purchased (*see* **Note 2**) or prepared in house using ultrapure water and analytical grade reagents.
4. Liquid nitrogen.
5. Centrifugal ultrafiltration unit (EMD Millipore, Darmstadt, Germany) for concentrating protein.
6. Cryoprotectants such as glycerol, ethylene glycol, methylpentanediol.

2.2 Experimental Equipment

1. Dewar (*see* **Note 2**).
2. Magnetic crystal mounting bases, cryoloops, and cryotools (cryotongs, magnetic wands) (*see* **Note 2**).
3. Stereomicroscope (*see* **Note 2**).

3 Methods

3.1 Crystallization

1. While several intact immunoglobulin structures have been determined [5–9] (Fig. 1), the flexibility between the Fab and Fc arms can make crystallization of these large molecules difficult. Thus, Fab fragments are usually prepared by cleaving immunoglobulin with pepsin, papain [10–12], IdeS [13] or LysC [14], or Fab can be produced recombinantly. Fab should be >95 % pure as estimated by SDS gel chromatography (*see* **Note 1**).
2. To prepare a complex with a large antigen (protein or large DNA or RNA fragment), the antigen should be mixed with Fab so that the smaller of the two components is in excess, and the complex purified by size exclusion chromatography and then concentrated in a centrifugal ultrafiltration unit with an appropriate molecular weight cutoff is somewhere between approximately 5–20 mg/ml (*see* **Note 3**).

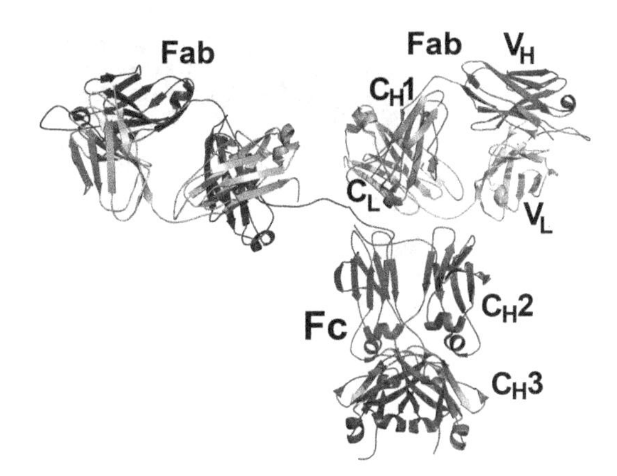

Fig. 1 Immunoglobulin G architecture. An intact immunoglobulin is shown, with light and heavy chains colored in *light green* and *light purple* (light chains) and *dark green* and *dark purple* (heavy chains). The two Fab domains are attached to the Fc domain by flexible linkers, and there is further domain flexibility between the Fab variable (V_L–V_H) and constant (C_L–C_H) domains that needs to be considered when determining the crystal structure

3. To prepare a complex with a small antigen (peptide, small DNA or RNA fragment, or small molecule hapten) the antigen should be added in excess (usually 5:1–10:1 M excess of antigen) to pre-concentrated Fab and that material directly screened for crystallization. Very small haptens (*see* **Note 4**) can also sometimes be soaked into preexisting crystals of the unliganded Fab, but peptidic antigens (or anything larger) are usually too big to soak into pre-formed crystals and thus need to be co-crystallized with the Fab.
4. Crystallization can be carried out in many different formats (sitting drops, hanging drops, under oil, etc.), manually or with robotic dispensing [15]. For simplicity we describe here the use of a 24-well vapor diffusion sitting drop plate (Fig. 2a) with manual reagent and protein dispensing. Prepare or purchase crystallization reagents. Antibody Fab fragments and their complexes with antigen crystallize under many different conditions, and primary crystallization screens are no different from those used for other soluble proteins (*see* **Note 5**). Add 0.5–1.0 ml crystallization screening reagent to each well. When all the well solutions have been added, protein drops can be dispensed. Using a manual 0.1–2 μl pipet, 1 μl drops can easily be formed, and smaller drops are possible with practice (*see* **Note 6**). Choose your drop size, dispense protein drops across ½ the plate, and then add the same amount of well solution to each protein droplet. The well solution can be mixed with or layered over the drop. Seal the completed wells with optically clear sealing tape (Fig. 2b). Then carry out the same

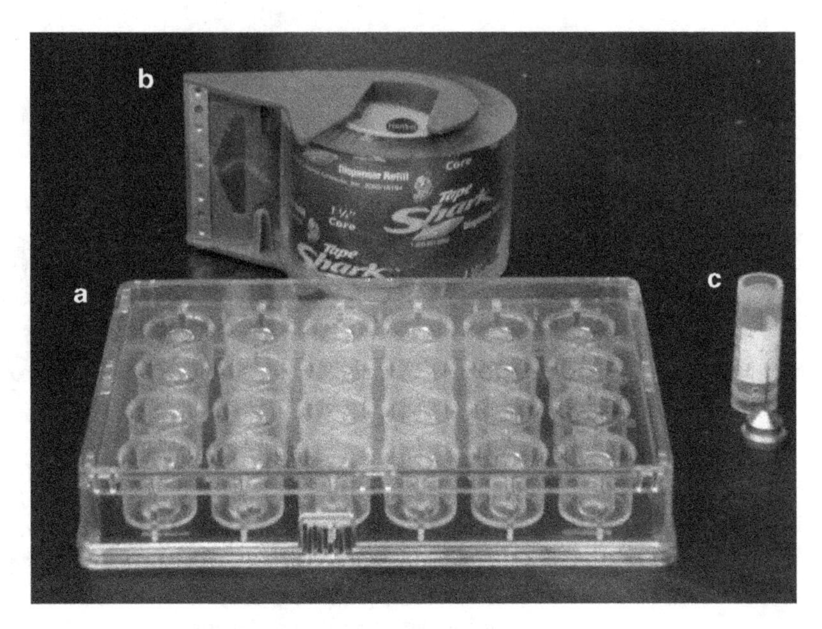

Fig. 2 Crystallization supplies. (**a**) 24-well Cryschem sitting drop plate, (**b**) optically clear sealing tape, (**c**) a magnetic cryopin and storage vial

procedure for the second ½ of the plate. Work quickly to minimize evaporation of the sample droplets.

5. When plates have been carefully sealed with tape, examine the wells under a stereomicroscope and make note of any precipitation or foreign materials (dust, fibers, etc.) in the drops, or if you have access to an automated imager take images of each drop directly after setup. Then store the crystallization plates in a constant-temperature incubator. If you have sufficient sample, prepare identical plates for incubation at two or more temperatures (for example 20 and 4 °C). If you do not have access to a constant temperature incubator, place plates in a Styrofoam cooler, on the bench, or in the cold room, away from any disturbances.
6. After one or more days, and every few days thereafter for a period of at least 1 month, inspect each plate under a microscope. Any conditions resulting in crystals can be optimized by setting up a fine screen of similar conditions, where pH and precipitant concentrations are varied in small increments around the initial condition. After precipitant concentrations and pH are optimized, further optimization can be carried out by testing chemical additives (*see* **Note 2**). Techniques such as streak-seeding [16] with a cat whisker or microseed matrix screening [17] can also be used to nucleate crystals in similar or totally different conditions. If no crystals are obtained, conditions resulting in granular precipitate can also be fine-screened, or

additional crystallization screens can be tested (*see* **Note 7**). If after extensive crystallization screening no crystals are obtained, changes to the protein sample may be necessary (*see* **Note 8**).

3.2 Crystal Mounting

1. When considering mounting and cryocooling, Fab crystals are no different from other macromolecular crystals. The crystal will be mounted in a cryoloop attached to a magnetic base (Fig. 2c). The total height of the loop/base assembly depends on the goniometer geometry, and this detailed information will be available from the X-ray facility where you plan to collect data.
2. Before crystal mounting, prepare a cryoprotectant solution, usually the well solution augmented with ~25 % glycerol, ethylene glycol, or methylpentanediol (*see* **Note 9**). Crystals should be mounted in a cryoloop (*see* **Note 2**) by placing the loop under the crystal and slowly lifting the crystal from the crystallization drop, ending with the crystal suspended by surface tension in the liquid within the loop. The crystal should then be dipped briefly into a cryoprotecting solution and then lifted from that solution and flash-cooled by quickly plunging into liquid nitrogen. Alternately, the cryosolution can be layered over the crystallization drop and the crystal pulled slowly through this cryoprotecting layer during the mounting procedure.
3. Transfer mounted crystal into CryoTongs (Hampton) under liquid nitrogen, then place crystal on goniometer under nitrogen gas cryostream (*see* **Note 10**).

3.3 Data Collection and Processing

1. As for crystal mounting, there are no special rules or tricks for collecting data from a Fab or a Fab–antigen complex crystal as compared with other macromolecular crystals. As Fabs crystallize in many different crystallization conditions, the resulting crystals also belong to many different space groups with different unit cell dimensions. The highest resolution data will likely be collected at a synchrotron light source, where the data collection software will be specifically tailored for the beamline and/or detector, so that training in the use of that software will of necessity be provided by the beamline staff (*see* **Note 11**) who can also suggest strategies to achieve maximum data quality. Data processing can be carried out with many different programs, such as HKL-2000 [18], XDS [19], Mosflm [20], or d*TREK [21] (*see* **Note 11**).
2. Resolution of diffraction from Fab and Fab–antigen samples is variable, with some crystals diffracting to very high resolution and others diffracting poorly. The highest resolution Fab structure currently in the PDB is that of HyHel-10 in complex with

lysozyme (3D9A) [22] that has been refined to 1.2 Å resolution. If crystals diffract poorly and need to be optimized, first make sure that the starting protein sample is extremely pure and homogeneous. If the sample is of an antibody–antigen complex, alterations in the antigen may drastically affect the quality of the resulting crystals. For example, different lengths of peptide antigens can alter the resulting crystals, or reducing glycan heterogeneity of a glycosylated antigen can improve crystal quality. Sometimes Fabs cleaved from IgG using different enzymes or Fab produced recombinantly may result in different quality crystals.

3.4 Structure Determination and Refinement

1. The method of choice for structure determination of Fab or Fab–antigen complexes is molecular replacement (MR) [23]. There are well over 1,000 Fab structures deposited in the PDB representing a multitude of suitable MR models. The MR model can be either used as the intact Fab or split into the variable (V_L–V_H) and constant (C_L–C_H1) domains (Fig. 1). There is a great deal of variability in the angle between the variable and constant domains (the elbow angle) [24], and multiple Fabs in the asymmetric unit may even have different elbow angles. Thus, when using an intact Fab as the MR model, many models with different elbow angles must be tested. To circumvent this problem, we usually select several Fab models with the highest sequence homology to the target Fab, split those models into separate variable and constant domains, and carry out MR using Phaser [25], AMoRe [26], EPMR [27], or any other available MR software. Doing the MR search with separate variable and constant domains is also extremely useful for validating the MR process, as a correct solution will pack forming biologically correct, intact Fab molecules. However, difficult structures such as those with multiple Fabs in the asymmetric unit may require numerous, iterative MR searches with a large number of different, intact Fab models.
2. After an initial solution is obtained, carry out rigid body refinement, letting the individual V_L, V_H, C_L, and C_H1 subunits move as rigid bodies, as there can be some flexibility in the arrangement of these domains within the variable and constant regions.
3. If a small antigen, such as a hapten or a peptide, is bound to Fab, the electron density for that hapten should be readily visible after phasing with just the Fab model. If a larger antigen is bound, and a model is available for that antigen, MR should also be carried out with the antigen model. If no appropriate model is available, depending on the size of the antigen, it may be possible to iteratively build the antigen model from the electron density obtained with phasing from just the Fab portion of the structure; however, a very large antigen with an unknown

structure may require additional phasing information from heavy atoms. In this case, the partial phase information from MR with the Fab portion of the structure can still be very useful in the structure determination process, and MR phases can be combined with heavy atom phases. If no clear density for bound ligand is apparent, new crystallization conditions may be necessary (for example with lower salt concentration), or if the antigen has only weak affinity for Fab, a larger ratio of antigen to Fab may be required.

4. After initial structure determination by MR or heavy atom methods, the Fab or the Fab/antigen model can then be rebuilt and refined in iterative cycles using model building software such as Coot [28] and refinement programs such as REFMAC5 [29], PHENIX [30], CNS [31], and BUSTER-TNT [32]. As there are so many high-resolution Fab structures available in the PDB, it is useful to superimpose several of these with your Fab model and refer to them during the rebuilding process, especially to help interpret flexible loop regions. The complementarity-determining region (CDR) loops are the most variable part of the Fab and will require special attention during rebuilding, especially CDR H3 that does not have a conserved structure except at its base [33].

3.5 Structure Validation and Analysis

1. As for any macromolecular structure, the final model should be carefully validated [34], to ensure good geometry, good fit of calculated to observed data, and strong, clear, well-ordered density for any antigen included in the model. Commonly used validation tools are available for use in the Validation Server from the Joint Center for Structural Genomics (http://smb.slac.stanford.edu/jcsg/QC/).
2. The final Fab model should be numbered by a standardized numbering system such as the Kabat or the IMGT systems [35, 36] (*see* **Note 12**). Specific residues from a Fab model numbered in this standard way are easily compared to those from the many other structures available. Standard calculations specific for Fabs that can be carried out are the calculation of the Fab elbow angle (http://proteinmodel.org/AS2TS/RBOW/index.html) [24] and analysis of the CDR loop conformations to see if they fall into one of the known canonical structure classes [37–40].

4 Notes

1. Fabs or other antibody fragments should be purified by affinity chromatography, followed by size exclusion chromatography, and ion exchange chromatography if necessary. Do not purify

samples in phosphate-containing buffers as the phosphate can form insoluble complexes (salt crystals) with divalent cations.

2. Crystal plates, sealing tape, crystallization and additive screens, and mounts and tools for crystal mounting and cryocooling can be purchased from many sources, with some of the major suppliers listed here: Hampton Research, Aliso Viejo, CA; Qiagen, Germantown, MD; Emerald Biosystems, Bainbridge Island, WA; Molecular Dimensions, Altamonte Springs, FL; Microlytic, Burlington, MA; Jena Bioscience, Jena, Germany; and MiTeGen, Ithaca, NY. The suppliers also provide online crystallization screen compositions, and many have excellent online guides relating to crystal growth and cryoprotection. Several of the suppliers also carry stereomicroscopes for protein crystallography or see major microscope manufacturers including Meiji Techno, Santa Clara, CA; Leica Microsystems, Buffalo Grove, IL; Olympus America, Center Valley, PA; and Carl Zeiss, Thornwood, NY.

3. If a Fab–antigen complex does not remain intact during size exclusion chromatography, it may be difficult to crystallize the complex, but crystallization can be attempted by simply mixing the two components in approximate 1:1 ratio before crystallization trials. Appropriate protein concentration values for crystallization can vary widely from protein to protein but usually range from around 5 to 20 mg/ml. Commercial kits can be used to determine if the sample is concentrated enough for crystallization (PCT Pre-Crystallization Test, Hampton Research, Aliso Viejo, CA), or we often use conditions D1–D6 of Stura Footprint Screen I [41] (Molecular Dimensions, Altamonte Springs, FL) to evaluate the sample. If protein precipitate appears after 24 h in most of these six Footprint conditions the protein is probably sufficiently concentrated. However, if no precipitation appears, the sample should be concentrated further.

4. Small haptens or peptides that are not water soluble may need to be dissolved in DMSO, DMF, or other organic solvent. Avoid exposing Fab to high concentrations of the organic solvents, and keep the final organic solvent concentration low (<5 % (v/v)) in the final complex with Fab by adding a very small amount of highly concentrated (50–100 mg/ml) hapten or peptide in organic solvent to buffer and then adding that diluted hapten/organic solvent/buffer mixture to the Fab sample.

5. Good screens to start with are the Stura Footprint Screens [12, 41], Hampton Peg-Ion screen, or the JCSG Core Suites. If trying to crystallize a Fab complex with antigen, avoid high-salt conditions as these may disrupt the complex formation.

6. Small drops are often used during the initial screening process to minimize the amount of protein sample used, but larger

drops can be used once productive crystallization conditions are obtained. Pipetting very small volumes requires some practice. Before attempting with precious protein samples, practice pipetting water drops onto parafilm until the drops are well formed and of uniform size. For someone setting up his or her first crystallization plate, first try crystallizing an easily obtainable and crystallizable protein such as hen egg white lysozyme (several crystallization recipes are available at hamptonresearch.com). When refining crystallization conditions, crystal size and quality may vary with different drop sizes, and also with the sort of crystallization setup used, for example sitting drop versus hanging drop.

7. Large and well-formed crystals are easy to identify, but if you obtain only small or no crystals, help and advice from someone experienced in crystal growth are invaluable in identifying microcrystals or precipitate suitable for optimization.
8. If you cannot obtain crystals of a Fab or a Fab–antigen complex, try changing the way the Fab was prepared (i.e., try cleavage with a different enzyme, or try making the Fab recombinantly). For a peptide antigen, try longer or shorter peptides. For a protein antigen, try with and without tags, cleavage of flexible loops, deglycosylation, or reductive methylation of lysines [42].
9. First test your crystals for diffraction at room temperature, by mounting in a glass or a quartz capillary or mounting in a MicroRT mount (Mitegen). By comparing the room-temperature diffraction to that obtained at cryogenic temperatures, you can see if cryocooling has damaged the crystal in any way. Then test crystals with several different cryosolutions made with different cryoprotecting compounds (for example glycerol or ethylene glycol), as different compounds can give very different results, depending on the crystal. Choose a cryosolution that freezes clear, that does not give ice diffraction when exposed to X-rays, and that does not adversely affect the diffraction properties of your crystal. In addition to commonly used glycerol, ethylene glycol, and methylpentanediol, crystals can be cryoprotected with low-molecular-weight polyethylene glycols, some sugars, oils, and "cryosalts" [43] such as the highly soluble sodium malonate. For a good starting point, this list of glycerol concentrations necessary to cryoprotect some common crystallization conditions is very helpful [44] as is this excellent guide [45]. Some crystallization conditions are suitable as cryoprotectants with no modification, such as those already containing a cryoprotecting chemical.
10. Alternately, the crystal can be placed directly on the goniometer head in the path of the nitrogen gas cryostream for cryocooling, bypassing the liquid nitrogen step. Crystals to be sent to the synchrotron can be stored by placing in cryovials (Fig. 2c)

and then into cryocanes in liquid nitrogen storage dewars. The crystals can be shipped by overnight delivery in dry shippers (Taylor-Wharton) at cryogenic temperatures.

11. Many excellent guides to data collection are available [46], and many different software packages are available for data processing, structure determination, refinement, and analysis [47]. Synchrotron beamlines for macromolecular crystallography will have many of these packages installed for their users, so that what you use will likely depend on what is installed at the beamline or in your home laboratory.

12. While it is extremely helpful to renumber your Fab model by the Kabat convention for structure analysis and publication, many of the available refinement packages do not accommodate the special insertions and deletions used in the numbering system and will distort the geometry of the model at those positions. Thus, use sequential numbering for your model during all phases of refinement, and only renumber using the Kabat system when refinement is complete. Any MR model from the PDB that used Kabat numbering should be renumbered sequentially before starting refinement. Kabat numbering for the variable regions (V_L and V_H) can be easily implemented with the Abnum server (http://www.bioinf.org.uk/abs/abnum/) [48].

Acknowledgements

This is manuscript number 21975 from the Scripps Research Institute.

References

1. Nelson AL, Dhimolea E, Reichert JM (2010) Development trends for human monoclonal antibody therapeutics. Nat Rev Drug Discov 9:767–774
2. Burton DR, Poignard P, Stanfield RL et al (2012) Broadly neutralizing antibodies present new prospects to counter highly antigenically diverse viruses. Science 337:183–186
3. Griffin L, Lawson A (2011) Antibody fragments as tools in crystallography. Clin Exp Immunol 165:285–291
4. Bernstein FC, Koetzle TF, Williams GJ et al (1977) The Protein Data Bank: a computer-based archival file for macromolecular structures. J Mol Biol 112:535–542
5. Marquart M, Deisenhofer J, Huber R et al (1980) Crystallographic refinement and atomic models of the intact immunoglobulin molecule Kol and its antigen-binding fragment at 3.0 Å and 1.9 Å resolution. J Mol Biol 141:369–391
6. Harris LJ, Larson SB, Hasel KW et al (1992) The three-dimensional structure of an intact monoclonal antibody for canine lymphoma. Nature 360:369–372
7. Harris LJ, Larson SB, Hasel KW et al (1997) Refined structure of an intact IgG2a monoclonal antibody. Biochemistry 36:1581–1597
8. Harris LJ, Skaletsky E, McPherson A (1998) Crystallographic structure of an intact IgG1 monoclonal antibody. J Mol Biol 275:861–872
9. Saphire EO, Parren PW, Pantophlet R et al (2001) Crystal structure of a neutralizing human IGG against HIV-1: a template for vaccine design. Science 293:1155–1159
10. Wilson IA, Rini JM, Fremont DH et al (1991) X-ray crystallographic analysis of free and

antigen-complexed Fab fragments to investigate structural basis of immune recognition. Methods Enzymol 203:153–176
11. Smith TJ (1993) Purification of mouse antibodies and Fab fragments. Methods Cell Biol 37:75–93
12. Stura EA, Fieser GG, Wilson IA (1993) Crystallization of antibodies and antibody-antigen complexes. Immunomethods 3: 164–179
13. Wenig K, Chatwell L, von Pawel-Rammingen U et al (2004) Structure of the streptococcal endopeptidase IdeS, a cysteine proteinase with strict specificity for IgG. Proc Natl Acad Sci U S A 101:17371–17376
14. Ofek G, Tang M, Sambor A et al (2004) Structure and mechanistic analysis of the anti-human immunodeficiency virus type 1 antibody 2 F5 in complex with its gp41 epitope. J Virol 78:10724–10737
15. Chayen NE, Saridakis E (2008) Protein crystallization: from purified protein to diffraction-quality crystal. Nat Methods 5:147–153
16. Stura EA, Wilson IA (1991) The streak seeding technique in protein crystallization. J Cryst Growth 110:270–282
17. Ireton GC, Stoddard BL (2004) Microseed matrix screening to improve crystals of yeast cytosine deaminase. Acta Crystallogr D Biol Crystallogr 60:601–605
18. Otwinowski Z, Minor W (1997) Processing of X-ray diffraction data collected in oscillation mode. Methods Enzymol 276A:307–326
19. Kabsch W (2010) XDS. Acta Crystallogr D Biol Crystallogr 66:125–132
20. Leslie AGW, Powell HR (2007) Processing diffraction data with Mosflm. In: Read RJ, Sussman JL (eds) Evolving methods for macromolecular crystallography, vol 245, Springer, Netherlands, pp 41–51
21. Pflugrath JW (1999) The finer things in X-ray diffraction data collection. Acta Crystallogr D Biol Crystallogr 55:1718–1725
22. Acchione M, Lipschultz CA, DeSantis ME et al (2009) Light chain somatic mutations change thermodynamics of binding and water coordination in the HyHEL-10 family of antibodies. Mol Immunol 47:457–464
23. Rossmann MG (1972) The molecular replacement method. Gordon & Breach, New York
24. Stanfield RL, Zemla A, Wilson IA et al (2006) Antibody elbow angles are influenced by their light chain class. J Mol Biol 357:1566–1574
25. McCoy AJ, Grosse-Kunstleve RW, Adams PD et al (2007) Phaser crystallographic software. J Appl Crystallogr 40:658–674
26. Navaza J (1994) AMoRe: an automated package for molecular replacement. Acta Crystallogr A 50:157–163
27. Kissinger CR, Gehlhaar DK, Fogel DB (1999) Rapid automated molecular replacement by evolutionary search. Acta Crystallogr D Biol Crystallogr 55:484–491
28. Emsley P, Cowtan K (2004) Coot: model-building tools for molecular graphics. Acta Crystallogr D Biol Crystallogr 60:2126–2132
29. Murshudov GN, Skubak P, Lebedev AA et al (2011) REFMAC5 for the refinement of macromolecular crystal structures. Acta Crystallogr D Biol Crystallogr 67:355–367
30. Afonine PV, Grosse-Kunstleve RW, Echols N et al (2012) Towards automated crystallographic structure refinement with phenix.refine. Acta Crystallogr D Biol Crystallogr 68:352–367
31. Brunger AT (2007) Version 1.2 of the crystallography and NMR system. Nat Protoc 2:2728–2733
32. Blanc E, Roversi P, Vonrhein C et al (2004) Refinement of severely incomplete structures with maximum likelihood in BUSTER-TNT. Acta Crystallogr D Biol Crystallogr 60:2210–2221
33. Shirai H, Kidera A, Nakamura H (1996) Structural classification of CDR-H3 in antibodies. FEBS Lett 399:1–8
34. Kleywegt GJ (2009) On vital aid: the why, what and how of validation. Acta Crystallogr D Biol Crystallogr 65:134–139
35. Kabat EA, Wu TT, Perry HM et al (1991) Sequences of proteins of immunological interest, vol 1, 5th edn. U.S. Department of Health and Human Services, Bethesda, MD
36. Lefranc MP, Giudicelli V, Ginestoux C et al (2009) IMGT, the international ImMunoGeneTics information system. Nucleic Acids Res 37:D1006–D1012
37. Al-Lazikani B, Lesk AM, Chothia C (1997) Standard conformations for the canonical structures of immunoglobulins. J Mol Biol 273:927–948
38. Chothia C, Lesk AM (1987) Canonical structures for the hypervariable regions of immunoglobulins. J Mol Biol 196:901–917
39. Martin AC, Thornton JM (1996) Structural families in loops of homologous proteins: automatic classification, modelling and application to antibodies. J Mol Biol 263:800–815
40. North B, Lehmann A, Dunbrack RL Jr (2011) A new clustering of antibody CDR loop conformations. J Mol Biol 406:228–256
41. Stura EA, Nemerow GR, Wilson IA (1992) Strategies in the crystallization of glycoproteins and protein complexes. J Cryst Growth 122:273–285
42. Rypniewski WR, Holden HM, Rayment I (1993) Structural consequences of reductive methylation of lysine residues in hen egg white lysozyme: an X-ray analysis at 1.8-Å resolution. Biochemistry 32:9851–9858

43. Rubinson KA, Ladner JE, Tordova M et al (2000) Cryosalts: suppression of ice formation in macromolecular crystallography. Acta Crystallogr D Biol Crystallogr 56:996–1001
44. Garman EF, Mitchell EP (1996) Glycerol concentrations required for cryoprotection of 50 typical protein crystallization conditions. J Appl Crystallogr 29:584–587
45. Garman E, Owen RL (2007) Cryo-crystallography of macromolecules: practice and optimization. Methods Mol Biol 364: 1–18
46. Garman E, Sweet RM (2007) X-ray data collection from macromolecular crystals. Methods Mol Biol 364:63–94
47. Jain D, Lamour V (2010) Computational tools in protein crystallography. Methods Mol Biol 673:129–156
48. Abhinandan KR, Martin AC (2008) Analysis and improvements to Kabat and structurally correct numbering of antibody variable domains. Mol Immunol 45:3832–3839

Chapter 24

Affinity Maturation of Monoclonal Antibodies by Multi-Site-Directed Mutagenesis

Hyung-Yong Kim, Alexander Stojadinovic, and Mina J. Izadjoo

Abstract

High-affinity antibodies are crucial for development of monoclonal antibody (MAb)-based therapeutics for human diseases. Many new detailed methods for affinity maturation have been developed to improve MAb qualities by site-directed mutagenesis, chain shuffling, and error-prone PCR. Site-directed mutagenesis on hotspots in variable heavy (VH) complementary-determining region (CDR) 3 is a commonly used method for improving therapeutic potency and efficacy of targeted MAbs. Strategies for affinity maturation via multi-site-directed mutagenesis in VH-CDR3 described here are for valuable technical tool in the armamentarium of immunologists for development of fast-performance MAbs. Our strategy includes (1) selection of targeted MAb, (2) replacement of certain amino acid residues (e.g., negative or neutral charge to positive amino acids) in VH-CDR3, and (3) determination of binding activity to a target antigen.

Key words Monoclonal antibodies, Affinity maturation, Complementary-determining region, Site-directed mutagenesis

1 Introduction

Monoclonal antibodies (MAbs) have been used extensively for targeted therapeutics. Currently, there are 27 therapeutic MAbs approved by the US Food and Drug Administration (FDA) [1]. The majority of these therapeutic MAbs are used for the treatment of human diseases, such as cancer and autoimmune syndromes. Significant research efforts are under way for developing high-affinity MAbs to a specific target antigen. In designing therapeutic MAbs, it is desirable to create super-active MAbs with high specificity and binding affinity to the targeted antigen. Generation of high-affinity MAbs is important for achieving desirable detection limits, dissociation half times, and therapeutic dosages and increasing the overall efficacy of these therapeutic agents. However, affinity maturation in vivo often fails to produce MAbs of targeted potency. Therefore, making further affinity improvement in vitro by site-directed mutagenesis, chain shuffling, and error-prone polymerase

Vincent Ossipow and Nicolas Fischer (eds.), *Monoclonal Antibodies: Methods and Protocols*, Methods in Molecular Biology, vol. 1131, DOI 10.1007/978-1-62703-992-5_24,

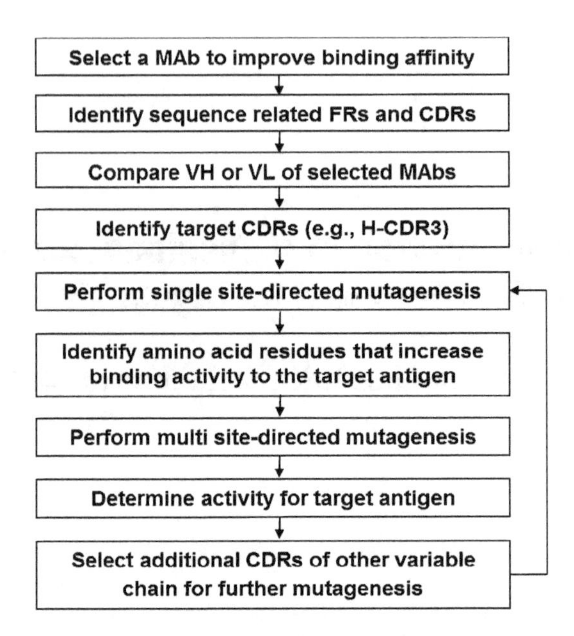

Fig. 1 Flow chart showing MAb affinity maturation using the site-directed mutagenesis. This is a commonly used procedure for generation of high-affinity MAbs to a target antigen via site-directed mutagenesis in target CDRs

chain reaction (PCR) is highly desirable. These technologies enable a rapid and powerful strategy to drive the next generation of MAb-based therapeutics for numerous important clinical applications [2, 3]. Affinity maturation strategies for therapeutic MAbs involve several steps including selection and sequencing of target complementary-determining regions (CDRs), mutation, MAb production in mammalian cells, and examination of MAb binding activity (Fig. 1).

Site-directed mutagenesis on hotspots in CDRs, mainly VH-CDR3, is promising and has often proven to be a successful approach [3, 4]. The essence of MAb maturation approach is to mimic the in vitro natural somatic hypermutation process that underlies affinity maturation of MAbs in vivo [5]. During somatic hypermutation, mutations in antibody genes encoding the variable domains are preferentially focused in "hotspot" regions. The consensus sequence of hotspots is the tetranucleotide, RGYW (in which R can be A or G, Y can be C or T, and W can be A or T). The AGY serine codons may also constitute hotspots [3, 6].

Replacement of certain amino acid residues in the target CDRs (e.g., negative or neutral charge to positive amino acids) can be carried out by four different strategies: (1) only the amino acid residues that differ between the first MAb and related MAb in the target region are modified; (2) only the amino acid residues that are the same between the first MAb and the related MAb in the target region are modified; (3) all of the amino acid residues in

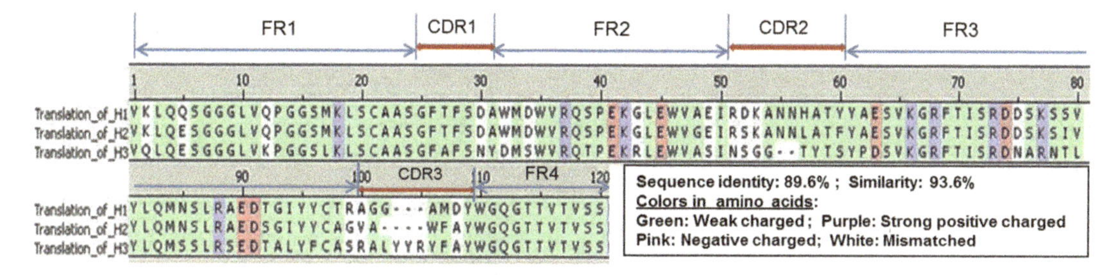

Fig. 2 Multiple sequence alignment of three related MAbs. To compare VHs of three related MAbs and identify target CDRs, sequences of each heavy chain (H1, H2, and H3) are identified by frame regions (FRs) and CDRs, indicated on the basis of charge difference of amino acids in each chain. Affinity maturation is created through the combination of mutations that change electrostatic charges (e.g., neutral or negative to positive) at specific amino acid positions in VH3-CDR3. Here we demonstrate the substitution of amino acids (Y101R_Y104K) in the VH3-CDR3 to improve MAb affinity. GeneBank access numbers: H1 (GU350787), H2 (GU350788), and H3 (GU350789). doi: 10.1371/journal.pone.0019867.g002

the target region are modified; and (4) scanning amino acid residues different from the wild-type sequence (alanine, threonine, proline, and glycine). Generally speaking, high-affinity MAb is obtained by the combination of single or multiple mutations on CDRs of either VH or VL chains. The target CDR mutations of the mutant clones should be confirmed each time by sequencing. In addition, the binding activity of the mutant MAb should be compared with the original MAb by enzyme-linked immunosorbent assay (ELISA), Western blot (WB) analysis, and potentially affinity and functional studies [7, 8].

Herein, we describe a method for improving MAb affinity using the QuikChange Lightning Multi Site-Directed Mutagenesis Kit (Agilent Technologies, La Jolla, CA, USA), focusing on the variable heavy-chain (VH3)-CDR3 domain genes of the targeted MAb (Fig. 2). This kit offers rapid and reliable method for site-directed mutagenesis of plasmid DNA for up to five different sites simultaneously using advanced high-fidelity enzyme technology; hence, this protocol can accelerate the process of affinity maturation of targeted MAb. Successful affinity maturation results in the generation of high-performance MAbs with increased and improved activity for a target antigen compared to the original MAb.

2 Materials

Please note that all disposables, plasmid DNAs, and reagents used for affinity improvement must be sterile. Molecular and culture procedures are conducted under a biosafety cabinet. Extreme care is taken to prevent cross contamination of mutant MAb clones.

2.1 Bacterial Cultures

1. NZY+broth: 10 g of NZ amine (casein hydrolysate), 5 g of yeast extract, 5 g of NaCl; add deionized water to a final volume of 1 l. Adjust to pH 7.5 using NaOH, autoclave, and add the following filer-sterilized supplements prior to use: 12.5 ml of 1 M $MgCl_2$, 12.5 ml of 1 M $MgSO_4$, 20 ml of 20 % (w/v) glucose (or 10 ml of 2 M glucose).
2. LB–ampicillin agar (per liter): 1 l of LB agar, autoclave, cool to 55 °C. Add 10 ml of 10 mg/ml filter-sterilized ampicillin. Pour into petri dishes (~25 ml/10 cm plate).
3. 10 mM Isopropyl-1-thio-β-D galactopyranoside (IPTG) (per 10 ml): 24 mg of IPTG, 10 ml of sterile deionized water (store at −20 °C). Spread 100 μl per LB agar plate. 2 % X-Gal (per 10 ml): 0.2 g of 5-bromo-4-chloro-3-indolyl-β-D galactopyranoside (X-Gal), 10 ml of dimethylformamide (store at −20 °C). Spread 100 μl per LB agar plate. For blue-white color screening, add 80 μg/ml of 5-bromo-4-chloro-3-indolyl-β-D-galactopyranoside (X-Gal), 20 mM IPTG, and the appropriate antibiotic to the LB agar. Do not mix the IPTG and the X-gal before pipetting them onto the plates because these chemicals may precipitate.

2.2 Site-Directed Mutagenesis

1. Plasmid vectors (50 ng/μl) harboring the antibody heavy- and light-chain gene to be mutated.
2. QuikChange Lightning Multi Site-Directed Mutagenesis Kit (Agilent Technologies, La Jolla, CA, USA), including QuikChange multi enzyme blend, 10× QuikChange buffer, dNTP mix, *Dpn* I restriction enzyme, QuikSolution, multi-control template (50 ng/μl), control primer mix (100 ng/μl), XL10-Gold ultracompetent cells, XL10-Gold β-mercaptoethanol (ME) mix, and pUC18 control plasmid [0.1 ng/μl in TE buffer (10 mM Tris–HCl and 1 mM ethylenediamine tetra-acetic acid (EDTA)), pH 7.5].
3. Restriction enzymes and reaction buffers (New England Biolabs, Ipswich, MA, USA).
4. Plasmid preparation kit (Qiaprep spin miniprep kit, Qiagen, Valencia, CA, USA).
5. Luria broth (LB): 10 g of tryptone, 5 g of yeast extract, and 10 g of NaCl made up to 1 l in deionized water; autoclave at 121 °C for 15 min.
6. Antibiotic stock (100 mg/ ml) for resistance provided by vector.
7. LB-antibiotic plates (10 g of tryptone, 5 g of yeast extract, 10 g of NaCl, and 15 g of agar made up to 1 l in deionized water; autoclave at 121 °C for 15 min, let cool to 50 °C, and add appropriate antibiotic [100 μg/ml ampicillin (carbenicillin is more stable) or kanamycin]; pour 20 ml into a 10-cm petri dish; once cool, store at 4 °C).

8. Mutagenic primers [high-performance liquid chromatography (HPLC)-purified oligonucleotides] for mutagenesis (final concentration: 0.4 μM each) (*see* **Note 1**).
9. Double-distilled water, sterile-filtered through a 0.22-μm filter.
10. Agarose gel.
11. Geneclean (BIO 101, Inc., Vista, CA, USA), and appropriate gel electrophoresis equipment.
12. TAE stock (50×) buffer (2 M Tris, 1 M acetic acid, and 50 mM EDTA, pH 8.3): Weighing out 242 g Tris base and dissolving in approximately 750 ml deionized water. Carefully add 57.1 ml glacial acid and 100 ml of 0.5 M EDTA (pH 8.0), and adjust the solution to a final volume of 1 l in deionized water.
13. Ethidium bromide solution (10 mg/ml, Sigma, St. Louis, MO, USA).

2.3 Cells and Culture Media

For transient expression of mutant MAb, Chinese hamster ovarian cells [CHO-K1, American Type Culture Collection (ATCC), Manassas, VA, USA] are maintained in F-12 medium (Invitrogen, Grand Island, NY, USA) supplemented with 10 % fetal bovine serum (FBS).

2.4 Expression of Mutant MAbs

1. Culture dishes (10 cm).
2. Lipofectamine-2000 (Invitrogen).
3. Opti-MEM I medium (Invitrogen).
4. PBS (137 mM NaCl, 10 mM Na_2HPO_4, 2.7 mM KCl, and 1.8 mM KH_2PO_4, pH 7.2).

2.5 ELISA

1. ELISA plates.
2. PBST washing buffer (PBS: 2.7 mM KCl, 1.8 mM KH_2PO_4, 137 mM NaCl, 10 mM Na_2HPO_4, and 0.05 % Tween-20, pH 7.2).
3. Coating buffer [50 mM sodium bicarbonate buffer ($NaHCO_3$): 15 mM Na_2CO_3, 35 mM $NaHCO_3$, pH 9.6].
4. Blocking buffer [1 % bovine serum albumin (BSA) in coating buffer]. Make freshly only the amount needed before use.
5. Substrate 2,2′-azino-bis(3-ethylbenzothiazoline-6-sulphonic acid) (ABTS, KPL Inc., Gaithersburg, MD, USA).
6. Positive and negative control sera.
7. MAb dilution buffer (1 % BSA in 1× PBS). Make freshly only the amount needed before use.
8. Secondary antibody, horseradish peroxidase (HRP)-conjugated secondary antibody (IgG and IgM, KPL Inc., Gaithersburg, MD, USA).

2.6 SDS-PAGE and Western Blotting

1. Sodium dodecyl sulfate (SDS)-10 % polyacrylamide gel electrophoresis (PAGE) gels: Prepare 25 ml of 10 % SDS gel [6.25 ml of 40 % acrylamide/bis, 6.25 ml of 1.5 M Tris–HCl (pH 8.8), 250 μl of 10 % SDS, 12.15 ml of deionized water, and add 12.5 μl tetramethylethylenediamine (TEMED), 0.125 ml of 10 % ammonium persulfate (AP)] and prepare 12.5 ml of 4 % acrylamide/bis stacking gel [1.25 ml of 40 % acrylamide/bis, 3.15 ml of 0.5 M Tris–HCl (pH 6.8), 0.125 ml of 10 % SDS, 7.5 ml of deionized water, and finally add 12.5 μl TEMED, 62.5 μl of 10 % AP] to make two mini-gels.
2. 4× tank buffer (make 1 l: 12 g Tris, 57.6 g glycine, and 4 g SDS). Use 500 ml (take 125 ml of stock solution, and make 500 ml with deionized water) to run SDS-10 % PAGE gels.
3. 2× sample buffer [2.5 ml of 0.5 M Tris (pH 6.8), 4 ml of 10 % SDS, 2 ml of glycerol, 1 ml of 2-ME, 0.01 % bromophenol blue, and store at room temperature].
4. Antigen samples (20 μg) in 2× sample buffer by boiling at 100°C for 4 min.
5. PAGE chamber (Bio-Rad, Hercules, CA, USA).
6. Antigen samples (10 μg/ml/well for each sample).
7. Molecular weight standards (Bio-Rad).
8. 3 M filter papers (cut 12 filter papers at 9×9 cm^2 with a cutter) (6 sheets of filter papers are required for each membrane).
9. Clean petri dishes.
10. Nitrocellulose membranes (9×9 cm^2, BA 85, pore size: 0.45 μm, Schleicher & Schuell, Keene, NH, USA).
11. Transfer buffers [anode buffer 1 (0.3 M Tris, 20 % methanol, pH 10.4), anode buffer 2 (0.025 M Tris, 20 % methanol, pH 10), and cathode buffer (0.025 M Tris, 40 mM acetic acid, 20 % methanol, pH 9.4)].
12. Semidry transblotter.
13. Destaining solution (40 % methanol, 10 % acetic acid, 50 % deionized water).
14. Culture supernatants containing target MAbs (1:10 or 1:100) in 1 % BSA–PBS.
15. TBS wash buffer (20 mM Tris–HCl, 150 mM NaCl, 2 mM NaN_3, pH 7.4).
16. HRP-conjugated secondary antibody (KPL, Gaithersburg, MD, USA) in 1 % BSA–PBS solution at 1:2,000.
17. Color developing 4CN substrate reagent (KPL Inc., Gaithersburg, MD, USA).

3 Methods

To select targeted MAb to improve binding affinity, an alignment of variable heavy chains of three related MAbs against a target antigen is created for each loop against its best hits from the BLAST searches. Multiple sequences of each VH chain are aligned by frame regions (FRs) and CDRs using the ClustalW multiple sequence alignment software (Fig. 2). Affinity maturation is created through the combination of mutations that change electrostatic charges (e.g., neutral or negative to positive) at specific amino acid positions in VH3-CDR3. Here we demonstrate the substitution of amino acids (Y101R_Y104K) in the VH3-CDR3 to improve MAb affinity.

3.1 Site-Directed Mutagenesis

QuikChange Lightning Multi Site-Directed Mutagenesis Kit [9] delivers results up to three times faster than the original multi-site mutagenesis kit (*see* **Note 2**).

1. Prepare the double-strand (ds) DNA plasmid (5 > kb) template by standard miniprep protocols (e.g., Qiaprep spin miniprep kit).
2. Prepare the mutant strand synthesis reactions for thermal cycling. Add 25-μl reaction components in the order [2.5 μl 10× QuikChange Lightning Multi reaction buffer, X μl double-distilled water, 0.5 μl QuikSolution, X μl plasmid DNA (50 ng), X μl mutagenic primers (100 ng each)] (Figs. 3 and 4), 1 μl dNTPs, 1 μl QuikChange Lightning Multi enzyme blend, then mix gently by pipetting or tapping the reaction tube.
3. Cycle the reactions using the cycling parameters [95 °C for 2 min, 30 cycles (95 °C for 20 s, 55 °C for 30 s, and 65 °C for 30 s/kb of plasmid length), and 65 °C for 5 min; for the control reaction, use a 2.5-min extension time].
4. Add 1 μl of *Dpn* I restriction enzyme directly to each PCR reaction.
5. Gently and thoroughly mix each reaction mixture by pipetting the solution up and down several times.
6. Spin down the reaction mixtures in a microcentrifuge for 1 min, and then immediately incubate each reaction at 37 °C for 5 min to digest the parental ds DNA.
7. Gently thaw the XL10-Gold ultracompetent cells on ice. For each mutagenesis reaction, aliquot 45 μl of the ultracompetent cells to a pre-chilled 14-ml BD Falcon polypropylene round-bottom tube.
8. Add 2 μl of the β-ME mix to the 45 μl of cells (*see* **Note 3**).
9. Swirl the contents of the tube gently. Incubate the cells on ice for 10 min, swirling gently every 2 min.

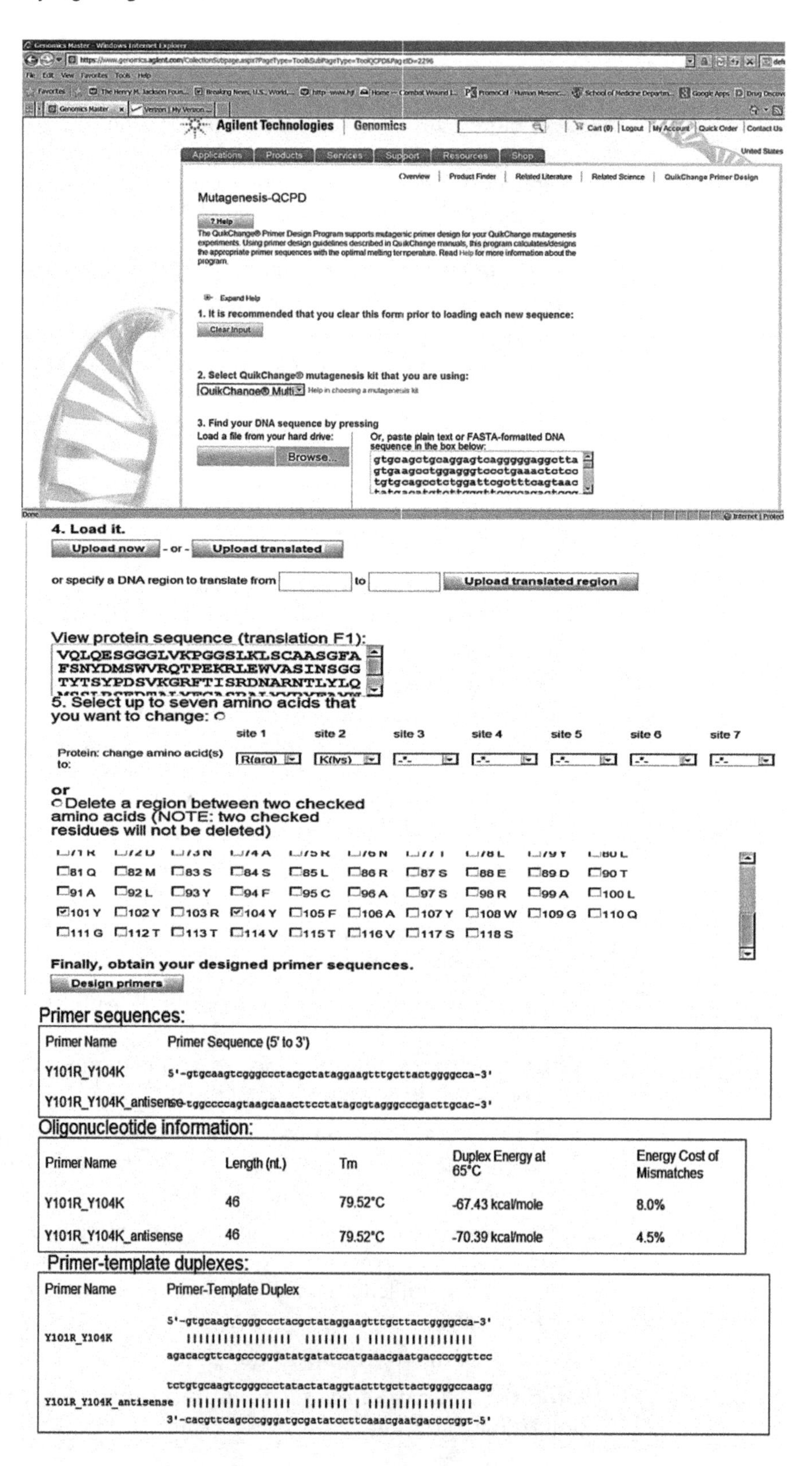

Primer sequences:

Primer Name	Primer Sequence (5' to 3')
Y101R_Y104K	5'-gtgcaagtcgggccctacgctataggaagtttgcttactggggcca-3'
Y101R_Y104K_antisense	5'-tggccccagtaagcaaacttcctatagcgtagggcccgacttgcac-3'

Oligonucleotide information:

Primer Name	Length (nt.)	Tm	Duplex Energy at 65°C	Energy Cost of Mismatches
Y101R_Y104K	46	79.52°C	-67.43 kcal/mole	8.0%
Y101R_Y104K_antisense	46	79.52°C	-70.39 kcal/mole	4.5%

Primer-template duplexes:

Primer Name	Primer-Template Duplex
Y101R_Y104K	5'-gtgcaagtcgggccctacgctataggaagtttgcttactggggcca-3' agacacgttcagcccgggatatgatatccatgaaacgaatgaccccggttcc
Y101R_Y104K_antisense	tctgtgcaagtcgggccctatactataggtactttgcttactggggccaagg 3'-cacgttcagcccgggatgcgatatccttcaaacgaatgaccccggt-5'

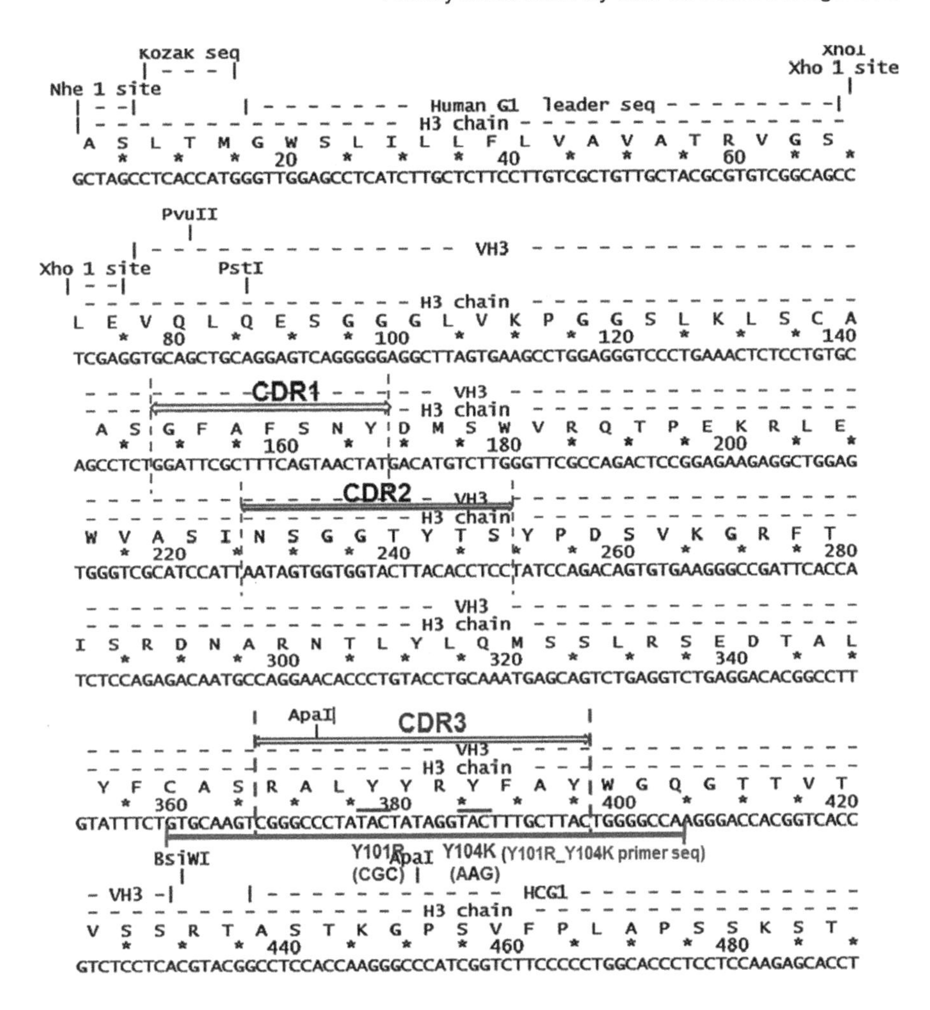

Fig. 4 Identification of target mutation positions (Y101R_Y104K) in the VH3-CDR3. To improve MAb affinity without affecting antibody activity to the target antigen, amino acid substitution at specific positions in the VH3-CDR3 is dependent on changes of electrostatic charge (e.g., neutral or negative to positive). The position of Y101R_Y104K primer synthesized by the web-based QuikChange primer design program is indicated by the *blue bar*

10. Transfer 1.5 μl of the *Dpn* I-treated DNA from each mutagenesis reaction to a separate aliquot of the ultracompetent cells (*see* **Note 4**).
11. Swirl the transformation reactions gently to mix, and then incubate the reactions on ice for 30 min.

Fig. 3 An example of web-based QuikChange primer design program (http://www.genomics.agilent.com). To design mutagenic primers for mutating tyrosine 101 to arginine (Y101R) and tyrosine 104 to lysine (Y104K) in the VH3-CDR3, program selection can be one of these six choices using the left-side menu bar. DNA sequences can be entered by copy-pasting to the provided text area. Amino acid change must be specified by checking the "radio" button in the gray area and selecting the target amino acid from the drop-down menu. After clicking "design primers" button, the recommended "sense" and "antisense" primers with melting temperature close to 80 °C are shown

12. Preheat NZY + broth in a 42 °C water bath. Transformation of XL10-Gold ultracompetent cells has been optimized using NZY + broth (*see* **Note 5**).
13. Heat-pulse the tubes in a 42 °C water bath for 30 s (*see* **Note 6**).
14. Incubate the tubes on ice for 2 min.
15. Add 0.5 ml of preheated (42 °C) NZY + broth to each tube, and incubate the tubes at 37 °C for 1 h with shaking at 225–250 rpm.
16. Plate the appropriate volume (1 μl, 10–100 μl) of each transformation reaction on agar plates containing the appropriate antibiotic for the plasmid vector. For the mutagenesis and transformation controls, spread cells on LB–ampicillin agar plates that have been prepared with 80 μg/ml X-Gal and 20 mM IPTG for color screening (optional).
17. Incubate the transformation plates at 37 °C for >16 h (*see* **Note 7**).
18. After positive clones are identified by colony PCR, plasmid DNAs from approximate ten positive clones are purified by Qiaprep spin miniprep kit (Qiagen). Target mutation is confirmed by sequencing of plasmid DNAs in the hypervariable VH3-CDR3 of each MAb.

3.2 Expression of Mutant MAbs

1. Two days prior to scheduled transfection, CHO-K1 cells are seeded at 15 ml in a 10-cm dish at a density of 5×10^5 cell/ml in culture media containing 10 % FBS to compare binding activity of mutant clones with wild type.
2. For each culture dish to be transfected, the plasmid DNAs (10 μg) at molar ratios (H:L = 1:1) are diluted into 800 μl of Opti-MEM I without serum.
3. In a separate tube, 80 μl of Lipofectamine 2000 are diluted into 720 μl Opti-MEM I medium and incubated for 5 min at room temperature.
4. The diluted DNAs are combined with the diluted Lipofectamine 2000 reagent and incubated at room temperature for 30 min to allow DNA-Lipofectamine 2000 reagent complexes to form.
5. After adding additional 6 ml Opti-MEM I medium to DNA–Lipofectamine 2000 complexes (~1,600 μl), the tubes are mixed gently and plated onto the 10-cm culture dish prepared.
6. The cells are incubated at 37 °C in a CO_2 incubator for 6 h.
7. Six hours post exposure with the DNA–Lipofectamine 2000 complex in the Opti-MEM I medium, the cells are washed once with PBS and changed with the regular culture medium containing 10 % FBS.

8. After 16-h recovery with the regular medium supplemented with 10 % FBS, the media is replaced with the serum-free culture medium.
9. After 72-h incubation, the culture supernatants are harvested, and the binding properties and MAb levels are examined at each condition.

3.3 Determination of Binding Activity

Binding activity of the mutant MAbs is compared with the original MAb by ELISA and WB analysis [7, 8].

3.3.1 ELISA

1. Dissolve antigen stock at 10 μg/ml in 10 ml of 50 mM coating buffer (pH 9.6) for one 96-well ELISA plate.
2. Add 100 μl (100 ng/well) to each well of ELISA plate by multichannel pipet.
3. Incubate at 37 °C for 1 h, and then shake off the antigen.
4. Rinse with PBST once by shaking.
5. Add 200 μl of 2 % BSA in sodium bicarbonate buffer to each well to block the plate. Leave plates at 4 °C overnight.
6. Wash with PBST three times.
7. Add 100 μl of culture supernatant in each well. Add 100 μl of positive and negative control sera at 1:10 dilution in 1 % BSA in 1× PBS.
8. Incubate at 37 °C for 2 h, and then wash off the plate.
9. Rinse with PBST three times.
10. Dilute HRP-conjugated secondary antibody to 1:1,000 in 1 % BSA in 1× PBS. Add 100 μl to each well.
11. Incubate at 37 °C for 1 h.
12. Rinse with PBST three times.
13. Add 100 μl of substrate ABTS solution to each well and keep in the dark for 15 min at room temperature.
14. Measure optical density (OD) at 405 nm using plate reader (e.g., Molecular Devices, Sunnyvale, CA, USA).

3.3.2 SDS-PAGE and Western Blotting

1. Twenty-five micrograms of antigens are dissolved in the sample buffer.
2. The samples, including 10 μl of diluted (1:20) broad-range molecular weight standards (Bio-Rad) per gel, are boiled at 100 °C for 4 min and separated by SDS-10 % PAGE (e.g., Bio-Rad Mini-gel Apparatus Separation System). The gels are stained for 15 min with Coomassie brilliant blue solution (0.2 % Coomassie brilliant blue R-250, 50 % methanol, and 10 % acetic acid).

3. Electrophoretic separation of each antigen is performed at room temperature at 90 V for the stacking gel and 125 V for the separating gel.
4. Separated 10 % gels are transferred to the pure nitrocellulose membranes (Schleicher & Schuell) at room temperature at 135 mA for 40 min on semidry transblotter.
5. To determine the molecular sizes of antigens recognized on each blot, one strip of molecular marker is cut and stained for 10 min in Amido black solution (Sigma).
6. The other blots are immersed in blocking buffer (5 % nonfat dry milk in 1× PBS) at 4 °C overnight to saturate protein-binding sites, then incubated at room temperature with diluted supernatants (1:10) from mutant clones and wild-type and control antibodies in blocking buffer for 2 h, and washed three times with TBST buffer (20 mM Tris–HCl, 150 mM NaCl, 0.05 % Tween-20, pH 7.4).
7. The blots are incubated at room temp for 2 h with peroxidase-conjugated secondary antibody at 1:2,000 dilution in blocking buffer.
8. The blots are washed three times with TBST buffer for 5 min, and the peroxidase-positive bands are detected by immersing the blots in a color-developing 4CN substrate reagent for 5 min.
9. The enzyme reaction is terminated by washing the blots in 0.1 M H_2SO_4.
10. Throughout the experiments, molecular sizes are determined based on broad-range molecular weight standards (e.g., Bio-Rad).

4 Notes

1. QuikChange primer designer is a part of the Agilent LabTools website (http://www.genomics.agilent.com). This software designs efficient primers for site-directed mutagenesis by calculating free energy of the mismatched primer–template duplex and comparing it to the energy of perfect duplex. The following items should be considered when designing mutagenic primers: (1) Primers should be between 25 and 45 bases in length, with a melting temperature (Tm) of ≥75 °C. Primers longer than 45 bases may be used, but using longer primers increases the likelihood of secondary structure formation, which may affect the efficiency of the mutagenesis reaction. (2) The desired point mutation or degenerate codon should be close to the middle of the primer with approximately 10–15 bases of template-complementary sequence on both sides. (3) Optimal primers have a minimum GC content of 40 % and terminate in one or more C or G bases at the 3′-end. (4) PAGE

purification of primers is not necessary in all cases but beneficial for high mutagenesis efficiency. Using this program (Fig. 3), mutagenic primers are designed for mutating (Y101R) and (Y104K) in the VH3-CDR3 (Fig. 4).

2. Three major steps of the QuikChange Lightning Multi Site-Directed Mutagenesis Kit: (1) faster mutant strand synthesis by a DNA thermal cycler (Bio-Rad); (2) faster *Dpn* I digestion of methylated and hemimethylated DNAs with new *Dpn* I enzyme; and (3) fast transformation of mutated ssDNA into XL10-Gold ultracompetent cells.
3. XL10-Gold cells are resistant to tetracycline and chloramphenicol. If the mutagenized plasmid contains only the tet^R or the cam^R resistance marker, an alternative strain of competent cells must be used. Using an alternative source of β-ME may reduce transformation efficiency.
4. (Optional) Verify the transformation efficiency of the XL10-Gold ultracompetent cells by adding 1 μl of 0.01 ng/μl pUC18 control plasmid (dilute the provided pUC18 DNA 1:10) to another 45-μl aliquot of cells.
5. Transformation reaction plating volumes: experimental mutagenesis 1, 10, and 100 μl; mutagenesis control 10 μl; and transformation control (pUC 18) 5 μl. When plating less than 100 μl from the transformation reaction, place a 100-μl pool of NZY+broth on the agar plate, pipet the cells into the pool, and then spread the mixture.
6. The duration of the heat pulse is critical for obtaining the highest efficiencies. Do not exceed 42 °C. This heat pulse has been optimized for transformation in 14-ml BD Falcon polypropylene round-bottom tubes.
7. The expected colony number from the mutagenesis control transformation is between 50 and 800 colonies. The mutagenesis efficiency (ME) for the QuikChange Multi control plasmid is calculated by the following formula: ME = {Number of blue colony-forming units (cfu)/total number of cfu} × 100 %. The XL10-Gold ultracompetent cells should be placed at −80 °C directly from the dry ice shipping container. It is essential that the BD Falcon polypropylene tubes are placed on ice before the cells are thawed and that the cells are aliquoted directly into the pre-chilled tubes. It is important that 14-ml BD Falcon polypropylene round-bottom tubes are used for the transformation protocol, since other tubes may be degraded by the β-ME during transformation. In addition, the duration of the heat-pulse step is critical and has been optimized specifically for the thickness and shape of these tubes. Optimal efficiencies are observed when cells are heat-pulsed for 30 s. Efficiencies decrease when incubating for <30 s or for >40 s. Do not exceed 42 °C.

Acknowledgment

We thank Tommy Kim for editorial assistance.

Disclaimer. The views expressed in this manuscript are those of the authors and do not reflect the official policy of the Department of the Army, the Department of Defense, the US Government, or the Henry M Jackson Foundation for the Advancement of Military Medicine.

References

1. Kim H-Y, Stojadinovic A, Weina PJ et al (2011) Monoclonal antibody-based therapeutics for melioidosis and glanders. Am J Immunol 7:39–53
2. Brady K, Lo BKC (2004) Direct mutagenesis of antibody variable domains. In: Lo BKC (ed) Antibody engineering—methods and protocols, vol 248, Methods in molecular biology. Humana Press, Totowa, NJ, pp 319–326
3. Wark KL, Hudson PJ (2006) Latest technologies for the enhancement of antibody affinity. Adv Drug Deliv Rev 58:657–670
4. Chowdhury PS, Wu H (2005) Tailor-made antibody therapeutics. Methods 36:11–24
5. Neuberger MS, Milstein C (1995) Somatic hypermutation. Curr Opin Immunol 7:248–254
6. Yau KY, Dubuc G, Li S, Hirama T et al (2005) Affinity maturation of a V(H)H by mutational hotspot randomization. J Immunol Methods 297:213–224
7. Kim HY, Rikihisa Y (1998) Characterization of monoclonal antibodies to the 44-kilodalton major outer membrane protein of the human granulocytic ehrlichiosis agent. J Clin Microbiol 36:3278–3284
8. Kim H-Y, Tsai S, Lo S-C et al (2011) Production and characterization of chimeric monoclonal antibodies against *Burkholderia pseudomallei* and *Burkholderia mallei* using the DHFR expression system. PLoS One 6:e19867
9. http://www.genomics.agilent.com/

Chapter 25

Epitope Mapping with Membrane-Bound Synthetic Overlapping Peptides

Terumi Midoro-Horiuti and Randall M. Goldblum

Abstract

Epitope mapping with synthetic overlapping peptides is used for identifying epitopes of monoclonal antibodies (mAbs) and antibodies from patient sera (Midoro-Horiuti et al. Mol Immunol 43:509–518, 2006; Ivanciuc et al. J Agric Food Chem 51:4830–4837, 2003; Midoro-Horiuti et al. Mol Immunol 40:555–562, 2003; Wang et al. J Allergy Clin Immunol 125:695–702, 702.e1–702.e6, 2010). When the mAbs recognize epitopes that are also recognized by patients of interest, they may be useful as surrogates for patient antibodies.

Key words Antigen, Epitope, Monoclonal antibody, Overlapping peptides, Patient serum

1 Introduction

The peptides are typically synthesized directly on the cellulose membrane starting from their C-termini and thus have free N-termini. The size of peptides used to identify epitopes is typically 8–15 amino acids. These are either linear or can be circularized by oxidizing cysteins added at designated positions in the peptides.

2 Materials

1. Overlapping synthetic peptide: Obtain or "print" overlapping synthetic peptides based on the amino acid sequence of the antigen of interest. Store in −20 °C freezer.
2. Specific monoclonal antibodies (mAbs).
3. Secondary antibodies, enzyme (e.g., peroxidase)-labeled antibodies against the first (primary) antibody.
4. MilliQ water (*see* **Note 1**).
5. Blocking buffer (GENOSYS Cat. No. SU-07-250): Add and mix 10 mL of Genosys concentrated blocking buffer, 90 mL of Tris-buffered saline/Tween (T-TBS), pH 8.0, and 5 g of

Vincent Ossipow and Nicolas Fischer (eds.), *Monoclonal Antibodies: Methods and Protocols*, Methods in Molecular Biology, vol. 1131, DOI 10.1007/978-1-62703-992-5_25, © Springer Science+Business Media New York 2014

sucrose to make 100 mL of blocking buffer. Prepare it just before use. Do not store.

6. Tris-buffered saline (TBS), pH 8.0: Add and mix NaCl (8.0 g), KCl (0.2 g), Tris base (6.1 g), and 800 mL of MilliQ water. Adjust pH to 8.0 with HCl. Make up to 1 L with MilliQ water. Store at room temperature.
7. 0.05 % T-TBS, pH 8.0: Add 0.5 mL of Tween20 to 1 L of TBS. Store at room temperature.
8. PBS (137 mM NaCl, 8.1 mM $Na_2HPO_4{\cdot}12H_2O$, 2.68 mM KCl, 1.47 mM KH_2PO_4, pH 7.4): Add and mix 8 g NaCl, 2.9 g $Na_2HPO_4{\cdot}12H_2O$, 0.2 g KCl, and 0.2 g KH_2PO_4 with 800 mL MilliQ water. This will be pH 7.2–7.6. If the pH is not within this range, adjust with HCl or NaOH. Make up to 1 L with MilliQ water (*see* **Note 2**).
9. Regeneration buffer I: Restore Western Blot Stripping Buffer (Thermo Scientific), store in 4 °C [5, 6].
10. Regeneration buffer II (62.5 mM Tris, 2 % SDS, pH 6.7, 100 mM 2-mercaptoethanol): Dissolve 7.57 g Tris base and 20 g SDS in 800 mL MilliQ water. Adjust pH with HCl to 6.7. Make up to 1 L with MilliQ water. Add 70 μL 2-mercaptoethanol per 10 mL regeneration buffer before use (*see* **Note 3**).
11. Regeneration buffer IIIA (8 M urea, 1 % SDS, 0.1 % 2-mercaptoethanol): Dissolve 480.5 g urea and 10 g SDS in 800 mL MilliQ water. Make up to 1 L with MilliQ water. Store at room temperature. Add 100 μL of 2-mercaptoethanol to 100 mL of regeneration buffer A in a fume hood just before use (*see* **Note 4**).
12. Regeneration buffer IIIB (50 % ethanol, 10 % acetic acid): Mix 400 mL MilliQ water with 500 mL ethanol and add 100 mL acetic acid. Do not mix ethanol and acetic acid directly. Store at room temperature.
13. Plastic bag and sealer.
14. Transparent plastic film (e.g., Saran wrap).
15. Chemiluminescent substrate (e.g., ECL Western blotting detection reagents, Amersham Pharmacia Biotech).
16. Film and film developer.

3 Methods

Solution volumes indicated below are for about 3×8 cm membrane. This size of membrane can contain about 120 peptides.

3.1 Testing the Nonspecific Antibody Binding

1. Remove the membrane from the freezer, allow to warm to room temperature, and rinse with 5 mL of methanol in polypropylene container for 1 min.
2. Wash the membrane three times with 10 mL TBS for 10 min with shaking. The membrane should be covered by the solution.
3. Block the membrane with 1 mL blocking buffer in the sealed plastic bag overnight at room temperature with gentle shaking. Plastic container with lid can be used, instead of plastic bag. You need 10 mL blocking buffer if you use a plastic container. Do not stack membranes (*see* **Note 5**).
4. Wash the membrane in the plastic container once with 10 mL T-TBS for 10 min with shaking.
5. Incubate the membrane with 1 mL peroxidase-labeled second antibody (antibody directed against first antibody) in blocking buffer (1:1,000–1:2,000 dilution) for 2 h at room temperature with shaking.
6. Wash the membrane three times with 10 mL T-TBS for 10 min with shaking.
7. Incubate the membrane with 1 mL ECL solution for 1 min in the plastic container in the darkroom. Make sure that the ECL reagent covers the membrane.
8. Wrap the membrane in the transparent plastic film, and insert the membrane, with the peptide side facing the film in the film cassette. Expose the membrane to the film for 15 s, 30 s, 1 min, 5 min, and 30 min.
9. Develop the film.
10. If you see spots, you will need to use a other secondary antibody system to avoid nonspecific antibody binding. If you see no spots, go to the epitope mapping experiments.

3.2 Epitope Mapping

1. Remove the membrane from the freezer, allow to warm to room temperature, and rinse with 5 mL of methanol in polypropylene container for 1 min.
2. Wash the membrane three times with 10 mL TBS for 10 min with shaking. The membrane should be covered by the solution.
3. Block the membrane with 1 mL of blocking buffer in the plastic bag overnight at room temperature with shaking. Plastic container with lid can be used instead of plastic bag. You will need 10 mL blocking buffer if you use a plastic container. Do not overlay membranes.
4. Wash the membrane in the plastic container once with 10 mL T-TBS for 10 min with shaking.
5. Incubate membrane with first antibody diluted in blocking buffer for 2 h to overnight (concentration is about 1 ng/mL–1 μg/mL) (*see* **Note 4**).

6. Wash the membrane three times with 10 mL T-TBS for 10 min with shaking.
7. Incubate the membrane with 1 mL peroxidase-labeled second antibody (antibody directed against first antibody) in blocking buffer (typically 1:1,000–1:2,000 dilution) for 2 h at room temperature with shaking (*see* **Note 4**).
8. Wash the membrane three times with 10 mL T-TBS for 10 min with shaking.
9. Incubate the membrane with 1 mL ECL solution for 1 min in the plastic container in the darkroom. Make sure that the ECL reagent covers the membrane (*see* **Note 6**).
10. Wrap the membrane in the transparent plastic film, and insert the membrane with the peptide side facing the film in the film cassette. Expose the membrane to the film for 15 s, 30 s, 1 min, 5 min, and 30 min (*see* **Note 7**).
11. Develop the film.

3.3 Regeneration of Peptide Membrane (Stripping Antibodies from the Membrane) Protocol I

1. Wash the membrane with 10 mL MilliQ water for 10 min with shaking.
2. Incubate the membrane in Restore Western Blot Stripping Buffer for 5–15 min at room temperature with shaking or incubate at 37 °C for high-affinity antibodies.
3. Wash the membrane with 10 mL T-TBS for 10 min with shaking.
4. If the membrane was incubated with directly labeled antibody, check the success of the regeneration by incubating the membrane with ECL and exposing to the film at least as long as the original exposure.
5. If the membrane was incubated with a primary antibody followed by labeled second antibody, incubate the membrane with an enzyme-labeled secondary antibody, ECL, and expose to the film at least as long as the original exposure.
6. If the membrane has antibody signal after procedure 4 or 5, repeat **steps 1–3** and test the membrane as in **step 4** or **5**. Alternatively, regeneration protocol II or III can be used at this step (*see* **Note 8**).

3.4 Regeneration of Peptide Membrane (Stripping Antibodies from the Membrane) Protocol II

1. Wash the membrane three times with 10 mL water for 10 min with shaking.
2. Wash the membrane at least four times with 10 mL regeneration buffer II for 30 min at 50 °C with shaking.
3. Wash the membrane at least three times with 10 mL PBS for 20 min at room temperature with shaking.
4. Wash the membrane three times with 10 mL T-TBS for 20 min at room temperature with shaking.

5. Wash the membrane three times with 10 mL TBS for 10 min at room temperature with shaking.
6. If the membrane was incubated with directly labeled antibody, check the success of the regeneration by incubating the membrane with ECL and exposing to the film at least as long as the original exposure.
7. If the membrane was incubated with a primary antibody with labeled second antibody, incubate the membrane with an enzyme-labeled secondary antibody, ECL, and expose to the film at least as long as the original exposure.
8. If the membrane has antibody (luminescent) signal after procedure 6 or 7, repeat **steps 1–5** and test the membrane as in **step 6** or 7. Alternatively, regeneration protocol I or III can be used at this step.

3.5 Regeneration of Peptide Membrane (Stripping Antibodies from the Membrane) Protocol III

1. Wash the membrane three times with 10 mL water for 10 min with shaking.
2. Incubate the membrane three times with regeneration buffer IIIA for 10 min with shaking.
3. Incubate the membrane three times with regeneration buffer IIIB for 10 min with shaking.
4. Wash the membrane with water for 10 min with shaking.
5. Wash the membrane three times with water for 10 min with shaking.
6. If the membrane was incubated with directly labeled antibody, check the success of the regeneration by incubating the membrane with ECL and exposing to the film at least as long as the original exposure.
7. If the membrane was incubated with a primary antibody with labeled second antibody, incubate the membrane with an enzyme-labeled secondary antibody, ECL, and expose to the film for at least as long as the original exposure.
8. If the membrane has antibody signal after procedure 6 or 7, repeat **steps 1–5** and test the membrane as in **step 6** or 7. Alternatively, regeneration protocol I or II can be used at this step.

3.6 Storage of Peptide Membrane

1. New membrane should be stored at –20 °C until use.
2. Used membrane, which will be used again within a few days, should be washed three times with T-TBS for 10 min and kept with a small volume of T-TBS in the plastic bag at 4 °C. Avoid drying out.
3. Used membrane which will be stored for a longer period should be regenerated, washed with methanol twice, air-dried, and kept at –20 °C.

4 Notes

1. MilliQ water should be used to make all the buffers in this protocol.
2. Other PBS can be used. Avoid NaN_3. This will inactivate peroxidase.
3. Store 2-mercaptoethanol in 4 °C and add just before each use.
4. Appropriate concentration and incubation time vary depending on the antibody. Adjust these based on the strength of signals from the first experiment results.
5. You can put two membranes in the same bag back to back to avoid overlaying them on their surface.
6. Add 0.5 mL each of ECL solution A and B in the darkroom just before use, mix well, and add the membrane.
7. Alternatively a bioimaging system (e.g., ChemiDoc-It®TS2 Imager, UVP) can be used.
8. The strength of regeneration protocol is I < II < III. You should try protocol I first, II next, and then III for most of the antibodies to avoid damaging the peptides on the membrane. If one procedure does not work, you can repeat the same procedure or try the next procedure.

   Commentary

   Synthetic peptides on the membrane are used to identify epitopes for mAbs and human serum antibodies.

References

1. Midoro-Horiuti T, Schein CH, Mathura V, Braun W, Czerwinski EW, Togawa A, Kondo Y, Oka T, Watanabe M, Goldblum RM (2006) Structural basis for epitope sharing between group 1 allergens of cedar pollen. Mol Immunol 43:509–518
2. Ivanciuc O, Mathura V, Midoro-Horiuti T, Braun W, Goldblum RM, Schein CH (2003) Detecting potential IgE-reactive sites on food proteins suing a sequence and structure database, SDAP-Food. J Agric Food Chem 51: 4830–4837
3. Midoro-Horiuti T, Mathura V, Schein CH, Braun W, Chin CCQ, Yu S, Watanabe M, Lee JC, Brooks EG, Goldblum RM (2003) Major linear IgE epitopes of mountain cedar pollen allergen Jun a 1 map to the pectate lyase catalytic site. Mol Immunol 40:555–562
4. Wang J, Lin J, Bardina L, Goldis M, Nowak-Wegrzyn A, Shreffler WG, Sampson HA (2010) Correlation of IgE/IgG4 milk epitopes and affinity of milk-specific IgE antibodies with different phenotypes of clinical milk allergy. J Allergy Clin Immunol 125:695–702, 702.e1-702.e6
5. Kaufmann SH, Ewing CM, Shaper JH (1987) The erasable Western blot. Anal Biochem 161:89–95
6. Kaufmann SH, Kellner U (1998) Erasure of western blots after autoradiographic or chemiluminescent detection. Methods Mol Biol 80: 223–235

Chapter 26

Epitope Mapping by Epitope Excision, Hydrogen/Deuterium Exchange, and Peptide-Panning Techniques Combined with In Silico Analysis

Nicola Clementi, Nicasio Mancini, Elena Criscuolo, Francesca Cappelletti, Massimo Clementi, and Roberto Burioni

Abstract

The fine characterization of protective B cell epitopes plays a pivotal role in the development of novel vaccines. The development of epitope-based vaccines, in fact, cannot be possible without a clear definition of the antigenic regions involved in the binding between the protective antibody (Ab) and its molecular target. To achieve this result, different epitope-mapping approaches have been widely described (Clementi et al. Drug Discov Today 18(9–10):464–471, 2013). Nowadays, the best way to characterize an Ab bound region is still the resolution of Ab–antigen (Ag) co-crystal structure. Unfortunately, the crystallization approaches are not always feasible. However, different experimental strategies aimed to predict Ab–Ag interaction and followed by in silico analysis of the results may be good surrogate approaches to achieve this result. Here, we review few experimental techniques followed by the use of "basic" informatics tools for the analysis of the results.

Key words B cell epitope, Epitope excision, Hydrogen/deuterium exchange, Peptide-panning techniques, Epitope-based vaccines

1 Introduction

The in silico study of the Ab–Ag interaction spans from the prediction of their docking through methods based solely on computational analysis to the three-dimensional (3D) rendering of the Ag regions involved in the binding. While the first approach requires the use of different scoring functions and/or bioinformatic algorithms calculated by computational tools, the second one depends on the generation of experimental data to be further elaborated [1]. For the sake of brevity, here we describe only the latter approaches by reviewing three possible epitope-mapping strategies,

Nicola Clementi and Nicasio Mancini contributed equally to the chapter.

Vincent Ossipow and Nicolas Fischer (eds.), *Monoclonal Antibodies: Methods and Protocols*, Methods in Molecular Biology, vol. 1131, DOI 10.1007/978-1-62703-992-5_26, © Springer Science+Business Media New York 2014

combining empirical data generation and in silico analysis. Two of them (epitope excision, hydrogen/deuterium (H/D) exchange) are characterized by generation of putative epitope amino acid sequence data through mass spectrometry. The third approach (peptide panning) allows to obtain a putative epitope consensus sequence after affinity selection on the Ab of interest of a phage-bound peptide library. In both cases the data generated from these approaches are amino acid sequences to be analyzed in silico.

2 Materials

2.1 Epitope Excision

Prepare all solutions using ultrapure water (prepared by purifying deionized water to attain a sensitivity of 18 MΩ cm at 25 °C) and analytical grade reagents. Prepare and store all reagents at room temperature (unless indicated otherwise).

1. CN-Br Sepharose beads (GE Healthcare): Weigh 0.2 g of beads and rinse in 10 mL HCl 1 mM in a 15 mL centrifuge tube; shake for 15 min, and equilibrate for 15 min.
2. Filter column, provided with the CN-Br Sepharose beads.
3. HCl: 1 mM solution in water.
4. $NaHCO_3$: 100 mM solution in water, pH 8.3.
5. Coupling buffer: $NaHCO_3$ 100 mM, pH 8.3 and NaCl 500 mM.
6. Quenching buffer: Tris–HCl 100 mM, pH 8.0.
7. Washing buffer: Sodium acetate 100 mM, pH 4 and NaCl 500 mM.
8. Phosphate-buffered saline (PBS): Sodium phosphate buffer 100 mM, NaCl 150 mM, pH 7.2.
9. Bissulfosuccinimidyl suberate (BS3): 10 mM solution in PBS (*see* **Note 1**).
10. Working solution: NH_4HCO_3 50 mM, pH 7.8.
11. Formic acid: 0.1 % solution in water.
12. Complete protease inhibitor (Complete EDTA-free, Roche): Prepare the stock solution 25× according to the producer's instructions.
13. Antibody immobilized on the beads: Store in sodium phosphate 10 mM and NaCl 250 mM, pH 7.6; the optimal ratio is 1–10 μM of ligand for each mL of beads (*see* **Note 2**).
14. Antibody of interest: Prepare a solution of 50 μg of protein and 100 μL of coupling buffer; adjust with PBS to a final volume of 200 μL (*see* **Note 3**).
15. Antigen: Prepare a 200 μL solution with 50 μg of protein diluted in PBS, pH 7.2.

16. Trypsin stock solution: 100 μg/mL solution in HCl 1 mM.
17. Trypsin working solution: Dilute just before use the Trypsin stock solution in working solution; use an enzyme/protein ratio (w/w) between 1:20 and 1:100.

2.2 Hydrogen/ Deuterium Exchange

Prepare all solutions using ultrapure water (prepared by purifying deionized water to attain a sensitivity of 18 MΩ cm at 25 °C) and analytical grade reagents. Prepare and store all reagents at room temperature (unless indicated otherwise).

Important: Diligently follow all waste disposal regulations when disposing waste materials.

1. POROS 20 AL media (Applied Biosystems).
2. Sodium cyanogenborohydride ($NaCNBH_3$).
3. Saturated sodium sulfate (Na_2SO_4) solution.
4. PBS: NaCl 137 mM, KCl 2.7 mM, Na_2HPO_4 10 mM, and KH_2PO_4 1.8 mM, pH 7.4.
5. Capping buffer: 2 mL of PBS containing ethanolamine 1 M and 8 mg/mL of $NaCNBH_3$.
6. Deuterated buffer: Phosphate 10 mM and NaCl 200 mM, pH 7.2 in 90 % D_2O.
7. Aqueous buffer: Phosphate 10 mM and NaCl 200 mM, pH 7.2.
8. Formic acid: 0.8 % solution in water.
9. Quenched solution: Guanidine hydrochloride (GuHCl) 1.6 M, 0.8 % formic acid.
10. Pepsin column (Hamuro Y, Coales SJ, Molnar KS, Tuske SJ, Morrow JA. *Rapid Commun. Mass Spectrom. 2008*; 22: 1041).
11. Trifluoroacetic acid (TFA): 0.05 % solution in water.
12. Reversed-phase trap column (4 μL bed volume; Applied Biosystems).
13. C18 column (Michrom BioResources, Inc.).
14. Solvent A: 0.05 % TFA solution in water.
15. Solvent B: 95 % acetonitrile and 5 % buffer A.
16. Antibody solution: Dilute in its buffer to a final concentration of 4.4 mg/mL.
17. Antigen solution: Dilute in its buffer to a final concentration of 1 mg/mL.

2.3 Peptide Panning

1. Phage Display Peptide Library 12-mer 100 μL, ~1×10^{13} pfu/mL (NEB—*New England BioLabs, Inc.*).
2. 96 gIII sequencing primer: 5′-CCC TCA TAG TTA GCG TAA CG-3′.
3. 28 gIII sequencing primer: 5′-GTA TGG GAT TTT GCT AAA CAA C-3′.

4. *E. coli* ER2738 host strain—F′ proA + B + lacIq Δ(lacZ) M15 zzf::Tn10(TetR)/fhuA2 glnV Δ(lac-proAB) thi-1 Δ(hsdS-mcrB)5.
5. SB medium—per liter: 35 g Bacto-Tryptone, 20 g yeast extract, 5 g NaCl. Autoclave, and store at room temperature.
6. LB medium—per liter: 10 g Bacto-Tryptone, 5 g yeast extract, 10 g NaCl. Autoclave, and store at room temperature.
7. Isopropyl-β-D-thiogalactoside (IPTG)/5-Bromo-4-chloro-3-indolyl-β-D-galactoside (Xgal) stock—Mix 1.25 g IPTG and 1 g Xgal in 25 mL dimethyl formamide (DMF). Store at −20 °C.
8. LB/IPTG/Xgal plates—1 l LB medium + 15 g/L agar. Autoclave, cool to <70 °C, add 1 mL IPTG/Xgal stock per liter, and pour. Store at 4 °C in the dark.
9. Top agar—per liter: 10 g Bacto-Tryptone, 5 g NaCl, 7 g Bacto-Agar.
10. Tetracycline stock (suspension)—20 mg/mL in 1:1 ethanol: water. Store at −20 °C.
11. LB + Tet plates—1 l LB medium + 15 g/L agar. Autoclave, cool to <70 °C, add 1 mL tetracycline stock, and pour. Store at 4 °C in the dark.
12. Blocking buffer—0.1 % PBS/BSA. Filter sterilize, and store at 4 °C.
13. PBS 1×. Autoclave, and store at room temperature.
14. PEG/NaCl—20 % (w/v) polyethylene glycol—8000, 2.5 M NaCl. Autoclave, and store at room temperature.
15. Washing solution—PBS 1× + 0.1 % [v/v] Tween-20.
16. Elution buffer—0.1 M HCl pH 2.2. Filter sterilize, and store at 4 °C.
17. Neutralizing buffer—1 M Tris base, pH 9. Filter sterilize, and store at 4 °C.

3 Methods

3.1 Epitope Excision

This epitope-mapping technique has been developed considering the differential accessibility of surface-exposed amino acid residues of the antigen in the presence or the absence of the antibody of interest [2]. Briefly, the Ab to be characterized is incubated with its target and digested. The epitope and the paratope complex will not be digested, and the amino acid composition of the eluted epitope will be evaluated through mass spectrometry by subtracting the unbound Ag control and the unbound Ab as well (Fig. 1).

Carry out all procedures at room temperature unless otherwise specified.

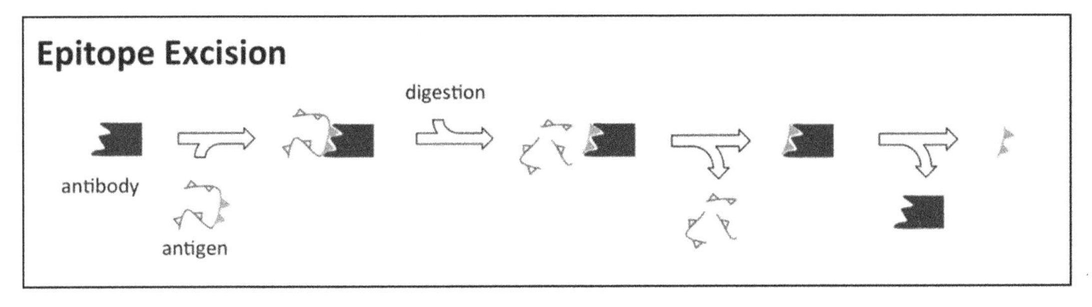

Fig. 1 Schematic representation of the experimental procedure of the Epitope Excision protocol: the antigen is bound to the immobilized antibody and digested with endoproteinases. Unbound peptides are washed off, and the affinity-bound peptides are subsequently eluted and collected to be further analyzed

Important: Opportune controls must be included in order to perform correctly the further mass spectrometric analysis since the exclusion of peptides obtained from the digestion of the unbound Ag or the unbound Ab of interest is essential.

For this purpose we suggest to perform the following digestions (a and b) in order to compare the differences between "a, b" resulting data and those obtained from the mAb-bound antigen (all the digestions must be performed in column).

(a) Digestion on antigen → add the Ag to a dedicated column without Ab bound to the beads.

(b) Digestion on the antibody → digest the Ab bound to the beads, without the presence of the antigen.

1. Rinse the CN-Br Sepharose beads as described, and take 200 μL of the slurry from the bottom of the centrifuge tube; place it in a 1.5 mL collection tube and spin down in a tabletop microcentrifuge.
2. Wash 6× with HCl solution: Add 800 μL each time, centrifuge in a tabletop microcentrifuge at 13,000 RCF for 1 min, and discard the supernatant.
3. Wash 6× with NaHCO3 solution, 400 μL each time as described above.
4. Centrifuge one more time at 13,000 RCF for 1 min, and discard the supernatant.
5. Add 100 μL of coupling buffer with the desired amount of antibody to be immobilized on the beads, and check that the pH value of the solution of beads and protein has not been altered.
6. Incubate for 2 h with slow agitation.
7. Centrifuge at 13,000 RCF for 1 min, and discard the supernatant.
8. Wash with 400 μL quenching buffer as described above.

9. Incubate for 2 h with 400 μL quenching buffer (*see* **Note 4**).
10. Wash 6× with alternating washing buffer and quenching buffer, 400 μL each time as described above.
11. Add the antibody of interest and incubate overnight (o.n.) at 4 °C (*see* **Note 5**).
12. Centrifuge at 13,000 RCF for 1 min, and discard the supernatant.
13. Wash 3× with PBS, 400 μL each time as described above.
14. Add 100 μL of BS3 to cross-link the antibody immobilized on the beads to the antibody of interest.
15. Incubate for 45 min with slow agitation.
16. Centrifuge at 13,000 × *g* for 1 min, and discard the supernatant.
17. Wash 2× with quenching buffer, 400 μL each time as described above.
18. Wash 3× with PBS, 400 μL each time as described above.
19. Add the antigen solution and incubate for 2 h with slow agitation.
20. Centrifuge at 13,000 RCF for 1 min, and discard the supernatant.
21. Wash 3× with PBS, 400 μL each time as described above.
22. Centrifuge at 13,000 RCF for 1 min, and discard the supernatant.
23. Add 100 μL of trypsin working solution and incubate o.n. at 37 °C.
24. Add the slurry on the filter column provided and centrifuge at 13,000 RCF for 1 min.
25. Collect the flow through, and add 4 μL of complete protease inhibitor stock solution.
26. Wash the column 3× with PBS, 400 μL each time as described above.
27. Add 500 μL of formic acid and incubate for 15 min with slow agitation.
28. Centrifuge at 13,000 RCF for 1 min, and collect the elution.
29. Store at 4 °C.
30. Follow proper sample preparation techniques depending on the mass spectrometric approach selected (MALDI-MS or ESI-MS) for further analysis.

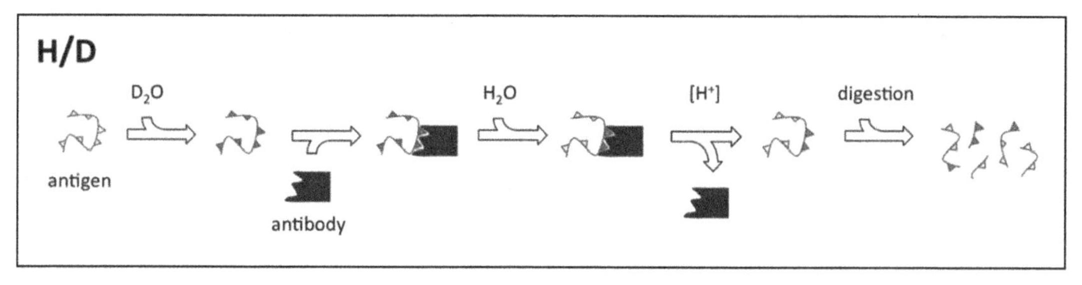

Fig. 2 The on/off-exchange protocol: Schematic representation of the experimental procedure. The antigen, firstly treated with a deuterated solution, is bound to the immobilized antibody. The complex is washed with a water solution allowing the ion exchange only on the unbound and exposed surfaces. Subsequently, the antigen is eluted and digested with endoproteinase. Finally, the MS measurement permits the identification of deuterated peptides

3.2 Hydrogen/Deuterium Exchange

Here is described the "on-solution/off-column" (on-exchange in solution followed by off-exchange in column) experiment where the antigen is first mixed with a deuterated buffer in solution and incubated for a predetermined duration ("on-solution" = on-exchange in solution) [3]. The on-exchanged antigen is next loaded onto the antibody column. The antigen-bound column is then washed with aqueous buffer allowing the ion exchange only on the unbound and exposed surfaces and incubated for half of the solution on-exchange time ("off-column" = off-exchange in column; the intrinsic D → H exchange reaction is twice as fast as the intrinsic H → D exchange reaction at the same pH reading due to isotopic effects). After the on/off-exchange reaction, the antigen is eluted out by a low pH buffer and digested by pepsin. Finally, the MS measurement permits the identification of deuterated peptides (Fig. 2).

Carry out all procedures at room temperature unless otherwise specified.

Important: A fully deuterated sample and a non-deuterated sample must be included in the experiment in order to fulfil the determination of deuteration level of each peptide after the on/off-exchange reaction.

3.2.1 Immobilization of the Antibody

1. Add 100 mg of POROS AL and 5 mg $NaCNBH_3$ to 600 μL of antibody solution.
2. Add 800 μL of saturated sodium sulfate (Na_2SO_4) solution and incubate *o.n.* with agitation.
3. Wash 5× with PBS.
4. Add the capping buffer and incubate for 2 h with slow agitation.

   (*Note*: This step is required in order to cap the unreacted aldehyde groups.)
5. Wash 5× with PBS, and resuspend in PBS.
6. Pack the resin in a column with 353 μL bed volume.

3.2.2 "On-Solution/Off-Column" Experiment

1. Dilute 5 μL of antigen solution with 45 μL of deuterated buffer.
2. Incubate at 38 °C for varying times (150, 500, 1,500, and 5,000 s).
3. Load the on-exchanged solution onto the antibody column (pre-equilibrated with deuterated buffer).
4. Wash the column with 500 μL of aqueous buffer.
5. Incubate for one-half of the preceding on-exchange duration (75, 250, 750, and 2,500 s).
6. Add 320 μL of formic acid (*see* **Note 6**).
7. Collect the last 40 μL of the eluent.
8. Add 20 μL of quenched solution.

3.2.3 Fully Deuterated Experiment

1. Add 4 μL of antigen solution to 36 μL of deuterated buffer and incubate at 60 °C *o.n.*
2. Load the sample onto the antibody column (pre-equilibrated with deuterated buffer).
3. Wash with 100 μL of deuterated buffer.
4. Add 320 μL of formic acid.
5. Collect the last 40 μL of the eluent.
6. Add 20 μL of quenched solution.

3.2.4 General Process for H/D-Exchanged Sample

1. Pump the quenched solution over a pepsin column (104 μL bed volume) with TFA (200 μL/min) for 3 min.
2. At the same time collect the proteolytic products by a reversed-phase trap column (4 μL bed volume).
3. Elute the peptide fragments and separate using a C18 column with a linear gradient of 13 % solvent B to 40 % solvent B over 23 min (flow rate 10 to 5 μL/min).
4. Follow proper sample preparation techniques depending on the mass spectrometric approach selected (MALDI-IRMPD or FI-TICR MS) for further analysis.

3.3 Peptide Panning

This epitope-mapping approach is based on the affinity selection of phage-displayed peptides against the Ab of interest (Fig. 3) [4–7]. Interestingly, this approach can lead to identification of mimotopes to be used in the rational design of epitope-based vaccine approaches [8]. This method can be followed using different phage libraries, spanning from commercially available libraries (containing linear peptides as well as loop-constrained peptides) to libraries deriving from the enzymatic digestion of the antigen of interest. As an example, we describe a "standard" panning procedure adapted from the screening protocol of Peptide Library 12-mer indicated by *New England BioLabs Inc.*

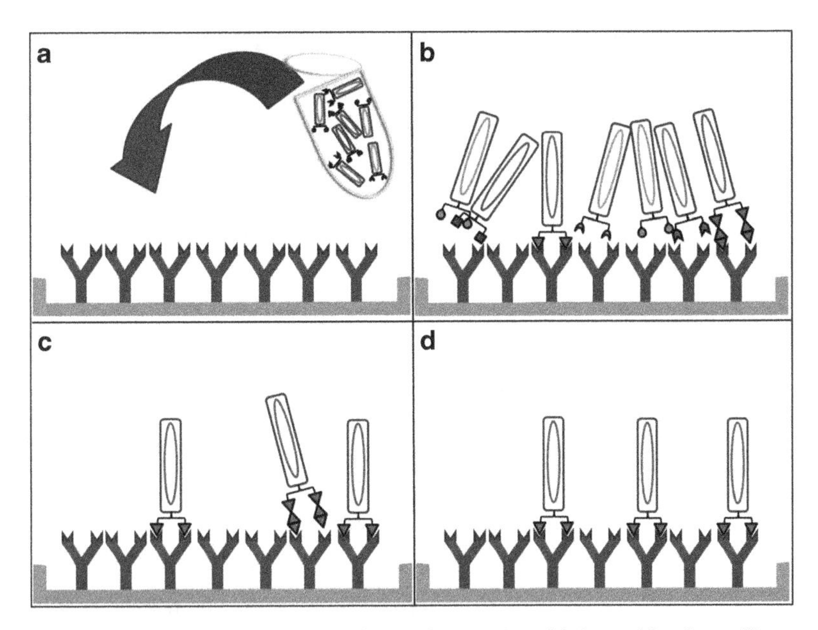

Fig. 3 Schematic representation of peptide panning. (**a**) A peptide phage library (commercially available or "homemade") is "panned" with the Ab to be characterized coated on a flat-bottom plate. (**b**) The library is then incubated at 37 °C. (**c**) Unbound phages are washed away. (**d**)After selection rounds, single high-affinity clones are picked up and sequenced

3.3.1 Phage Titering

1. Inoculate 5–10 mL of SB with ER2738 (single colony) and incubate at 37 °C with shaking, for 4–8 h. When the culture reaches mid-log phase, dispense 200 μL into microfuge tubes.
2. While cells are growing, melt top agar in microwave and dispense 3 mL into sterile culture tubes, one per expected phage dilution. Maintain tubes at 45 °C.
3. Prepare serial dilutions of phage in SB. Suggested dilution ranges: for amplified phage culture supernatants, 10^{-8} to 10^{-11}; for unamplified panning eluates, 10^{-1} to 10^{-4}.
4. Add 1 μL of each phage dilution to each tube to carry out infection, vortex quickly, and incubate for 15 min at 37 °C.
5. Transfer the infected cells to culture tubes containing 45 °C top agar. Shake and pour culture onto a pre-warmed LB/IPTG/Xgal plate. Tilt gently and rotate plate to spread top agar.
6. Allow the plates to cool for 5 min at room temperature and incubate overnight at 37 °C.
7. Phage plaques to be counted will appear in blue.

3.3.2 Peptide Panning

Day 1

1. Prepare Ab/PBS 1× solution (generally 300 ng of Ab per well), and coat four flat-bottom wells for each Ab to be characterized.
2. Add 25 μL of Ab-containing solution to each well and swirl gently until the well surfaces are completely covered by the coating solution.
3. Incubate overnight at 4 °C.

Day 2

4. Inoculate 10 mL of SB+Tet medium with a single colony of ER2738. This culture will be used for phage tittering (*see* **Note 7**).
5. Pour off the coating solution from each plate, and firmly slap it face down onto a clean paper towel to remove residual solution. Fill each plate or well completely with blocking buffer. Incubate for at least 1 h at 37 °C.
6. Discard the blocking solution as in **step 5**. Wash each plate rapidly six times with washing solution.
7. Dilute 100-fold a representation of the library with 40 μL of PBS/well. Pipette onto coated plate, and incubate at 37 °C for 2 h.
8. Discard nonbinding phage by pouring off and slapping plate face down onto a clean paper towel.
9. Wash plates ten times with the washing solution (**step 6**).
10. Elute bound phage with 50 μL of elution buffer. Rock gently for 10 min, scrape the wells, pipette eluate into a microcentrifuge tube, and neutralize with 3 μL/well of neutralizing buffer.
11. Titer the eluate generally using 1 μL diluted as described above. Plaques from the first- or second-round eluate titering can be sequenced if desired.
12. Inoculate 1 mL from the ER2738 culture performed at **step 4** in 100 mL SB. Add the eluate phages and incubate for 4.5 h at 37 °C with vigorous shaking.
13. Transfer the culture to a centrifuge tube (50 mL) and spin for 30 min at 4,000 ×*g* at 4 °C. Transfer the supernatant to a fresh tube and respin (discard the pellet).
14. Transfer the upper 80 % of the supernatant to a fresh tube, and add to it 1/6 volume of 20 % PEG/2.5 M NaCl. Allow the phage to precipitate at 4 °C for at least 2 h, preferably overnight.

Day 3

15. Spin the PEG precipitation at 12,000 ×*g* for 20 min at 4 °C. Decant and discard the supernatant, respin the tube briefly, and remove residual supernatant with a pipette.
16. Suspend the pellet in 1 mL of PBS. Transfer the suspension to a microcentrifuge tube and spin at maximum (18,000 ×*g*) for 5 min at 4 °C to pellet residual cells.

17. Transfer the supernatant to a fresh microcentrifuge tube and reprecipitate by adding 1/6 volume of 20 % PEG/ 2.5 M NaCl. Incubate on ice for 60 min. Microcentrifuge at 18,000 RCF for 10 min at 4 °C, discard the supernatant, respin briefly, and remove residual supernatant with a micropipet.
18. Suspend the pellet in 1 mL of PBS. Microcentrifuge for 1 min to pellet any remaining insoluble material. Transfer the supernatant to a fresh tube. This is the amplified eluate.
19. Titer the amplified eluate as described above, on LB/IPTG/ Xgal plates.
20. Coat the plate for the second round of panning as described in **steps 1–3**.

Days 4 and 5

21. Count blue plaques from the titering plates (**step 19**), and determine the phage titer, which should be on the order of $10^{13/14}$ pfu/ml. Succeeding rounds of panning can be carried out with as little as 10^9 pfu of input phage.
22. Carry out a second round of panning: Repeat **steps 4–18** using the first-round amplified eluate as input phage and raising the Tween concentration in the wash steps to 0.5 % (v/v).
23. Titer the resulting second-round amplified eluate on LB/ IPTG/Xgal plates.
24. Coat a plate or a well for the third round of panning as described above.

Day 6

25. Carry out a third round of panning.
26. Titer the unamplified third-round eluate as in **step 11** on LB/ IPTG/Xgal plates. Plaques from this titering can be used for sequencing.

    If there is no clear sequence consensus, carry out a fourth round of panning.

3.4 Data Analysis

The data of any epitope-mapping approach should be analyzed considering those previously obtained in the biological characterization of the mAb of interest, following "standard" experimental procedures (e.g., assays evaluating the Ab biological activity, western blotting, immunoprecipitation, Biacore). This will permit to focus better the data obtained from the three methods described above. The three epitope-mapping experimental approaches described above are just examples among epitope-mapping techniques [1, 7, 9–12]. We focused on these three approaches since the data resulting from all these methods are amino acidic sequences to be further analyzed in silico.

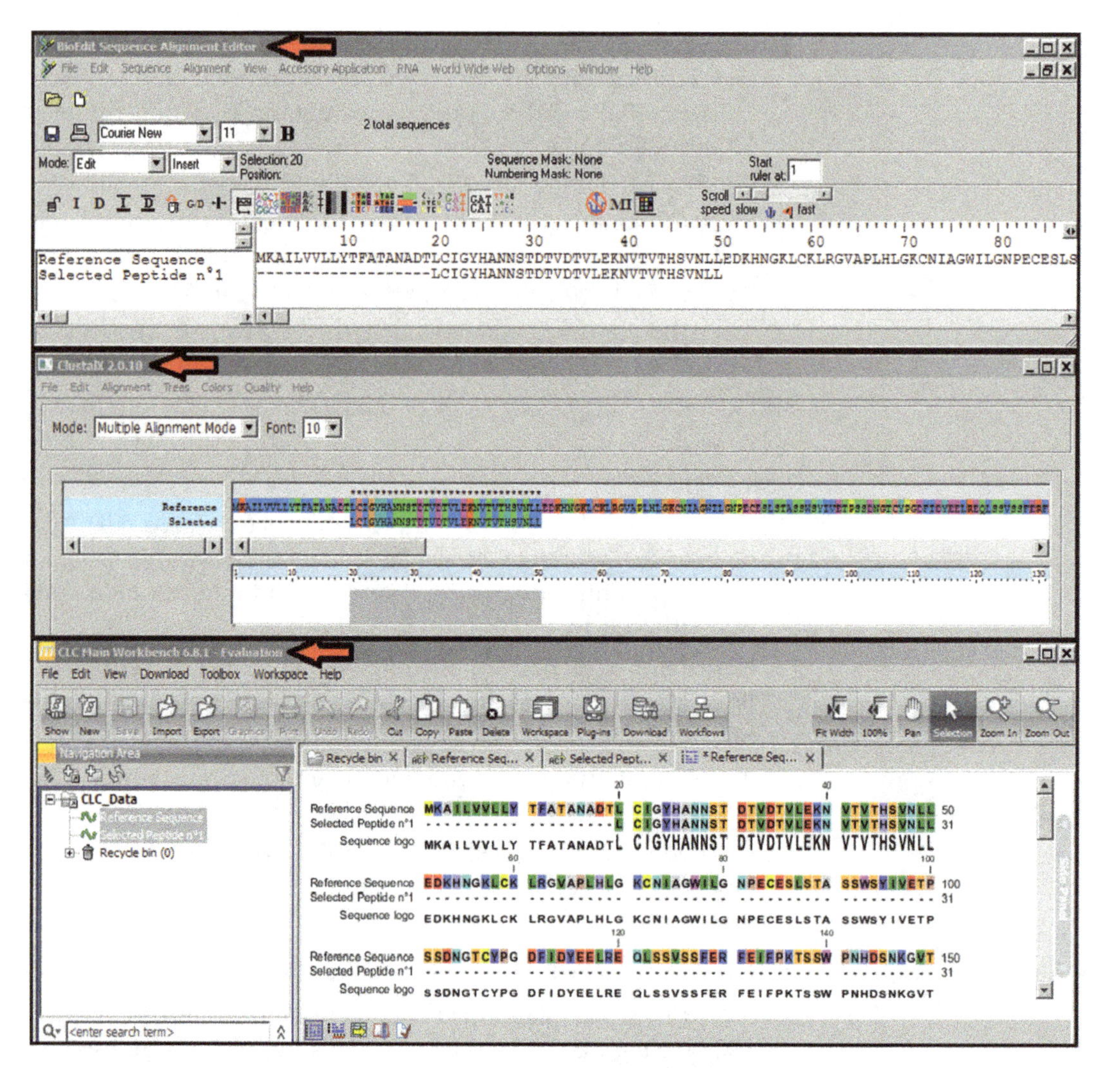

Fig. 4 Comparison between the reference sequence and the peptide sequence deriving from the two mass spectrometric approaches. *Red arrows* highlight three software commonly used for sequence analysis

3.4.1 Analysis of Data Deriving from the Mass Spectrometric Approaches

The amino acid sequences resulting from the mass spectrometric approaches described above should be analyzed by using dedicated database servers, software to "manage" the peptide sequences, and web servers or meta-web servers for the analysis of consensus motifs on the target antigen crystal structures (if available) [1]. In particular, the "custom" procedure should be performed as follows:

Sequence Alignment

The "output" sequence (sequence resulting from the experimental assays) must be checked by aligning it with a reference sequence (usually the amino acid sequence of the protein used to perform the assays). This step can be easily performed by using several programs such as ClustalX (http://www.clustal.org), Bio-Edit (http://www.mbio.ncsu.edu/bioedit/bioedit.html), or CLC-Workbench (http://www.clcbio.com) [13, 14] (Fig. 4). The same result can also be achieved by uploading the sequence(s) (.fasta format) into Protein-Blast server (http://blast.ncbi.nlm.nih.gov/Blast.cgi?PAGE=Proteins). The latter allows for the discovery of sequence analogies among different antigenic "variants" of the same antigen (Fig. 5).

Once the correspondence between the reference linear amino acid sequence and the "output" sequence is checked, it is possible to identify (in the "reference" sequence) the antigenic region involved, i.e., in the Ab–Ag binding. Moreover, by highlighting the different protein functional and/or structural domains (when possible or if they are present) it is possible to evaluate potential consistencies between the epitope "position" in the linear sequence and an eventual biological activity peculiar of the Ab previously characterized (Fig. 6).

Identification of the Putative Epitope on 3D Crystal Structures or Molecule Models

When a crystal structure of the Ag is available, it can be helpful to identify the amino acid residues of the putative Ab-bound protein region on the 3D structure of the protein targeted by the Ab. When more than one structure are available, it can be important to use crystal structures endowed with the highest resolution. To do this, structures of the Ag of interest can be downloaded from Internet databases (e.g., RCSB Protein Data Bank, http://www.rcsb.org/pdb/home/home.do) (Fig. 7).

In order to visualize, edit, and analyze the downloaded crystal structures, it can be useful to use molecular visualization and editing computer programs freely available "online" such as UCSF Chimera, DeepView/SPDBV, VMD (more complex computer program designed for visualizing and analyzing molecular dynamics (MD) simulation and biological systems), RasMol, or Cn3D (Table 1).

Once the crystal structure (usually downloaded with .pdb extension) is loaded onto a rendering program (Table 1) (Fig. 8), the residues identified in the linear amino acid sequence can now be highlighted on the crystal (or model), allowing to appreciate the 3D conformational motifs of the putative Ab-bound region containing the epitope (Fig. 9).

The putative recognized motif, highlighted in the crystal structure, can now be further analyzed. In particular, it can be measured, evaluated for its hydrophilic or hydrophobic proprieties, and compared to epitopes belonging to Ab possibly well characterized, already described, and available for comparative experimental assays.

3.4.2 Analysis of the Affinity-Selected Peptides

The amino acid sequences of the peptide-panning selected peptides can be analyzed in order to find possible consensus motifs on the 3D structure of the Ab-targeted Ag. Also in this case, prior to proceeding with further analysis, the antigen crystal structure (or model) must be downloaded from dedicated protein structure databases (see above). Nowadays, different informatics tools able

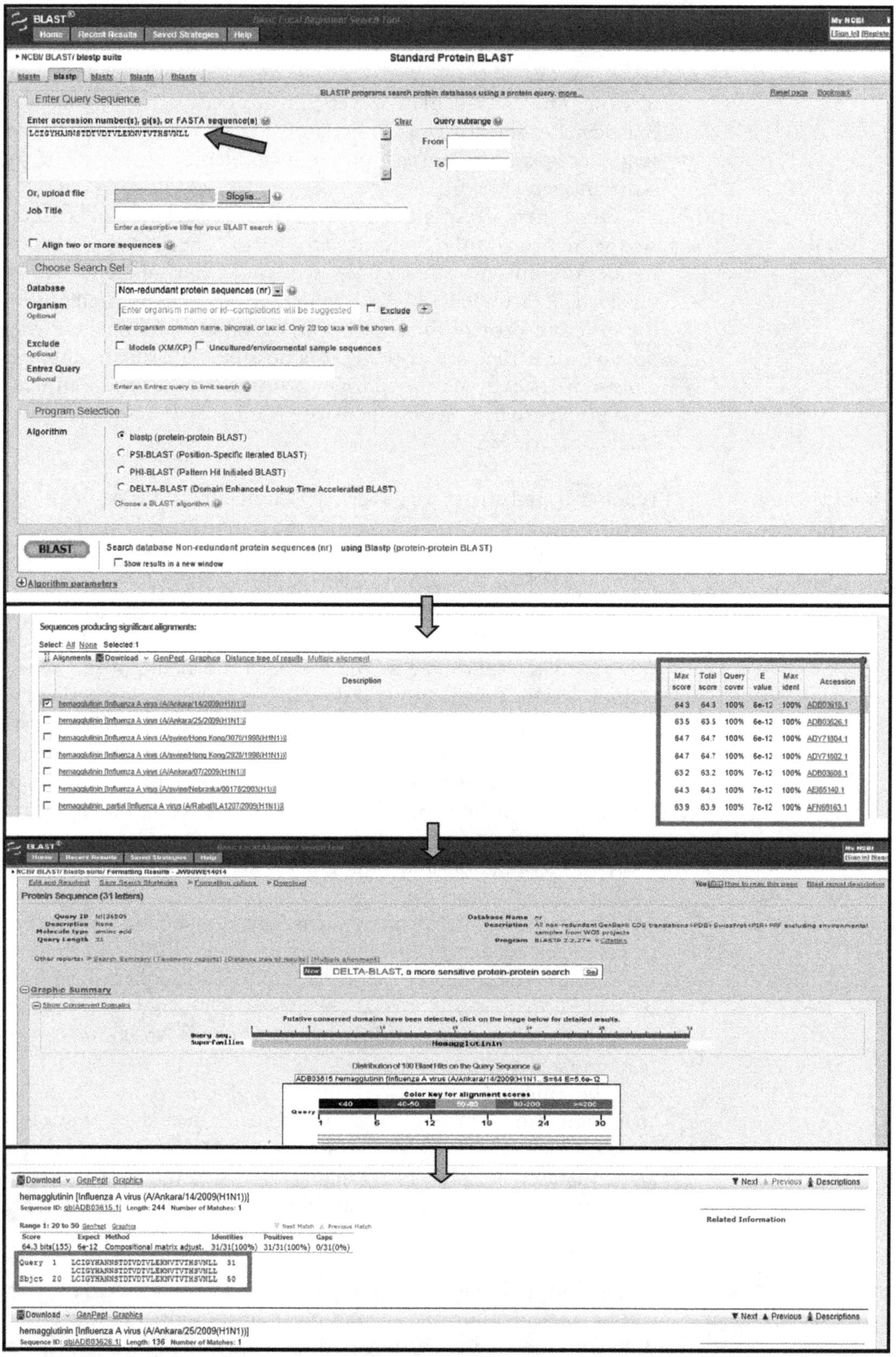

BLAST®

Home Recent Results Saved Strategies Help

NCBI/ BLAST/ blastp suite

Standard Protein BLAST

blastn blastp blastx tblastn tblastx

BLASTP programs search protein databases using a protein query. more...

Enter Query Sequence

Enter accession number(s), gi(s), or FASTA sequence(s)

LCIGYHANNSTDIVDTVLEKNVTVTHSVNLL

Query subrange From To

Or, upload file Sfoglia...

Job Title

Enter a descriptive title for your BLAST search

Align two or more sequences

Choose Search Set

Database: Non-redundant protein sequences (nr)

Organism (Optional): Enter organism name or id--completions will be suggested. Exclude

Enter organism common name, binomial, or tax id. Only 20 top taxa will be shown.

Exclude (Optional): Models (XM/XP) Uncultured/environmental sample sequences

Entrez Query (Optional)

Enter an Entrez query to limit search

Program Selection

Algorithm

- blastp (protein-protein BLAST)
- PSI-BLAST (Position-Specific Iterated BLAST)
- PHI-BLAST (Pattern Hit Initiated BLAST)
- DELTA-BLAST (Domain Enhanced Lookup Time Accelerated BLAST)

Choose a BLAST algorithm

BLAST Search database Non-redundant protein sequences (nr) using Blastp (protein-protein BLAST)

Show results in a new window

Algorithm parameters

Sequences producing significant alignments:

Select: All None Selected:1

Alignments Download GenPept Graphics Distance tree of results Multiple alignment

Description	Max score	Total score	Query cover	E value	Max ident	Accession
hemagglutinin [Influenza A virus (A/Ankara/14/2009(H1N1))]	64.3	64.3	100%	6e-12	100%	ADB03615.1
hemagglutinin [Influenza A virus (A/Ankara/25/2009(H1N1))]	63.5	63.5	100%	6e-12	100%	ADB03626.1
hemagglutinin [Influenza A virus (A/swine/Hong Kong/3070/1998(H1N1))]	64.7	64.7	100%	6e-12	100%	ADY71804.1
hemagglutinin [Influenza A virus (A/swine/Hong Kong/2928/1998(H1N1))]	64.7	64.7	100%	6e-12	100%	ADY71802.1
hemagglutinin [Influenza A virus (A/Ankara/07/2009(H1N1))]	63.2	63.2	100%	7e-12	100%	ADB03608.1
hemagglutinin [Influenza A virus (A/swine/Nebraska/00178/2003(H1))]	64.3	64.3	100%	7e-12	100%	AEI85140.1
hemagglutinin, partial [Influenza A virus (A/Rabat/ILA1207/2009(H1N1))]	63.9	63.9	100%	7e-12	100%	AFN69163.1

NCBI/ BLAST/ blastp suite/ Formatting Results - JW90WE14014

Protein Sequence (31 letters)

Query ID lcl|26505; Description None; Molecule type amino acid; Query Length 31

Database Name nr; Description All non-redundant GenBank CDS translations+PDB+SwissProt+PIR+PRF excluding environmental samples from WGS projects; Program BLASTP 2.2.27+

DELTA-BLAST, a more sensitive protein-protein search

Graphic Summary

Show Conserved Domains

Putative conserved domains have been detected, click on the image below for detailed results.

Query seq. / Superfamilies: Hemagglutinin

Distribution of 100 Blast Hits on the Query Sequence

ADB03615 hemagglutinin [Influenza A virus (A/Ankara/14/2009(H1N1)), S=64 E=5.6e-12

Color key for alignment scores: <40, 40-50, 50-80, 80-200, >=200

Query 1 6 12 18 24 30

Download GenPept Graphics

hemagglutinin [Influenza A virus (A/Ankara/14/2009(H1N1))]

Sequence ID: gb|ADB03615.1| Length: 244 Number of Matches: 1

Range 1: 20 to 50 GenPept Graphics

Score	Expect	Method	Identities	Positives	Gaps
64.3 bits(155)	6e-12	Compositional matrix adjust.	31/31(100%)	31/31(100%)	0/31(0%)

```
Query  1   LCIGYHANNSTDIVDTVLEKNVTVTHSVNLL  31
           LCIGYHANNSTDIVDTVLEKNVTVTHSVNLL
Sbjct  20  LCIGYHANNSTDIVDTVLEKNVTVTHSVNLL  50
```

Download GenPept Graphics

hemagglutinin [Influenza A virus (A/Ankara/25/2009(H1N1))]

Sequence ID: gb|ADB03626.1| Length: 136 Number of Matches: 1

Fig. 5 Protein Blast. An example BLAST search. Once the input sequence is loaded (*red arrow*), the server will process the sequence by questing different databases. The results (*green arrows*) will highlight the database sequences showing the highest identity and higher query coverage (*red box*). Finally, after the selection of the "right" sequence, it will be possible to visualize the peptide sequence aligned with the master sequence found by Blast server (*blue box*) (Color figure online)

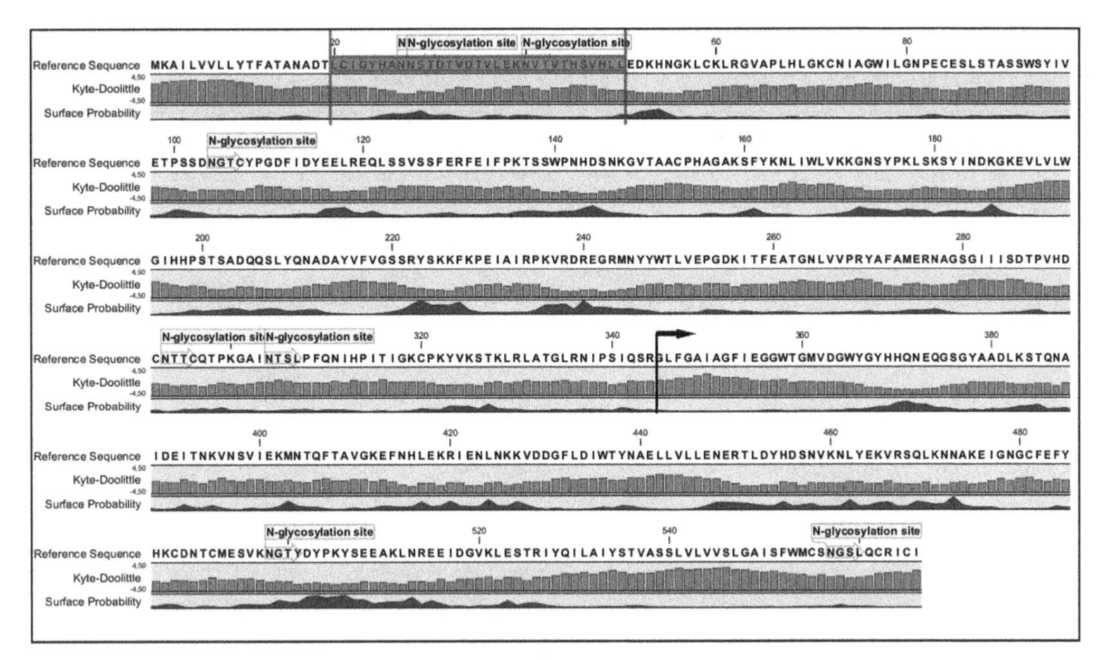

Fig. 6 As an example, the linear amino acid sequence of influenza hemagglutinin (HA)-coding region (reference sequence) has been analyzed. The output sequence deriving from the mass spectrometric approaches described above has been aligned as described in Subheading 3.4.1 and highlighted in *purple*. The HA2 domain, containing the highly hydrophobic fusion peptide, has been marked with a *black arrow*. Kyte–Doolittle analysis of the ref. sequence has been performed to evaluate the protein hydrophilic/hydrophobic regions (*boxes in grey*); moreover, the surface probability for each residues has been calculated for a first screening on the linear sequence (*graphs in green*). Finally the putative glycosylation sites have been highlighted (*yellow flags*). From this first analysis it is possible to draw possible correlations between the putative epitope and the Ab biological features (CLC Workbench) (Color figure online)

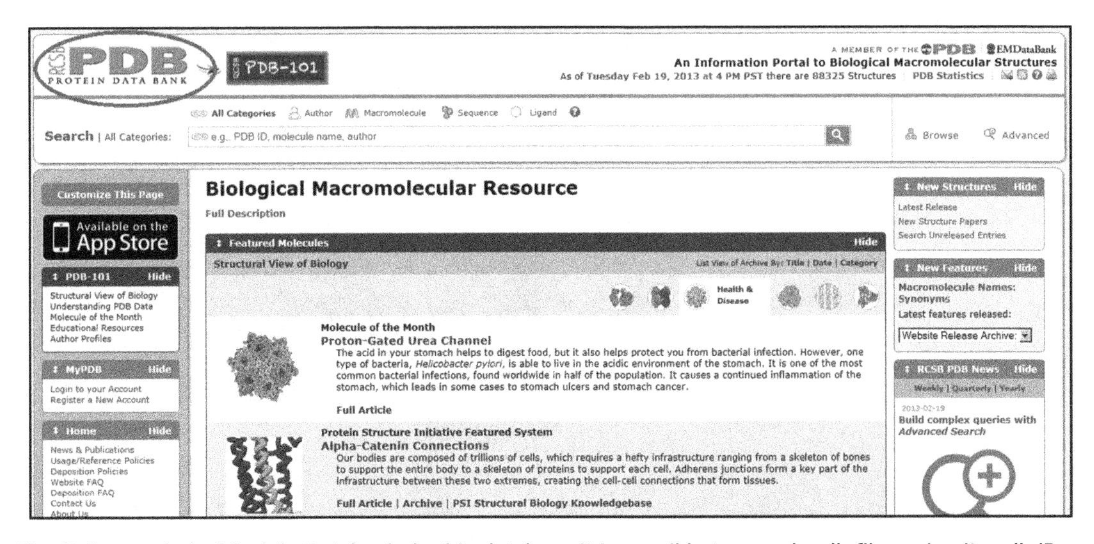

Fig. 7 Screenshot of Protein Databank. In this database it is possible to search pdb files using its pdb ID or alternatively using protein name filtering the results by categories. The file of interest supplies the references describing it and, usually, is possible to download in pdb and fasta format as well

Table 1
List of common molecular visualization and editing computer programs

Program name	Source (URLs)	References
UCSF Chimera	http://www.cgl.ucsf.edu/chimera/	Resource for Biocomputing, Visualization, and Informatics at the University of California, San Francisco, supported by the National Institutes of Health
Ras Mol	http://rasmol.org/	Based on RasMol 2.6 by Roger Sayle Biomolecular Structures Group, Glaxo Wellcome Research & Development Stevenage, Hertfordshire, UK
VMD	http://www.ks.uiuc.edu/Research/vmd/	Humphrey, W. et al.
Cn3D	http://www.ncbi.nlm.nih.gov/Structure/CN3D/cn3d.shtml	NCBI
DeepView/SPDBV	http://www.expasy.org/spdbv/	The Swiss Institute of Bioinformatics (SIB)

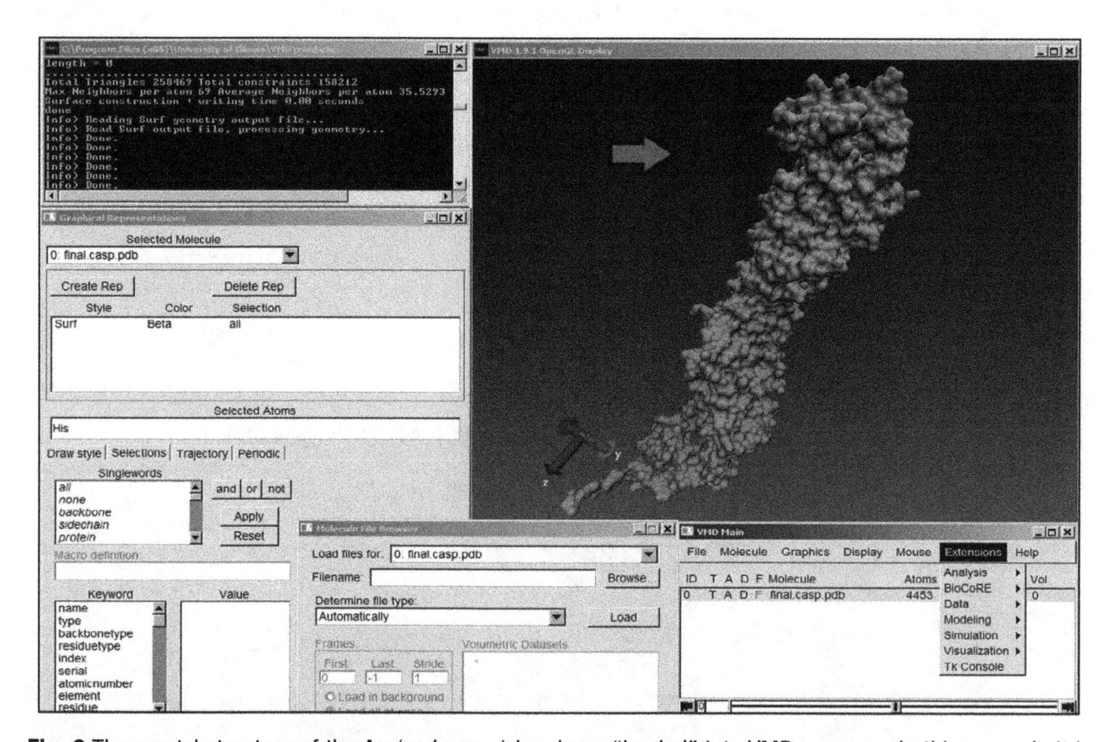

Fig. 8 The crystal structure of the Ag (*red arrow*) has been "loaded" into VMD program. In this example it is possible to appreciate the solvent-accessible surfaces (Color figure online)

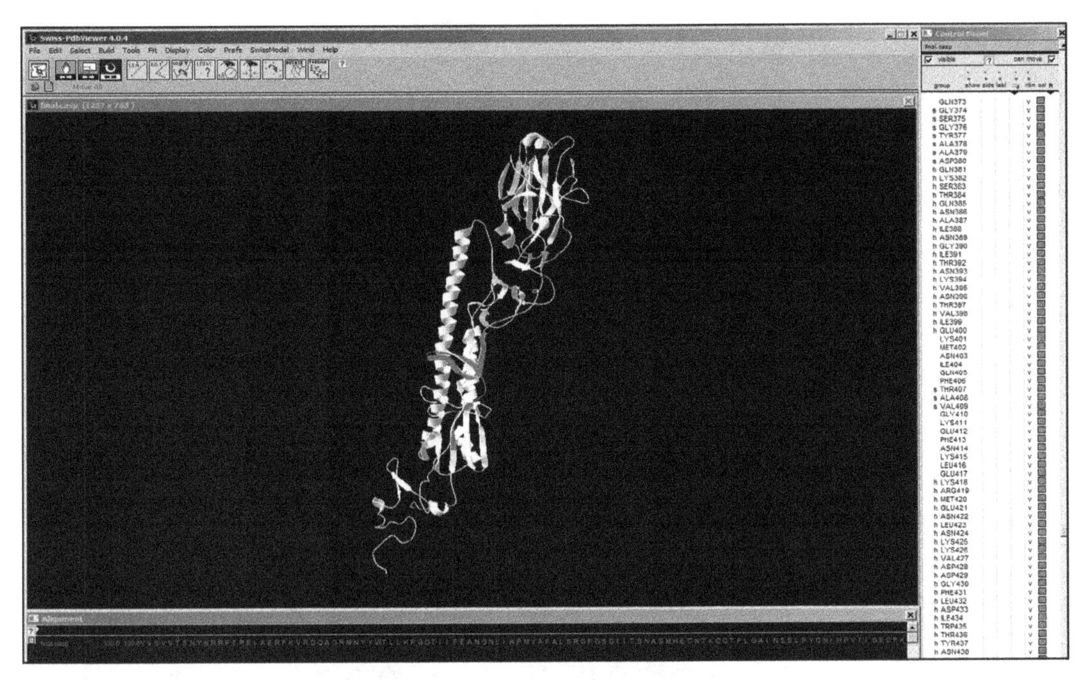

Fig. 9 The same structure (influenza hemagglutinin) edited using SPDBV. The sequence identified by the mass spectrometry experiments has been highlighted in *red*. From this picture it is possible to speculate that the epitope recognized by this Ab lies in a region not involved in the docking with the HA cellular receptor since the HA region involved in the receptor recognition lies on the "top" of this molecule (the so-called globular head). This can be easily confirmed by experimental approaches (Color figure online)

to perform the analysis of affinity-selected peptides have been developed (e.g., Mimox, http://immunet.cn/mimox; Pepitope Server, http://pepitope.tau.ac.il) [15, 16]. Generally, these tools allow for the identification of putative consensus motifs on the basis of the principal characteristics of the key consensus residues shared among the different affinity-selected peptides, by applying different algorithms. Here, we report the example of "user-friendly" server named Pepitope Server. This server is generally used to computationally predict epitopes based on selected peptide sequences extracted from a phage display library or to align a linear peptide sequence onto a three-dimensional protein structure. In brief, the first program output is the alignment of each affinity-selected phage-displayed peptide to the 3D structure of the Ag. Once few peptides have been aligned, the server adds on a clustering algorithm to find one or more patches of residues, corresponding to putative epitopes, on the surface of the 3D structure. This second output is the predicted Ab-interacting surface Ag region. Interestingly, Pepitope eliminates the non-solvent-exposed amino acid residues, assuming that the affinity-selected peptides mimic solvent-exposed amino acid residues [17]. Finally, the server allows to run three different epitope-mapping algorithms (PepSurf, Mapitope, or a combined analysis) (Fig. 10).

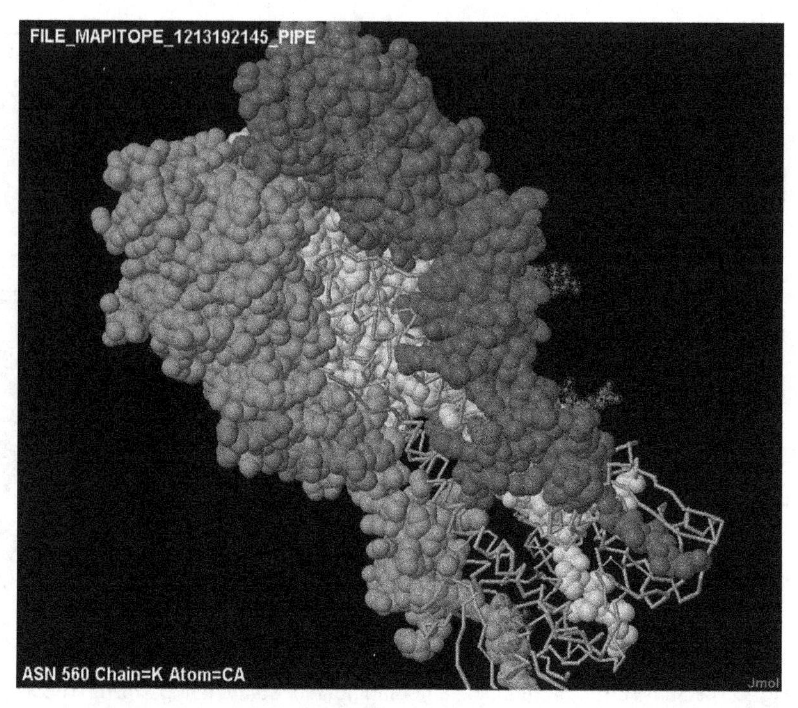

Fig. 10 Pepitope server screenshot. As described above, the Pepitope input files are Protein Data Bank files (".pdb" file extension) and the set of affinity-selected phage-displayed peptide sequences (".fasta" file format). *Red arrow* indicates the three algorithms purposed for the analysis. *Black box* indicates that it is possible to load a .pdb file or alternatively enter the Ag pdb ID. The affinity-selected peptides must be uploaded in .fasta format (*green arrow*). Several advanced settings can be also selected on the basis of the phage-displayed library used in the experimental session (Color figure online)

Once the proper algorithm is selected, the server will generate output files showing the putative epitope positions (Fig. 11).

3.4.3 Final Notes

As previously described [1, 8], the data generated by the examples of in silico analysis reported here must not leave aside the simultaneous generation of empirical data. Experimental results are essential to show a relationship between the epitope-mapping results and the biological features of the antibody being characterized. In addition, preliminary data generated as above can be further confirmed by experimental techniques such as alanine scanning or Pep-scan techniques.

4 Notes

1. Prepare it just before use: This step is required only for the indirect immunosorption approach.
2. For the indirect immunosorption approach, an Fc-specific antibody is required in this step in order to obtain a defined

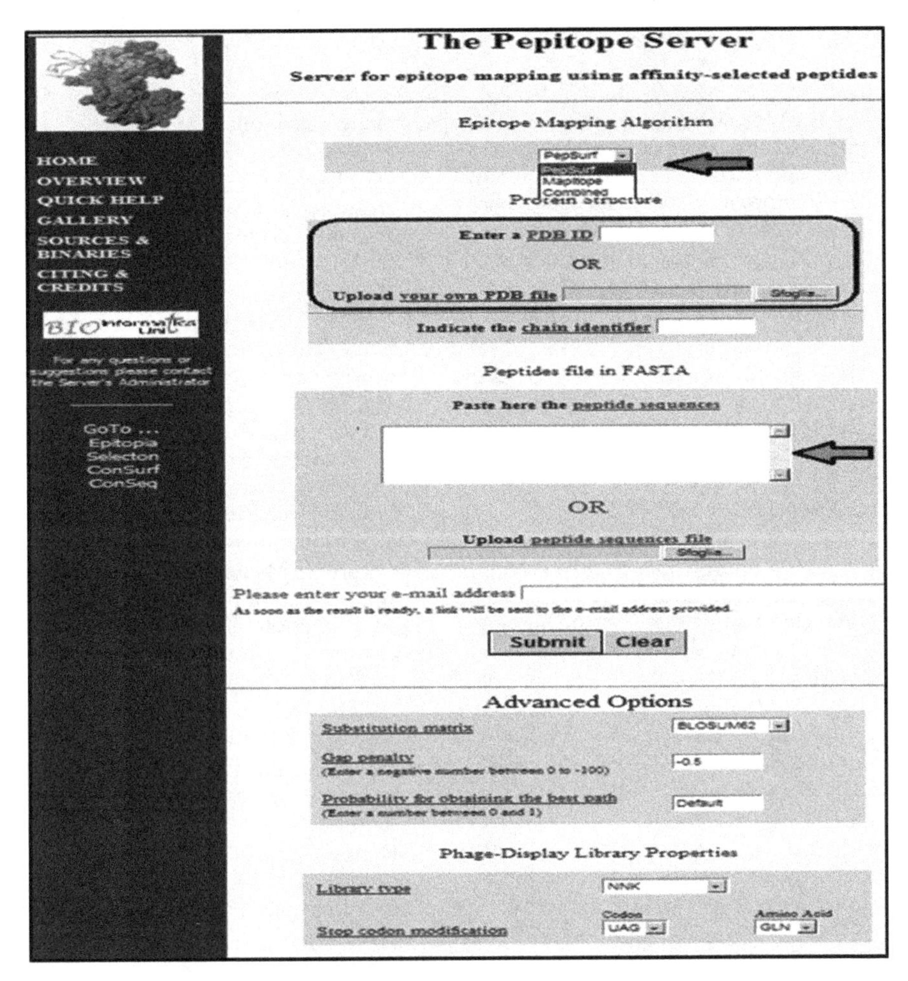

Fig. 11 Epitope (*red* and *purple*) predicted by Pepitope server's Mapitope algorithm on the influenza hemagglutinin trimeric form. Data visualized using Jmol: an open-source Java viewer for chemical structures in 3D (http://www.jmol.org/) (Color figure online)

orientation of the antibody of interest in which the paratopes are exposed to the solution.

3. This step is required only for the indirect immunosorption approach.
4. This step is required in order to block any reactive sites on the beads.
5. **Steps 11–18** are required only for the indirect immunosorption approach.
6. This step quenches the exchange reactions and elutes out the antigen from the antibody column.
7. If amplifying the eluted phage on the same day (*see* **step 12**), also inoculate 10 mL of SB medium. Incubate both cultures at 37 °C with vigorous shaking.

References

1. Clementi N, Mancini N, Castelli M, Clementi M, Burioni R (2013) Characterization of epitopes recognized by monoclonal antibodies: experimental approaches supported by freely accessible bioinformatic tools. Drug Discov Today 18(9–10):464–471
2. Hager-Braun C, Tomer KB (2005) Determination of protein-derived epitopes by mass spectrometry. Expert Rev Proteomics 2: 745–756
3. Coales SJ, Tuske SJ, Tomasso JC, Hamuro Y (2009) Epitope mapping by amide hydrogen/deuterium exchange coupled with immobilization of antibody, on-line proteolysis, liquid chromatography and mass spectrometry. Rapid Commun Mass Spectrom 23:639–647
4. Birkenmeier G, Osman AA, Kopperschlager G, Mothes T (1997) Epitope mapping by screening of phage display libraries of a monoclonal antibody directed against the receptor binding domain of human alpha2-macroglobulin. FEBS Lett 416:193–196
5. Dore JM, Morard F, Vita N, Wijdenes J (1998) Identification and location on syndecan-1 core protein of the epitopes of B-B2 and B-B4 monoclonal antibodies. FEBS Lett 426:67–70
6. Rowley MJ, Scealy M, Whisstock JC, Jois JA, Wijeyewickrema LC, Mackay IR (2000) Prediction of the immunodominant epitope of the pyruvate dehydrogenase complex E2 in primary biliary cirrhosis using phage display. J Immunol 164:3413–3419
7. Clementi N et al (2011) A human monoclonal antibody with neutralizing activity against highly divergent influenza subtypes. PLoS One 6:e28001
8. Clementi N, Criscuolo E, Castelli M, Mancini N, Clementi M, Burioni R (2012) Influenza B-cells protective epitope characterization: a passkey for the rational design of new broad-range anti-influenza vaccines. Viruses 4: 3090–3108
9. De Marco D et al (2012) A non-VH1-69 heterosubtypic neutralizing human monoclonal antibody protects mice against H1N1 and H5N1 viruses. PLoS One 7:e34415
10. Mancini N et al (2009) Hepatitis C virus (HCV) infection may elicit neutralizing antibodies targeting epitopes conserved in all viral genotypes. PLoS One 4:e8254
11. Mancini N et al (2006) Cloning and molecular characterization of a human recombinant IgG Fab binding to the Tat protein of human immunodeficiency virus type 1 (HIV-1) derived from the repertoire of a seronegative patient. Mol Immunol 43:1363–1369
12. Cardoso RM et al (2005) Broadly neutralizing anti-HIV antibody 4E10 recognizes a helical conformation of a highly conserved fusion-associated motif in gp41. Immunity 22:163–173
13. Larkin MA et al (2007) Clustal W and Clustal X version 2.0. Bioinformatics 23:2947–2948
14. Tippmann HF (2004) Analysis for free: comparing programs for sequence analysis. Brief Bioinform 5:82–87
15. Huang J, Gutteridge A, Honda W, Kanehisa M (2006) MIMOX: a web tool for phage display based epitope mapping. BMC Bioinformatics 7:451
16. Mayrose I et al (2007) Pepitope: epitope mapping from affinity-selected peptides. Bioinformatics 23:3244–3246
17. Tsodikov OV, Record MT Jr, Sergeev YV (2002) Novel computer program for fast exact calculation of accessible and molecular surface areas and average surface curvature. J Comput Chem 23(6):600–609. PMID: 11939594

Chapter 27

Fine Epitope Mapping Based on Phage Display and Extensive Mutagenesis of the Target Antigen

Gertrudis Rojas

Abstract

The residues contributing to the formation of the epitope recognized by a monoclonal antibody can be defined in several ways. Mutagenesis on the target antigen, followed by screening of the reactivity of the new variants with the antibody, is particularly powerful to reveal the functional contribution of each amino acid in the context of the native antigen. The current protocol provides a relatively simple procedure to study the surface of the target antigen in the search for residues involved in recognition. If the antigen is successfully displayed on the surface of filamentous bacteriophages, it can be quickly scanned by simultaneous mutagenesis of multiple solvent-exposed residues and high-throughput screening of the new variants with the antibody to be mapped. Once a few amino acids critically involved in recognition are defined, they can be used as starting points for a comprehensive exploration of the antigenic region by randomization of their whole neighborhood. The analysis of binding and sequence data allows delineating a detailed picture of the epitope recognized by the antibody under investigation.

Key words Antigen, ELISA screening, Epitope mapping, Kunkel mutagenesis, Monoclonal antibody, Phage display, Randomization

1 Introduction

Monoclonal antibodies (mAbs) recognizing different epitopes on the same target antigen can have different (and even opposite) biological effects [1]. That is why deciphering the molecular details of the antibody/antigen interaction helps to understand antibody-mediated functions and to target biological systems in the desired way [2]. Several epitope mapping approaches have been developed in order to precisely locate the antigenic region that is being recognized by a given antibody. These methods can be divided into two main categories: direct structural studies of the antigen/antibody complexes (mainly by X-ray crystallography) [3] and functional mapping techniques. The latter procedures, based on characterizing the impact of antigen changes on the recognition by the antibody,

Vincent Ossipow and Nicolas Fischer (eds.), *Monoclonal Antibodies: Methods and Protocols*, Methods in Molecular Biology, vol. 1131, DOI 10.1007/978-1-62703-992-5_27, © Springer Science+Business Media New York 2014

are useful to assess the individual contributions of antigen residues to the epitope [4]. Functional or energetic epitopes are thus defined.

Functional mapping studies are limited by the number of antigen variants that can be tested. Combinatorial biology procedures like yeast display have greatly expanded their potential, since large libraries derived from the original antigen through genetic diversification by error-prone polymerase chain reaction (PCR) can be constructed and screened [5]. Phage display is faster and allows the manipulation of even more molecules [6]. The use of random phage-displayed peptide libraries has indeed become very popular for a different mapping approach, based on the isolation of peptide sequences recognized by the antibody, presumably mimicking the epitope [7]. Finding of the similarities between the antigenic peptides and surface patches of the antigen is then used to predict the location of the original epitope. Unfortunately, selected peptides do not always resemble the nominal antigen in terms of the binding details [8]. Sometimes they establish a new network of interactions with the paratope [9], so there is no structural or even functional antigen mimicry by peptides.

The current protocol combines the high-throughput potential of phage display with the advantages of directly exploring the contribution of residues by mutagenesis in the context of the antigen itself. The first critical step to use this mapping approach is the ability to obtain a phage-displayed form of the target antigen which folds properly and is recognized by the antibody under investigation. Once this goal is achieved, the next step is to find a discrete set of solvent-exposed residues that can be individually replaced in order to assess their contribution to the epitope formation. When the antibody is not cross-reactive with a second protein homologous to the antigen (sharing certain degree of sequence identity with it and originating from a different species), the first mutagenesis round can be easily designed. Replacing every individual residue in the target antigen by its counterpart in the non-cross-reactive protein should reveal at least one amino acid which contributes to the epitope formation and is associated with the lack of cross-reactivity.

Enzyme-linked immunosorbent assay (ELISA) screening of this initial panel of variants constructed through modified Kunkel mutagenesis [10, 11] thus highlights the importance of a few residues that are critically involved in recognition, pointing to the recognized antigenic region, which can be more comprehensively explored in a second mutagenesis round by randomizing all the accessible residues in the whole area. The complete profile of tolerated and non-tolerated replacements at multiple positions delineates a functional picture of the epitope. The current protocol has been successfully used in our laboratory to map the epitopes recognized by several antibodies [2].

2 Materials

2.1 Microorganisms

1. Use TG1 *Escherichia coli* (*E. coli*) strain (K12_(*lac-pro*), *sup*E, *thi*, *hsd*D5/F′ *tra*D36, *pro*A$^+$B$^+$, *lac*I^q, *lac*Z_M15) to rescue phagemid-containing viral particles displaying the original antigen and its mutated variants.
2. Use CJ236 *E. coli* strain (*dut*$^-$ *ung*$^-$ *thi*-1 *relA1 spoT1 mcrA*/pCJ105 (F′ *cam*^{r})) to obtain single-stranded DNA template for site-directed mutagenesis and/or randomization.
3. Use XL-1 Blue *E. coli* strain (*rec*A1 *end*A1 *gyr*A96 *thi*-1 *hsd*R17 *sup*E44 *rel*A1 *lac* F′ *pro*AB *lac*IqZ_M15 Tn10 Tet^r) to obtain plasmid DNA for sequencing.

2.2 Helper Phage

Use M13KO7 helper phage to rescue phagemid-containing viral particles.

2.3 Culture Media for E. coli

1. Prepare 2XTY by dissolving tryptone (16 g/L), yeast extract (10 g/L), and sodium chloride (5 g/L) in water. Sterilize by autoclaving during 20 min at 120 °C and 1 atm.
2. Add 1.5 % (w/v) of bacteriological agar to liquid 2XTY to obtain solid 2XTY. Sterilize by autoclaving during 20 min at 120 °C and 1 atm.
3. Prepare stock solutions of the following additives in sterilized water and filter through 0.2 μm disposable filters:
   (a) Glucose (40 %, w/v).
   (b) Ampicillin (100 mg/mL).
   (c) Kanamycin (70 mg/mL).
   (d) Chloramphenicol (5 mg/mL).
   (e) Uridine (0.25 mg/mL).
4. Dissolve tetracycline in ethanol (5 mg/mL) to obtain a stock solution.
5. Supplement 2XTY and solid 2XTY with the additives described above immediately before use (at the final concentrations described in the protocol).
6. Prepare 2XTY/AG by supplementing 2XTY with ampicillin (100 μg/mL) and glucose (2 % w/v).
7. Prepare 2XTY/AK by supplementing 2XTY with ampicillin (100 μg/mL) and kanamycin (70 μg/mL).

2.4 Plastic Materials and Glassware

1. Use always aerosol-resistant filtered pipette tips to manipulate samples containing phages. This applies to M13KO7 helper phage, phage-containing supernatants, and purified phagemid-containing viral particles.

2. Leave all the glassware (like Erlenmeyer flasks) and non-disposable plastic materials (such as centrifuge bottles) that have been in contact with phage samples (M13KO7 helper phage, phage-containing supernatants, or purified phagemid-containing viral particles) in 2 % bleach overnight to decontaminate.
3. Use only disposable gamma-irradiated plastic materials for *E. coli* culture, DNA purification, phage production and purification, and site-directed mutagenesis:
   (a) Pipette tips, aerosol-resistant filtered pipette tips, vials, and PCR tubes/plates.
   (b) 50 mL conical bottom tubes and round-bottom 96-deep-well polypropylene plates (2 mL maximal volume/well).
   (c) Gas-permeable adhesive lids and freezing-resistant adhesive lids for 96-well plates.
   (d) Petri dishes, 12-well (flat bottom) and 96-well (U bottom) tissue culture plates.
4. Sterilize decontaminated glassware and non-disposable plastic materials during 20 min at 120 °C and 1 atm before using for *E. coli* culture, DNA purification, and phage production and purification.

2.5 Solutions for Phage Purification

1. Prepare PEG/NaCl precipitation solution by dissolving polyethylene glycol 6000 (20 % w/v) and sodium chloride (2.5 mol/L) in water. Sterilize by autoclaving during 20 min at 120 °C and 1 atm.
2. Prepare 10× phosphate-buffered saline (PBS), by dissolving sodium chloride (1.5 mol/L), dibasic sodium phosphate (80 mmol/L), and potassium monobasic phosphate (20 mmol/L) in water. Adjust pH to the range 7.2–7.4. Sterilize by autoclaving during 20 min at 120 °C and 1 atm.
3. Prepare PBS by diluting 100 mL of 10× PBS up to 1 L with sterilized water in a previously sterilized bottle.
4. Prepare 60 % glycerol solution (v/v) in water. Sterilize by autoclaving during 20 min at 120 °C and 1 atm.

2.6 Reagents, Materials, and Solutions for ELISA Screening

1. Use polyvinyl chloride microplates as the solid support.
2. Use 4 % (w/v) skim powder milk in PBS (M-PBS) to block the plates and to dilute phage samples as well as the conjugated antibody.
3. Prepare washing solution by diluting Tween 20 in water at a final concentration of 0.1 % (v/v).

4. The protocol is optimized to use the anti-M13/horseradish peroxidase conjugate provided by GE Healthcare (1/5,000 working dilution) to detect bound phages.
5. Prepare a solution of dibasic sodium phosphate (200 mmol/L) in water. Prepare a second solution of citric acid (100 mmol/L) in water. Adjust the pH of the phosphate solution to 5.0 by adding the citric acid solution to obtain peroxidase substrate buffer. Keep the buffer at 4 °C.
6. Prepare peroxidase substrate solution immediately before use by adding 5 mg of *o*-phenylenediamine and 5 μL of hydrogen peroxide (30 % v/v) to 10 mL of peroxidase substrate buffer.
7. Dilute HCl (10 % v/v) in water to obtain the stop solution.

2.7 Kits for DNA Purification

1. Use QIAprep Spin M13 kit solutions and columns (Qiagen) to purify single-stranded DNA template, with a modified procedure described in the protocol.
2. Use QIAprep Spin minikit (Qiagen) to purify plasmid DNA for sequencing, according to the manufacturer's instructions.

2.8 Reagents and Solutions for Site-Directed Mutagenesis

1. The protocol is optimized to use the following DNA-modifying enzymes provided by New England Biolabs:
   (a) T4 Polynucleotide kinase (10,000 U/mL).
   (b) T7 DNA polymerase (unmodified) (10,000 U/mL).
   (c) T4 DNA ligase (400,000 U/mL).
2. Use solutions of ATP (10 mmol/L) and dNTPs (10 mmol/L each) also provided by New England Biolabs.
3. Prepare 10× TM buffer by dissolving Tris (500 mmol/L) and magnesium chloride (100 mmol/L) in water. Adjust pH to 7.5 with HCl. Sterilize by autoclaving during 20 min at 120 °C and 1 atm.
4. Prepare 100 mmol/L dithiothreitol (DTT) solution in sterilized water. Filter through 0.2 μm disposable filters. Make individual aliquots in vials and keep at −20 °C.
5. Use sterilized water to dilute mutagenic oligonucleotides and to prepare the mutagenesis reactions.

2.9 Software

The software CLUSTALX, Swiss PDB Viewer, and Pymol can be downloaded from:

http://www.clustal.org/download/current/

http://spdbv.vital-it.ch/download/binaries/SPDBV_4.10_PC.zip

http://sourceforge.net/projects/pymol/files/latest/download

3 Methods

3.1 Displaying the Target Antigen on Filamentous Phages

1. Select the antigen (or antigen fragment) to be displayed (*see* **Note 1**).
2. Synthesize the gene coding for the target antigen, flanked by *Apa*LI and *Not*I unique restriction sites. Optimize the DNA sequence according to *E. coli* codon usage, and remove any unpaired Cys by replacing the corresponding codon by a triplet encoding a Ser residue (*see* **Note 2**).
3. Clone the gene coding for the target antigen between *Apa*LI and *Not*I restriction sites into pCSM phagemid vector (Fig. 1) using standard DNA cloning procedures (*see* **Note 3**).
4. Transform TG1 *E. coli* competent cells with the resulting genetic construct, and grow transformed cells on solid 2XTY/AG during 16–20 h at 37 °C.
5. Pick an isolated colony, and inoculate it into a 50 mL tube containing 10 mL of 2XTY/AG (*see* **Note 4**). Grow at 37 °C with shaking at 250 rpm until the culture reaches an absorbance at 600 nm in the range 0.4–0.8 (*see* **Note 5**).
6. Add 10^{11} plaque-forming units (pfu) of M13KO7 helper phage (*see* **Note 6**) to grown *E. coli* and incubate at 37 °C during 30 min without shaking.

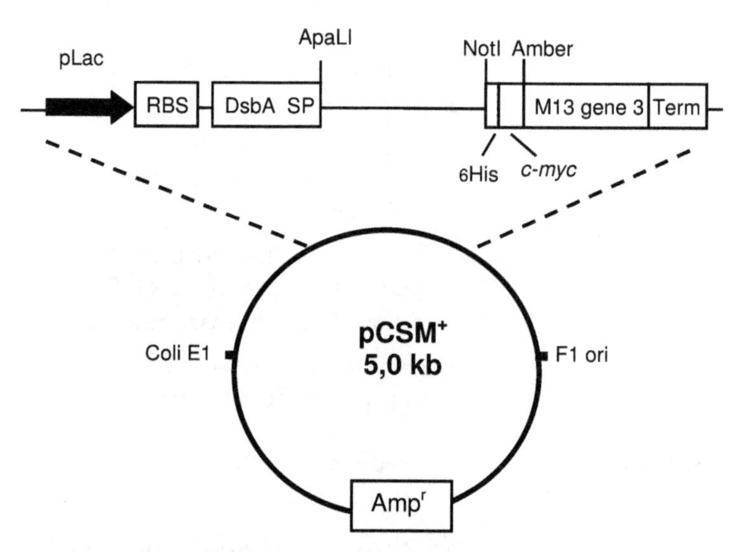

Fig. 1 Schematic representation of the pCSM+ phagemid vector. The vector includes DNA sequences coding for DsbA signal peptide (MKKIWLALAGLVLAFSASA) and hexahistidine and *c-myc* tags. Additional elements are shown: *lac* promoter, ribosomal binding site, amber stop codon, *Apa*LI and *Not*I restriction sites, M13 gene 3, transcription terminator sequence, *E. coli* and phage replication origins, and ampicillin resistance gene

7. Centrifuge at 2,000 × *g* during 15 min. Carefully remove the supernatant.
8. Resuspend the cell pellet in 40 mL of 2XTY/AK contained in a 250 mL Erlenmeyer flask.
9. Grow the cells at 28 °C during 16–20 h with shaking at 250 rpm.
10. Centrifuge at 4,000 × *g* during 15 min at 4 °C.
11. Collect the supernatant, and mix it with 10 mL of PEG/NaCl solution. Incubate during 1 h on ice to precipitate phages (*see* **Note 7**).
12. Centrifuge at 4,000 × *g* during 15 min at 4 °C. Carefully remove the supernatant.
13. Resuspend the phage pellet in 1 mL of PBS and transfer to a vial.
14. Centrifuge at 10,000 × *g* during 10 min at 4 °C in a microcentrifuge to remove the remaining *E. coli* and cell debris.
15. Mix the supernatant with 250 μL of PEG/NaCl solution in a new vial (*see* **Note 8**). Incubate during 20 min on ice to precipitate phages.
16. Centrifuge at 10,000 × *g* during 10 min at 4 °C in a microcentrifuge. Carefully remove the supernatant.
17. Gently resuspend the phage pellet in 1 mL of PBS (*see* **Note 9**). Add 500 μL of 60 % glycerol solution to the vial and mix.
18. Keep purified phages at −20 °C, and characterize the antigenicity of the phage-displayed target antigen as described in Subheading 3.2.

3.2 Confirming the Antigenicity of the Phage-Displayed Target Antigen

1. Prepare coating solutions by diluting separately the following antibodies in PBS at a final concentration of 10 μg/mL: anti-*c-myc* tag 9E10 mAb, unrelated control mAb, and mAbs against the target antigen (*see* **Note 10**).
2. Coat different sections of the polyvinyl chloride ELISA microplates with 100 μL/well of each coating solution. Cover the plates with a lid and incubate them during 16–20 h at 4 °C (*see* **Note 11**).
3. Discard the coating solutions by inverting the plates several times.
4. Add 200 μL/well of 4 % M-PBS to block the plates. Cover the plates with a lid and incubate for 30 min at room temperature (RT).
5. Discard the blocking solution by inverting the plates several times.
6. Add 100 μL/well of the phage samples (previously diluted in M-PBS) to the wells coated with the different antibodies (*see* **Note 12**). Add M-PBS only to some coated and blocked wells

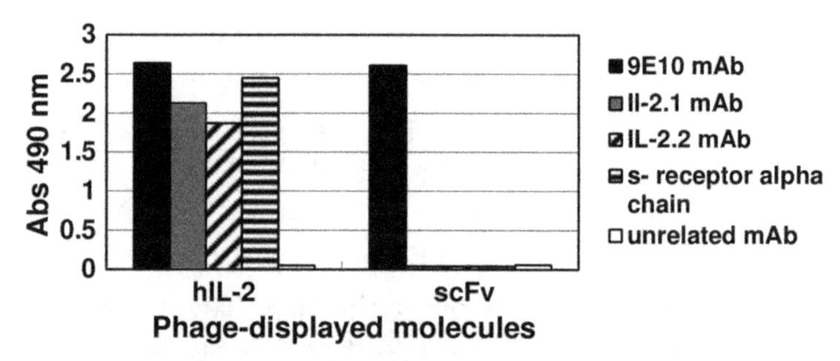

Fig. 2 Typical ELISA results confirming the antigenicity of the phage-displayed target antigen. Phages displaying either human interleukin-2 (hIL-2) or a control protein (unrelated single-chain Fv antibody fragment, scFv) were added to polyvinyl chloride plates coated with the anti-*c-myc* tag 9E10 mAb, anti-IL-2 mAbs (IL-2.1 and IL-2.2), an unrelated antibody, and soluble alpha chain of hIL-2 receptor. Bound phages were detected with an anti-M13 mAb conjugated to horseradish peroxidase. Specific recognition of phage-displayed hIL-2 by both anti-IL2 mAbs and by the receptor chain was shown

in order to assess the level of background. Coat the plates with a lid and incubate during 1 h at RT.

7. Discard the samples by inverting the plates several times, and wash the plates at least five times by filling the wells with washing solution with a washing bottle and discarding it (*see* **Note 13**).
8. Add 100 μL/well of the anti-M13/peroxidase conjugate (diluted 1/5,000 in M-PBS). Coat the plates with a lid and incubate during 1 h at RT.
9. Wash the plates at least eight times as described in **step** 7.
10. Add 100 μL/well of peroxidase substrate solution. Incubate during 15 min at RT. A yellow/orange color should develop in some wells, while the wells containing no phage samples and all the wells coated with the unrelated antibody should remain without any color (indicating low background levels).
11. Stop the reaction with 50 μL/well of the stop solution.
12. Read the absorbances at 490 nm with a microplate reader.
13. Analyze background levels (*see* **Note 14**) and the results with control phages (*see* **Note 15**) in order to assess the usefulness of the assay.
14. Analyze the results with the phage-displayed target antigen in order to confirm its antigenicity (*see* **Note 16**). Typical results showing proper display of a model antigen are shown in Fig. 2.

3.3 Producing a Single-Stranded DNA Template for Antigen Mutagenesis

1. Transform CJ236 *E. coli* competent cells with the genetic construct obtained after **step 3** of Subheading 3.1, and grow transformed cells in solid 2XTY/AG during 16–20 h at 37 °C (*see* **Note 17**).

2. Pick an isolated colony (*see* **Note 18**), and inoculate it into a 50 mL tube containing 10 mL of 2XTY/AG supplemented with 5 μg/mL chloramphenicol (*see* **Note 19**). Grow at 37 °C with shaking at 250 rpm until the culture reaches an absorbance at 600 nm in the range 0.4–0.8 (*see* **Note 20**).
3. Add 10^{11} pfu of M13KO7 helper phage (*see* **Note 6**) and incubate at 37 °C during 30 min without shaking.
4. Centrifuge at 2,000 × *g* during 15 min. Carefully remove the supernatant.
5. Resuspend the cell pellet in 40 mL of 2XTY/AK supplemented with 2 % glucose and 0.25 μg/mL uridine, contained in a 250 mL Erlenmeyer flask (*see* **Note 21**).
6. Grow the cells at 37 °C during 20 h with shaking at 250 rpm (*see* **Note 22**).
7. Centrifuge at 4,000 × *g* during 15 min at 4 °C.
8. Collect the supernatant, and mix it with 10 mL of PEG/NaCl solution. Incubate during 1 h on ice to precipitate phages (*see* **Note 23**).
9. Centrifuge at 4,000 × *g* during 15 min at 4 °C. Carefully remove the supernatant.
10. Resuspend the phage pellet in 0.5 mL of PBS and transfer to a vial.
11. Centrifuge at 10,000 × *g* during 10 min in a microcentrifuge to remove the remaining *E. coli* and cell debris.
12. Mix the supernatant in a new vial with 10 μL of MP solution from the QIAprep Spin M13 kit. Incubate for 2 min at RT (*see* **Note 24**).
13. Add the mixture to a QIA column. Centrifuge for 1 min at 6,000 × *g*. Discard the flow-through.
14. Add 0.7 mL of MLB solution from the QIAprep Spin M13 kit to the column. Centrifuge for 1 min at 6,000 × *g*. Discard the flow-through.
15. Add 0.7 mL of MLB solution from the QIAprep Spin M13 kit to the column. Incubate for 1 min at RT.
16. Centrifuge for 1 min at 6,000 × *g*. Discard the flow-through.
17. Add 0.75 mL of PE solution from the QIAprep Spin M13 kit (already containing ethanol as recommended by the manufacturer) to the column. Centrifuge for 1 min at 6,000 × *g*. Discard the flow-through.
18. Repeat **step 17**.
19. Transfer the column to an empty vial and centrifuge again for 1 min at 6,000 × *g* to remove residual PE.

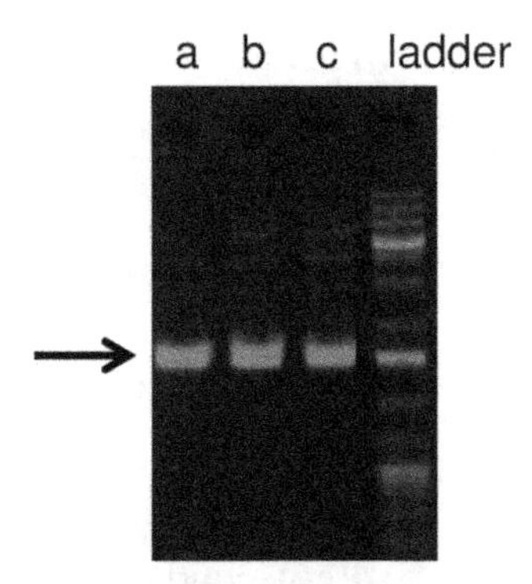

Fig. 3 Typical results of agarose gel electrophoresis of purified single-stranded DNA. A main band (indicated by the *arrow*) was observed after running single-stranded DNA samples (a, b, and c) on an agarose gel (1 %). Less intense bands with lower electrophoretic mobility were seen in some of the samples. GeneRuler 1 kb Plus DNA ladder was included

20. Transfer the column to a new vial. Add 100 μL of EB solution from the QIAprep Spin M13 kit to the center of the column and incubate for 10 min at RT in order to elute DNA.
21. Centrifuge for 1 min at 6,000 × *g*. Collect the eluted DNA in the vial.
22. Visualize the eluted single-stranded DNA (1 μL) in an agarose gel (1 %) electrophoresis in the presence of ethidium bromide according to standard procedures (*see* **Note 25**). Typical electrophoresis results are shown in Fig. 3.
23. Determine the single-stranded DNA concentration in a Nanodrop quantitation machine. Yields vary widely between 50 ng/μL and 1 μg/μL, depending on the particular genetic construct.

3.4 Site-Directed Mutagenesis to Construct Antigen Variants with Individual Replacements

1. Align the protein sequences of the target antigen and a related antigen from a different species (non-cross-reactive with the antibody to be mapped) with the CLUSTALX software. Select the positions occupied by residues differing in the two related antigens (*see* **Note 26**).
2. Analyze solvent accessibility of the amino acids in these positions in the 3D structure (or a suitable model) of the target antigen by using the Swiss PDB Viewer accessibility tools. Choose those residues with ≥20 % solvent accessibility.
3. Design and order the synthesis of the mutagenic oligonucleotides required to replace every individual solvent-exposed residue of the target antigen by the one found in the equivalent position of the non-cross-reactive antigen. These are antisense oligonucleotides (*see* **Note 27**) annealing with the target antigen gene from 15 nucleotides before the triplet coding for

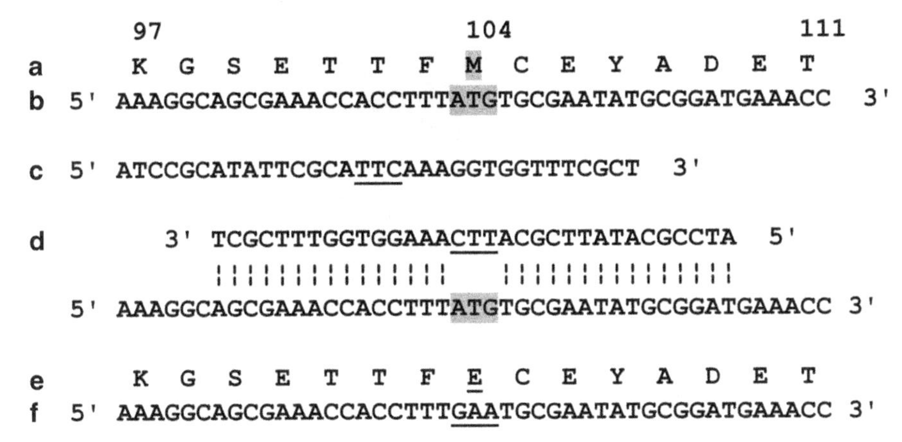

Fig. 4 Design of mutagenic oligonucleotides to replace individual residues in a model antigen. The original protein (**a**) and DNA (**b**) sequences of the segment comprised between Lys 97 and Thr 111 of human interleukin-2 are shown. The residue to be replaced (Met 104) and its coding triplet are highlighted in *grey*. The mutagenic oligonucleotide designed to replace Met 104 by Glu appears in (**c**). The mutagenic triplet is *underlined*. Annealing between the antisense mutagenic oligonucleotide and the template strand is shown in (**d**). Both protein (**e**) and DNA (**f**) sequences of the segment of the interleukin-2 variant to be obtained after mutagenesis are also represented. Differences from the original sequences are *underlined*

the residue to be replaced to 15 nucleotides after this position. In the complementary position to this triplet, the mutagenic oligonucleotide contains an antisense triplet complementary to the one coding for the residue to be introduced in the mutated target antigen (*see* Fig. 4 for an example).

4. Prepare stocks of mutagenic oligonucleotides (dissolved in water) at 330 ng/μL.
5. Prepare a phosphorylation mix, containing the following components per each mutagenic oligonucleotide to be phosphorylated (*see* **Note 28**):
   (a) 2 μL 10× TM buffer.
   (b) 2 μL ATP (10 mmol/L).
   (c) 1 μL DTT (100 mmol/L).
   (d) 11 μL water.
   (e) 2 μL Polynucleotide kinase (10,000 U/mL).
6. Add 2 μL of each oligonucleotide stock to 18 μL of the mix described above. Incubate for 1 h at 37 °C.
7. Prepare an annealing mix consisting of the following components per each mutagenic oligonucleotide to be used (*see* **Note 29**):
   (a) 2.5 μL 10× TM buffer.
   (b) xμL single-stranded template DNA (1 μg DNA).
   (c) yμL water to complete a total volume of 23 μL.

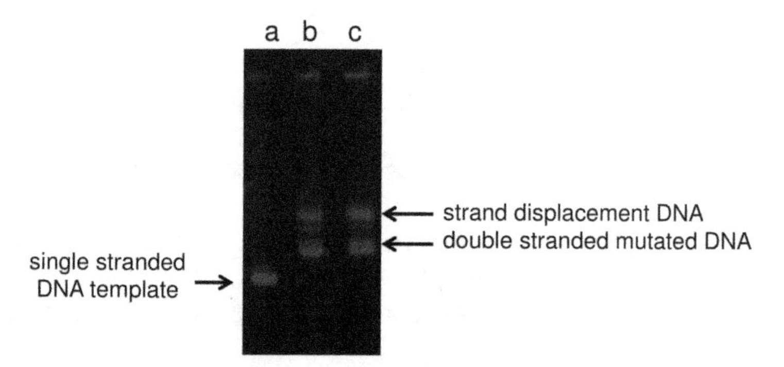

Fig. 5 Typical results of agarose gel electrophoresis of site-directed mutagenesis products. The starting single-stranded DNA template was run in (**a**). Site-directed mutagenesis products ((**b**) and (**c**)) exhibited the band corresponding to double-stranded mutated DNA and a band with lower electrophoretic mobility which should represent DNA derived from aberrant strand displacement synthesis (polymerization was not stopped after the emerging strand reached the double strand formed between the template and the mutagenic oligonucleotide). A less intense band between both presumably corresponds to double-stranded nicked DNA (the nick between the end of the emerging strand and the beginning of the oligonucleotide was not sealed)

8. Add 2 μL of the phosphorylated oligonucleotide from **step 6** to 23 μL of the above-described annealing mix in a PCR tube or a PCR microplate (*see* **Note 30**).
9. Incubate all the annealing reactions for 3 min at 90 °C, 3 min at 50 °C, and 5 min at 20 °C on a thermocycler.
10. Prepare a fill-in mix with the following components per each annealing reaction (*see* **Note 31**):
    (a) 1 μL ATP (10 mmol/L).
    (b) 2.5 μL dNTPs (10 mmol/L each).
    (c) 1.5 μL DTT (100 mmol/L).
    (d) 0.6 μL T4 DNA ligase (400,000 U/mL).
    (e) 0.4 μL T7 DNA polymerase (10,000 U/mL).
11. Add 6 μL of the above-described fill-in mix to each annealing reaction. Incubate during 16–20 h at RT.
12. Visualize 8 μL of the products of each reaction in an agarose gel (1 %) in the presence of ethidium bromide according to standard procedures. Use 1 μL of the single-stranded DNA template as a control. Figure 5 shows the typical electrophoretic pattern of successful reactions (*see* **Note 32**).
13. Transform competent XL-1 Blue *E. coli* cells with 5 μL of the products from each reaction, and grow transformed cells on solid 2XTY/AG during 16–20 h at 37 °C.
14. Pick isolated colonies (six from each mutagenesis reaction), and inoculate each one in 5 mL of 2XTY/AG.

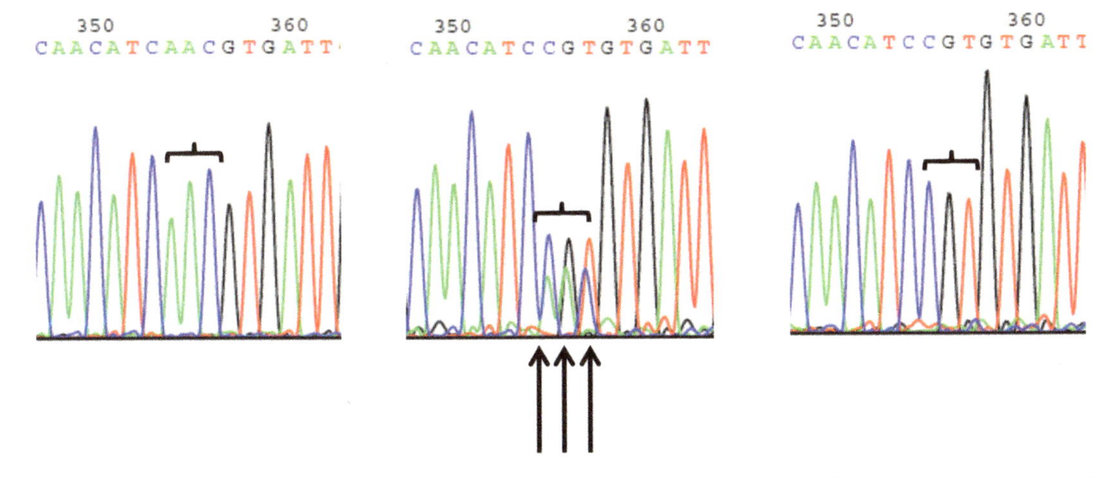

Fig. 6 Typical sequencing results after mutagenesis of the target antigen gene. *Left panel* shows a segment of the original antigen (human interleukin-2) gene sequence. The DNA segment in the *middle panel*, although apparently mutated, shows a clear overlapping of the mutated triplet CGT with the template triplet AAC. *Right panel* is an example of successful mutagenesis, in which the triplet coding for Asn 90 (AAC) has been replaced by CGT (coding for Arg). *Braces* indicate the targeted codon. *Arrows* point to overlapping peaks. Except for the targeted triplet, sequencing results show sharp peaks with no overlaps

15. Grow the cultures during 16–20 h at 37 °C with shaking at 250 rpm.
16. Purify plasmid DNA with QIAprep Spin minikit according to the manufacturer's instructions.
17. Send the plasmids for automated sequencing of the inserted target antigen genes.
18. Select one genetic construct having each desired mutation in the target antigen gene without additional undesired mutations along it. Verify that there is no overlap between the mutated sequence and the non-mutated original sequence at the mutation site (*see* **Note 33**). Examples of overlapped and non-overlapped sequences are shown in Fig. 6.

3.5 Identifying Critical Residues Contributing to Epitope Formation by Phage ELISA Screening

1. Transform competent TG1 *E. coli* cells with the selected plasmids containing mutated target antigen genes. Transform other cells with the original genetic construct containing the target antigen gene and with the one containing the control antigen gene.
2. Add 900 μL of 2XTY/AG to each well of a 96-deep-well polypropylene plate. Inoculate three isolated colonies obtained in **step 1** from each transformation reaction in individual wells of the plate. Inoculate six wells with colonies transformed with the plasmid containing the original target antigen gene (positive control) and six additional wells with colonies transformed with the plasmid containing the control antigen gene.

Leave several wells non-inoculated as negative controls in order to detect any contamination.

3. Coat the plate with a gas-permeable lid, and grow the cells at 37 °C with shaking at 250 rpm during at least 4 h (*see* **Note 34**).
4. Add 100 μL/well of M13KO7 helper phage diluted in 2XTY (at a final concentration of 10^{11} pfu/mL, *see* **Note 35**). Incubate during 30 min at 37 °C without shaking.
5. Centrifuge the plate at 1,000 × *g* during 15 min.
6. Quickly discard the supernatant by inverting the plate in a single movement. Be extremely careful to avoid cross-contamination between the wells (*see* **Note 36**).
7. Coat the plate with a new gas-permeable lid, and shake it at 250 rpm during 20 min at 28 °C to disrupt the cell pellets (*see* **Note 37**).
8. Add 1 mL of 2XTY/AK to each well. Coat the plate with a new gas-permeable lid, and grow the cells during 16–20 h at 28 °C with shaking at 250 rpm.
9. Prepare coating solutions by diluting separately the following antibodies in PBS at a final concentration of 10 μg/mL: anti-*c-myc* tag 9E10 mAb, unrelated control mAb, and mAbs against the target antigen (*see* **Note 38**).
10. Coat different sections of the polyvinyl chloride ELISA microplates with 100 μL/well of each coating solution (*see* **Note 39**). Cover the plates with a lid, and incubate them during 16–20 h at 4 °C (*see* **Note 40**).
11. Discard the coating solutions by inverting the plates several times.
12. Add 200 μL/well of 4 % M-PBS to block the plates. Cover the plates with a lid and incubate for 30 min at RT.
13. During the incubation, centrifuge the deep-well culture plates obtained after **step 8** during 15 min at 4,000 × *g* at 4 °C. Make sure that there is no cell pellet in the non-inoculated wells, indicating the absence of contamination.
14. Add 400 μL/well of M-PBS to an empty deep-well plate (dilution plate).
15. Transfer 100 μL of the phage-containing supernatants obtained after **step 13** from each well of the culture plate to the equivalent well in the dilution plate (fivefold final dilution). Transfer can be easily done with a multichannel pipette.
16. Discard the blocking solution by inverting the ELISA plates several times.
17. Transfer 100 μL of diluted phage-containing supernatants obtained after **step 15** from each well of the dilution plate to the corresponding wells of the blocked ELISA plates. Diluted supernatant from each culture well should be tested on the sections of

the ELISA plates coated with each antibody (*see* **Note 41**). Coat the plates with a lid and incubate during 1 h at RT.

18. Discard the samples by inverting the plates several times, and wash the plates at least five times by filling the wells with washing solution with a washing bottle and discarding it (*see* **Note 13**).
19. Add 100 μL/well of the anti-M13/peroxidase conjugate (diluted 1/5,000 in M-PBS). Coat the plates with a lid and incubate during 1 h at RT.
20. Wash the plates at least eight times as described in **step 18**.
21. Add 100 μL/well of peroxidase substrate solution. Incubate during 15 min at RT. A yellow/orange color should develop in some wells, while the wells containing supernatant from the non-inoculated wells of the culture plate and all the wells coated with the unrelated antibody should remain without any color (indicating low background levels).
22. Stop the reaction with 50 μL/well of stop solution.
23. Read the absorbances at 490 nm with a microplate reader.
24. Analyze background levels (*see* **Note 42**) and the results with control phages (*see* **Note 43**) in order to assess the usefulness of the assay.
25. Calculate the mean of the absorbance values obtained for replicas derived from colonies transformed with the same plasmid (either containing a given mutation in the target antigen gene or control plasmids containing the original target antigen gene/control antigen gene).
26. Calculate the normalized reactivity of each mutated antigen variant by dividing the mean absorbance detected on the wells coated with the antibody to be mapped by the mean absorbance of the same samples determined with 9E10 anti-*c-myc* tag coating antibody (*see* **Note 44**).
27. Calculate the relative reactivity (%) for each mutated antigen variant as the ratio between the normalized reactivity of a given variant and that of control non-mutated phage-displayed antigen.
28. Define the critical residues as those whose replacement results in mutated variants having less than 50 % relative reactivity with the antibody to be mapped as compared to the original phage-displayed antigen. Figure 7 shows typical results of the screening to identify critical residues contributing to a model epitope.

3.6 Randomizing Positions in the Neighborhood of the Critical Residues to Obtain a Complete View of the Epitope

1. Visualize with the Pymol software the identified critical residues (*see* Subheading 3.5) on the 3D target antigen structure (either determined by X-ray crystallography or predicted from modeling studies).
2. Determine the residues in the neighborhood of the critical ones using the suitable Pymol tools. Distances to define the neighborhood can vary between 8 and 12 Å around a critical residue.

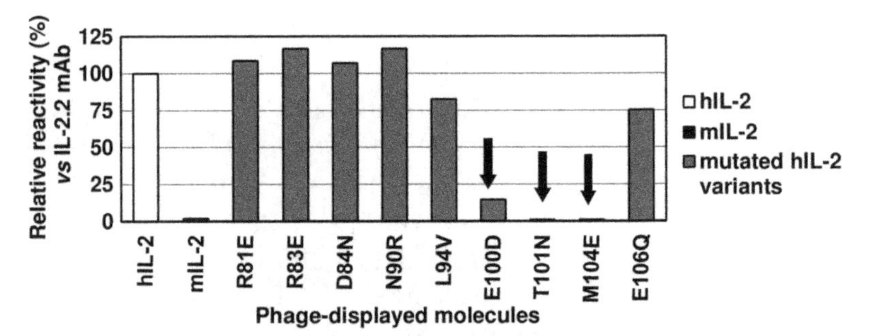

Fig. 7 Typical results of the screening to identify critical residues contributing to the formation of a model epitope. Multiple mutated phage-displayed variants of human interleukin-2 (hIL-2) were constructed by individually replacing one residue by the amino acid found in the equivalent position of mouse interleukin-2 (m-IL-2), which is not recognized by the anti-hIL-2 antibody under investigation (IL-2.2). Phages displaying all these variants were tested on polyvinyl chloride microtiter plates coated with IL-2.2 mAb and with the anti-*c-myc* tag 9E10 mAb. Phage-displayed hIL-2 and mIL-2 were included as controls. Bound phages were detected with an anti-M13 mAb conjugated to horseradish peroxidase. Normalized reactivity for each variant was estimated by dividing the signal obtained on IL-2.2 by the reference signal (measured with the anti-tag mAb). Relative reactivity (%) was calculated as the ratio between normalized reactivity of a given variant and that of the original hIL-2. Replacing Glu 100, Thr 101, and Met 104 resulted in more than 50 % decrease of the relative reactivity (indicated by *arrows*), showing that these are critical residues for the epitope recognized by IL-2.2. All the remaining mutated variants kept more than 75 % of the original reactivity. Only a sample of the tested variants is shown

3. Select those residues in the defined neighborhood that are solvent exposed (≥20 %) using the Swiss PDB Viewer accessibility tools.
4. Exclude from the previously identified residues those being Cys (which should not be replaced in order to avoid global folding defects and aggregation due to the absence of critical disulfide bonds and/or the formation of undesired intra- and intermolecular disulfide bonds mediated by unpaired Cys).
5. Design and order the synthesis of the mutagenic oligonucleotides required to individually randomize each of the already identified critical positions and their solvent-exposed neighbors in the target antigen gene. These are antisense degenerate oligonucleotides (*see* **Note 27**) annealing with the target antigen gene from 15 nucleotides before the triplet coding for the residue to be randomized to 15 nucleotides after this position. In the complementary position to this triplet, the mutagenic oligonucleotide contains the sequence MNN (antisense sequence complementary to the degenerate codon NNK, which codes for the mix of the 20 amino acids). Introduction of NNK in the target antigen gene thus results in every residue

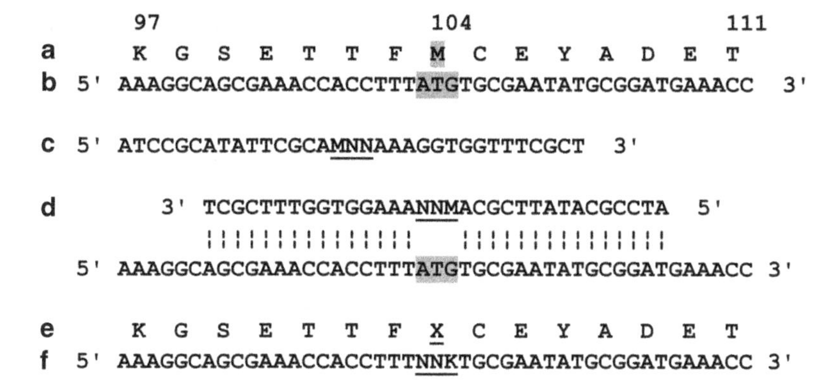

Fig. 8 Design of mutagenic oligonucleotides to randomize selected positions of a model antigen. The original protein (**a**) and DNA (**b**) sequences of the segment comprised between Lys 97 and Thr 111 of human interleukin-2 are shown. The residue to be replaced (Met 104) and its coding triplet are highlighted in *grey*. The mutagenic oligonucleotide designed to randomize residue 104 appears in (**c**). The mutagenic triplet is *underlined*. Annealing between the antisense mutagenic oligonucleotide and the template strand is shown in (**d**). Both protein (**e**) and DNA (**f**) sequences of the segment of the interleukin-2 panel of variants to be obtained after randomization are also represented. Differences from the original sequences are *underlined*. X indicates the introduction of any amino acid at this position

being coded at this position with no stop codons, except for the amber codon TAG, which codes for a Gln residue in amber-suppressor strains like TG1. Figure 8 shows an example of mutagenic primer design for randomization.

6. Randomize the desired positions by site-directed mutagenesis with the degenerate oligonucleotides previously designed. Follow **steps 4–12** described in Subheading 3.4 in order to do that.
7. Transform competent TG1 *E. coli* cells with 5 μL of the products from each mutagenesis reaction, and grow transformed cells in solid 2XTY/AG during 16–20 h at 37 °C. Transform other cells with the original genetic construct containing the target antigen gene and with the one containing the control antigen gene.
8. Add 900 μL of 2XTY/AG to each well of a 96-deep-well polypropylene plate. Inoculate individual colonies obtained after **step 7** from each transformation reaction in individual wells of the plate (*see* **Note 45**). Inoculate six wells with colonies transformed with the plasmid containing the original target antigen gene (positive control) and six wells with colonies transformed with the plasmid containing the control antigen gene. Leave several wells non-inoculated as negative controls in order to detect any contamination.

9. Continue following **steps 3–24** of the procedure described in Subheading 3.5 in order to screen by ELISA the reactivity of the new variants arising from randomization of selected positions of the target antigen.
10. Calculate the mean of the absorbance values obtained for replicas derived from colonies transformed with control plasmids containing either the original target antigen gene or the control antigen gene.
11. Calculate the normalized reactivity of each mutated antigen variant by dividing the detected absorbance on a well coated with the antibody to be mapped by the absorbance of the same sample determined with 9E10 anti-*c-myc* tag coating antibody (*see* **Note 46**).
12. Calculate the relative reactivity (%) for each mutated antigen variant as the ratio between the normalized reactivity of a given variant and that of the control non-mutated phage-displayed antigen.
13. Classify phage-displayed variants in three groups. Those variants showing more than 75 % relative reactivity as compared to the original non-mutated antigen are considered reactive. On the other hand, the variants with relative reactivities below 50 % that of the original phage-displayed antigen are nonreactive. Variants with relative reactivities in the range between 50 and 75 % are partially reactive (*see* **Note 47**).
14. Select variants corresponding to each of the three groups to be sequenced (*see* Subheading 3.7).
15. After the ELISA screening, coat the centrifuged culture deep-well plates containing phages obtained from colonies derived from the randomization reactions with freezing-resistant lids, and keep them at –20 °C as a source of the genes coding for the different variants.

3.7 Sequencing a Sample of Antibody-Reactive and Nonreactive Antigen Variants

1. Grow XL-1 Blue *E. coli* cells in 2XTY containing 10 μg/mL of tetracycline at 37 °C with shaking at 250 rpm until the culture reaches an absorbance at 600 nm in the range 0.4–0.8.
2. Defrost the deep-well plates stored at **step 15** of Subheading 3.6.
3. Add 150 μL/well of 2XTY to a tissue culture plate (U bottom).
4. Add 6 μL of phage-containing supernatant from each well in the deep-well plate corresponding to a variant selected for sequencing (*see* **step 14** of Subheading 3.6) to one of the wells in lane A of the tissue culture plate (*see* **step 3**).
5. Make serial dilutions by transferring with a multichannel pipette 6 μL from each well in lane A to the corresponding well in lane B and repeating the procedure until lane H (*see* **Note 48**).

6. Add 60 μL/well of the suspension of grown XL-1 Blue cells derived from **step 1** to each well of lane H (the one containing the highest phage dilutions) of the tissue culture plate. Incubate during 30 min at 37 °C without shaking.
7. Transfer 30 μL of infected cells from each well in lane H to a well of a 12-well tissue culture plate containing solid 2XTY/AG. Add 30 μL of non-infected control cells to an additional well. Add 100 μL of 2XTY to each well in order for the liquid to cover the surface of the well. Dry the plate, invert it, and grow at 37 °C during 16–20 h (*see* **Note 49**).
8. Pick isolated colonies (one from each selected variant), and inoculate each one in 5 mL of 2XTY/AG.
9. Grow the cultures during 16–20 h at 37 °C with shaking at 250 rpm.
10. Purify plasmid DNA with QIAprep Spin minikit according to the manufacturer's instructions.
11. Send the plasmids for automated sequencing of the inserted target antigen genes.
12. Deduce the protein sequences of the different variants. Select only those variants that display a residue different from the original one at the randomized position and no other replacements or undesired insertions/deletions along the antigen sequence (*see* **Note 50**).

3.8 Analyzing Binding and Sequence Data to Delineate a Detailed Picture of the Epitope

1. Classify the replacements found in the phage-displayed antigen variants in the following groups:
   (a) Tolerated replacements: those found in reactive mutated antigen variants.
   (b) Non-tolerated replacements: those contained in nonreactive antigen variants.
   (c) Partially tolerated replacements: those found in partially reactive variants.
2. Select the positions in the antigen that tend to exhibit several non-tolerated or partially tolerated replacements. The original residues occupying these positions in the antigen are likely to contribute to epitope formation and can be defined as part of the functional epitope.
3. Non-tolerated and partially tolerated replacements by Cys or Pro should be analyzed with caution, since both kinds of changes could affect the global antigen folding without necessarily disturbing the particular epitope recognized by the antibody to be mapped. If the only non-tolerated or partially tolerated replacements at a given position involve Cys or Pro, exclude the original residue from the functional epitope.

Table 1
Exploration of the whole neighborhood of the identified critical residues contributing to the formation of a given epitope on a model antigen by site-directed randomization

Original residue	Effect of replacements		Common property for tolerated residues
	Tolerated	Non-tolerated	
Glu 100[a]	–	A, D, H, N, P, R, S, T, V	–
Thr 101[a]	–	G, L, N, P, Q, R, Y	–
Thr 102	D, Q, S	A, C, F, I, L, P, W	Hydrophylic
Phe 103	W	C, D, L, N, P, Q, R, S, T, V	Aromatic
Met 104[a]	L	D, E, G, P, R, S, T	Hydrophobic (similar size)

[a]Glu 100, Thr 101, and Met 104 were identified as critical residues contributing to the formation of the epitope recognized by IL-2.2 mAb on human interleukin-2 in the first mutagenesis round (replacing human residues by their counterparts in mouse IL-2). Subsequent randomization of these residues and their solvent-exposed neighbors confirmed the relevance of E100, T101, and M104 through the emergence of a set of additional non-tolerated replacements (<50 % relative reactivity with IL-2.2 mAb as compared to the original non-mutated antigen) and also highlighted the contribution of the residues Thr 102 and Phe 103 to the epitope. Strict amino acid conservation was not always required for epitope formation. Position 103 needs to have an aromatic ring, position 104 must display a hydrophobic side chain of certain size, and several hydrophilic residues at position 102 are compatible with recognition by the antibody. The above-described data delineate a functional map of the epitope

4. Exclude from the epitope those residues that can be replaced by a variety of amino acids with different chemical and physical properties without affecting the antigenicity towards the antibody under investigation.
5. Search for shared properties between the residues involved in tolerated replacements and the corresponding original residues at positions belonging to the functional epitope. Such an analysis would lead to a definition at the molecular and even atomic level of the features of the residues in the epitope that contribute to interactions with the antibody (charges, aromatic rings, hydrophilic or hydrophobic groups). *See* Table 1 for an example.
6. Visualize the functional epitope on the known or the predicted antigen 3D structure using Pymol (*see* Fig. 9 for an example). A graphical representation of the epitope is quite informative to understand the biological consequences of recognition by the antibody.

4 Notes

1. The target antigen to be displayed on filamentous phages is usually the whole antigen recognized by the antibody to be mapped. If there is reliable information about a particular domain within the antigen being responsible for binding, an antigen fragment comprising only such domain can be displayed instead.

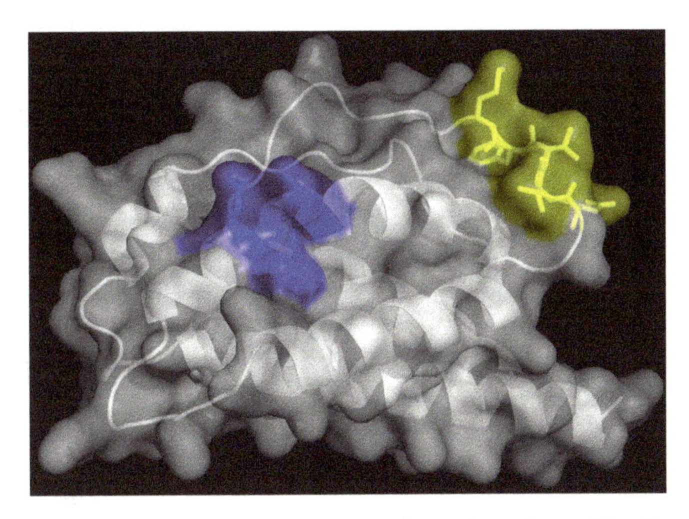

Fig. 9 Location of two functional epitopes on the surface of a model antigen. The cluster of residues recognized by IL-2.1 mAb on human IL-2 is shown in *blue*, while amino acids contributing to the epitope recognized by IL-2.2 mAb are shown in *yellow*. Side chains of the residues of both epitopes are represented as *sticks*. The rest of the molecule is shown as a cartoon in *white*. Surface of the whole hIL-2 is represented. The figure was obtained with Pymol

The second approach allows focusing the mapping effort on the relevant antigenic region. Although the procedure described in Subheading 3.1 refers to the display of the target antigen, the same steps should be followed in order to obtain phage-displayed proteins that can be used as negative controls in subsequent experiments. Totally unrelated proteins can fulfill this role, but it is particularly useful to display a related antigen (for instance the homologous protein from a different species), as long as it is not recognized by the antibody to be mapped. This strategy provides an ideal negative control closely resembling the phage-displayed target antigen.

2. Even though the use of synthetic genes allows optimizing sequences in terms of codon usage, replacing undesired residues, and removing restriction sites if necessary, another alternative is to amplify the gene coding for the target antigen from a natural source by PCR. Non-optimized genes of diverse origins have been successfully used to display proteins on filamentous phages.

3. Other phagemid vectors can be used in order to achieve the display of the target antigen instead of pCSM. An essential feature of the genetic constructs to be prepared is that the target antigen should always be produced as a fusion with a tag peptide that can be recognized by an available monoclonal antibody. Confirmation of the proper display of every antigen variant is critical for the success of the current mapping approach.

4. The procedure described in **steps 5–18** of Subheading 3.1 refers to the production and purification of phages displaying the target antigen starting from a single colony. It is highly recommendable to obtain simultaneously phages from several colonies (at least two per each displayed antigen) in order to use independent replicas in the ELISA screening.
5. It usually takes from 3 to 6 h for the 10 mL cultures to reach the desired absorbance level, starting with fresh transformed TG1 colonies.
6. M13KO7 helper phage stocks are labeled with the phage titer (measured as pfu/mL). 10^{11} pfu are usually contained in 10 μL of a typical helper phage stock obtained at the lab (10^{13} pfu/mL). Depending on the source of helper phage, the titers can vary widely, and the volume to be added needs to be calculated.
7. During the incubation of phage-containing supernatants with PEG/NaCl solution on ice, the formation of a white precipitate (composed mainly by phages) that tends to accumulate at the bottom of the conical tube is usually observed.
8. Since in the second phage precipitation step (in 1.5 mL vials) phages are more concentrated than in the first precipitation step, the formation of a white precipitate is immediately observed after mixing PBS containing phages with the PEG/NaCl solution. The mixture looks like milk. If this does not happen, that means that a failure in phage production/purification has occurred, the samples should be discarded, and the procedure should be started again (from **step 5** of Subheading 3.1).
9. Purified phage pellets (after *E. coli* and cell debris have been removed) are very difficult to resuspend. Try to avoid vortexing. Resuspend the pellets by gentle pipetting, and (if necessary) leave the pellet in contact with PBS at RT with slow shaking until it can be disrupted.
10. Confirming the display of the target antigen on the phage surface through capture by a coating anti-tag antibody that recognizes the *c-myc* tag fused to the antigen in our display system is the first prerequisite to use the current mapping approach. The phage-displayed antigen also needs to be recognized by the antibody to be mapped. Screening with both antibodies immobilized on the plates, together with an unrelated negative control mAb to discriminate specific recognition from background reactivity levels, is thus always required. Testing the recognition by other mAbs against the same antigen and also by additional proteins known to interact with the antigen (receptors, ligands, …), although not strictly necessary, is highly recommendable as it could provide additional information about the

global folding status of the antigen. That is why plates should be separately coated with all the available antibodies (or other proteins) able to recognize the target antigen.

11. Even though polyvinyl chloride microplates and PBS are properly suited for coating with multiple antibodies and other proteins and can routinely be used to start the experiments, different coating conditions could be required in some particular cases. This issue should be explored by the researcher if the ELISA results are not good enough.
12. The same array of phage samples should be tested on the wells coated with every antibody (or protein). Ideally, two independent preparations (separately obtained through the procedure described in Subheading 3.1) of the phages displaying the target antigen should be included as well as two preparations of the control phages (displaying either an unrelated protein fused to the same tag or the homologous control antigen from a different species). In order for an antigen to be useful as a control in this experimental setting, it cannot be recognized by the antibody to be mapped. Several dilutions of each phage preparation should be tested by duplicate on coated wells. Typical phage dilutions are in the range between 10- and 1,000-fold, starting with purified phages obtained in Subheading 3.1. If three dilutions are used, testing each phage preparation thus requires six wells coated with each protein.
13. When filling the ELISA plate wells with washing solution, make sure that all the wells are totally filled and no air burbles are preventing their surface be washed. Exhaustive washing of every well guarantees low background levels and reproducibility of results.
14. Low background levels are required for the antigenicity assay to be valid. The absorbance values determined for all the coated wells to which M-PBS was added instead of diluted phages (*see* **step 6** of Subheading 3.2) should be very low (usually less than 0.1). All the phage samples should also produce low absorbance values on the wells coated with the control antibody unrelated to the antigenic system under investigation. Sometimes, wells with the lowest phage dilutions (around tenfold) show higher background signals on the unrelated antibody due to nonspecific binding of concentrated phages. In these cases, such dilutions should not be taken into account. This is not a problem since nonspecific background readily disappears when using higher dilutions and all the subsequent analysis can be based on them. Low background levels guarantee the specificity of the positive signals in the assay.
15. Control phages displaying a protein different from the target antigen (either unrelated or related to it) should produce positive

signals (absorbances above twice the background level without phages) on the wells coated with 9E10 antibody recognizing the *c-myc* tag fused to every protein in our display system. On the other hand, they should be nonreactive with the antibody to be mapped.

16. Once low background levels and proper behavior of the control phages are assessed, positive signals of the phages displaying the target antigen (absorbances above twice the background level without phages) on the wells coated with the anti-tag antibody confirm successful display. Recognition in a similar context by the antibody to be mapped indicates that the target epitope is well conserved in the phage-displayed molecule and defines the suitability of the current mapping approach to investigate its nature. Recognition by a panel of additional available antibodies (or other proteins) known to interact with the target antigen can be interpreted as the proof of its correct folding on the surface of filamentous bacteriophages.
17. CJ236 is the preferred *E. coli* strain to obtain single-stranded DNA to be used as the template for site-directed mutagenesis due to the ability of this strain to produce uracil-containing template DNA. This feature allows the mutated DNA strand to be preferentially replicated in a conventional *E. coli* host over the non-mutated uracil-containing template and increases the efficiency of the mutagenesis process (*see* Subheading 3.4). Despite this advantage, it should be taken into account that some template genetic constructs cannot be produced when using CJ236 strain. If the procedure described in Subheading 3.3 repeatedly fails with a given genetic construct, replace CJ236 by XL-1 Blue *E. coli* strain. This strain renders high amounts of single-stranded DNA for virtually any template genetic construct, and mutagenesis efficiency levels on these templates are acceptable.
18. Phage production from different colonies of CJ236 *E. coli* transformed with the same genetic construct is less reproducible than the same procedure when using other *E. coli* strains, resulting in a failure to isolate single-stranded DNA from some colonies without any evident reason. That is why it is highly recommendable to start the procedure described in Subheading 3.3 with at least three or four colonies, in order to be successful with some of them.
19. When using XL-1 Blue *E. coli* strain instead of CJ236 for single-stranded DNA production, chloramphenicol should be replaced by tetracycline at 10 μg/mL.
20. It can take up to 8 h for the 10 mL cultures to reach the desired absorbance level, starting with fresh transformed CJ236 or XL-1 Blue colonies.

21. When using XL-1 Blue *E. coli* strain instead of CJ236 for single-stranded DNA isolation, uridine is not included during phage production.
22. When using XL-1 Blue *E. coli* strain instead of CJ236 for single-stranded DNA isolation, phage production is performed at 28 °C instead of 37 °C.
23. Low phage production by the CJ236 *E. coli* strain usually results in the absence of any visible precipitate during the incubation of phage-containing supernatants with PEG/NaCl solution on ice, in contrast to what is observed for other *E. coli* strains like TG1 or XL-1 Blue.
24. During phage precipitation with MP solution, the formation of a white precipitate is sometimes seen. While it can be observed or not when using CJ236 *E. coli* strain for phage production (and this fact is not a useful indicator to predict the subsequent success of single-stranded DNA isolation), precipitate formation resulting in the solution becoming completely white must be seen when using XL-1 Blue strain.
25. Successful single-stranded DNA purification results in a main band that can be clearly observed on an agarose gel electrophoresis. Sometimes it is accompanied by several minor bands with less electrophoretic mobility (*see* Fig. 3).
26. The mapping approach described in this chapter takes advantage of the lack of cross-reactivity of the antibody to be mapped with a second antigen closely related to the target one but originating from a different species. Such knowledge (when available) provides a good starting point to scan the surface of the target antigen in the search for critical residues. In the absence of this kind of information, a different set of replacements can be designed, on the basis of other experimental evidences that suggest a possible location for the epitope.
27. Since pCSM^{+} phagemid vector is used in the current protocol, the phage replication origin is in the sense strand of the plasmid, resulting in packaging of this strand into filamentous phages and isolation of sense single-stranded DNA template. Antisense mutagenic oligonucleotides are thus required to anneal with the template. When using other phagemid vectors, it is important to know whether they are $^{+}$ or $^{-}$ versions. In the second case, the antisense template DNA would be obtained through the protocol described in Subheading 3.3, and sense mutagenic oligonucleotides directly including the triplet to be introduced and flanking sequences of the target antigen gene should be designed.
28. Even though individual phosphorylation mixes for each mutagenic oligonucleotide could be prepared, it is advantageous to make a master mix with all the common components and

distribute it in different vials to be mixed with each oligonucleotide. This is particularly useful when doing multiple simultaneous mutagenesis reactions. If n phosphorylation reactions are going to be performed, add enough components to the master mix for an excess of reactions ($>n$) in order to make sure that any lack of accuracy in pipetting will not result in the mix to be exhausted before preparing all the required phosphorylation reactions.

29. Even though individual annealing mixes for each mutagenic oligonucleotide could be prepared, it is advantageous to make a master mix with template DNA and buffer, to be mixed with each phosphorylated oligonucleotide. This is particularly useful when doing multiple simultaneous mutagenesis reactions. If n annealing reactions are going to be performed, add DNA and buffer for an excess of reactions ($>n$) in order to make sure that the mix will be enough to prepare all the required annealing reactions.
30. The use of a PCR microplate instead of individual tubes is particularly advantageous to perform multiple simultaneous mutagenesis reactions.
31. Even though individual fill-in reactions could be prepared, it is advantageous to make a master mix to be distributed among all the annealing reactions. This is particularly useful when doing multiple simultaneous mutagenesis reactions. If n fill-in reactions are going to be performed, add enough components to the mix for an excess of reactions ($>n$) in order to make sure that the mix will be enough for all the mutagenesis reactions.
32. Typically, at least two bands are observed after the mutagenesis reaction: one with slightly less electrophoretic mobility than the template DNA, that corresponds to the hybrid double-stranded DNA formed by the template strand and the newly synthesized mutated strand (the desired mutagenesis product), and a second one with even lower electrophoretic mobility. The latter comes from strand displacement, an artifact arising during the fill-in reaction, when the mutagenic oligonucleotide primes the synthesis of a new strand as expected, but the synthesis is not stopped when the emerging chain reaches the double strand formed between the template and the mutagenic oligonucleotide. Synthesis thus continues, displacing the mutagenic oligonucleotide and the newly formed strand and giving rise to high-molecular-weight aberrant DNA species (*see* Fig. 5). Sometimes a third, and very weak, band is observed between the two previously described bands.
33. Since the product of a successful mutagenesis reaction is a hybrid molecule composed by the original template strand and a new mutated strand, both strands can in principle be replicated in *E. coli.*

Usually one of them is preferentially replicated after transformation of *E. coli* with mutagenesis products, resulting in sequences indicative of either successful mutation at the target position or no mutation at all. In some cases, however, sequencing reveals the coexistence of mutated and non-mutated DNA in the same bacteria (shown by overlapping sequencing peaks at the target position, *see* Fig. 6).

34. Growth of transformed *E. coli* cells in deep-well plates results in a visible difference between inoculated wells, which become cloudy, and non-inoculated control wells, which should remain totally clear. That difference should be detectable after approximately 4 h of culture or more, indicating that grown cells are ready for the next step. If non-incoculated wells become cloudy as well, plates should be discarded due to media contamination.
35. M13KO7 helper phage stocks are labeled with the phage titer (measured as pfu/mL). A typical helper phage stock obtained at the lab (10^{13} pfu/mL) should be diluted 100-fold before infecting *E. coli* in the deep-well plates. Depending on the source of helper phage, the titers can vary widely and the dilution needs to be calculated.
36. Supernatant from each well of deep-well plates can alternatively be removed by pipetting after infection with helper phage. This reduces the chances of cross-contamination between wells. But removing the supernatants by simply inverting the plates as described, if properly done, is very simple and particularly advantageous when working simultaneously with several plates.
37. Shaking the deep-well plates without media is an easy way to disrupt the pellets before adding the final media to produce phages and allows simultaneous manipulation of multiple plates. Alternatively, this step can be omitted and each pellet can be resuspended by pipetting after adding 1 mL of 2XTY/AK (*see* **step 8** of Subheading 3.5).
38. Every target antigen variant should be tested on wells coated with the antibody to be mapped in order to discriminate recognized from non-recognized variants. Confirming the display of every variant through capture by a coating anti-tag antibody that recognizes the *c-myc* tag fused to all of them in our display system is also necessary in order to exclude a deleterious effect of a given mutation on protein folding and processing. Such gross defects could affect protein display itself and cause the lack of recognition even when the mutations do not directly compromise the target epitope. Screening with both antibodies immobilized on the plates, together with an unrelated negative control mAb to discriminate specific recognition from background reactivity levels, is thus always required.

Testing the recognition by other mAbs against different epitopes on the same antigen, although not strictly necessary, is highly recommendable as it could confirm the specificity of the effects of mutations on recognition by the antibody under investigation.

39. Even though polyvinyl chloride microplates and PBS are properly suited for coating with multiple antibodies, different coating conditions can be required in some particular cases. Use coating conditions already validated when testing the display of the target antigen on phages (*see* Subheading 3.2).
40. Coat the ELISA plates for screening the same day you start phage production in deep-well plates, since both procedures take 16–20 h. Fresh phage-displayed antigen variants could thus be tested the next day.
41. Diluted phage-containing supernatant from each well should be tested on wells coated with at least three different antibodies (the one to be mapped, the anti-tag 9E10 mAb, and an unrelated control antibody) and possibly with other antibodies against different epitopes on the same antigen. Diluted supernatants from each column of the dilution plate should thus be transferred to several columns in coated ELISA plates. It is important to carefully record the distribution of phage-containing supernatants in ELISA plates in order to be able to match ELISA results with the location of the corresponding phage-displayed antigen variants in the original deep-well plate.
42. Low background levels are required for the assay to be valid. The absorbance values obtained for supernatants of the non-inoculated wells (*see* **step 2** of Subheading 3.5) should be very low (usually less than 0.1). All the phage samples should also produce low absorbance values on the wells coated with a control unrelated antibody. Low background levels guarantee the specificity of the positive signals in the assay.
43. Control phages displaying a protein different from the target antigen (either unrelated or related to it) should produce positive signals (absorbances above twice the background level obtained from the non-inoculated deep-wells) on wells coated with 9E10 antibody recognizing the *c-myc* tag fused to every protein in our display system. On the other hand, they should be nonreactive with the antibody to be mapped. Phage-displayed original antigen (positive control) should produce positive signals on both coating antibodies.
44. Normalization of the recognition of every single-mutated antigen variant by the antibody to be mapped takes the display level of the same variant (assessed with the anti-tag antibody 9E10) as the reference. Differences in recognition due to different display levels are thus excluded from subsequent analysis, and the

attention can be focused on the effects of replacements on antigenicity (probably related to epitope disruption).

45. Since each colony obtained after direct transformation of TG1 cells with the products of the site-directed randomization reaction should have a unique sequence (either mutated or not), screening at this step has no replicas. Several dozens of individual clones from randomization of each position should be screened.

46. Normalization of the recognition of every antigen variant by the antibody to be mapped takes the display level of the same variant (assessed with the anti-tag antibody 9E10) as the reference. Differences in recognition due to different display levels are thus excluded from subsequent analysis. Emerging variants that are not recognized by the anti-tag antibody (signals around the background level) should not be considered, because they probably harbor mutations affecting protein display itself.

47. Depending on the contribution of the original residue at the randomized position to epitope formation and the efficiency of the mutagenesis procedure, randomization can result in the appearance of only nonreactive, partially reactive, or reactive variants or in a mixture of variants with different properties. When only reactive variants are detected in phage ELISA screening, this can be due to either a failure in the mutagenesis procedure (the original non-mutated target antigen gene coming from the template is always present) or the lack of influence of the mutated residue in the interaction with the antibody to be mapped. The mixed effects of mutating a given residue could be a composite of the presence of non-mutated and mutated molecules and the different effects of distinct replacing residues on epitope formation. Since these possibilities cannot be distinguished before sequencing the emerging antigen variants, all the successfully phage-displayed variants (as assessed with the anti-tag antibody), independently of their recognition level by the antibody to be mapped, may be picked for sequencing (*see* Subheading 3.7).

48. Serial dilutions of phage-containing supernatants obtained after randomization are required in order to obtain isolated colonies that can be used as sources of unique mutated antigen variants. The procedure described to make dilutions does not require changing the aerosol-resistant filtered pipette tips between lanes, but it is crucially important to remove any residual media from the tips before loading the volume of 6 μL that will be transferred to the next lane. Otherwise, individual colonies carrying unique antigen sequences would be difficult to pick after infecting XL-1 Blue cells with diluted phage-containing supernatants.

49. Although phage-infected XL-1 Blue cells could be grown on conventional Petri dishes containing solid 2XTY/AG, the use of multi-well tissue culture plates filled with solid media is advantageous to work simultaneously with multiple phage-displayed antigen variants for sequencing. 12-well plates are easy to handle, and the surface area of each well is convenient to pick isolated colonies. Alternatively, 6-well or 24-well plates could be used, depending on the number of antigen variants to be sequenced and the separation between colonies that is observed after growth in solid 2XTY/AG.
50. Plasmid DNA to be sequenced in Subheading 3.7 does not come directly from transformation with site-directed randomization products. Phage production from transformed colonies and infection of XL-1 Blue cells with the resulting phages are intermediate steps that guarantee in this case that a single antigen sequence will be found after sequencing, since only one DNA strand (either mutated or not) can be packaged into a phage particle. That is why overlapping sequences are a very rare finding at the current step.

Acknowledgement

Prof. Sachdev Sidhu contributed to the design of this mapping approach with helpful advice.

References

1. Boyman O, Kovar M, Rubinstein MP et al (2006) Selective stimulation of T cell subsets with antibody-cytokine immune complexes. Science 311:1924–1927
2. Rojas G, Pupo A, Leon K et al (2013) Deciphering the molecular bases of the biological effects of antibodies against Interleukin-2: a versatile platform for fine epitope mapping. Immunobiology 218:105–113
3. Davies DR, Cohen GH (1996) Interactions of protein antigens with antibodies. Proc Natl Acad Sci U S A 93:7–12
4. Benjamin DC, Perdue SS (1996) Site-directed mutagenesis in epitope mapping. Methods 9: 508–515
5. Chao G, Cochran JR, Wittrup KD (2004) Fine epitope mapping of anti-epidermal growth factor receptor antibodies through random mutagenesis and yeast surface display. J Mol Biol 342:539–550
6. Smith GP, Petrenko VA (1997) Phage display. Chem Rev 97:391–410
7. Böttger V, Böttger A (2009) Epitope mapping using phage display peptide libraries. Methods Mol Biol 524:181–201
8. Deroo S, Muller CP (2001) Antigenic and immunogenic phage displayed mimotopes as substitute antigens: applications and limitations. Comb Chem High Throughput Screen 4:75–110
9. Saphire EO, Montero M, Menendez A et al (2007) Structure of a high affinity "mimotope" peptide bound to HIV-1 neutralizing antibody b12 explains its inability to elicit gp120 cross-reactive antibodies. J Mol Biol 369:696–709
10. Kunkel TA (1985) Rapid and efficient site-specific mutagenesis without phenotypic selection. Proc Natl Acad Sci U S A 82:488–492
11. Fellouse FA, Sidhu SS (2007) Making antibodies in bacteria. In: Howard GC, Kaser MR (eds) Making and using antibodies. A practical handbook. CRC Press, Boca Raton, FL, pp 157–177

Chapter 28

Epitope Mapping with Random Phage Display Library

Terumi Midoro-Horiuti and Randall M. Goldblum

Abstract

Random phage display library is used to map conformational as well as linear epitopes. These libraries are available in varying lengths and with circularization. We provide here a protocol conveying our experience using a commercially available peptide phage display library, which in our hands provides good results.

Key words Allergen, Conformational epitope, Monoclonal antibody, Random phage display library

1 Introduction

It is believed that B cell epitopes, for instance aeroallergens, are generally conformational. Conformational epitopes are formed when several regions on the surface of proteins that are separated on the unfolded primary amino acid strands come together in the process of protein folding. Using random phage display libraries one can identify these epitopes, which are often not identified in overlapping peptide array experiments [1].

2 Materials

1. Random phage display library: Ph.D. Phage Display Libraries (New England Biolabs, NEB). Random peptides of 7 and 12 or circularized with cysteine residues are available from NEB. Additional protocols for these libraries are available from NEB.
2. Protein G magnetic beads (NEB), if monoclonal antibody (mAb) of interest is not mouse IgG, biotinylate it and use streptavidin magnetic beads.
3. Specific mAbs.
4. Isotype control antibody for preincubation of phage.
5. ER2738 *E. coli* strain (NEB).

Vincent Ossipow and Nicolas Fischer (eds.), *Monoclonal Antibodies: Methods and Protocols*, Methods in Molecular Biology, vol. 1131, DOI 10.1007/978-1-62703-992-5_28, © Springer Science+Business Media New York 2014

6. LB medium.
7. SOC medium.
8. IPTG/X-gal stock: Mix 1.25 g isopropyl-β-D-galactoside (IPTG) and 1 g 5-bromo-4-chloro-3-indolyl-β-D-galactoside (X-gal) in 25 mL dimethyl formamide (DMF), Store at −20 °C.
9. LB/X-gal plates.
10. LB/IPTG/X-gal plates.
11. Top agar: Mix 10 g Bacto-Trypone, 5 g yeast extract, 5 g NaCl, 7 g Bacto-Agar, and MilliQ water to make up 1 L. Autoclave, and aliquot in 50 mL. Store at room temperature.
12. Tetracycline stock: 20 mg/mL in 1:1 ethanol:water. Store at −20 °C. Vortex before use.
13. LB/Tet plates.
14. Tris-buffered saline (TBS), pH 7.5: Add and mix NaCl (8.0 g), KCl (0.2 g), Tris base (6.1 g), and 800 mL of MilliQ water. Adjust pH to 7.5 with HCl. Make up to 1 L with MilliQ water. Autoclave, and store at room temperature.
15. Polyethylene glycol-8000 (PEG-8000)/NaCl: 20 % (w/v) PEG-8000, 2.5 M NaCl, autoclave, mix well to combine separated layers while still warm. Store at room temperature.
16. Iodide buffer: 10 mM Tris–HCl, pH 8.0, 1 mM EDTA, 4 M sodium iodide (NaI). Store at room temperature in the dark.
17. Streptavidin stock solution: 1.5 mg lyophilized streptavidin (supplied in NEB kit) in 1 mL 10 mM sodium phosphate, pH 7.2, 100 mM NaCl, 0.02 % NaN3. Store at 4 °C or −20 °C. Avoid freezing and thawing.
18. TTBS: 0.01 % TBS–Tween20 (50 mM Tris–HCl, pH 7.5, 150 mM NaCl).
19. Blocking buffer: 0.1 M $NaHCO_3$, pH 8.6, 5 mg/mL bovine serum albumin (BSA), 0.02 % NaN_3, 0.2 μm filter and store at 4 °C.
20. Sequencing primer (NEB).
21. Peroxidase labeled mouse anti-M13 phage antibody (Amersham).
22. 3,3,5,5-Tetramethylbenzidine (TMB) substrate (KPL, Pierce or Sigma).
23. Aerosol-resistant pipette.
24. Microcentrifuge tubes.
25. Toothpicks or gel-loading pipette tips, autoclaved.

3 Methods

3.1 Pre-cleaning Phage (See Note 1)

1. Take 50 μL of protein G or the appropriate beads in a 1.5 mL microcentrifuge tube.
2. Add 1 mL of TTBS. Suspend resin by tapping the tube.
3. Centrifuge tube for 30 s or use magnetic stand to pellet resin.
4. Remove supernatant.
5. Suspend resin in 1 mL blocking buffer. Incubate for 1 h at 4 °C with gentle rocking.
6. Pellet resin and discard supernatant.
7. Wash resin with 1 mL TTBS. Wash will be done with adding 1 mL TTBS to the tube, suspending resin by tapping the tube, centrifuging the tube for 30 s or using magnetic stand to pellet resin, and removing supernatant. Wash four times.
8. Dilute 1.5×10^{11} phage and 300 ng of isotype antibody to a final volume of 0.2 mL with TTBS. Final concentration of antibody is 10 nM.
9. Incubate at room temperature for 20 min.
10. Transfer the phage antibody mixture to the tube containing washed resin (Subheading 3.1, **step** 7).
11. Mix gently and incubate at room temperature for 15 min, mixing occasionally.
12. Wash ten times with 1 mL TTBS.
13. Elute bound phage by suspending pelleted resin in 1 mL 0.2 M glycine–HCl, pH 2.2, and in 1 mg/mL BSA. Incubate at room temperature for 10 min.
14. Centrifuge elution mixture for 1 min.
15. Carefully transfer the supernatant to a new microcentrifuge tube. Do not disturb the resin.
16. Immediately neutralize the eluate with 0.15 mL 1 M Tris–HCl, pH 9.1.
17. Titer the eluate.

3.2 Titering the Phage

1. Incubate 5 mL of LB with ER2738 with shaking for 4–8 h to OD 600 ~0.5.
2. Heat top agar in microwave and dispense 3 mL into sterile culture tubes, 1 tube/expected phage dilution. Maintain tubes at 45 °C.
3. Pre-warm LB/IPTG/X-gal plate/expected dilution for at least 1 h at 37 °C until ready to use.

4. Dilute serially to produce $10–10^3$-fold dilution of phage in LB. Use the $10–10^4$ dilutions for unamplified panning eluates and the $10^8–10^{11}$ dilutions for amplified culture supernatants.
5. When the culture reaches to OD_{600} ~0.5, dispense 0.2 mL into microcentrifuge tubes, 1 for each phage dilution.
6. Add 10 μL of each phage dilution to individual tube to infect, vortex quickly, and incubate at room temperature for 1–5 min.
7. Transfer the infected cells, one infection at a time, to culture tubes containing 45 °C top agar.
8. Vortex briefly, and immediately pour culture onto a pre-warmed LB/IPTG/X-gal plate.
9. Gently tilt and rotate plate to spread top agar evenly.
10. Allow the plates to cool for 5 min.
11. Invert and incubate overnight at 37 °C.
12. Count plaques on those plates with about 100 plaques/plate.
13. Multiply each number by dilution factor to get the phage titer in plaque-forming unit (pfu).

3.3 Panning

1. Take 50 μL of protein G or other appropriately labeled beads in a 1.5 mL microcentrifuge tube.
2. Add 1 mL of TTBS. Suspend resin by tapping the tube.
3. Centrifuge tube for 30 s or use magnetic stand to pellet resin.
4. Remove supernatant.
5. Suspend resin in 1 mL of blocking buffer. Incubate for 1 h at 4 °C with gentle rocking.
6. Pellet resin, and discard supernatant.
7. Wash resin with 1 mL TTBS four times.
8. Dilute 1.5×10^{11} pre-cleaned phages (eluate phages from Subheading 3.1, **step 17**) and 300 ng of antibody of interest to a final volume of 0.2 mL with TTBS. Final concentration of antibody is 10 nM.
9. Incubate at room temperature for 20 min.
10. Transfer the phage/antibody mixture to the tube containing washed resin.
11. Mix gently, and incubate at room temperature for 15 min, mixing occasionally.
12. Wash ten times with 1 mL TTBS.
13. Elute bound phage by suspending pelleted resin in 1 mL 0.2 M glycine–HCl, pH 2.2, and 1 mg/mL BSA.
14. Incubate at room temperature for 10 min.
15. Centrifuge elution mixture for 1 min.

16. Carefully transfer the supernatant to a new microcentrifuge tube. Do not disturb the resin.
17. Immediately neutralize the eluate with 0.15 mL 1 M Tris–HCl, pH 9.1.
18. Titer the eluate (*see* Subheading 3.2).
19. Panning will be repeated three times (procedure Subheadings 3.3 and 3.4).

3.4 Amplification for Next Panning

1. Incubate 20 mL of LB with ER2738 with shaking for 4–8 h to OD 600 ~0.5.
2. Add the remaining phage from Subheading 3.3, **step 17**.
3. Incubate at 37 °C for 4.5–5 h with shaking.
4. Transfer the culture to a centrifuge tube and centrifuge for 10 min at 12,000 ×*g* at 4 °C.
5. Transfer the supernatant to a new centrifuge and centrifuge for 10 min at 12,000 ×*g* at 4 °C.
6. Transfer the upper 80 % to the new centrifuge tube, and add 1/6 volume of 20 % PEG/2.5 M NaCl.
7. Incubate at 4 °C for 2 h to overnight.
8. Centrifuge PEG precipitation at 12,000 ×*g* for 15 min at 4 °C. Discard the supernatant. Double centrifuge for a short time, and remove supernatant with pipette.
9. Resuspend the pellet in 1 mL TBS by pipetting.
10. Transfer the suspension to a microcentrifuge tube and centrifuge at 14,000 rpm for 5 min at 4 °C.
11. Transfer the supernatant to a new microcentrifuge tube and reprecipitate with 1/6 volume of 20 % PEG/2.5 M NaCl.
12. Incubate for 15–60 min on ice.
13. Centrifuge at 14,000 rpm for 10 min at 4 °C. Double centrifuge for a short time, and remove supernatant with pipette.
14. Resuspend the pellet in 200 μL of TBS.
15. Centrifuge at 14,000 rpm for 1 min.
16. Transfer the supernatant to a new microcentrifuge tube. Amplified eluate can be stored for several weeks at 4 °C or stored at −20 °C by adding one volume of sterile glycerol.
17. Titer the amplified eluate on LB/IPTG/X-gal plate (Subheading 3.2).
18. Perform the next panning (Subheading 3.3).

3.5 Amplification and Purification of the Selected Phages

1. Culture ER2738 in 5 mL of LB with tetracycline (20 μg/mL) overnight.
2. Dilute overnight culture of ER2738 1:100 in LB.

3. Aliquot diluted culture in 14 mL tubes.
4. Select about 20–40 individual plaques from the plates after third panning (Subheading 3.2, **step 9**, *see* **Note 2**).
5. Stab a blue plaque by toothpick or gel-loading pipette tip and add to LB with ER2738 tube.
6. Incubate the tubes at 37 °C for 4.5–5 h with shaking.
7. Transfer 1.5 mL from the cultures into microcentrifuge tubes.
8. Centrifuge at 14,000 rpm for 3 min.
9. Transfer the supernatant to a new microcentrifuge tube and centrifuge again at 14,000 rpm for 3 min.
10. Transfer the upper 80 % of supernatant to a new microcentrifuge tube.
11. Transfer 500 μL of supernatant to a new microcentrifuge tube.
12. Add 200 μL of 20 % PEG-8000/2.5 M NaCl.
13. Mix gently by inverting tubes several times.
14. Let the microcentrifuge tube stand for 10–20 min at room temperature.
15. Centrifuge at 10,000 rpm for 20 min.
16. Discard supernatant.
17. Centrifuge quickly, and remove the remaining supernatant by pipetting.
18. Resuspend pellet in 1 mL TBS.
19. Store at 4 °C.

3.6 Phage ssDNA Extraction

1. Select amplified phage (1 mL) from individual clone to a new microcentrifuge tube.
2. Add 1/6 volume of 20 % PEG/2.5 M NaCl.
3. Incubate at room temperature for 10 min.
4. Centrifuge at 10,000 rpm for 10 min.
5. Discard the supernatant.
6. Suspend phage pellet thoroughly in 100 μL of iodide buffer followed by 250 μL 100 % ethanol.
7. Incubate the mixture for 10 min at room temperature to precipitate the single-stranded DNA.
8. Centrifuge at 10,000 rpm for 10 min.
9. Discard supernatant.
10. Add 0.5 mL of 70 % ethanol.
11. Centrifuge at 10,000 rpm for 10 min.
12. Discard supernatant.
13. Dry the pellet.

14. Suspend pellet in 30 μL of TE buffer (10 mM Tris–HCl [pH 8.0], 1 mM EDTA).
15. Quantify the product (5 μL) by agarose gel electrophoresis.

3.7 Nucleotide Sequencing

1. Sequence purified phage DNA with -96 g III primer 5′-HOCCC TCA TAG TTA GCG TAA CG -3′ (NEB). For instance while mapping for the major cockroach allergens Bla g 2 epitope, we sequenced 40 clones and obtained 32 clean sequences. Twenty-four out of 32 were SMMKADFDEEPR, and other 8 were SMMKADFEEEPR [3] (*see* **Note 3**).

   The small number of sequences recovered in this case suggested that the original library had limited phage sequences that interacted with the antibody 7C11 and that these were cleanly separated from relatively large number of other phage by this technique.

3.8 Phage ELISA (See Note 4)

1. Coat the plate with the antibody of interest (10 μg/ml) according to the plate manufacturer.
2. Add three dilutions (1:1, 1:10, 1:100) of phage diluted in TTBS to the wells. A nonspecific streptavidin-binding phage clone can be used as a negative control.
3. Incubate at 4 °C overnight.
4. Wash microtiter plate three times with TTBS.
5. Add peroxidase-labeled mouse anti-M13 phage antibody (Amersham).
6. Incubate at room temperature for 1 h.
7. Wash microtiter plate with TTBS four times.
8. Add TMB substrate.
9. Measure absorbance at 492 nm.

In our experiments, we tested for Bla g 2 epitope with 1×10^{8}–1×10^{15} PFU/mL. Significant inhibition of selected phages was observed in a dose–response manner [3].

3.9 Inhibition ELISA with Antibody-Specific Phage Clones (See Note 5)

1. Coat microtiter plate with antibody (10 μg/ml in 100 μL borate buffer).
2. Incubate at 4 °C overnight.
3. Wash microtiter plate with TTBS three times.
4. Add phage clones at concentrations ranging from 10^{7} pfu/mL to 10^{14} pfu/ml diluted in TTBS.
5. Incubate at 4 °C overnight. Unlabelled antigen is used as a negative control, and a nonspecific streptavidin-specific phage is used as a positive control.
6. Wash microtiter plate with TTBS three times.

7. Add biotin-labeled allergen at a concentration of 20 ng/mL.
8. Incubate microtiter plate ice for 30 min.
9. Wash microtiter plate with TTBS four times.
10. Add peroxidase-conjugated streptavidin diluted 1:1,000 in TTBS.
11. Incubate microtiter plate at room temperature for 1 h.
12. Wash microtiter plate with TTBS six times.
13. Add TMB substrate.
14. Measure absorbance at 492 nm.

4 Notes

1. This is for removing the nonspecific phage binding to antibody using isotype antibody.
2. Plates should be <3 days old, stored in 4 °C, and should have <100 plaques.
3. To map the epitope on the molecule structure, there are several computational programs available. EpiSeach (http://curie.utmb.edu/episearch.html) has been recently modified, and we successfully used it in combination with crystal structure analysis [2, 3].
4. After three rounds of selection, enrichment and specificity of the phage clones are confirmed by ELISA.
5. Inhibition ELISA is performed to determine the degree of inhibition of allergen binding to antibody by the phage clones.

References

1. Jensen-Jarolim E, Leitner A, Kalchhauser H, Zurcher A, Ganglberger E, Bohle B, Scheiner O, Boltz-Nitulescu G, Breiteneder H (1998) Peptide mimotopes displayed by phage inhibit antibody binding to bet v 1, the major birch pollen allergen, and induce specific IgG response in mice. FASEB J 12:1635–1642
2. Negi SS, Braun W (2009) Automated detection of conformational epitopes using phage display peptide sequences. Bioinform Biol Insights 3:71–81
3. Tiwari R, Negi SS, Braun B, Braun W, Pomes A, Chapman MD, Goldblum RM, Midoro-Horiuti T (2012) Validation of a phage display and computational algorithm by mapping a conformational epitope of Bla g 2. Int Arch Allergy Immunol 157:323–330

Chapter 29

Epitope Mapping of Monoclonal and Polyclonal Antibodies Using Bacterial Cell Surface Display

Anna-Luisa Volk, Francis Jingxin Hu, and Johan Rockberg

Abstract

The unique property of specific high-affinity binding to more or less any target of interest has made antibodies tremendously useful in numerous applications. Hence knowledge of the precise binding site (epitope) of antibodies on the target protein is one of the most important features for understanding its performance and determining its reliability in immunoassays. Here, we describe a high-resolution method for mapping epitopes of antibodies based on bacterial surface expression of antigen fragments followed by antibody-based flow cytometric sorting. Epitopes are determined by DNA sequencing of the sorted antibody-binding cells followed by sequence alignment back to the antigen sequence. The method described here has been useful for the mapping of both monoclonal and polyclonal antibodies with varying sizes of epitopes.

Key words Cell surface display, Epitope mapping, Antibody, *S. carnosus*, FACS, Gram-positive

1 Introduction

Antibodies are invaluable tools for a vast number of applications including protein purification and characterization, medical diagnosis, and therapeutic treatment. Despite the different aims of these various applications, the antibody's binding characteristics and its epitope on the target protein will most likely affect the assay's reliability. Several methods for determination of an antibody's epitope are available, including peptide scanning involving chemically synthesized peptides [1] and X-ray crystallography of the antibody–antigen complex [2].

Flow cytometry is a technique for counting and examining microscopic cells by suspending them in a stream of fluid that passes by a detection device. Fluorescence-activated cell sorting (FACS) is a type of flow cytometry, enabling sorting and collecting heterogeneous mixtures of cells based on specific light scattering

The authors Anna-Luisa Volk and Francis Jingxin Hu have contributed equally to this work.

Vincent Ossipow and Nicolas Fischer (eds.), *Monoclonal Antibodies: Methods and Protocols*, Methods in Molecular Biology, vol. 1131, DOI 10.1007/978-1-62703-992-5_29,

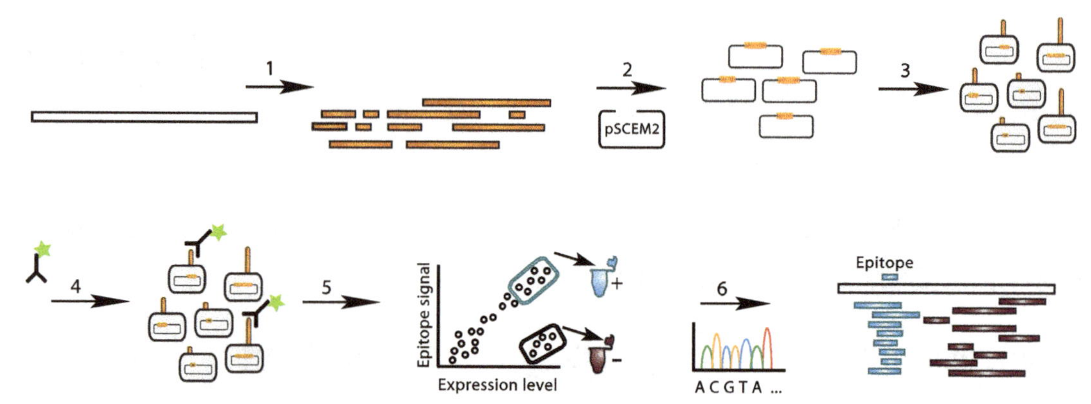

Fig. 1 The gene encoding the target antigen to be mapped is amplified by PCR and fragmented (1) followed by cloning into the staphylococcal display vector (pSCEM2) (2). The library containing plasmids are transformed into *Staphylococcus carnosus*, where in-frame fragments are displayed on the bacterial surface (3). After incubation with antibody (4), binding and non-binding cells are sorted using FACS to isolate epitope- and non-epitope-containing cells (5). Colonies derived from individual cells are sequenced (6), and sequences from epitope (*blue*)- and non-epitope (*red*)-binding cells are aligned back to the antigen in order to together determine the antibody epitope

and/or fluorescent characteristics of each cell [3]. Although this technique is mainly used in mammalian research, there is a potential in biotechnology to utilize it as a powerful tool for library screening [4].

Here, we describe a protocol for determination of antibody-binding epitopes by using an antigen-focused, library-based approach in conjunction with flow cytometric sorting (Fig. 1). The library members are generated by fragmentation of the target antigen DNA, and the Gram-negative bacterium *E. coli* is first used for preparation of the library containing plasmids. The library is then presented as peptides on the cell surface of the nonpathogenic Gram-positive bacterium *Staphylococcus carnosus*. The cell surface anchoring domain is located downstream of the epitope peptides in the expression cassette; hence, only in-frame fragments are analyzed since out-of-frame regions cannot bind to the membrane nor be detected by labeled HSA. Epitope expression level is monitored by measuring labeled HSA binding to the albumin-binding domain (ABD) downstream of the epitope sequence, on the cell surface just before the membrane-anchoring region. This organism allows for high multivalent surface expression, rapid library screening, and sorting of antibody-binding cells using flow cytometric devices thanks to its rigid cell structure. In contrast to conventional epitope mapping methods, this method offers a powerful and efficient way to map continuous epitopes by screening both long and shorter antigen fragments in one library suitable for both monoclonal and polyclonal antibodies.

2 Materials

2.1 Generation of Antigen-Specific DNA-Fragment Library

1. Vector DNA containing the antigen gene as template plus corresponding primers and additional reagents needed for PCR amplification of the template DNA.
2. 15-mL conical polypropylene tubes (BD Falcon).
3. Sterile, DNase-free water.
4. Sonicator with 6-mm micro tip (Vibra cell 750 W; Sonics and Materials, http://www.sonicsandmaterials.com/).
5. 1 % agarose gel.
6. Amicon Ultra-15 centrifugal concentrators (MWCO 10 kDa; Millipore).
7. Centrifuge with swinging-bucket rotor.
8. Low DNA Mass Ladder (Invitrogen).
9. T4 DNA polymerase (Promega).
10. T4 polynucleotide kinase (PNK) and 10× PNK buffer (New England Biolabs).
11. 10 mM dNTP mix (10 mM each dNTP; New England Biolabs).
12. Display vector pSCEM2 (modified pSCEM1; [5]; this vector is not commercially available but may be obtained by contacting Prof. Ståhl, School of Biotechnology, Royal Institute of Technology (KTH), 10691, Stockholm, Sweden).
13. JETSTAR Maxiprep kit (Genomed; http://www.genomed-dna.com/).
14. High DNA Mass Ladder (Invitrogen).
15. EcoRV restriction endonuclease (New England Biolabs).
16. 10× NEB Buffer 3 (New England Biolabs).
17. 100× BSA (New England Biolabs).
18. Antarctic Phosphatase and buffer (New England Biolabs).
19. QIAquick PCR Purification Kit (Qiagen).
20. T4 DNA ligase and 10× ligation buffer (Fermentas).
21. 10× PEG (Fermentas).
22. 25:24:1 phenol:chloroform:isoamyl alcohol (VWR Scientific).
23. 96 and 70 % ethanol, ice cold.
24. 3 M sodium acetate, pH 5.5.
25. 10 mM Tris–Cl buffer, pH 8.5.

2.2 Amplification of the Target-Specific Vector Library in E. coli

1. DNA library (Subheading 2.1).
2. Electrocompetent *E. coli* (e.g., XL1-Blue; Stratagene).
3. 2-mm electroporation cuvettes (Cell Projects).

4. Sterile pipet tips with aerosol-barrier filters (Sarstedt).
5. Electroporation instrument (e.g., MicroPulser, BioRad).
6. Plastic Pasteur pipettes (Cell Projects).
7. TSB + Y medium: For 1 L of medium dissolve 30 g tryptic soy broth (Merck) and 5 g yeast extract (Merck) in distilled water and autoclave.
8. SOC medium: Mix 47 mL TSB + Y medium with 2.5 mL 20 % (w/v) sterile-filtered glucose and 0.5 mL sterile-filtered salt solution consisting of 1 M $MgCl_2$, 1 M $MgSO_4$, 1 M NaCl, and 0.25 M KCl.
9. 15-mL round-bottom tubes (BD Falcon).
10. Tryptose blood agar base (TBAB) plates: Per liter of agar dissolve 40 g TBAB (Merck) in distilled water and autoclave. Let cool to 50 °C before adding antibiotics (e.g., ampicillin) as indicated in the protocol.
11. PCR primers:

    Forward: SAPA23 5′-GGCTCCTAAAGAAAATACAACGGC-3′.

    Reverse:SAPA245′-TGTTGAATTCTTTAAGGGCATCTGC-3′.
12. Ampicillin (VWR Scientific).
13. JETSTAR Maxiprep kit (Genomed; http://www.genomed-dna.com).
14. 10 mM Tris–Cl buffer, pH 8.5.
15. Shaking incubator.
16. 1-L Erlenmeyer flask.

2.3 Generation of a *S. carnosus* Library Expressing the Antigen for Epitope Mapping

2.3.1 Preparation of Electrocompetent *S. carnosus*

1. *S. carnosus* TM 300 cells ([6]; may be obtained from Prof. F. Götz, University of Tübingen, 72076, Tübingen, Germany).
2. B2 medium: Dissolve 20 g casein hydrolysate (peptone, Sigma), 50 g yeast extract (Merck), 50 g NaCl, and 2 g $K_2HPO_4 \cdot 2H_2O$ (VWR Scientific) in 1.9 L distilled water (dH_2O); adjust the pH to 7.5; fill up to 1.95 L with dH_2O; and autoclave. Dissolve 10 g glucose (VWR Scientific) in 50 mL of dH_2O, sterile filtrate, and add aseptically to the autoclaved medium.
3. 5-L and 500-mL sterile shake flasks.
4. Spectrophotometer and cuvettes.
5. Six sterile GSA centrifuge tubes (Sorvall).
6. Sorvall centrifuge with GSA rotor.
7. 3 L sterile distilled water.
8. 1 L 10 % (v/v) glycerol in distilled water.

9. 5-mL Stripette serological pipets (Corning).
10. PipetBoy pipetting aid (Integra Biosciences; http://www.pipetboy.info/).
11. Aerosol-barrier pipet tips (Sarstedt).
12. Sterile 1.5 mL microcentrifuge tubes.

2.3.2 Generation of Staphylococcal Display Library

1. Electrocompetent *S. carnosus* (*see* Subheading 2.3.1).
2. Library plasmid (~4–6 μg/μL; *see* Subheading 2.2).
3. Heat block.
4. 0.5 M sucrose + 10 % (w/v) glycerol in distilled water, sterile filtered.
5. Electroporation instrument (e.g., MicroPulser, BioRad).
6. 1-mm electroporation cuvettes (1 mm; Cell Projects).
7. Disposable plastic Pasteur pipettes (Cell Projects).
8. 100 μL and 1 mL aerosol-barrier pipet tips (Sarstedt).
9. B2 medium: Dissolve 20 g casein hydrolysate (peptone, Sigma), 50 g yeast extract (Merck), 50 g NaCl, and 2 g $K_2HPO_4 \cdot 2H_2O$ (VWR Scientific) in 1.9 L distilled water (dH_2O); adjust the pH to 7.5; fill up to 1.95 L with dH_2O; and autoclave. Dissolve 10 g glucose (VWR Scientific) in 50 mL of dH_2O, sterile filtrate, and add aseptically to the autoclaved medium.
10. Sterile 50 mL shake flask.
11. Shaking incubator.
12. TBAB plates supplemented with chloramphenicol: Per liter of agar dissolve 40 g TBAB (Merck) in distilled water and autoclave. Let cool to 50 °C before adding 10 μg/mL chloramphenicol.
13. 85 % glycerol, sterile (VWR Scientific).
14. Sterile 1.5 mL microcentrifuge tubes.

2.4 Antibody Labeling and Flow Cytometric Analysis of the *S. carnosus* Antibody Library

1. Antigen-expressing *S. carnosus* (*see* Subheading 2.3.2).
2. TSB + Y medium: For 1 L of medium dissolve 30 g tryptic soy broth (Merck) and 5 g yeast extract (Merck) in distilled water and autoclave. After the medium has cooled down to approx. 50 °C supplement it with 10 μg/mL chloramphenicol.
3. Sterile 100 mL shake flasks.
4. PBSP buffer: Add 10 mL Pluronic F108 NF surfactant (BASF Corp.) to 990 mL phosphate buffer saline (PBS), pH 7.2, to yield 1 L 1 % (v/v) PBSP. Filtrate through a 0.45 μm filter (Millipore).
5. Refrigerated centrifuge.

6. Primary antibodies to be epitope mapped.
7. End-over-end rotator.
8. Alexa 488-labeled secondary antibody appropriate for detecting primary antibody.
9. Alexa 647-labeled human serum albumin: Label 20 mg human serum albumin (Sigma) with Alexa 647 carboxylic acid succinimidyl ester (Invitrogen) according to the manufacturer's instructions, and determine the concentration in a spectrophotometer.
10. Flow cytometer (e.g., MoFlo Astrios, Beckman Coulter).
11. Flow cytometer tubes.
12. TBAB plates supplemented with chloramphenicol: Per liter of agar dissolve 40 g TBAB (Merck) in distilled water and autoclave. Let cool to 50 °C before adding 10 μg/mL chloramphenicol.
13. PCR and sequencing primers:

    Forward:SAPA235′-GGCTCCTAAAGAAAATACAACGGC-3′.

    Reverse:SAPA245′-TGTTGAATTCTTTAAGGGCATCTGC-3′.
14. BigDye® Terminator v3.1 Cycle Sequencing Kit (Applied Biosystems).
15. 96-well PCR plate.

3 Methods

3.1 Amplification, Fragmentation, and Ligation of the Target DNA into Display Vector

1. Amplify the antigen DNA in 96 parallel 25 μL/well PCR reactions using 5 pmol of forward and reverse primer, respectively, and 5 ng target DNA per reaction.
2. Collect the PCR products in a 15-mL conical centrifuge tube, and dilute them to a final volume of 5 mL with sterile, DNase-free water.
3. To yield random molecules of antigen DNA, fragment the diluted PCR product in a sonicator with a 6 mm microtip (Vibra cell 750 W) at 21 % amplitude and constant sonication for 75 min (*see* **Note 1**).
4. Verify successful fragmentation with molecule sizes of 50–150 bp (*see* **Note 2**) by gel electrophoresis on a 1 % agarose gel.
5. Concentrate the fragmented DNA by ultrafiltration using an Amicon Ultra-15 concentrator (MWCO 10 kDa, Millipore) in a centrifuge with swinging-bucket rotor according to the manufacturer's recommendations.
6. Determine the DNA concentration either by preparing serial dilutions (1:1, 1:5, 1:10, 1:50, 1:100) of the fragmented DNA and analyzing it in comparison to Low DNA Mass ladder

(Invitrogen) on a 1 % agarose gel stained with ethidium bromide or by using a NanoDrop spectrophotometer (Thermo Fisher Scientific).

7. Blunt and 5′ phosphorylate the DNA fragments by preparing the following reaction mixture:

   3 μg fragmented DNA (5–15 μL).

   3 U T4 DNA polymerase.

   10 U T4 PNK.

   2.5 μL 10× PNK buffer.

   2.5 μL 10 mM dNTP.

   Add sterile H_2O to 25 μL.

8. Incubate at room temperature for 30 min.
9. Heat-inactivate the enzymes for 10 min at 70 °C. The products can be kept at −20 °C until use.
10. Prepare pSCEM2 vector DNA from an overnight culture as described in Rockberg et al. [7]. In brief, purify the plasmid from an *E. coli* culture using JETSTAR Maxiprep kit (Genomed) or equivalent according to the manufacturer's instructions, and determine concentration as described in paragraph 6 using High DNA Mass Ladder (Invitrogen) instead.
11. Digest 100 μg of pSCEM2 with *Eco*RV by incubating the following reaction mixture for 1 h at 37 °C:

    100 μg pSCEM2 vector

    25 μL 10× NEBuffer 3

    100 U EcoRV

    2.5 μL 100× BSA

    Add water to 250 μL.

12. Confirm complete restriction enzyme digestion on a 1 % agarose gel. In case of no linearized vector still being present continue with digestion for another 30 min and repeat this step.
13. Heat-inactivate *Eco*RV for 20 min at 80 °C.
14. Add 30 μL of Antarctic Phosphatase buffer (NEB) and 10 μL of Antarctic phosphatase to the digestion mixture and incubate for 30 min at 37 °C.
15. Purify digested and dephosphorylated pSCEM2 vector on ten QIAquick columns in parallel (using 28 μL of the reaction on each column) according to the protocol supplied with the QIAquick PCR Purification Kit (Qiagen).
16. Pool the ten eluates from the columns, and determine the vector concentration as in **step 10**. The vector can be stored at −20 °C until use.

17. Ligate blunted antigen DNA (**step 9**) with 10 μg of dephosphorylated vector pSCEM2 (**step 16**) at a molar ratio of vector to insert of 1:3. Calculate the amount of insert DNA needed based on the length of your fragments, and prepare the following reaction mixture:

    10 μg pSCEM2 vector DNA

    Calculated the amount of insert DNA

    20 μL 10× ligation buffer

    20 μL 10× PEG

    2 μL 5 U/μL T4 DNA ligase

    Add water to 200 μL.
18. Incubate overnight at room temperature.
19. Fill up the ligation mixture to a final volume of 400 μL with sterile water. Add 400 μL of the lower fraction of 25:24:1 phenol:chloroform:isoamyl alcohol to the diluted mixture and vortex intensively for 1 min. Perform further phase separation by microcentrifugation at maximum speed. Pipet the aqueous top phase to a new tube and add 400 μL of 25:24:1 phenol:chloroform:isoamyl alcohol. Repeat the described extraction step two times, and after the third extraction transfer the aqueous phase to a new tube.

    CAUTION: Phenol, chloroform, and isoamyl alcohol are harmful. Handle only with appropriate personal protection equipment, and perform all work under a fume hood.
20. Concentrate the extracted DNA by ethanol precipitation. Add 40 μL of 3 M sodium acetate and 1 mL of ice-cold 96 % ethanol to the extracted phase and incubate at −20 °C for 20 min. Pellet the precipitated DNA in a bench-top centrifuge at maximum speed for 15 min at 4 °C. Remove the supernatant without disturbing the pellet. Rinse the pellet with 500 μL 75 % ice-cold ethanol, and repeat centrifugation step. Discard the supernatant, and let the pellet air-dry at room temperature for 30 min. Dissolve the pellet in 15 μL 10 mM Tris–Cl buffer (pH 8.5), and determine plasmid concentration as described in **step 6**.
21. The DNA library can be stored at −20 °C until further use.

3.2 Amplification of the Target-Specific Vector Library in E. coli

1. Thaw the DNA library prepared in the previous chapter and electrocompetent *E. coli* cells on ice for 10 min.
2. Place electroporation cuvettes on ice.
3. Add 1 μL (around 5 μg typically) of library plasmids to a tube containing 50 μL competent cells using a sterile filter pipet tip. Mix carefully by pipetting up and down and incubate on ice for 10 min.

4. Transfer 50 μL of the cell suspension to a 2 mm electroporation cuvette. Carefully hit the cuvette on the bench a few times to remove air bubbles that might be trapped in the cuvette.
5. Wipe the cuvette dry, place it in the cuvette holder of the electroporator, and insert the holder.
6. Pipet 1 mL SOC medium using a disposable plastic Pasteur pipet, but do not add it to the cells before **step 8**.
7. Switch the BioRad Micropulser electroporator on, and transform the cells with program EC2 (V=2.5 kV).
8. Withdraw the cuvette promptly, and let the cells recover by instantly adding 1 mL SOC medium (**step 6**). Pipet carefully up and down to mix, and transfer the whole suspension to a 15 mL round-bottom tube.
9. Incubate the tube(s) horizontally, to avoid sedimentation, on a shaking table at 150 rpm and 37 °C for 1 h.
10. Pre-warm eight TBAB plates supplemented with 100 μg/mL ampicillin at room temperature for 1 h.
11. Prepare two dilutions, 1:1,000 and 1:10,000, of the cell suspension in SOC medium.
12. Distribute the undiluted cell suspension equally onto six TBAB plates. Plate out 100 μL per dilution on the remaining two plates, and incubate all plates at 37 °C for approx. 16 h overnight.
13. The library size calculated from the colony number on the dilution plates should be around 10^6 (*see* **Note 3**).
14. To evaluate the ligation efficiency and fragment length distribution perform a PCR screen on 24 colonies with primers SAPA23 and SAPA24, which amplify the vector insert (*see* **Note 4**).
15. Dissolve the colonies from the six plates containing undiluted sample with 500 μL TSB + Y per plate, and pool them in a 1 L shake flask holding 100 mL TSB + Y supplemented with 100 μg/mL ampicillin. Incubate at 150 rpm, 37 °C overnight.
16. Next day, pellet the cells by centrifugation for 8 min at 2,500 × g, 4 °C, and isolate plasmid with the help of JETSTAR Maxiprep kit according to the manufacturer's instructions. Resuspend the final DNA pellet in 50 μL Tris–Cl buffer.
17. Quantify the amount of vector with the methods described in Subheading 3.1, **steps 6** and **10**. Vector library concentrations can be expected to be >1 μg/mL. The vector library can be stored at −20 °C until further use.

3.3 Generation of a S. carnosus Library Expressing the Antigen for Epitope Mapping

3.3.1 Preparation of Electrocompetent S. carnosus

1. On the morning of the first day autoclave equipment needed on the following days:
   (a) Two 500 mL shake flasks holding 60 mL B2 medium
   (b) Two 5 L shake flasks holding 500 mL B2 medium
   (c) Six GSA centrifuge tubes
   (d) Fifty 1.5 mL microcentrifuge tubes
   (e) 1 L 10 % glycerol
   (f) 3 L distilled water.
2. Once autoclaved and cooled down, inoculate one of the 500 mL shake flasks with *S. carnosus* TM 300 cells from a glycerol stock stored at −80 °C. The second flask serves as blank. Incubate both flasks at 37 °C, 150 rpm, overnight. Allow the cultivation to go on for at least 20 h to yield sufficiently high cell concentration ($OD_{578\ nm} > 8$).
3. The next day, put the following items needed for subsequent steps into a cold room on ice:
   (a) Sterile 10 % glycerol
   (b) Sterile distilled water
   (c) GSA rotor
   (d) GSA centrifuge tubes
   (e) 1.5 mL microcentrifuge tubes.
4. Measure $OD_{578\ nm}$ for the overnight culture using 100-fold dilution and the second shake flask as blank. The OD is typically around 8–12.
5. Calculate the amount of overnight culture needed to yield a start OD_{578} of 0.5 in 500 mL medium. Inoculate both 5 L shake flasks (**step 1**) with that volume.
6. Measure the OD_{578} for both main cultures and adjust if needed.
7. Incubate at 37 °C, 150 rpm, until OD_{578} reaches 4 (approx. 3.5 h).
8. Make sure that the centrifuge is at 4 °C.
9. Once OD_{578} reached 4, stop growth by placing the cultures on ice for 15 min. All subsequent steps should be carried out in a cold room.
10. Distribute both cultures equally to the six precooled GSA centrifuge tubes. Place the tubes into the precooled rotor, and take it to the centrifuge.
11. Centrifuge at $3{,}000 \times g$, 4 °C, for 10 min.
12. Take rotor and tubes back to the cold room, and decant the supernatant. Resuspend the pellets in the residual medium by vortexing.

13. Make sure that the pellets are completely resolved before gradually adding 1 L of ice-cold, sterile water, i.e., add a small volume of water, then vortex, and repeat until 1 L is used.
14. Centrifuge at 4,000×*g*, 4 °C, for 10 min.
15. Resuspend the pellets in 1 L ice-cold water as described above (**step 13**).
16. Centrifuge at 4,500×*g*, 4 °C, for 10 min.
17. Resuspend the pellets in 540 mL ice-cold water in the manner described above (**step 13**) and pool into three GSA tubes.
18. Centrifuge at 5,000×*g*, 4 °C, for 10 min.
19. Resuspend the pellets in 540 mL 10 % glycerol in the manner described above (**step 13**).
20. Centrifuge at 5,000×*g*, 4 °C, for 10 min.
21. Resuspend the pellets in 180 mL 10 % glycerol in the manner described above (**step 13**) and pool into one GSA tube.
22. Centrifuge at 5,500×*g*, 4 °C, for 10 min.
23. Make sure that the supernatant is completely removed before resuspending the pellet in 6 mL 10 % glycerol using at 5 mL stripette and a pipette controller.
24. Apportion the cells in aliquots of 240 μL into 1.5 mL microcentrifuge tubes using a pipette and 1,000 μL sterile aerosol-barrier filter tips.
25. Store the electrocompetent cells at −80 °C.

3.3.2 Generation of Staphylococcal Display Library

1. Thaw electrocompetent *S. carnosus* (Subheading 3.3.1) and the library plasmid (Subheading 3.2) on ice for 5 min (*see* **Note 5**).
2. Incubate cells at room temperature for 20–30 min.
3. Mix cells gently by flipping and spin down for 1 s in a microcentrifuge.
4. Heat the tube with 240 μL electrocompetent cells in a heat block with water-filled wells at 56 °C for exactly 1.5 min (*see* **Note 6**).
5. Immediately add 1 mL sucrose–glycerol solution and mix gently by inverting the tubes two to three times.
6. Centrifuge at 4,500×*g* for 7 min at room temperature in a bench-top centrifuge.
7. Completely remove the supernatant by aspiration.
8. Resuspend the cells in 140 μL 0.5 sucrose +10 % glycerol per tube. Make sure that the pellets are completely dissolved before proceeding with electroporation (*see* **Note 7**).
9. Add 4 μL of plasmid library (around 5 μg typically) per tube using a sterile filter tip. Mix gently by pipetting up and down

several times and flipping the tube with a finger. Spin down in a microcentrifuge for 1 s.

10. Incubate at room temperature for 10 min.
11. In the meantime, set the electroporator (BioRad MicroPulser) to 2.3 kV and 1.1 ms as time constant, and prepare the cuvettes and disposable plastic pipettes.
12. Transfer 55 μL of the cell suspension to a 1 mm electroporation cuvette with a sterile filter tip. Carefully hit the cuvette on the bench a few times to remove air bubbles that might be trapped in the cuvette. The cell suspension in one tube is sufficient for four electroporations.
13. Place it in the cuvette holder of the electroporator, and insert the holder.
14. Pipette 1 mL B2 medium into a plastic Pasteur pipet.
15. Transform the cells with the settings stated in **step 11**.
16. Withdraw the cuvette promptly, and let the cells recover by instantly adding 1 mL B2 medium (**step 14**). Pipet carefully up and down to mix, and transfer the entire suspension to a 50 mL shake flask (*see* **Note 8**).
17. For library transformation, a total of 10 electroporations is recommended. Repeat **steps 12–16**, and pool all transformations in the same shake flask.
18. Incubate the cells at 37 °C for 2 h while shaking with 150 rpm.
19. Pre-warm two TBAB plates supplemented with 10 μg/mL chloramphenicol at room temperature for 1 h.
20. Prepare a 1:600 dilution of the cell suspension in B2 medium in duplicates.
21. Plate out 100 μL of the dilution on the two plates and incubate at 37 °C over two nights (at least 36 h).
22. After the incubation (**step 18**), transfer the culture from the 50 mL flask to a 5 L shake flask. Rinse the 50 mL flask with B2 medium to recover as many cells as possible. Cultivate the cells in a total volume of 500 mL B2 medium supplemented with 10 μg/mL chloramphenicol for 16–24 h at 37 °C while shaking at 150 rpm.
23. Next day, pellet the cells by centrifugation at 2,000 × *g* for 10 min.
24. Decant the supernatant, and resuspend the cells in the residual medium by vortexing.
25. To prepare glycerol stocks determine the volume of the cell suspension and add sterile 85 % glycerol to a final concentration of 15 %.
26. Aliquot the cell suspension into 1.5 mL microcentrifuge tubes and store at −80 °C until further use.

3.4 Antibody Labeling and Flow Cytometric Analysis of the *S. carnosus* Antibody Library

1. Inoculate 10 mL TSB + Y medium supplemented with 10 μg/mL chloramphenicol with 10 μL *S. carnosus* library from a glycerol stock (Subheading 3.3.2). Cultivate overnight at 37 °C and 150 rpm in a 100 mL shake flask.
2. Next day (after approx. 16 h), take 2 μL of overnight culture to a microcentrifuge tube. Wash the cells by adding 200 μL PBSP and pipetting up and down several times.
3. Centrifuge at 3,500 × *g*, 4 °C, for 6 min, and discard the supernatant.
4. Dissolve the pellet in 200 μL PBSP, and repeat the centrifugation step.
5. Resuspend the pellet in 75 μL PBSP supplemented with primary antibody, which is to be epitope mapped, at a final concentration of ~1 μg/mL (*see* **Note 9**).
6. Incubate at room temperature for 45 min in an end-over-end rotator.
7. Pellet and wash the cells as described in **steps 3** and **4**.
8. Resuspend the cells in 75 μL PBSP containing Alexa-488-labeled secondary and Alexa-647-labeled human serum albumin (HSA) at a final concentration of 1 μg/mL and 40 nM, respectively.
9. Incubate on ice in the dark for 30 min.
10. Repeat **step 7**.
11. Dissolve the pellet in 600 μL PBSP.
12. Analyze antibody binding and surface expression level simultaneously in a flow cytometer using appropriate excitation lasers and emission filters centered around 520 and 670 nm, respectively (*see* **Note 10**). Gate cells showing surface expression and antibody binding, i.e., high signal at both 520 and 670 nm (*see* Fig. 2), and sort >1,000 cells into 200 μL TSB + Y medium. Epitope expression level is monitored by measuring labeled HSA binding to the ABD downstream of the epitope sequence on the cell surface located just before the membrane-anchoring region. Only in-frame fragments will be possible to analyze since out-of-frame regions will not be able to bind to the membrane nor be detected by labeled HSA.
13. Incubate sorted cells for 2 h at 37 °C in an end-over-end rotator.
14. Transfer the cells into a 100 mL shake flask containing 10 mL TSB + Y medium supplemented with 10 μg/mL chloramphenicol and cultivate overnight at 37 °C while shaking at 150 rpm.
15. Perform a second round of sorting by repeating **steps 2–13**, this time gating both positive and negative cells (*see* Fig. 2).
16. Plate the sorted cells on TBAB plates supplemented with 10 μg/mL chloramphenicol, and incubate the plates over two nights (~48 h) at 37 °C. After this step plates can be kept

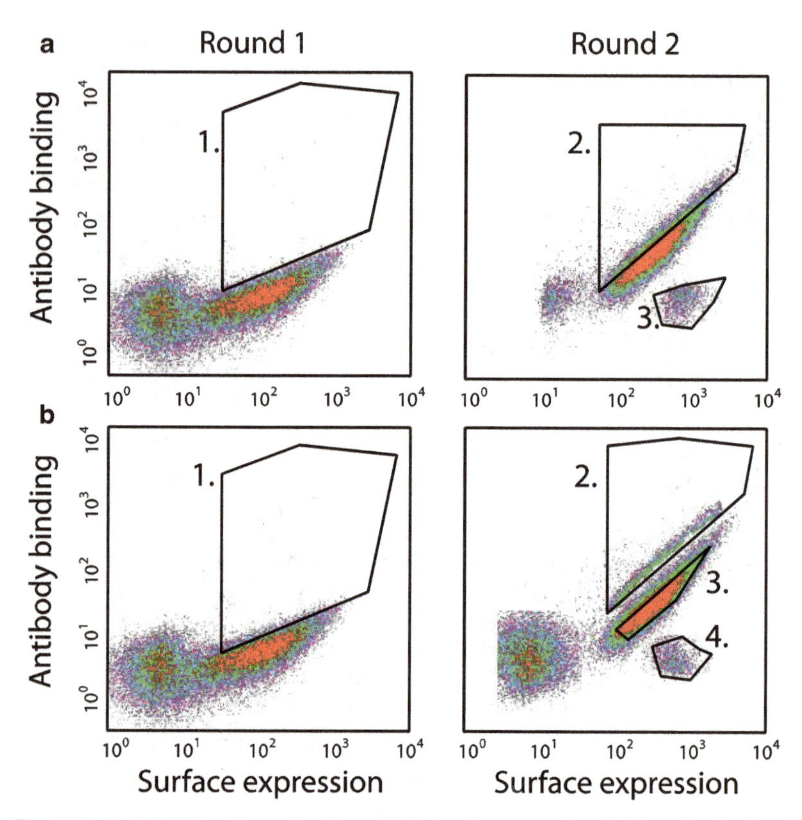

Fig. 2 Typical FACS gating of epitope libraries for monoclonal (**a**) and polyclonal (**b**) antibodies. Examples of positive gates can be seen in A (1–2) and B (1–3); negative populations are gated as seen in A3 and B4

refrigerated at 4–8 °C and sealed with parafilm until further use for a couple of weeks.

17. Pick 96 gated colonies into separate wells on a 96-well PCR plate holding 10 μL sterile water. Dissolve the colonies by pipetting up and down several times.
18. Transfer 1 μL from each well into a new 96-well PCR plate.
19. Amplify the vector inserts by PCR in a total volume of 25 μL using 5 pmol of forward and reverse primer.
20. Sequence the PCR products using the Big Dye sequencing kit (Applied Biosystems) according to the manufacturer's instructions.
21. Determine the shortest consensus sequences within a gated population by aligning the sequences obtained from the cycle sequencing to the original antigen sequence to identify regions of antibody binding. In case of mapping a polyclonal antibody alignment may result in several antibody-binding sites, i.e., several consensus sequences. Proper gating ought to be confirmed by checking that no negative clones align to the identified binding sites (*see* **Note 11**).

4 Notes

1. If foam formation is observed during sonication, dilute the PCR product or reduce the sonication amplitude.
2. The sonication step controls the length of the peptides later to be cloned and expressed in *S. carnosus*. It can easily be adapted for antibodies with known longer or shorter epitopes by plainly adjusting the fragmentation time. If longer unwanted fragments are observed after 75 min of sonication, let fragmentation continue for another 30 min.
3. If few transformants are observed, possible explanations are a too low concentration of vector DNA or too old competent cells. Determine the concentration of the library plasmid and control on a gel that vector was successfully circularized. Electrocompetent cells older than 6 months should not be used for library preparations.
4. For efficient ligation a threefold molar excess of insert to vector is essential. Thus the concentration of insert and vector DNA should be determined as accurately as possible. If still an insert-to-relegated vector ratio below 50 % is observed, prepare two ligations with three times higher or lower insert:vector ratio, respectively, additionally to the original ratio. However, in case the number of inserts is low while a sufficiently high transformation efficiency is maintained, EcoRV digestion and subsequent dephosphorylation of the vector library prior to transformation into *S. carnosus* are advisable to eliminate empty vectors from library since they have shown to possess a growth advantage compared to insert-baring plasmids. In addition, gel purification of the digested plasmid can also eliminate undigested vector, thereby improving the ratio of vectors containing an insert.
5. Do not keep the cells on ice longer than 5 min. Prolonged incubation at this step reduces transformation frequency.
6. Do not handle more than three tubes at once as prolongated handling and incubation decrease transformation frequency. Make sure that the heat treatment is precisely for 1 min 30 s and that the glycerol solution is added instantly after that. It might be advisable to place the tubes in the heat block successively with 30-s interval.
7. Insufficient resuspension can cause arcing during electroporation. Thoroughly resuspend the cells pipetting up and down, but avoid the introduction of air bubbles into your cell suspension.
8. Do not wash the cuvette with an additional volume of B2 medium in an attempt to recover the small amount of residual cells. Total volumes above 1 mL/electroporation during phenotyping result in drastically reduced transformation frequency.

9. In general epitope mappings are easiest done in singleplex (one monoclonal or polyclonal antibody per tube), but it is also possible to analyze mixture of antibodies [8].
10. If no antibody-binding signal can be detected during FACS, raise the antibody concentration. A library based on longer DNA fragments (>200 bp) might prove useful in case the lacking signal is suspected to result from possible structural epitopes.
11. Failure of validation, i.e., sequences of positive and negative clones align to the same site, indicates that the negative and positive gate during FACS were insufficiently separated. Try to lower the negative gate to decrease the number of false-negative clones.

References

1. Geysen HM, Meloen RH, Barteling SJ (1984) Use of peptide synthesis to probe viral antigens for epitopes to a resolution of a single amino acid. Proc Natl Acad Sci U S A 81:3998–4002
2. Sheriff S, Silverton EW, Padlan EA et al (1987) Three-dimensional structure of an antibody-antigen complex. Proc Natl Acad Sci U S A 84:8075–8079
3. Herzenberg LA, Parks D, Sahaf B et al (2002) The history and future of the fluorescence activated cell sorter and flow cytometry: a view from Stanford. Clin Chem 48:1819–1827
4. Mattanovich D, Borth N (2006) Applications of cell sorting in biotechnology. Microbial Cell Fact 5:12
5. Rockberg J, Löfblom J, Hjelm B et al (2008) Epitope mapping of antibodies using bacterial surface display. Nat Methods 5
6. Augustin J, Götz F (1990) Transformation of Staphylococcus epidermidis and other staphylococcal species with plasmid DNA by electroporation. FEMS Microbiol Lett 54:203–207
7. Rockberg J, Löfblom J, Hjelm B et al (2010) Epitope mapping using gram-positive surface display. Curr Protoc Immunol Chapter 9:Unit9.9. doi: 10.1002/0471142735.im0909s90
8. Hudson EP, Uhlen M, Rockberg J (2012) Multiplex epitope mapping using bacterial surface display reveals both linear and conformational epitopes. Sci Rep 2:706

Chapter 30

Ion Exchange-High-Performance Liquid Chromatography (IEX-HPLC)

Marie Corbier, Delphine Schrag, and Sylvain Raimondi

Abstract

The ion exchange-high-performance liquid chromatography is a high-throughput analytical method that allows to determine the charge profile of purified antibodies. Here, we describe the preparation of the samples, chromatographic conditions to be used (buffer preparation, salt gradient, quantity injected, flow rate, run time, column suitability, etc.), validity of the analysis, and integration of the chromatogram in order to calculate the proportion of the different isoforms.

Key words Ion exchange-high-performance liquid chromatography, Isoforms, Basic/acidic/main peak integrations

1 Introduction

Ion exchange-high-performance liquid chromatography (IEX-HPLC) allows the separation of ions and polar molecules based on their charge [1]. The stationary phase surface contains ionic functional groups which interact with analyte ions of opposite charge (coulombic interactions). IEX-HPLC methods are further subdivided into cation and anion exchange chromatography, where cation exchange HPLC retains cations (positively charged ions) and anion exchange HPLC retains anions (negatively charges ions) [2, 3]. IEX-HPLC separates proteins based on their net charge, which is dependent on the composition of the mobile phase. By adjusting the pH or the ionic concentration of the mobile phase, various proteins can be separated [4].

Here we describe a weak cation exchange-high-performance liquid chromatography (WCX-HPLC) method to determine the charge homogeneity of antibodies [5, 6]. This WCX-HPLC method has been developed to allow for the separation of the acidic, basic, and main variants according to their charge distribution. The posttranslational chemical and enzymatic modifications result in the formation of various charge variants and heterogeneity

Vincent Ossipow and Nicolas Fischer (eds.), *Monoclonal Antibodies: Methods and Protocols*, Methods in Molecular Biology, vol. 1131, DOI 10.1007/978-1-62703-992-5_30,

of antibodies, thus modifying their isoelectric pH (p*I*) [7]. The acidic peaks elute first followed by the main peak and finally the basic peaks. The three isoform populations are identified and analyzed by their percentage relative area [8].

2 Materials

2.1 Equipment

The IEX analysis is performed using a high-performance liquid chromatography system (e.g., Waters Alliance HPLC) with a 214 nm UV detector and appropriate software (e.g., Empower software) to allow for accurate chromatogram integration. The samples are analyzed on a suitable column (e.g., a cation exchange BioMab NP5-SS, 250 × 4.6 mm from Agilent).

2.2 Buffer Preparation

1. Stock solution 1: 100 mM sodium phosphate (dibasic). Dissolve 14.20 g dibasic sodium phosphate in a 1 L beaker containing approximately 900 mL ultrapure water (*see* **Note 1**). Transfer to a 1 L graduated cylinder, and add ultrapure water to a final volume of 1 L. Filter through a 0.22 μm filter (*see* **Note 2**).
2. Stock solution 2: 100 mM sodium phosphate (monobasic). Dissolve 12.00 g monobasic sodium phosphate in a 1 L beaker containing approximately 900 mL ultrapure water (*see* **Note 1**). Transfer to a 1 L graduated cylinder, and add ultrapure water to a final volume of 1 L. Filter through a 0.22 μm filter (*see* **Note 2**).
3. Running buffer: 10 mM sodium phosphate, pH 6.5. Pour 100 mL of stock solution 1 into a 1 L graduated cylinder containing approximately 800 mL ultrapure water. Add ultrapure water to a final volume of 1 L to obtain a 10 mM dibasic sodium phosphate solution. Pour 100 mL of stock solution 2 into a 1 L graduated cylinder containing approximately 800 mL ultrapure water. Add ultrapure water to a final volume of 1 L to obtain a 10 mM monobasic sodium phosphate solution. Pour the necessary volume of 10 mM dibasic sodium phosphate in a 2 L beaker containing 1 L 10 mM monobasic sodium phosphate to reach the target pH of 6.5 (*see* **Note 3**). Filter through a 0.22 μm filter (*see* **Note 4**).
4. Elution buffer: 10 mM sodium phosphate, 500 mM NaCl, pH 6.5. Pour 100 mL of stock solution 1 into a 1 L graduated cylinder containing approximately 800 mL ultrapure water. Add ultrapure water to a final volume of 1 L to obtain a 10 mM dibasic sodium phosphate solution. Pour approximately 900 mL of this solution into a 1 L beaker and dissolve in 29.22 g sodium chloride. Transfer to a 1 L graduated cylinder,

and add 10 mM dibasic sodium phosphate to a final volume of 1 L to obtain a 10 mM dibasic sodium phosphate and 500 mM NaCl solution. Pour 100 mL of stock solution 2 into a 1 L graduated cylinder containing approximately 800 mL ultrapure water. Add ultrapure water to a final volume of 1 L to obtain a 10 mM sodium phosphate monobasic solution. Pour approximately 900 mL of this solution into a 1 L beaker and dissolve in 29.22 g sodium chloride. Transfer to a 1 L graduated cylinder, and add 10 mM monobasic sodium phosphate to a final volume of 1 L to obtain a 10 mM monobasic sodium phosphate and 500 mM NaCl solution. Pour the required volume of 10 mM monobasic sodium phosphate and 500 mM NaCl into a 2 L beaker containing 1 L 10 mM dibasic sodium phosphate and 500 mM NaCl in order to reach the target pH of 6.5 (*see* **Note 5**). Filter through a 0.22 μm filter (*see* **Note 4**).

2.3 Sample Preparation

Samples must be prepared on the day of the experiment to avoid any dilution effects which could influence the protein charge distribution. Vortex the sample vials well, and ensure that no air bubbles are present. Although the WCX-HPLC method typically shows satisfactory repeatability, it is recommended to inject samples in duplicates.

1. Blank samples: Prepare 100 μL of blank (*see* **Note 6**) directly in the appropriate HPLC vial.
2. Samples: Prepare 100 μL of sample by diluting with the running buffer to 1 mg/mL (*see* **Note 7**) directly in the appropriate HPLC vial.
3. Reference sample: Prepare 100 μL of sample reference in duplicate (as described in the previous step) (*see* **Note 8**).

3 Method

3.1 IEX Run

1. Connect and equilibrate the column on the HPLC system using a flow rate of 0.8 mL/min with the running buffer for at least 20 min (*see* **Note 9**) and at a column temperature of 25 °C (*see* **Note 10**).
2. Perform the analysis using a gradient method (*see* Table 1) with a 0.8 mL/min flow rate and a total analysis time of 60 min and with an injection volume of 50 μL (i.e., 50 μg). The sample order is typically run as follows: blanks, reference standard sample duplicate 1, samples, and reference standard sample duplicate 2.
3. At the end of the analysis, rinse and store the column with running buffer (*see* **Note 11**).

Table 1
Description of the linear gradient of 10 mM sodium phosphate and 500 mM NaCl, pH 6.5 (elution buffer) used in order to separate the different isoforms

Time (min)	% of Running buffer	% of Elution buffer
0	95	5
10	95	5
50	79	21
50.1	0	100
55	0	100
55.1	95	5
60	95	5

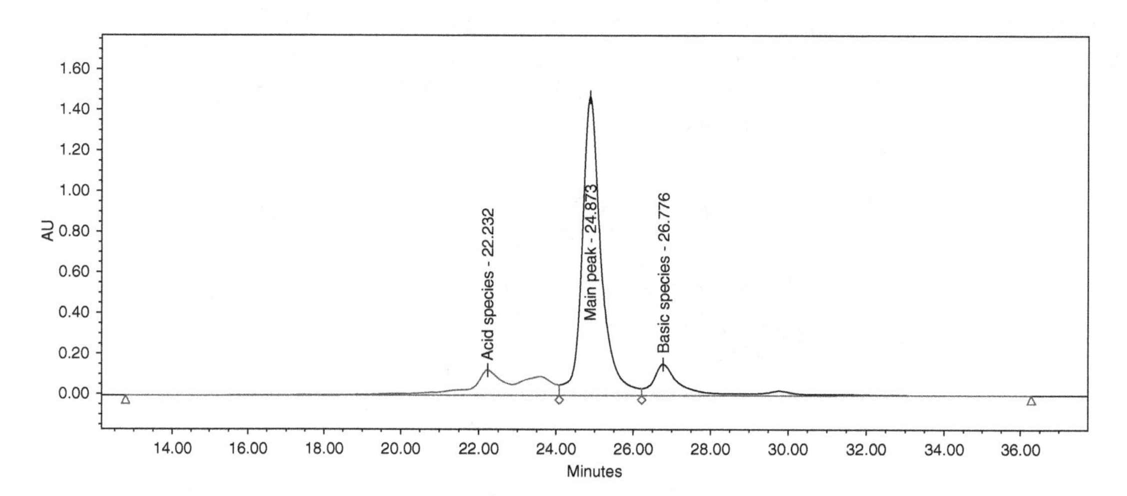

Fig. 1 Example of chromatogram manual integration of the acidic species (which elute first), the main component, and the basic species (which elute at the end) using the Empower software

3.2 IEX Results: Acceptance Criteria

1. The analysis is accepted and considered to be valid if the reference standard samples run at the beginning and at the end of the analysis have the same main peak retention time and the appropriate charge distribution in terms of acidic, main acidic peak, and basic isoform relative percentages.

3.3 IEX Results: Chromatogram Integration

1. Using the Empower software, identify the main component in reference sample and samples by its characteristic retention time and position (between the acidic and basic species).
2. Integrate the main component separately from the acidic species (which elute first) and the basic species (which elute at the end) in the reference standard and samples (*see* **Note 12**) (Fig. 1).

Table 2
Example of each sample component (main peak, acidic species, and basic species) characteristics obtained following the chromatogram integration with Empower software

	Peak name	RT	Height	Area	% Area
1	Acid species	22.232	126570	13437941	18.30
2	Main peak	24.873	1471406	50711431	69.06
3	Basic species	26.776	155353	9279278	12.64

The proportion of main peak, acidic species, and basic species is reported

3. Calculate the proportion of each sample component (main peak, acidic species, and basic species) by reporting the peak areas of each component relative to the total integrated peak area (Table 2).

4 Notes

1. To facilitate the dissolving process and to avoid crystallization, add the sodium phosphate slowly and mix well using a magnetic stir bar.
2. The stock solutions can be stored at 4 °C for 6 months. It is highly recommendable however to refilter the buffer (0.22 μm) when it is not regularly used.
3. Around 300 mL of 10 mM dibasic sodium phosphate is required to adjust the pH to 6.5. The pH has to be checked with a pH meter as the temperature-dependent variation may be observed.
4. The running buffer can be stored at 4 °C for 1 month. It is highly recommendable however to refilter the buffer (0.22 μm) before each use when it is not regularly used.
5. Around 800 mL of 10 mM monobasic sodium phosphate is necessary to adjust the pH to 6.5. Temperature-dependent variation may be observed.
6. Blanks include running buffer and sample buffer.
7. The volume of injection is 50 μL at a concentration of 1 mg/mL (i.e., 50 μg). If the sample concentration is below 1 mg/mL, the volume of injection can be increased up to the maximum volume allowed by the HPLC system in order to inject 50 μg. As the results are reported in terms of relative percentages of main acidic peak, acidic species, and basic species, the quantity injected can also be decreased according to the limit of quantification of the method (if determined). The areas would, however, not be comparable between samples.

8. If available, it is recommended to include an internal reference standard sample with a known relative percentage of main peak, acidic species, and basic species at the beginning and at the end of the sequence to verify the suitability of the column and the system.
9. To equilibrate the column, start at a flow rate of 0.2 mL/min and slowly increase the flow rate (0.1–0.2 mL/min steps) up to 0.8 mL/min whilst ensuring a stable pressure.
10. If available, a column oven is highly recommended to have a constant and accurate column temperature control for improved repeatability (i.e., retention time). Performing the analysis at room temperature is however acceptable.
11. For long-term column storage, refer to the column instructions provided by the supplier.
12. Different integration techniques may be used. The WCX-HPLC method described in this chapter uses the "drop" method. It involves placement of a vertical line from the valley between the peaks to the horizontal baseline, which is drawn between the start and stop points of the peak group [9]. Integration may be assessed using automated methods with parameters previously established for peak detection or manual integration may be used to distinguish the desired peaks and the baseline from noise.

References

1. Melter L, Ströhlein G, Butté A, Morbidelli M (2007) Adsorption of monoclonal antibody variants on analytical cation-exchange resin. J Chromatogr A 1154:121–131
2. Macchi FD, Shen FJ, Kwong M, Andya JD, Shire SJ, Bjork N, Totpal K, Chen AB (2001) Identification of multiple sources of charge heterogeneity in a recombinant antibody. J Chromatogr B Biomed Sci Appl 752:233–245
3. Vlasak J, Ionescu R (2008) Heterogeneity of monoclonal antibodies revealed by charge-sensitive methods. Curr Pharm Biotechnol 9:468–481
4. Rea JC, Moreno GT, Lou Y, Farnan D (2011) Validation of a pH gradient-based ion-exchange chromatography method for high-resolution monoclonal antibody charge variant separations. J Pharm Biomed Anal 54(2):317–323
5. Zhang T, Bourret J, Cano T (2011) Isolation and characterization of therapeutic antibody charge variants using cation exchange displacement chromatography. J Chromatogr A 1218(31):5079–5086
6. Liu H, Gaza-Bulseco G, Faldu D, Chumsae C, Sun J (2008) Heterogeneity of monoclonal antibodies. J Pharm Sci 97(7):2426–2447
7. Rozhkova A (2009) Quantitative analysis of monoclonal antibodies by cation-exchange chromatofocusing. J Chromatogr A 1216(32):5989–5994
8. Khawli LA, Goswami S, Hutchinson R, Kwong ZW, Yang J, Wang X, Yao Z, Sreedhara A, Cano T, Tesar D, Nijem I, Allison DE, Wong PY, Kao Y-H, Quan C, Joshi A, Harris RJ, Motchnik P (2010) Charge variants in IgG1. mAbs 2(6):613–624
9. 402 LCGC NORTH AMERICA (2006) 24(4) www.chromatographyonlime.com

Chapter 31

Size Exclusion-High-Performance Liquid Chromatography (SEC-HPLC)

Delphine Schrag, Marie Corbier, and Sylvain Raimondi

Abstract

The size exclusion-high-performance liquid chromatography is a high-throughput analytical method, through isocratic condition, that allows to determine and quantify the level of aggregates and fragments of purified antibodies. Here, we describe the preparation of the samples, chromatographic conditions to be used (buffer preparation, quantity injected, flow rate, run time, column suitability, etc.), validity of the analysis, and integration of the chromatogram in order to calculate the proportion of the different components.

Key words Size exclusion-high-performance liquid chromatography, Aggregates, Monomer, Fragments, Peak integration

1 Introduction

The size exclusion-high-performance liquid chromatography (SEC-HPLC) method separates protein components according to their size [1, 2]. The separation is achieved by the differential exclusion from the pores of the packing material, of the sample molecules as they pass through a bed of porous particles. The principle feature of size exclusion chromatography is its gentle non-adsorptive interaction with the sample, enabling high retention of biomolecular activity. This technique is frequently used for its ability to separate antibody monomer from both aggregates and fragments [3]. The largest components, i.e., protein aggregates, penetrate the matrix particles to a lesser extent and are therefore eluted from the column ahead of smaller components, such as protein monomers and fragments [4]. The fragments, being the smallest components, penetrate the matrix more readily and are therefore eluted after the monomer. Resolution is dependent on the Stokes radius of the components [5]. The components are identified by their typical retention time and position, relative to calibration markers (Fig. 1).

Vincent Ossipow and Nicolas Fischer (eds.), *Monoclonal Antibodies: Methods and Protocols*, Methods in Molecular Biology, vol. 1131, DOI 10.1007/978-1-62703-992-5_31, © Springer Science+Business Media New York 2014

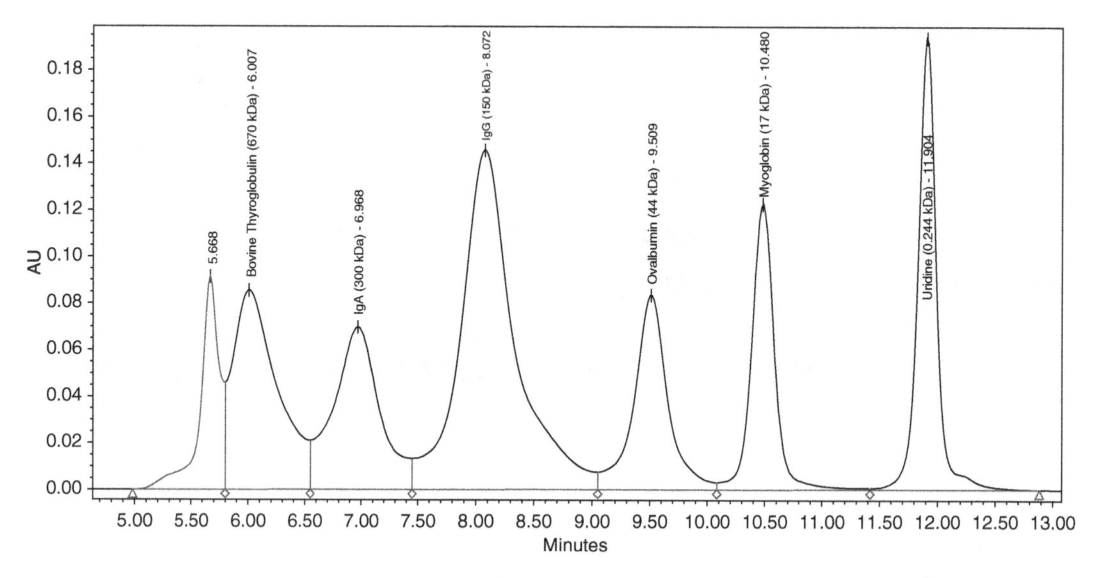

Fig. 1 Example of chromatogram integration of the calibration marker with the protein mixture components

Here we describe a SEC-HPLC method to determine the level of high- and low-molecular-weight species in purified antibody under isocratic conditions [6]. This SEC-HPLC method has been developed to allow for the separation of the aggregates, monomer, and fragment components according to their size. The aggregate peak elutes first followed by the monomer peak and finally the fragment peak [7]. The proportion of sample components is determined by calculation of the peak areas of each component relative to the total integrated peak area.

2 Materials

2.1 Equipment

The SEC-HPLC analysis is performed using a high-performance liquid chromatography system (e.g., Waters Alliance HPLC) with a 280 nm UV detector and appropriate software (e.g., Empower software) to allow for accurate chromatogram integration. Samples containing 100 μg of purified antibody are analyzed on a suitable column (e.g., TSKgel column G3000SWXL, 300 × 7.8 mm from Tosoh).

2.2 Buffer Preparation

1. Running buffer: 0.2 M sodium phosphate (dibasic), pH 7.0. Dissolve 28.39 g sodium phosphate in a 1 L beaker containing approximately 900 mL ultrapure water (*see* **Note 1**). Transfer to a 1 L graduated cylinder, and add ultrapure water to a final volume of 1 L. Adjust to pH 7.0 with 37 % HCl (*see* **Note 2**). Filter through a 0.22 μm filter (*see* **Note 3**).

2.3 Sample Preparation

Samples must be prepared on the day of the experiment to avoid any dilution effects which could influence the proportion of reversible aggregate populations. Vortex the sample vials well, and ensure that no air bubbles are present. Although the SEC-HPLC method typically shows satisfactory repeatability, it is recommended to inject samples in duplicates.

1. Blank sample: Prepare 150 μL of blank (*see* **Note 4**) directly in the appropriate HPLC vial.
2. Calibration marker: Prepare the calibration marker solution (Phenomenex) by combining 100 μL of reconstituted protein mixture with 6 μL reconstituted uridine (*see* **Note 5**).
3. Samples: Prepare 150 μL of sample by diluting with the running buffer to 1 mg/mL (*see* **Note 6**) directly in the appropriate HPLC vial.
4. Reference sample: Prepare 150 μL of reference sample in duplicate (as described in the previous step) (*see* **Note 7**).

3 Method

3.1 SEC Run

1. Connect and equilibrate the column (*see* **Note 8**) at a flow rate of 1 mL/min with the running buffer for at least 20 min (*see* **Note 9**) and at column temperature of 25 °C (*see* **Note 10**).
2. Perform the analysis using an isocratic method with a 1 mL/min flow rate and a total analysis time of 15 min, with an injection volume of 100 μL (i.e., 100 μg), except for the calibration marker (10 μL). The sample order is typically run as follows: blanks, calibration marker, reference standard sample duplicate 1, samples, and reference standard sample duplicate 2.
3. At the end of the analysis, rinse and store the column with running buffer (*see* **Note 11**).

3.2 SEC Results: Acceptance Criteria

1. The analysis is accepted and considered to be valid if the calibration marker profile is in accordance with the certificate provided by the supplier.
2. The analysis is accepted and considered to be valid if the reference standard samples run at the beginning and at the end of the analysis have the same monomer retention time and the appropriate aggregate, monomer, and fragment relative percentages.

3.3 SEC Results: Chromatogram Integration

1. Using the Empower software, integrate the calibration marker and identify the protein mixture components according to the certificate provided by the supplier (Fig. 1).
2. Identify the monomeric component in reference sample and samples by its characteristic retention time and position, relative to the calibration marker.

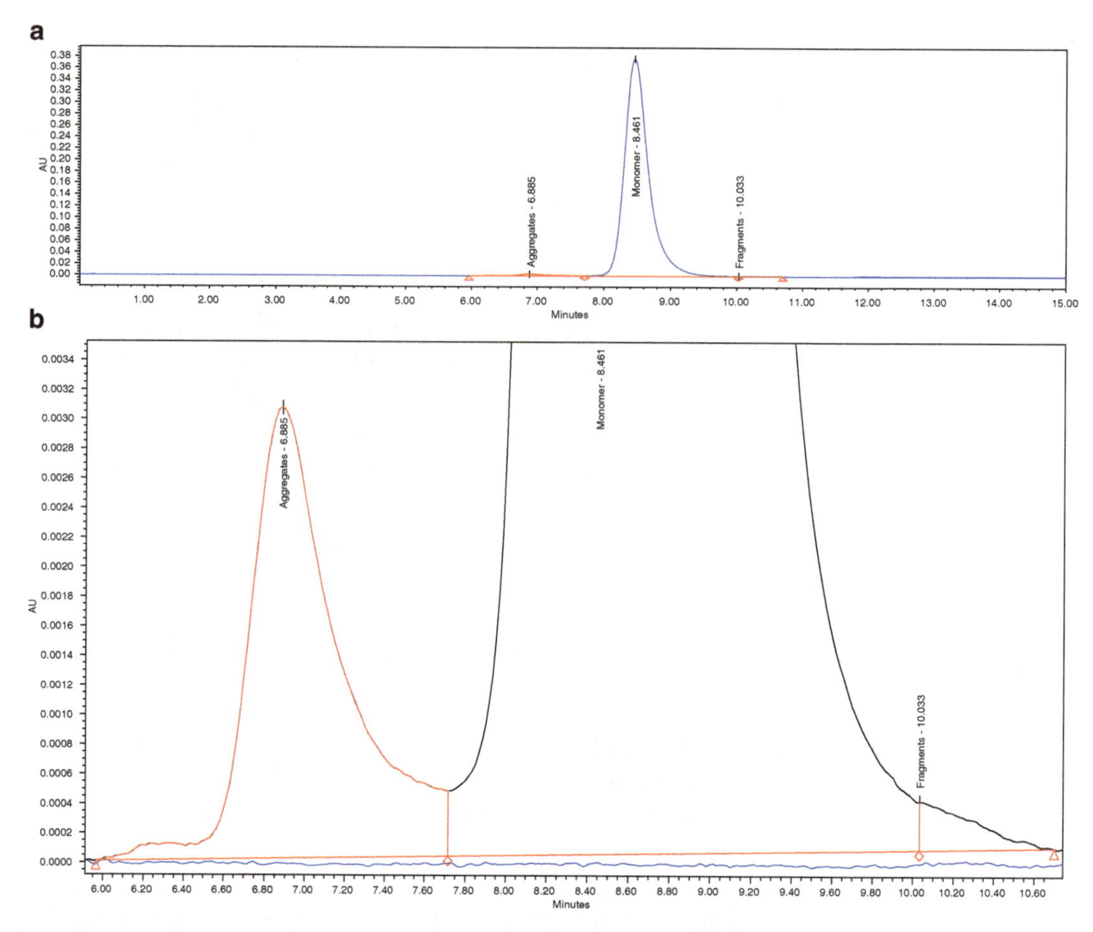

Fig. 2 (**a**) Example of manual chromatogram integration (full chromatogram view) of the high-molecular-weight species (i.e., aggregates, which elute first), monomeric component, and low-molecular-weight species (i.e., fragments, which elute at the end) using the Empower software. (**b**) Chromatogram zoom on the manual integration of aggregates, monomer, and fragments

3. Integrate the monomeric component from the high-molecular-weight species (i.e., aggregates) and the low-molecular-weight species (i.e., fragments) in the reference standard and samples (*see* **Note 12**) (Fig. 2).
4. Calculate the proportion of each sample component (monomer, aggregates, and fragments) by reporting the peak areas of each component relative to the total integrated peak area (Table 1).

4 Notes

1. To facilitate the dissolving process and to avoid crystallization, add the sodium phosphate slowly and mix well using a magnetic stir bar.

Table 1
Example of each sample component (aggregates, monomer, and fragments) characteristics obtained following the chromatogram integration with Empower software

	Peak name	RT	Height	Area	% Area
1	Aggregates	6.885	3048	95664	1.00
2	Monomer	8.461	377584	9502931	98.94
3	Fragments	10.033	331	6206	0.06

The proportion of each component is reported

2. When using 37 % HCl, avoid any skin contact and wear an appropriate mask or use a fume hood. The required volume of 37 % HCl to reach the target pH is about 4.5 mL for 1 L of buffer.
3. The running buffer can be stored at 4 °C for 1 month. It is highly recommendable however to refilter the buffer (0.22 μm) when it is not regularly used.
4. Blanks include running buffer and sample buffer.
5. The protein and uridine mixtures (Phenomenex kit) are both reconstituted with 1 mL of running buffer. Take care to vortex the protein mixture until completely dissolved. It is highly recommended to always inject the calibration marker when performing an SEC-HPLC analysis. This allows to both check the column separation performance and to identify sample protein components relative to their molecular weight.
6. The volume of injection is 100 μL at a concentration of 1 mg/mL (i.e., 100 μg). If the sample concentration is below 1 mg/mL, the volume of injection can be increased up to the maximum volume allowed by the HPLC system in order to inject 100 μg. As the results are reported in terms of relative percentages of monomer, aggregate, and fragment, the quantity injected can be decreased according to the limit of quantification of the method (if determined). The areas would, however, not be comparable between samples.
7. If available, it is recommended to include an internal reference standard sample with a known relative percentage of aggregates, monomer, and fragments at the beginning and at the end of the sequence to verify the suitability of the column and the system.
8. To protect the HPLC system and column shelf life when injecting unpurified or precipitated samples, it is highly recommended to use an HPLC security guard cartridge system (Phenomenex) which attaches directly to the top of the column.

9. To equilibrate the column, start at a flow rate of 0.2 mL/min and slowly increase the flow rate up to 1 mL/min whilst ensuring a stable pressure.
10. If available, a column oven is recommended to have a constant and accurate column temperature control for improved repeatability (i.e., retention time). Performing the analysis at room temperature is however acceptable.
11. For long-term column storage, refer to the column instructions provided by the supplier.
12. Different integration techniques may be used. The SEC-HPLC method described in this chapter uses the "drop" method. It involves placement of a vertical line from the valley between the peaks to the horizontal baseline, which is drawn between the start and stop points of the peak group [8]. Integration may be assessed using automated methods with parameters previously established for peak detection, or manual integration may be used to distinguish the desired peaks and the baseline from noise.

References

1. Jiskoot W, Beuvery EC, de Koning AA, Herron JN, Crommelin DJ (1990) Analytical approaches to the study of monoclonal antibody stability. Pharm Res 7:1234–1241
2. Park S, Cho H, Kim Y, Ahn S, Chang TJ (2007) Fast size-exclusion chromatography at high temperature. J Chromatogr A 1157(1–2):96–100
3. Diederich P, Hansen SK, Oelmeier SA, Stolzenberger B, Hubbuch JJ (2011) A sub-two minutes method for monoclonal antibody-aggregate quantification using parallel interlaced size exclusion high performance liquid chromatography. J Chromatogr A 1218(50):9010–9018
4. Arosio P, Barolo G, Müller-Späth T, Wu H, Morbidelli M (2011) Aggregation stability of a monoclonal antibody during downstream processing. Pharm Res 28(8):1884–1894
5. Farnan D, Moreno GT, Stults J, Becker A, Tremintin G, van Gils MJ (2009) Interlaced size exclusion liquid chromatography of monoclonal antibodies. J Chromatogr A 1216(51):8904–8909
6. Conte P, Piccolo A (1999) High Pressure Size Exclusion Chromatography (HPSEC) of humic substances: molecular sizes, analytical parameters, and column performance. Chemosphere 38(3):517–528
7. Arakawa T, Ejima D, Li T, Philo JS (2010) The critical role of mobile phase composition in size exclusion chromatography of protein pharmaceuticals. J Pharm Sci 99(4):1674–1692
8. 402 LCGC NORTH AMERICA (2006) 24(4) www.chromatographyonlime.com

Chapter 32

N-Glycosylation Characterization by Liquid Chromatography with Mass Spectrometry

Song Klapoetke

Abstract

Three methods are introduced here to fully characterize glycosylation at protein, peptides, and glycan levels by LC-MS (ESI TOF MS). At protein level, glycosylation could be detected by comparing protein masses with and without glycan. At peptide level, glycosylation sites could be detected by removing glycan and site specific glycosylation could be examined with glycopeptides. At glycan level, glycan profiling and identification could be achieved by analyzing released glycans labeled with procainamide. All these methods are MS based and able to provide a whole picture of glycosylation of a glycoprotein. These methods are also simple to implement in industrial or academic laboratory with proper training and instruments.

Key words Glycan characterization, Glycan labeling, Glycan identification, Glycoprotein, Glycopeptide, LCMS, LC-ESI TOF MS

In this chapter, we focus on how to characterize N-glycosylation in recombinant monoclonal antibodies by high-resolution mass spectrometry in-line with HPLC system (LCMS). We use IgG1s as model glycoprotein for the procedures. IgG1 is a tetrameric glycoprotein (approximately 150 kDa) composed of two identical heavy chains and two identical light chains linked to each other by disulfide bonds with one N-glycosylation site on each heavy chain.

We provide three methods for the purpose. The first method is to determine N-glycosylation site number by monitoring molecular weights of four different forms of the protein, namely, the intact, deglycosylated (after glycan being removed), reduced, and deglycosylated/reduced forms. The second method is to confirm glycan attachment sites and site-specific glycan structure information by peptide mapping with or without glycan. The third method is to profile and identify glycan removed from the protein. These three methods can be used together for full glycan characterization or used individually depending on the user's objective. These methods can be used as start point for other glycoproteins.

Vincent Ossipow and Nicolas Fischer (eds.), *Monoclonal Antibodies: Methods and Protocols*, Methods in Molecular Biology, vol. 1131, DOI 10.1007/978-1-62703-992-5_32, © Springer Science+Business Media New York 2014

1 Introduction

For glycoprotein analysis, monoclonal antibodies, for example IgG1s, are analyzed without any treatment by LCMS to obtain intact mass information. They are also reduced to heavy chain and light chain by breaking interchain disulfide bonds by reducing agents (for example: dithiothreitol or DTT) in order to examine more structure variants or modification.

N-glycans can be removed by N-glycanase (for example: PNGase F) without damaging the protein backbone. After removing N-glycan, proteins can be examined for other modifications.

By analyzing IgG1s in intact, reduced, deglycosylated, and deglycosylated/reduced states, we can predict numbers of glycosylation site and major glycoforms. As an added benefit, we should be able to verify the antibody's primary structure and predict sequence variants, posttranslational modification (for example: oxidation), and terminal amino acid truncation or modification.

Glycopeptide analysis is carried out by treating IgGs with proteases, for example, trypsin, which fragments the protein into peptides. When analyzing trypsin digestion products, the potential glycopeptides should be easily identified by comparing the peptide's molecular weight with expected value or theoretical value (amino acid sequence without glycan attached). The glycosylation site could also be identified from deglycosylated trypsin digestion products because the potential N-glycan attachment site will have a +1 Da mass shift due to conversion of asparagine to aspartic acid.

LCMS/MS^E experiment is used to verify glycosylation site from deglycosylated trypsin digestion sample and to verify glycan composition from trypsin digestion sample. Site-specific glycoform information can also be obtained from trypsin digestion sample if more than one glycosylation site contained in the glycoprotein.

The results from trypsin digestion sample and deglycosylated trypsin digestion sample should be examined together, and these results should be consistent with the results from glycoprotein analysis. Amino acid sequence coverage and other degradation or modification products such as deamidation and oxidation can also be revealed by analyzing digestion products.

For glycan profiling and identification, glycans are removed from the antibody. Because glycans lack chromophore, they are not detected by means relying on UVs. Therefore, an aromatic amine is usually linked through a reductive amination reaction [1] to enable fluorescent detection. We introduce procainamide as a labeling agent to improve both fluorescent detection and ESI MS ionization [2–4].

To label the glycan, sample needs to be vacuum-dried and react with procainamide after deglycosylation. The labeled glycan then is purified by a HILIC plate [5]. The HILIC plate retains glycan in high organic solvent, and the salt and proteins are not retained with the resin and washed out from the sample. The purified sample is eluted by high aqueous solvent. The glycan then is analyzed by a normal-phase UPLC system for profiling and ESI MS for glycan identification. The glycan assignment is made according to molecular weight, MS^E fragmentation pattern, and glycan maturation pathway [6].

The data acquired from glycan analysis should also be consistent with data from glycoprotein and glycopeptide analysis.

2 Materials

Prepare all solutions using LCMS-grade water and analytical grade reagents. Prepare and store all reagents at room temperature (unless indicated otherwise). Use LCMS-grade organic solvent to make mobile phase.

2.1 Glycoprotein Analysis

1. Hydrochloride (34–37 %).
2. PNGase F (≥2.5 mU/μL): A sterile-filtered solution in 20 mM Tris–HCl, 50 mM NaCl, 1 mM EDTA, pH 7.5 (Prozyme, Hayward, CA, USA).
3. Reaction buffer (5×): 100 mM sodium phosphate, 0.1 % sodium azide, pH 7.5 (Prozyme).
4. Rapigest solution (5 μg/μL in water): Add 200 μL of water to a vial containing 1 mg of Rapigest (Waters, Milford, MA, USA).
5. Diluent (50 mM ammonium bicarbonate): Weigh 400 mg ammonium bicarbonate to a 100 mL bottle, and add 100 mL water to the bottle.
6. DTT (100 mM): Weigh 15.43 mg DTT to a 1.5 mL centrifugal vial, and add 1 mL diluent to the vial (*see* **Note 1**).
7. Mobile phase A: 0.05 % formic acid in water.
8. Mobile phase B: 0.05 % formic acid in acetonitrile (ACN).
9. Mobile phase C: 0.05 % Trifluoroacetic acid (TFA) in water.
10. Mobile phase D: 0.05 % TFA in ACN.
11. Lockmass solution (10 % in ACN) (Agilent Technologies, Santa Clara, CA, USA) (*see* **Note 2**).
12. Heat block and/or oven.
13. Vortex mixer.
14. Micro-centrifuge: Compatible with 1.5 mL centrifuge vial.
15. Column: 2.1 × 5 mm desalting cartridge (Waters).
16. Pressure digester: Barocycler NEP2320 (PBI, South Easton, MA, USA [7]).

17. LC system: Acquity classic UPLC (Waters or equivalent).
18. ESI-TOF MS: QTOF Premier mass spectrometry (Waters or equivalent).

2.2 Glycopeptide Analysis

1. Tris–HCl buffer: 1 M, pH 8.0.
2. PNGase F (≥2.5 mU/μL): A sterile-filtered solution in 20 mM Tris–HCl, 50 mM NaCl, 1 mM EDTA, pH 7.5 (Prozyme).
3. Reaction buffer (5×): 100 mM sodium phosphate, 0.1 % sodium azide, pH 7.5 (Prozyme).
4. Trypsin (1 μg/10 μL): Add 160 μL water to a vial containing 20 μg trypsin/40 μL (Promega, Madison, WI, USA).
5. Rapigest solution (5 μg/μL in water): Add 200 μL of water to a vial containing 1 mg of Rapigest (Waters).
6. TFA: 20 % in water.
7. Mobile phase A: 0.05 % TFA in water.
8. Mobile phase B: 0.05 % TFA in ACN.
9. Lockmass solution: 10 % in ACN (Agilent Technologies).
10. Heat block and/or oven.
11. Vortex mixer.
12. Micro-centrifuge: Compatible with 1.5 mL centrifuge vial.
13. Column: Acquity BEH300 C18, 1.7 μm 100×2.1 mm (Waters).
14. Pressure digester: Barocycler NEP2320 (PBI or equivalent).
15. LC system: Acquity classic UPLC (Waters or equivalent).
16. ESI-TOF MS: QTOF Premier mass spectrometry (Waters or equivalent).

2.3 Glycan Profiling and Identification

1. Dimethyl sulfoxide (DMSO, >99.9 %).
2. Acetic acid (≥99 %).
3. Procainamide hydrochloride: Reagent grade.
4. Reductant: Sodium cyanoborohydride, reagent grade.
5. Formic acid (FA, >99 %).
6. Ammonium hydroxide (28–30 %).
7. PNGase F (≥2.5 mU/μL): A sterile-filtered solution in 20 mM Tris–HCl, 50 mM NaCl, 1 mM EDTA pH 7.5 (Prozyme).
8. Reaction buffer (5×): 100 mM sodium phosphate, 0.1 % sodium azide, pH 7.5 (Prozyme).
9. 95 % ACN.
10. 20 % ACN.
11. FA: 0.1 % in ACN.
12. Mobile phase A: 100 % ACN.

13. Mobile phase B: 50 mM Ammonium formate, pH 4.4: Measure 2 L of water to a 2 L bottle, add 3,770 μL of formic acid to the bottle, and adjust pH to 4.4 with ammonium hydroxide.
14. Lockmass solution: 1:10 in ACN (Agilent Technologies).
15. Heat block and/or oven.
16. Vortex mixer.
17. Micro-centrifuge: Compatible with 1.5 mL centrifuge vial.
18. Standard laboratory vacuum pump.
19. Centrifugal evaporator.
20. Pressure digester: Barocycler NEP2320 (PBI, USA).
21. pH meter.
22. Clean-up station (Prozyme).
23. HILIC 96-well plate (Waters).
24. 96-well collection plate (Waters).
25. Column: Waters Acquity BEH Glycan, 100 × 2.1 mm, 1.7 μm (Waters).
26. LC system: Acquity classic UPLC with fluorescent detector (Waters or equivalent).
27. ESI-TOF MS: QTOF Premier mass spectrometry (Waters or equivalent).

3 Methods

3.1 Glycoprotein Analysis

1. Mix about 100 μg of sample with water to about 1 mg/mL in a 1.5 mL vial (*see* **Note 3**). Mix well. This is the intact sample.
2. Mix about 100 μg of sample with 16 μL Rapigest (5 μg/μL) and dilute with water to about 1 mg/mL (*see* **Note 3**). Heat the solution at 100 °C for 5 min (*see* **Note 4**). Reduce the sample with 100 mM DTT (1:1; v/v) at 37 °C for 1 h (*see* **Note 1**). Add 2 μL of HCl (34–37 %) to the sample and heat at 37 °C for 1 h (*see* **Note 5**). This is the reduced sample.
3. Mix about 100 μg of sample, 10 μL reaction buffer, and 5 μL PNGase F and dilute with water to a total volume of 50 μL in a 1.5 mL vial (*see* **Note 3**). Mix well. Heat at 37 °C for 1 h in a pressure digester (setting: 30,000 psi, 60 cycles, holding time: 50 s, release time: 10 s). After the digestion, dilute with diluent to about 1 mg/mL. This is the deglycosylated sample.
4. Deglycosylate sample as described in **step 3**. After digestion, add 16 μL Rapigest (5 μg/μL) and dilute with water to about 1 mg/mL. Heat the solution at 100 °C for 5 min (*see* **Note 4**). Reduce the sample with 100 mM DTT (1:1; v/v) at 37 °C for

Table 1
Instrument setting for intact and deglycosylated samples

UPLC-UV instrument setting			
Mobile phase A	0.1 % formic acid in water		
Mobile phase B	0.1 % formic acid in ACN		
Gradient program	Time (min)	%B	Flow rate (mL/min)
	0	5	0.5
	2.00	5	0.5
	2.01	5	0.2
	3.50	90	0.2
	6.00	90	0.2
	6.10	5	0.5
	6.70	90	0.5
	6.80	5	0.5
	7.40	90	0.5
	7.50	5	0.5
	9.00	5	0.5
Column	Waters MassPREP online desalting cartridge, 2.1 × 5 mm		
Column temperature	80 °C		
Autosampler temperature	5 °C		
Injector volume	2 μL		
UV detection	254 nm		
QTOF premier mass spectrometer settings			
Ionization	ESI positive V mode		
Scan type	MS scan		
Scan time	2–6 min		
Mass range (Da)	500–4,000 *m/z*		

Note: With lockspray on, the lockmass solution is pumped into the ion source at a flow rate from 5 to 15 μL/min, depending on the intensity and stability of the signal. The reference channel will be sampled at 10-s intervals. The MS settings for reference channel are the same as those for analyte channel

1 h. Add 2 μL of HCl (34–37 %) to the sample and heat at 37 °C for 1 h (*see* **Note 5**). This is the deglycosylated/reduced sample.

5. Analyze intact and deglycosylated samples on a UPLC-ESI TOF MS system (*see* Tables 1 and 2 for instrument setting).
6. Data analysis:
   MaxEnt 1 in MassLynx (Waters) will be used to calculate molecular weight with gain correction (*see* **Note 2**).

3.1.1 Process Parameters for Intact and Deglycosylated Samples

Output mass range: 130,000–160,000.

Resolution: 1.0 Da/channel.

Damage model: Uniform Gaussian width at half height 0.75 Da.

Minimum intensity ratios: Left 33 %, right 33 %.

Completion options: Iterate to convergence.

Table 2
Instrument setting for reduced and deglycosylated/reduced samples

UPLC-UV instrument setting			
Mobile phase C	0.05 % TFA in water		
Mobile phase D	0.05 % TFA in ACN		
Gradient program	Time (min)	%D	Flow rate (mL/min)
	0	5	0.2
	1.00	5	0.2
	1.01	10	0.2
	8.11	50	0.2
	8.50	90	0.2
	8.60	5	0.5
	9.10	90	0.5
	9.20	5	0.5
	9.70	90	0.5
	9.80	5	0.5
	11.00	5	0.5
Column	Waters MassPREP online desalting cartridge, 2.1 × 5 mm		
Column temperature	60 °C		
Autosampler temperature	5 °C		
Injector volume	5 μL		
UV detection	254 nm		
QTOF premier mass spectrometer settings			
Ionization	ESI positive V mode		
Scan type	MS scan		
Scan time	2–8.5 min		
Mass range (Da)	500–4,000 *m/z*		

Note: With lockspray on, the lockmass solution is pumped into the ion source at a flow rate from 5 to 15 μL/min, depending on the intensity and stability of the signal. The reference channel will be sampled at 10-s intervals. The MS settings for reference channel are the same as those for analyte channel

3.1.2 Process Parameters for Reduced and Deglycosylated/Reduced Samples

Output mass range: 20,000–40,000 for light chain, 40,000–60,000 for heavy chain.

Resolution: 0.50 Da/channel for light chain and 1.0 Da/channel for heavy chain.

Damage model: Uniform Gaussian width at half height 0.75 Da.

Minimum intensity ratios: Left 33 %, right 33 %.

Completion options: Iterate to convergence.

The molecular weights can be used to predict glycosylation sites and confirm the amino acid composition (if the amino acid sequence is known) in both intact and reduced states (*see* **Note 7**). All data should be consistent to each other. The data can also be used to confirm total disulfide bond number.

Table 3
Instrument setting for trypsin-digested samples

UPLC-UV instrument setting		
Mobile phase A	0.05 % TFA in water	
Mobile phase B	0.05 % TFA in ACN	
Gradient program	Time (min)	% of Mobile phase B
	0	2
	35	18
	80	35
	97	60
	100	98
	102	98
	104	2
	106	2
Flow rate	0.2 mL/min	
Column	Waters Acquity BEH300 C18, 1.7 μm 100 × 2.1 mm	
Column temperature	35 °C	
Autosampler temperature	5 °C	
Injector volume	10 μL	
UV detection	214 nm	
QTOF premier mass spectrometer setting		
Ionization	ESI positive V mode	
Scan type	MS scan	
Scan time	3–102 min	
Mass range (Da)	100–3,000 *m/z*	

Note: With lockspray on, the lockmass solution is pumped into the ion source at a flow rate from 5 to 15 μL/min, depending on the intensity and stability of the signal. The reference channel will be sampled at 10-s intervals. The MS settings for reference channel are the same as those for analyte channel

3.2 Glycopeptide Analysis

1. Add 100 μg of sample, 10 μL of 1 M Tris buffer (pH 8.0), and 16 μL of Rapigest (5 μg/μL) to a 1.5 mL vial and dilute to a total volume of 100 μL with water. Heat the solution at 100 °C for 5 min (*see* **Note 3**).
2. Add 40 μL volume of the denatured sample (1 mg/mL) from **step 1**, 20 μL of trypsin (1 μg/10 μL), and 10 μL Tris buffer (1 M, pH 8.0) to a 1.5 mL vial, and bring the total volume to 100 μL with water. Heat the sample at 37 °C overnight (16–18 h). Add 2.8 μL of 20 % TFA to the sample solution (*see* **Note 8**). This is the trypsin digestion sample.
3. Add 40 μL volume of the deglycosylation sample (1 mg/mL) from **step 3** of Subheading 3.1, 20 μL of trypsin (1 μg/10 μL), and 10 μL Tris buffer (1 M, pH 8.0) to a 1.5 mL vial, and bring the total volume to 100 μL with water. Heat the sample at 37 °C overnight (16–18 h). Add 2.8 μL of 20 % TFA to the sample solution (*see* **Note 8**). This is deglycosylated trypsin digestion sample.
4. Analyze trypsin-digested samples on a UPLC-ESI TOF MS system (*see* Table 3 for instrument setting).

5. Data analysis:
   Verify glycosylation sites by comparing glycopeptide in trypsin digestion and deglycosylated trypsin digestion samples. Positive glycosylation site should be confirmed if the peptide molecule weight is expected mass plus glycan in trypsin digestion sample and 1 Da higher than expected mass in deglycosylated trypsin digestion sample (*see* **Note 9**).

3.3 Glycan Profiling and Identification

1. Add 200 μg of sample to a 1.5 mL micro-centrifuge tube and dilute to 40 μL with water (*see* **Note 3**).
2. Add 10 μL of reaction buffer and 10 μL PNGase F to the sample.
3. Incubate at 37 °C for 1 h with pressure digester (setting: pressure = 30,000 psi, T1 = 50S, T2 = 10S, cycles = 60) (*see* **Note 6**).
4. Completely dry the reaction mixture in a centrifugal evaporator (*see* **Note 10**).
5. Add 350 μL of DMSO to a 1.5 mL vial, add 150 μL of acetic acid to the vial, and mix well.
6. Weigh about 10 mg procainamide hydrochloride to a 1.5 mL vial.
7. Weigh about 6 mg reductant (sodium cyanoborohydride) to a different 1.5 mL vial.
8. Add 100 μL of the DMSO–acetic acid mixture to the vial containing 10 mg procainamide hydrochloride and mix well.
9. Add the mixture to the vial containing 6 mg reductant and mix well. If the reductant is difficult to dissolve, gently warm the vial for up to 3 min in an oven or a heat block set to 65 °C and then mix by pipetting. This is labeling reagent (*see* **Note 11**). Once prepared, the labeling reagent must be used within 1 h.
10. Add 5 μL of the labeling reagent from **step 9** to each sample. Vortex to mix thoroughly, and briefly spin down.
11. Incubate the reaction vials in an oven or a heat block set to 65 °C for 3 h.
12. Clean up sample by placing the waste reservoir tray and the Waters HILIC μElution Plate in the Glycan Clean-up Station. Connect the Glycan Clean-up Station to the vacuum pump.
13. Wash the μElution Plate wells with 200 μL of water. Allow the water to drain slowly by vacuum. Condition the wells with 600 μL 95 % ACN. Allow the 95 % ACN to drain slowly by vacuum.
14. Reconstitute each sample from **step 11** in 200 μL of 95 % ACN. Vortex vigorously, and centrifuge briefly.
15. Transfer the sample to the μElution Plate wells prepared in **step 13**. Leave the solutions in the plate wells for 15 min while

Table 4
Instrument setting for N-glycan samples

UPLC-UV instrument setting			
Mobile phase A	100 % ACN		
Mobile phase B	50 mM Ammonium formate, pH 4.4		
Gradient program	Time (min)	%B	Flow rate (mL/min)
	0	20	0.5
	46.5	40	0.5
	48	100	0.25
	49	100	0.25
	58	20	0.5
	63	20	0.5
Column	Waters Acquity BEH Glycan, 100 × 2.1 mm, 1.7 μm		
Column temperature	60 °C		
Autosampler temperature	5 °C		
Injector volume	10 μL		
Fluorescent detector settings	Excitation 305 nm Emission 360 nm		
QTOF premier mass spectrometer settings			
Ionization	ESI positive V mode		
Scan type	MSE scan		
Scan time	2–48 min		
Mass range (Da)	100–3,000 *m/z*		
Collision energy			

Note: With lockspray on, the lockmass solution is pumped into the ion source at a flow rate from 5 to 15 μL/min, depending on the intensity and stability of the signal. The reference channel will be sampled at 10-s intervals. The MS settings for reference channel are the same as those for analyte channel

the glycans are adsorbed onto the sorbent. Drain the remaining solution slowly by vacuum.

16. Wash each well with 200 μL of 95 % ACN. Allow the 95 % ACN to drain slowly by vacuum.
17. Replace the waste reservoir tray with a 96-well collection plate. Discard the liquid in the waste reservoir tray.
18. Elute by pipetting 50 μL of 20 % ACN into each well to elute the glycan. Leave the solutions in the plate wells for 5 min, and then drain the 20 % ACN slowly by vacuum.
19. Pipette 150 μL of 0.1 % FA in ACN into each well with eluted glycan (*see* **Note 12**).
20. Transfer to HPLC sample vials for UPLC analysis.
21. Analyze sample on a UPLC-ESI TOF MS system (*see* Table 4 for instrument setting).
22. Data analysis:

    Peak with %area ≥0.5 will be identified by MS according to mass value and glycan maturation pathway [2, 6].

4 Notes

1. DTT is not stable; you need to prepare it at the time of use.
2. We use lockmass from reference channel to correct mass in analyte channel through "modify calibration" in Masslynx software.
3. If the sample concentration is too low, concentrate sample first with a centrifugal evaporator.
4. We usually heat sample to 65 °C for 30 min to denature or 100 °C for 5 min. If sample precipitates when heating, using Rapigest can prevent sample precipitation.
5. This step is to decompose Rapigest. If you did not add Rapigest during denaturation, this step can be omitted.
6. This procedure is for the sample deglycosylated under native state; traditional method can be used without pressure digester by denaturing the sample first and then incubating it at 37 °C for 3–18 h.
7. Using mass difference between native and deglycosylated in either intact or reduced state, you can predict how many potential N-glycosylation sites exist in tested sample [8].
8. The purpose of adding TFA is to stop the digestion and also to decompose Rapigest.
9. Site-specific glycan and major species can be detected with this method and easily tracked by comparing UV chromatograms of glycopeptide and deglycopeptide [8].
10. It is critical to completely remove the solvent for a successful labeling procedure. Visual verification is required.
11. The labeling reagent may look cloudy. This will not affect the labeling.
12. Because normal-phase separation will be used to analyze glycan, additional 0.1 % FA in ACN is necessary to adjust the solvent strength to match the initial condition (water is a strong solvent here).

References

1. Bigge JC, Patel TP, Bruce JA, Goulding PN, Charles SM, Parekh RB (1995) Non-selective and efficient fluorescent labeling of glycans using 2-aminobenzamide and anthranilic acid. Anal Biochem 230:229–238
2. Klapoetke S, Zhang J, Becht S, Gu X, Ding X (2010) The evaluation of a novel approach for the profiling and identification of N-linked glycan with a procainamide tag by HPLC with fluorescent and mass spectrometric detection. J Pharm Biomed Anal 53:315–324
3. Pabst M, Kolarich D, Poltl G, Dalik T, Lubec G, Holfinger A, Altmann F (2009) Comparison of fluorescent labels for oligosaccharides and introduction of a new postlabeling purification method. Anal Biochem 384:263–273
4. Harvey DJ (2000) Electrospray mass spectrometry and fragmentation of N-linked carbohydrates

derivatized at the reducing terminus. J Am Soc Mass Spectrom 11:900–915
5. Yu YQ, Gilar M, Kaska J, Gebler JC (2005) Deglycosylation and sample clean up method for mass spectrometry analysis of N-linked glycans. Waters (The applications book, march 2005), pp 34–36
6. Freeze HH (2000) Current protocols in protein science. Wiley, New York, pp 12.4.2–12.4.5
7. Alvarado R, Tran D, Ching B, Phinney BS (2010) A comparative study of In-gel digestions using microwave and pressure-accelerated technologies. J Biomol Tech 21:148–155
8. Klapoetke SC, Zhang J, Becht S (2011) Glycosylation characterization of human IgA1 with differential deglycosylation by UPLC-ESI TOF MS. J Pharm Biomed Anal 56:513–520

Chapter 33

Fc Engineering of Antibodies and Antibody Derivatives by Primary Sequence Alteration and Their Functional Characterization

Stefanie Derer, Christian Kellner, Thies Rösner, Katja Klausz, Pia Glorius, Thomas Valerius, and Matthias Peipp

Abstract

Therapeutic antibodies used in the treatment of cancer patients are able to mediate diverse effector mechanisms. Dependent on tumor entity, localization, and tumor burden different effector mechanisms may contribute to the in vivo antitumor activity to a variable degree. Especially Fc-mediated effector functions such as antibody-dependent cell-mediated cytotoxicity (ADCC) and complement-dependent cytotoxicity (CDC) have been suggested as being important for the in vivo activity of therapeutic antibodies like rituximab or trastuzumab. In recent years, several strategies have been pursued to further optimize the cytotoxic potential of monoclonal antibodies by modifying their Fc part (Fc engineering) with the ultimate goal to enhance antibody therapy.

Since Fc engineering approaches are applicable to any Fc-containing molecule, strategies to enhance CDC or ADCC activity of full antibodies or scFv-Fc fusion proteins by altering the primary Fc sequence are described.

Key words Antibody engineering, Fc engineering, ADCC, CDC, NK cell

1 Introduction

Despite the overall convincing clinical activity, not all patients optimally benefit from antibody therapy. Animal models and clinical data suggest important roles for Fc-mediated effector mechanisms such as antibody-dependent cell-mediated cytotoxicity (ADCC) and complement-dependent cytotoxicity (CDC) in antibody therapy [1–3]. Based on these findings, engineering approaches designed to improve antibodies' interaction with distinct components of the patients' immune system [4] became evident and in the future may overcome shortcomings of conventional antibody therapy [5].

Since IgG_1 represents the most commonly used antibody isotype in cancer immunotherapy, different Fc engineering approaches

Vincent Ossipow and Nicolas Fischer (eds.), *Monoclonal Antibodies: Methods and Protocols*, Methods in Molecular Biology, vol. 1131, DOI 10.1007/978-1-62703-992-5_33, © Springer Science+Business Media New York 2014

Table 1
Selected engineered IgG_1 Fc variants with enhanced ADCC activity

Variant	FcγRIIIa binding	FcγRIIb binding	IIIa/IIb profile	ADCC induction	Fold reduction in EC50 value	Complement activation	References
Wild type	↑	↑	1	↑	–	+	
S298A-E333A-K334A	↑↑	↓	10	↑↑↑	10–100	n.d.a.	[9]
S239D-I332E	↑↑↑	↑↑	4	↑↑↑	10–100	+	[10]
S239D-I332E-A330L	↑↑↑↑	↑↑↑	9	↑↑↑	10–100	–	[10]
F243L-R292P-Y300L-V305I-P396L	↑↑↑	↑(↑)	7	↑↑↑	10–100	n.d.a.	[12]

IIIa/IIb = Fold FcγRIIIa binding/FcγRIIb binding; EC50 = Half-maximum effective concentration; ↑ = Enhanced activity/binding compared to wt; ↓ = Reduced activity/binding compared to wt; n.d.a. = No data available (adapted from [19]; reprinted with permission)

of this isotype have frequently been described. Several studies demonstrated that Fc glycosylation critically affected Fc-mediated effector functions [6–8]. Especially lack of Fc fucosylation led to higher affinity binding to the activating Fc receptor (FcR) FcγRIIIa resulting in enhanced natural killer (NK) cell-mediated ADCC. As an alternative approach to glyco-engineering, Fc variants with amino acid substitutions have been identified using different technologies [9,10]. Several variants with enhanced FcγRIIIa binding affinity demonstrated stronger NK cell-mediated ADCC activity as well as improved phagocytosis ([9–12]; Table 1). While several approaches were designed to enhance predominantly Fc binding to FcγRIIIa, recently amino acid exchanges around the glycosylation motif at amino acid position N297 were identified that led to aglycosylated antibody variants produced in mammalian cells with very unique FcR binding profiles and biologic functions ([13]; Table 2). Similar protein engineering approaches have been described to either diminish or enhance CDC ([14]; Table 3). Fc engineering technologies were frequently applied to whole antibody molecules [5] but probably are also applicable to any Fc-containing fusion protein or antibody derivative (e.g., scFv-Fc fusion proteins) [15].

In this chapter Fc engineering strategies enhancing CDC or ADCC activity of full antibodies or scFv-Fc fusion proteins by altering the primary amino acid sequence are described (Figs. 1 and 2). Detailed information is exemplarily given for ADCC- or CDC-optimized CD20 antibody variants (Figs. 1, 3, 4).

Table 2
Miscellaneous engineered IgG_1 Fc variants with altered biological function

Variant	Species	Isotype	Expression system	Glycosylation status	FcR binding	References
N297Q/A	Human	IgG_1	Mammalian	agly.	Residual FcγRI binding	[13]
E382V-M428I	Human	IgG_1	Bacterial	agly.	Selective binding to FcγRI	[20]
			Mammalian	gly.	Binding similar to wild type	
S298G-T299A	Human	IgG1	Mammalian	agly.	Selective binding to FcγRI, FcγRIIa, FcγRIIb; no binding to FcγRIIIa or C1q	[13]
G236R-L328R	Human	IgG_1	Mammalian		No FcγR interactions	[21]
S267E-L328F	Human	IgG_1	Mammalian		400-fold enhanced FcγRIIb binding	[21]
G236A	Human	IgG_1	Mammalian		Enhanced FcγRIIa binding	[22]

agly. aglycosylated, *gly.* glycosylated

Table 3
Selected engineered IgG_1 Fc variants with altered CDC activity (adapted from [19]; reprinted with permission)

Variant	ID	C1q binding (KD in nM)	C1q (fold) binding	CDC (fold) potency	ADCC (fold) potency	References
Wild type		48	1	1	1	
K326W			3	2	0	[14]
K326A-E333A	E3		2.5	2	1.0–1.5	[14]
K326W-E333S			5	2	0	[14]
S267E-H268F-S324T	EFT	1.0	47	6.9	0.045	[23]
S267E-H268F-S324T-G236A-I332E	EFT+AE			23	1.2	[23]

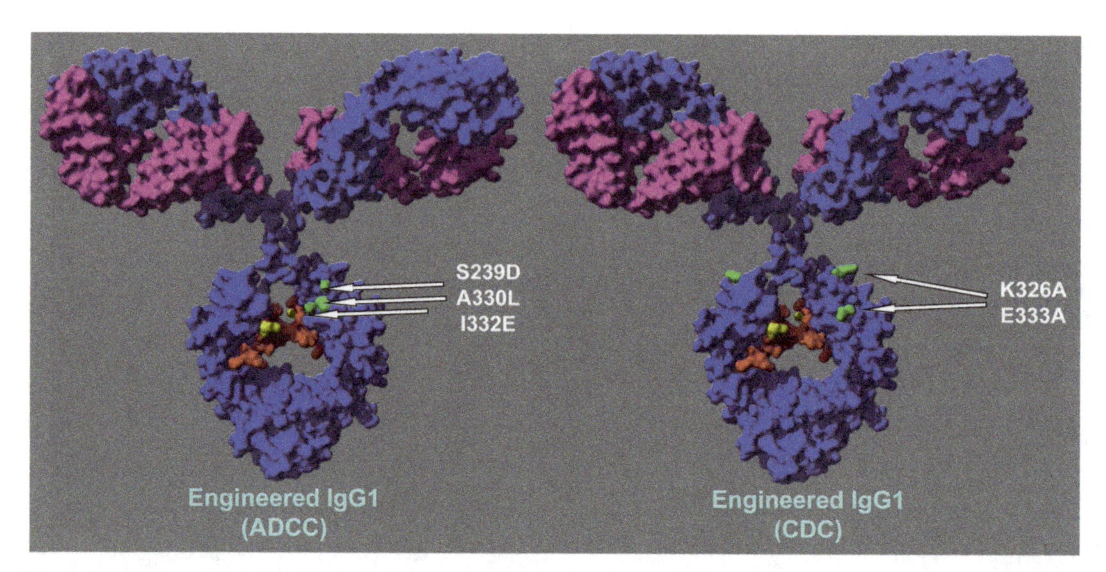

Fig. 1 Selected antibody models illustrating exchanged amino acid positions. IgG_1 model structure pdb file by M. Clark ([18]; adapted from [19]; reprinted with permission). *Blue* = heavy chain; *purple* = light chain; *red/yellow* = oligosaccharides; *yellow* = fucose residue; *green* = respective amino acid substitution. The picture was generated using YASARA structure software (available at http://www.YASARA.org)

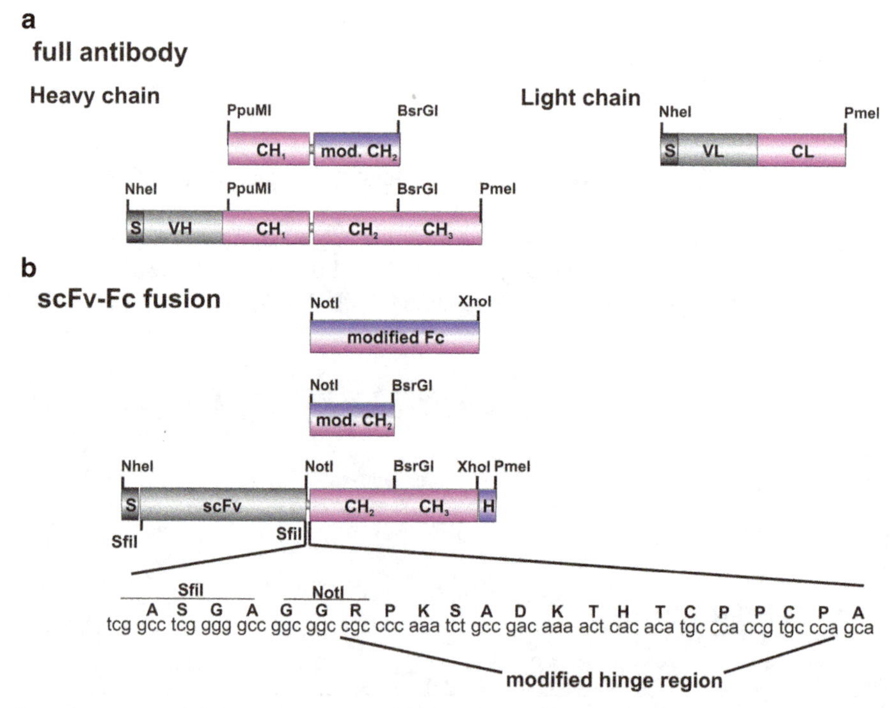

Fig. 2 Design of engineered IgG_1 antibodies or scFv-Fc fusion proteins. Scheme of the expression cassettes designed for the expression of (**a**) full-length antibodies and (**b**) scFv-Fc fusion proteins. Proteins are expressed using pSEC-Tag2-Hygro backbone (Invitrogen). *S* secretion leader, *H* hexahistidine tag

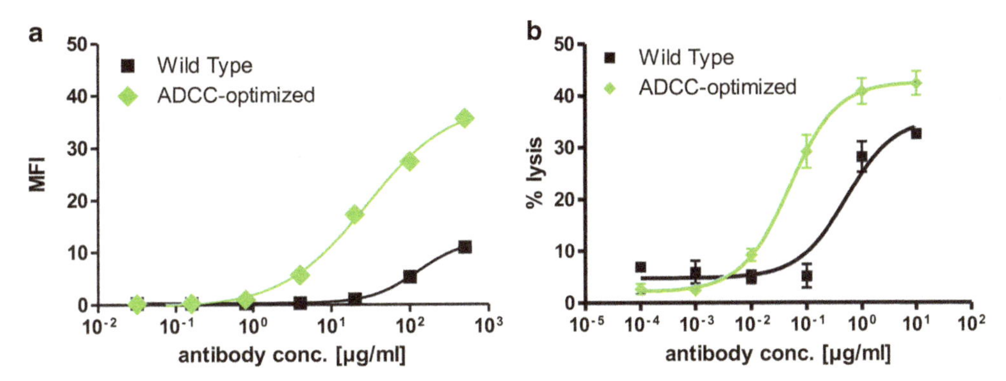

Fig. 3 Fc protein engineering enhances FcγRIIIa binding and extend of NK cell-mediated ADCC. (**a**) An Fc-engineered antibody and its non-engineered counterpart were tested for FcγRIIIa binding by immunofluorescence staining and flow cytometry using isolated NK cells. (**b**) To analyze the cytotoxic activity of the protein-engineered antibody variant, chromium release assays were performed. Isolated MNC served as effector cells and Raji cells as targets. Effector-to-target ratio = 40:1. Data are presented as mean ± SEM

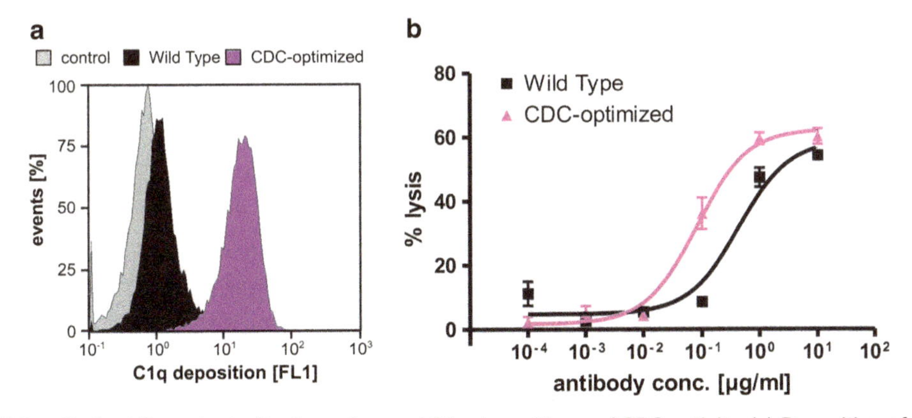

Fig. 4 CDC-optimized Fc variants display enhanced C1q deposition and CDC activity. (**a**) Deposition of complement component C1q on the surface of Ramos cells by a CDC-optimized antibody and its wild-type counterpart. (**b**) To analyze the complement-dependent cytotoxicity of the CDC-optimized antibody variant, chromium release assays were performed using human serum as a source of complement and Ramos cells as targets. Data are presented as mean ± SEM

2 Materials (Key Reagents)

2.1 Cloning of Expression Vectors

1. IgG_1 constant heavy-chain and kappa light-chain constructs (GenBank: J00228; J00241).
2. Antibody v-region resources: Phage display, database information, immunized animals, human B cells.
3. Eukaryotic expression vectors, e.g., GS expression system (Lonza Biologics, Slough, UK) or pSEC series of vectors (Invitrogen).
4. Restriction endonucleases, DNA-modifying enzymes (e.g., New England Biolabs, Frankfurt, Germany).
5. Standard equipment for DNA cloning and gel electrophoresis.

2.2 Expression of IgG_1

2.2.1 Culture of CHO-K1 and HEK293T Cells

1. CHO-Kl cell line (DSMZ, the German Resource Centre for Biological Material, Braunschweig).
2. HEK293T cell line (ATCC; or LentiX, Clontech).
3. DMEM medium-Glutamax I (Invitrogen) supplemented with 10 % FCS and 1 % penicillin, streptomycin (CHO-Kl cells; HEK293T).
4. Pen/Strep: Penicillin, Streptomycin 100× (PAA, Pasching, Austria).

2.2.2 Transfection of CHO-K1 and HEK293T Cells

1. Opti-MEM (Invitrogen).
2. Amaxa Nucleofection kit V.
3. Lipofectamine 2000 (Invitrogen).
4. 2× HBS buffer: 1.5 mM Na_2HPO_4, 50 mM HEPES, 280 mM NaCl; pH 7.05.
5. Chloroquine (Sigma-Aldrich).

2.2.3 Screening for Cell Clones with High Production Rates

1. Polyclonal anti-human κ-light chain antibody (Caltag, Buckingham, UK).
2. Penta-His antibody (Qiagen).
3. HRP-conjugated polyclonal anti-human IgG_1 antibody (Jackson Immunoresearch, Newmarket, UK).
4. Washing buffer: 0.05 % Tween 20, 3 % BSA in TBS:150 mM NaCl, 10 mM Tris–HCl, pH 7.6.
5. SigmaFAST o-Phenylenediamine-dihydrochloride (OPD) (Sigma-Aldrich Corp.).
6. Sunrise absorbance reader (Tecan, Groeding, Austria).

2.3 Antibody Purification

1. HiTrap Protein A columns (GE Healthcare) or protein A beads (Sigma-Aldrich).
2. Protein A washing buffer: 100 mM Tris–HCl; pH 8.0.
3. Protein A elution buffer: 0.1 M Na–citrate; pH 3.0.
4. Protein A neutralization buffer: 1 M Tris–HCl; pH 8.0.

2.4 Protein Quantification

1. Capillary electrophoresis, Experion Pro260 analysis kit, and Chip (Biorad).

2.5 Reagents and Materials for SDS-PAGE/Western Blot Analysis/Coomassie Staining

1. 4–12 % Tris–Glycine gels (Invitrogen).
2. Brilliant blue staining solution (Carl Roth GmbH).
3. PVDF membranes (GE Healthcare).
4. Non-fat dry milk (Carl Roth GmbH).
5. Tris-buffered saline: 150 mM NaCl, 10 mM Tris–HCl, pH 7.6.

6. Goat-anti-human-IgG–HRP conjugate (Sigma-Aldrich Corp.).
7. ECL detection system (Pierce Biotechnology Inc., Rockford, IL, USA).

2.6 Reagents for Cytotoxicity Assays (Chromium Release Assay)

1. Percoll (Biochrom, Holliston, MA, USA).
2. 1× PBS (Invitrogen).
3. 10× PBS (Invitrogen).
4. 1× Hank's Buffered Salt Solution (HBSS) without Ca/Mg (PAA, Pasching, Austria).
5. Distilled water (Invitrogen), ice cold.
6. ^{51}Chromium (Hartmann Analytic GmbH, Braunschweig, Germany).
7. Triton X-100 (Merck).
8. Round bottom 96-well microtiter plate (Sarstedt).
9. β/γ-counter instrument.

2.7 Immuno-fluorescence Analysis

1. FACS buffer: 1× PBS, 1 % BSA, 0.1 % sodium azide.
2. FITC-conjugated anti-human IgG (Dako, Hamburg, Germany).
3. Human serum (HS).
4. FITC-conjugated anti-human C1q (Dako).
5. Flow cytometer instrument.

3 Methods

3.1 Construction of Expression Vectors for Full Antibodies/scFv-Fc Fusion Proteins

1. Since various vector systems are available/suitable for mammalian expression of antibodies and have been used successfully, here more general considerations are given for the design of expression constructs.
2. Variable regions from various sources can be used as starting points: hybridoma cell lines, immunized animals, human B cells, or v-regions isolated by phage display.
3. The choice of the secretion leader is critical for efficient antibody production. If available, the v-regions' original secretion leader usually results in acceptable production yields and secures correct N-terminal processing of the molecule. Alternatively v-regions can be cloned into appropriate expression vectors such as pUC-HAVT20 vector [16] or pSEC-Tag2-based vectors to incorporate a generally well-transcribed/translated leader sequence such as the HAVT20 leader sequence or a murine kappa light-chain leader (*see* **Note 1**).
4. v-regions can be cloned into vectors encoding human constant regions (e.g., pNUT or pEE series of vectors).

5. For antibody expression in an "industry-compatible" vector system, GS expression vectors (Lonza, Inc., Allendale, NJ, USA) can be used.
6. For the expression of full antibodies, strategies using double-gene vectors, encoding both light- and heavy-chain coding sequences or co-transfections of separate light- and heavy-chain constructs, can be applied.
7. Fc-engineered scFv-Fc fusion proteins can be designed similarly with the exception that only one polypeptide chain is required, simplifying the design and expression process (Fig. 2).
8. Since with some exceptions most amino acid exchanges, that have been described to enhance ADCC or CDC activity, are located in the lower hinge and CH_2 domain, often, exchanging the complete Fc part is not necessary. Corresponding short cloning cassettes can be de novo synthesized with appropriate flanking restriction sites (Fig. 2). As an alternative, amino acid exchanges can be realized by site-directed mutagenesis [17].
9. The Fc variant S239D-I332E described by Lazar and colleagues has been demonstrated to enhance NK cell-mediated ADCC in the background of various specificities, thereby representing a good starting point to optimize NK cell-mediated ADCC. To date, CDC-enhancing variants have less frequently been applied; therefore, it could not easily be predicted which variant would be the best starting point in combination with a given antibody.

3.2 Expression of Antibodies and Antibody Derivatives in Mammalian Cell Lines

Complete antibodies or antibody derivatives can be produced either in a transient expression system or, if continuously higher amounts of antibody are needed, by establishing stable cell clones. For transient expression, a vector/cell line system allowing vector amplification is recommended (e.g., SV40 origin of replication coding vector (pSEC series) and large T antigen-expressing production cell lines (e.g., HEK293T)).

3.2.1 Transient Transfection of HEK293T Cells (Ca Phosphate Transfection)

1. Seed 2–5 × 10^6 HEK 293T cells in 100 mm dishes and grow overnight.
2. After 24 h, remove medium and add 8 ml of prewarmed culture medium (*see* **Note 2**).
3. Prepare DNA mix (n = number of tissue culture plates):

   (n+1) × 20 μg plasmid DNA in 895 μl sterile water (10 μg heavy-chain construct + 10 μg light-chain construct).

   (n+1) × 100 μl 2.5 M $CaCl_2$.

   (n+1) × 5 μl 100 mM Chloroquine (final 50–100 μM) (optional).
4. Continuously bubble air through (n+1) × 1 ml of 2×HBS buffer in a 50 ml reaction tube and add the DNA/Ca/chloroquine solution dropwise.

5. Vortex shortly and add 2 ml of the transfection solution dropwise to each tissue culture plate.
6. Incubate for 9–10 h in a humidified incubator at 37 °C and 5–6 % CO_2.
7. Change transfection media by prewarmed tissue culture media.
8. Collect supernatant every 24 h for 5–7 days.

3.2.2 Stable Transfection of CHO-K1 by Nucleofection

1. Use 2×10^6 CHO-K1 cells and 5 μg of plasmid DNA for stable transfection. Transfection could be performed using the Amaxa Nucleofection System and transfection kit V according to the manufacturer's instructions for CHO cells.
2. After 48 h exchange medium with culture medium containing 500 μg/ml hygromycin B if pSEC-Tag2-Hygro-based vectors were used.
3. Passage cells regularly for 2 weeks under selective pressure to eliminate cell clones not expressing the selection marker.
4. Supernatant of the mixed cell clones often contains sufficient recombinant protein for initial characterization. If higher amounts of protein are required, single-cell subclones can be established by limiting dilution.

3.2.3 Screening of Stably Transfected Cell Clones for High Antibody Production

1. Coat 96-well plates with 25–50 μl of polyclonal anti-human κ-light-chain antibody (screening of full antibodies) or anti-penta-his antibody (scFv-Fc fusion proteins) or the appropriate antibody diluted in 1× PBS (final conc. 1–5 μg/ml) overnight at 4 °C.
2. Wash twice with 125 μl washing buffer between all incubation steps.
3. Add 100 μl 3%BSA/1× PBS for 1 h at RT for blocking coated wells.
4. Remove blocking solution and add 50 μl tissue culture supernatant from 96-well plates established by limiting dilution.
5. Incubate for 30–60 min at RT.
6. Incubate with polyclonal anti-human IgG–HRP-conjugated antibodies or the appropriate antibody at a dilution of 1:1,000–1:5,000 for 30–60 min at RT.
7. Develop ELISA by adding 62.5 μl staining solution, and stop reaction by adding 37.5 μl of 3M HCl after significant staining is visible. Dilutions of commercially available antibodies can be used for quantitative analysis/calibration.
8. Readout is absorption at 492 nm (reference 670 nm).
9. Usually screening 100–300 cell clones is sufficient to identify one to five clones with moderate-to-high expression levels (*see* **Note 3**).

3.3 Purification of Proteins

IgG_1 antibodies and scFv-Fc fusion proteins can be purified from tissue culture supernatant by affinity chromatography using protein A columns (use low-salt conditions for human IgG_1; *see* **Note 4**).

1. Collect supernatant from stably transfected mixed or single-cell clones with high expression levels and store at 4 °C until use (*see* **Note 5**).
2. Add 1/10 volume of 1 M Tris–HCl pH 8.0 to collected supernatant.
3. Filtrate through 0.2–0.45 μm Steritop filter units or similar equipment.
4. Perform purification of IgG_1 antibodies or scFv-Fc fusion proteins according to protocols described for human IgG_1 antibodies by either using an automated chromatography system or batch purification using protein A beads.
5. Neutralize elution fractions immediately by adding 1/5 volume neutralization buffer.
6. Dialyze elution fractions extensively against 1× PBS (typically with a volume 1,000× higher than the eluted protein) or perform additional IMAC purification (6× His-Tag) using standard conditions for scFv-Fc fusion proteins.

3.4 Size Exclusion Chromatography

To separate correctly assembled IgG_1 or scFv-Fc fusion proteins from higher molecular weight aggregates, size exclusion chromatography can be performed.

1. Perform size exclusion on HPLC or FPLC chromatography system, e.g., an ÄKTAPurifier system.
2. Equilibrate a Superdex 200 26/60 column with 3 column volumes of 1× PBS at a flow rate of 1 ml/min.
3. Load 2 ml loop with equal volume of antibody solution (>1 mg/ml).
4. Collect peak fractions by measuring absorption at 280 nm. (IgG1 antibodies typically display one major peak with only minor peaks deriving from contaminating proteins or higher molecular mass aggregates. scFv-Fc fusion proteins may contain higher amounts of aggregates, largely depending on the respective scFv used in the construct.)
5. For analytical reevaluation, use Superdex 200 16/300 columns and load up to 100 μl of antibody solution, perform the run at a flow rate of 0.3–0.5 ml/min, and compare elution profile with appropriate molecular mass markers.

3.5 SDS-PAGE, Coomassie Staining, Western Blotting

1. Separate 1–5 μg of purified recombinant protein by SDS-PAGE using 4–12 % Tris–glycine gels under reducing or non-reducing conditions. (Adding DTT/beta-mercaptoethanol

results in reducing interchain disulfide bonds of heavy and light chains. For full antibodies the typical heavy- and light-chain pattern of 50–55 and 25–30 kDa bands will be detectable. scFv-Fc fusion proteins display one protein band of about 50–60 kDa. Using nonreducing conditions full IgG1 antibodies migrate with a mobility of 150 kDa and scFv-Fc fusion proteins with a molecular mass of about 110 kDa corresponding to the size of the proposed homo-dimeric protein.)

2. Stain gels with colloidal Coomassie brilliant blue staining solution or blot to PVDF membranes according to standard procedures.
3. Block membranes using 5 % nonfat dry milk in Tris-buffered saline for 1 h at RT.
4. Perform immunoblotting against human IgG-Fc using goat–anti-human IgG–HRP conjugate (diluted 1:2,000), and incubate blot for 1 h at RT.
5. Wash blots using TBS buffer containing 0.05 % Tween 20 and 0.2 % Triton X-100 for 5 min. Repeat this washing step three times.
6. Develop blots using ECL detection system and analyze by a digital imaging system.

3.6 Analysis of Fc/FcR Interaction on FcR-Positive Cells

To analyze binding of antibody variants to FcγRIIIa in a cellular background, enrich NK cells from mononuclear cells (MNC) (*see* Subheading 3.8.1) using magnetic activated cell sorting (MACS) and negative selection according to the manufacturer's instructions (e.g., Miltenyi Biotec).

1. Prepare aliquots of 1×10^5 cells.
2. Wash once with 500 μl of FACS buffer, pellet cells by centrifugation, and discard supernatant.
3. Resuspend cell pellet in 20–30 μl of serial dilutions of the respective antibody variants (range 0.1–500 μg/ml).
4. Incubate on ice for 30–60 min.
5. Wash three times by adding 1 ml ice-cold FACS buffer, and pellet cells by centrifugation.
6. Resuspend cell pellet in 20 μl FITC-conjugated anti-human IgG antibody (1:20 diluted in FACS buffer).
7. Incubate at 4 °C for 30–60 min.
8. Wash cells three times as described above.
9. Resuspend cell pellet in 500 μl FACS buffer and measure immediately or fix cells by adding fixation buffer (1× PBS, 1 % paraformaldehyde).
10. Analyze samples by flow cytometry (Fig. 3a). (In typical experiments, antibody variants harboring wild-type Fc, binding to

effector cells, could only be observed at higher antibody concentrations (>100–300 μg/ml). Molecules containing Fc variants with high-affinity FcγRIIIa binding activity demonstrate improved binding at significantly lower antibody concentrations.)

3.7 C1q Binding Analysis

To analyze functional consequences of amino acid substitutions, inserted into the CH_2 domain of IgG_1 antibodies, indirect immunofluorescence and FACS analyses can be performed using human serum (*see* **Note 6**).

1. Add 1×10^5 target cells/well/94 μl medium in a v-shaped 96-well microtiter plate.
2. Add 5 μl of antigen-specific or control antibody (stock 200 μg/ml, end concentration 10 μg/ml).
3. Incubate at RT for 15 min.
4. Add 1 μl human serum (*see* Subheading 3.8.2).
5. Incubate at 37 °C for 10 min.
6. Add 100 μl ice-cold FACS buffer. Centrifuge at $600 \times g$ and 4 °C for 3 min, and aspirate the supernatant avoiding the cell pellet.
7. Repeat **step 6** two times.
8. Add 10 μl of polyclonal FITC-conjugated anti-human C1q antibody (1:10 diluted in FACS buffer).
9. Incubate at 4 °C for 1 h.
10. Perform **step 6** three times.
11. Resuspend cell pellet in 500 μl FACS buffer.
12. Analyze samples by flow cytometry (Fig. 4a). (Typically CDC-optimized Fc variants demonstrate more efficient C1q deposition, resulting in stronger fluorescence signals.)

3.8 Measuring Cytotoxicity

To analyze whether Fc protein engineering led to enhanced cytolytic activity, chromium release assays can be performed. For Fc variants with enhanced FcγRIIIa binding, MNC can be used as effector cells. For testing CDC-optimized variants, human serum should be prepared from freshly drawn, peripheral blood as a source of complement.

3.8.1 Preparation of Effector Cells

1. Prepare 70 % Percoll solution (dilute with 1× PBS) and 63 % Percoll (dilute with 1× HBSS solution).
2. Overlay 3 ml of 70 % Percoll solution with 3 ml of 63 % Percoll solution.
3. Overlay with 5–9 ml freshly drawn peripheral blood (anticoagulated with citrate).
4. Centrifuge for 20 min at $1,300 \times g$ at RT without break.

5. Collect MNC from the plasma/63 % Percoll interface (upper ring) and polymorphonuclear (PMN) cells from the interface between the two Percoll layers.
6. Dilute transferred cell solution 1:10 in 1× PBS.
7. Centrifuge for 5 min PMN at 1,300 × *g* and RT without break and MNC at 600 × *g* and RT with break.
8. Decant the supernatant, and carefully resuspend the cell pellet avoiding excess pipetting.
9. If necessary, perform lysis of erythrocytes by adding 45 ml ice-cold distilled water to the cell suspension for 30 s (*see* **Note 7**).
10. Immediately add 5 ml of 10× PBS and slowly invert the tube several times.
11. Centrifuge for 5 min at 600 × *g* and RT with break.
12. Carefully decant the supernatant.
13. If still high numbers of erythrocytes are visible, repeat **steps 9–13** one more time.
14. Wash the cell pellet with 1× PBS three times to remove platelets, and finally resuspend cell pellet in 5 ml cell culture medium.
15. Count cells, and adjust cell concentration for the desired ratio in the following cytotoxicity assays (*see* **Note 8**).

3.8.2 Preparation of Human Serum (HS)

1. Collect freshly drawn peripheral blood into tubes containing coagulants.
2. Place collecting tubes on ice for 30 min.
3. Centrifuge collecting tubes at 850 × *g* and 4 °C for 20 min.
4. Transfer serum into 15 ml tubes and use immediately or store at −80 °C.

3.8.3 ^{51}Chromium Release Assay (ADCC and CDC Assay)

1. Harvest target cell line expressing the desired target antigen, and determine cell count (calculate 5,000 cells/well).
2. Incubate 1×10^6 target cells with 100 μCi ^{51}chromium in 200–500 μl of 1× PBS for 2 h at 37 °C and 5 % CO_2.
3. Prepare effector cells like MNC and human serum from peripheral blood to analyze ADCC or CDC, respectively.
4. After 2-h incubation, wash target cells three times with 7–10 ml of medium, and finally resuspend cell pellet (0.1×10^6 cells/ml).
5. Incubate labelled target cells (5,000 cells/well) and MNC or serum in the presence of antigen-specific antibodies or control antibodies in a 96-well microtiter plate for 3–4 h at 37 °C and 5 % CO_2 (= experimental counts per minute (cpm), see the reaction setup below). For maximum target cell destruction (= maximal release, see the reaction setup below) use Triton

X-100 in combination with target cells alone. Basal ^{51}chromium release (= basal release target cells, see the reaction setup below) can be determined by incubating target cells without adding an effector population. In the presence of effector cells/serum antibody-independent lysis mediated by the respective effector population could be analyzed (= antibody-independent effector cell/serum-mediated release, see the reaction setup below). Experimental release:

50 μl target cells (5,000 cells/well).

50 μl effector cells/human serum (variable cell counts).

X μl antibody.

X μl medium.

Σ 200 μl/well.

Antibody-independent effector cell/serum-mediated release:

50 μl target cells (5,000 cells/well).

50 μl effector cells/human serum (variable cell counts).

100 μl medium.

Σ 200 μl/well.

Basal release target cells:

50 μl target cells (5,000 cells/well).

150 μl medium.

Σ 200 μl/well.

Maximal release:

50 μl target cells (5,000 cells/well).

100 μl 2 % Triton X-100 (in 1× PBS or culture medium).

50 μl medium.

Σ 200 μl/well.

6. After 3-h incubation, centrifuge microtiter plate for 5 min at 850×*g* and use 25 μl supernatant to determine ^{51}chromium release (cpm) in a γ-counter instrument using cell culture supernatant.
7. Alternatively mix 25 μl supernatant of each sample with 150 μl OptiPhase Supermix Cocktail (Perkin-Elmer, Waltham, MA) and incubate for 15 min at RT and continuous shaking. ^{51}Cr release can then be measured in a scintillation/luminescence counter (1450 LSC and Luminescence Counter, Perkin-Elmer) according to the manufacturer's instructions.
8. Calculate target cell lysis using the following formula (Figs. 3b, c and 4b):

$$\%\ lysis = \frac{(experimental\ cpm - basal\ cpm)}{(maximal\ cpm - basal\ cpm)} \times 100.$$

4 Notes

1. Using antibodies' endogenous secretion leader may positively influence antibody secretion.
2. 50–70 % confluent plates are optimal for transfection. Depending on the expressed antibody 30–50 plates usually represent a good starting point to generate enough recombinant protein for initial testing.
3. Production yields vary largely between different constructs, probably depending on the respective v-regions and the secretion leaders used.
4. Proteins can be alternatively purified using anti-human kappa or anti-human Fc matrices (both from BAC BV, Leiden, The Netherlands).
5. Usually, it is not necessary to add protein inhibitors to collected cell culture supernatants for protein purification procedures.
6. For immunofluorescence analyses, perform all staining procedures at 4 °C to avoid antigen internalization.
7. Time is crucial for lysis of erythrocytes.
8. Avoid extended "storage" time especially of PMN.

References

1. Nimmerjahn F, Ravetch JV (2005) Divergent immunoglobulin g subclass activity through selective Fc receptor binding. Science 310(5753):1510–1512
2. Musolino A et al (2008) Immunoglobulin G fragment C receptor polymorphisms and clinical efficacy of trastuzumab-based therapy in patients with HER-2/neu-positive metastatic breast cancer. J Clin Oncol 26(11):1789–1796
3. Weng WK, Levy R (2003) Two immunoglobulin G fragment C receptor polymorphisms independently predict response to rituximab in patients with follicular lymphoma. J Clin Oncol 21(21):3940–3947
4. Carter PJ (2006) Potent antibody therapeutics by design. Nat Rev Immunol 6:343–357
5. Desjarlais JR et al (2007) Optimizing engagement of the immune system by anti-tumor antibodies: an engineer's perspective. Drug Discov Today 12(21–22):898–910
6. Jefferis R, Lund J, Goodall M (1995) Recognition sites on human IgG for Fc gamma receptors: the role of glycosylation. Immunol Lett 44(2–3):111–117
7. Lund J et al (1996) Multiple interactions of IgG with its core oligosaccharide can modulate recognition by complement and human Fc gamma receptor I and influence the synthesis of its oligosaccharide chains. J Immunol 157(11): 4963–4969
8. Jefferis R (2009) Recombinant antibody therapeutics: the impact of glycosylation on mechanisms of action. Trends Pharmacol Sci 30(7): 356–362
9. Shields RL et al (2001) High resolution mapping of the binding site on human IgG1 for Fc gamma RI, Fc gamma RII, Fc gamma RIII, and FcRn and design of IgG1 variants with improved binding to the Fc gamma R. J Biol Chem 276(9):6591–6604
10. Lazar GA et al (2006) Engineered antibody Fc variants with enhanced effector function. Proc Natl Acad Sci U S A 103(11):4005–4010
11. Presta LG et al (2002) Engineering therapeutic antibodies for improved function. Biochem Soc Trans 30(4):487–490
12. Stavenhagen JB et al (2007) Fc optimization of therapeutic antibodies enhances their ability to kill tumor cells in vitro and controls tumor expansion in vivo via low-affinity activating Fcgamma receptors. Cancer Res 67(18): 8882–8890
13. Sazinsky SL et al (2008) Aglycosylated immunoglobulin G1 variants productively engage

activating Fc receptors. Proc Natl Acad Sci U S A 105(51):20167–20172

14. Idusogie EE et al (2001) Engineered antibodies with increased activity to recruit complement. J Immunol 166(4):2571–2575
15. Repp R et al (2011) Combined Fc-protein- and Fc-glyco-engineering of scFv-Fc fusion proteins synergistically enhances CD16a binding but does not further enhance NK-cell mediated ADCC. J Immunol Methods 373(1–2):67–78
16. Boel E et al (2000) Functional human monoclonal antibodies of all isotypes constructed from phage display library-derived single-chain Fv antibody fragments. J Immunol Methods 239(1–2):153–166
17. Derer S et al (2012) Fc engineering: design, expression, and functional characterization of antibody variants with improved effector function. Methods Mol Biol 907:519–536
18. Clark MR (1997) IgG effector mechanisms. Chem Immunol 65:88–110
19. Peipp M, van de Winkel JG, Valerius T (2011) Molecular engineering to improve antibodies' anti-lymphoma activity. Best Pract Res Clin Haematol 24(2):217–229
20. Jung ST et al (2010) Aglycosylated IgG variants expressed in bacteria that selectively bind FcgammaRI potentiate tumor cell killing by monocyte-dendritic cells. Proc Natl Acad Sci U S A 107(2):604–609
21. Horton HM et al (2011) Antibody-mediated coengagement of FcgammaRIIb and B cell receptor complex suppresses humoral immunity in systemic lupus erythematosus. J Immunol 186(7):4223–4233
22. Richards JO et al (2008) Optimization of antibody binding to FcgammaRIIa enhances macrophage phagocytosis of tumor cells. Mol Cancer Ther 7(8):2517–2527
23. Moore GL et al (2010) Engineered Fc variant antibodies with enhanced ability to recruit complement and mediate effector functions. MAbs 2(2):181–189

Part IV

Applications of Monoclonal Antibodies

Chapter 34

Labeling and Use of Monoclonal Antibodies in Immunofluorescence: Protocols for Cytoskeletal and Nuclear Antigens

Christoph R. Bauer

Abstract

Antibodies are widely used to target and label specifically extra- or intracellular antigens within cells and tissues. Most protocols follow an indirect approach implying the successive incubation with primary and secondary antibodies. In these protocols the primary antibodies are specifically targeted against the antigen in question and are normally not labeled. The secondary antibodies come from a different species and are in contrast fluorescently labeled. The idea is that the primary antibodies specifically bind to their targets but cannot be visualized directly. Only binding of the secondary (fluorescent) antibodies to the constant region of the primary antibodies allows consecutively the visualization in a fluorescent microscope.

Primary antibodies can be either of monoclonal (normally produced in mouse) or of polyclonal origin (normally produced in rabbit, goat, sheep, or donkey). Using (primary) monoclonal antibodies has the clear advantage that all antibodies used are identical in origin and behavior and should thus give a more clear-cut labeling result. On the other hand the demands towards labeling protocols might be concomitantly higher: Binding of primary antibodies will only occur if fixation and labeling protocols preserve the antigen sufficiently to keep its specific and unique target structure available. One could imagine that for polyclonal antibodies this demand is slightly lower as there is a pool of antibodies with varying specificities against multiple parts of their target antigens. Certain fractions of this pool might thus tolerate a larger variety of conditions, and consequently a larger variety of protocols might still result in successful labeling.

Each step in a labeling protocol can be decisive for the outcome of an experiment especially if monoclonal antibodies are used. Especially critical are choice of buffer and fixation and permeabilization parameters of the protocol.

In this chapter we discuss and detail proven protocols using monoclonal antibodies for two key targets within a cell, namely, (a) the cytoskeleton with emphasis on microtubules (Cramer and Mitchison, J Cell Biol 31:179–189, 1995; Traub et al., Biol Cell 90:319–337, 1998; Desai et al., Methods Cell Biol 61:385–412, 1999) and (b) the nuclear pore complex (Davis and Blobel, Cell 45:699–709, 1986; Kimura et al., Mol Cell Biol 23:1304–1315, 2003). The two protocols which differ substantially in using either a chemical fixation/permeabilization approach versus a one-step coagulation protocol are chosen to offer the reader tools to design their own procedures and adapt them accordingly to fit their individual experimental setup and the biological question asked.

Key words Immunofluorescence, Antibody, Labeling, Fixation, Protocol, Permeabilization, Cytoskeleton, Nuclear, Antigens

Vincent Ossipow and Nicolas Fischer (eds.), *Monoclonal Antibodies: Methods and Protocols*, Methods in Molecular Biology, vol. 1131, DOI 10.1007/978-1-62703-992-5_34, © Springer Science+Business Media New York 2014

1 Introduction

Protocols for immunofluorescence labeling involve several consecutive steps, namely, fixation, permeabilization, blocking of nonspecific binding sites, incubations with primary and secondary antibodies, and finally mounting of specimen on microscope slides.

For a perfect fixation the antigens should be immobilized while retaining their cellular and subcellular architecture. For antibodies to reach their target within a cell additional permeabilization steps are often needed. The choice of fixative depends (a) on the nature of the antigen to be examined and (b) on the properties of the antibody used. The two fixation methods described in this chapter use two different approaches, namely, organic solvents and cross-linking reagents. Organic solvents s remove lipids and dehydrate the cells while precipitating and fixing proteins. Cross-linking reagents (such as formaldehyde) form intermolecular bridges, normally through free amino groups, thus creating a network of linked antigens. Cross-linkers preserve cell structure better than organic solvents but most often require the addition of a permeabilization steps. Organic solvents act very fast and do not need permeabilization but will often denature protein antigens. This last point might actually be an advantage when the primary antibody has been produced against a denatured protein, for example cut out of an electrophoretic gel.

The two protocols to be presented in this chapter differ mainly in the way fixation and permeabilization are achieved:

(a) For the cytoskeleton protocol the fixation step uses formaldehyde to chemically fix and cross-link the specimen. The following permeabilization is done with Triton X-100.

(b) For the nuclear antigen protocol a mixture of cold acetone/methanol coagulates and permeabilizes the specimen in one step.

After these first steps both protocols are nearly identical and involve the same buffers, solutions, and incubation procedures. Similarly the preparation and culture of cells are identical for both protocols.

2 Materials

2.1 Cell Culture

1. Fibroblast cells; we use NIH 3T3 cells.
2. DMEM medium (for example Sigma, D5796 with glucose, L-glutamine, pyroxidine, and $NaHCO_3$).
3. Collagen (for example Sigma, No C8919 1 mg/ml solution diluted freshly in sterile tissue culture water).
4. Sterile PBS.
    (a) 6-well cell culture dish, cover slips (18 × 18 mm), and slides.

2.2 Cytoskeleton Protocol

1. PEM buffer: 100 mM piperazine-N,N′-bis[ethanesulfonic acid] (PIPES), pH 7.0, 5 mM EGTA (*see* **Note 1**), 2 mM $MgCl_2$.
2. PBS.
3. TBS: 0.15 M NaCl, 0.02 M Tris–Cl, pH 7.4 (*see* **Note 2**).
4. TBST: TBS plus 0.1 % Triton X-100.
5. 4 % formaldehyde in PEM (*see* **Note 3**).
   (a) 4 % formaldehyde in PBS (*see* **Note 3**).
6. 1 % Triton X-100 in PEM.
7. AbDil (antibody-diluting solution): TBS plus 0.1 % Triton X-100, 2 % BSA, 0.1 % sodium azide; store at 4 °C.
8. Primary antibodies: Monoclonal anti-alpha-tubulin (Sigma No. T5168), dilute 1:1,000 in AbDil.
9. Secondary antibodies: Invitrogen Alexa Fluor® 488 Goat Anti-Mouse IgG (H + L), highly cross-adsorbed, diluted 1:1,000 in AbDil.
10. Phalloidin TRITC: Sigma No. P1951, diluted 1:500 in AbDil (*see* **Note 4**).
11. Mounting medium: Prolong plus DAPI (Invitrogen No. P36935).

2.3 Nuclear Antigen Protocol

1. Methanol/acetone (1:1).
2. PBS, TBST, AbDil, Prolong (*see* Subheading 2.2).
3. Primary antibodies: Anti-Nuclear Pore Complex Proteins antibody [Mab414] from Abcam (No. ab24609), diluted 1:1,000 in AbDil.
4. Secondary antibodies: Invitrogen Alexa Fluor® 488 Goat Anti-Mouse IgG (H + L), highly cross-adsorbed, diluted 1:1,000 in AbDil.

3 Methods

3.1 Cell Culture

1. Sterilize cover slips with 100 % ethanol and place into 6-well plate.
2. When dry coat cover slips for 10 min with collagen.
3. Wash two times with sterile PBS.
4. Plate fibroblast cells at rather low density to have nice morphology.
5. Let cells attach overnight in the cell culture incubator.

3.2 Cytoskeleton Protocol

1. Wash cells 2 × 1 min with PBS at 37 °C (*see* **Note 5**).
2. Fix with 4 % formaldehyde in PEM buffer for 10 min (solution should be at 37 °C when added).

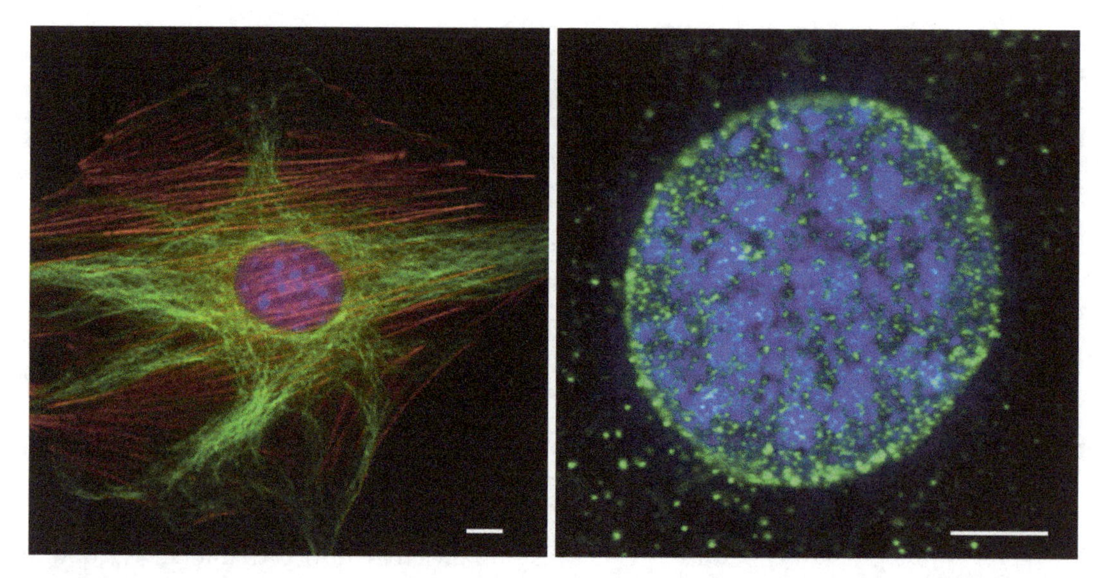

Fig. 1 Images showing the result of cytoskeleton (*left*) and nuclear antigen (*right*) labeling. *Left image* shows labeling of DNA (DAPI, *blue*), microtublues (Alexa 488, *green*), and actin filaments (phalloidin TRITC, *red*). Image was taken with a fluorescence widefield microscope (Zeiss AxioZ1). *Right image* shows labeling of DNA (DAPI, *blue*) and nuclear pore complex proteins (Alexa 488, *green*). Images were taken with a confocal microscope (Zeiss LSM 710). Bars represent 5 μm

3. Permeabilize with 1 % Triton X-100 in PEM for 5 min at room temperature (RT).
4. Wash 4 × 3 min in PBS at RT.
5. Block for 20 min in 3 ml AbDil at RT.
6. Incubate with primary antibodies for 60 min at RT (*see* **Note 6**).
7. Take cover slips back into 6-well dish (cell side up) and wash 4 × 5′ with TBST at RT.
8. Incubate as above with secondary antibodies and phalloidin TRITC for 60 min at RT (*see* **Note 6**).
9. Wash 4 × 5 min with TBST.
10. Rinse 2 × 2 min in PBS.
11. Postfix with 4 % formaldehyde in PBS for 10 min at RT.
12. Wash 4 × 5 min with PBS.
13. Rinse briefly with distilled water.
14. Drain and mount in Prolong plus DAPI (*see* **Note** 7).
15. Leave overnight in a flat dark place.
16. Seal the cover slip edges onto the glass slide with nail polish and analyze results with fluorescence or confocal microscope (*see* Fig. 1).

3.3 Nuclear Antigen Protocol

1. Fix cells in −20 °C methanol/acetone (1:1) for 3 min (*see* **Note 8**).
2. Rehydrate in TBST 3 × 5 min.

3. Block for 20 min in 3 ml AbDil.
4. Incubate with primary antibodies for 60 min at RT (*see* **Note 6**).
5. Take cover slips back into 6-well dish (cell side up) and wash 4 × 5′ with TBST at RT.
6. Incubate as above with secondary antibodies for 60 min at RT (*see* **Note 6**).
7. Follow **steps 9–16** as described above in Subheading 3.2.

4 Notes

1. Dissolve EGTA (MW 380.35 g) in H_2O at a concentration of 0.5 M, and adjust pH to 8.0 with NaOH (will only go into solution after pH adjustment). Sterilize by filtering, dispense into aliquots, and store sterile solution at room temperature.
2. Can make 10× TBS stock.
3. Use paraformaldehyde to make 16 % stock in H_2O: Warm 100 ml water to 70 °C, add 16 g paraformaldehyde, and add NaOH while stirring until the solution becomes clear (do not use too much NaOH). Formaldehyde stock solutions can be stored at −20 °C. They need to be warmed to 70 °C to dissolve after thawing. Commercial formaldehyde stock solutions (for example 37 %) can be used, but they contain methanol.
4. The secondary antibodies and the phalloidin TRITC can be added into the same solution, and the incubation can be done simultaneously.
5. Microtubules depolymerize in the cold, so make sure to use pre-warmed solutions. You can do the first steps within the 6-well dish (until antibody incubation).
6. Prepare clean parafilm sheet, and place cover slips (cell side down) on 100 μl drops of antibody solution. Before placing the cover slips hold and drain them briefly vertically onto a filter paper to absorb excessive liquid.
7. Prepare a clean microscope slide; add a small drop of Prolong with DAPI onto the slide; place the cover slip with the cells facing down onto the slide in a gliding movement to avoid air bubbles.
8. Use small (5 cm in diameter) glass petri dishes. Precool them on ice. Take the precooled acetone/methanol solution out of the −20 °C freezer and put into cold dish. Place cover slips inside dish with cells facing upwards. After fixation put cover slips back into 6-well dish.

References

1. Cramer LP, Mitchison TJ (1995) Myosin is involved in postmitotic cell spreading. J Cell Biol 31:179–189
2. Traub P, Bauer C, Hartig S, Grueb S, Stahl S (1998) Colocalization of single ribosomes with intermediate filaments in puromycin-treated and serum-starved mouse embryo fibroblasts. Biol Cell 90:319–337
3. Desai A, Murray A, Mitchison TJ, Walczak CE (1999) The use of Xenopus egg extracts to study mitotic spindle assembly and function in vitro. Methods Cell Biol 61:385–412
4. Davis LI, Blobel G (1986) Identification and characterization of a nuclear pore complex protein. Cell 45:699–709
5. Kimura T, Ito C, Watanabe S, Takahashi T, Ikawa M, Yomogida K, Fujita Y, Ikeuchi M, Asada N, Matsumiya K, Okuyama A, Okabe M, Toshimori K, Nakano T (2003) Mouse germ cell-less as an essential component for nuclear integrity. Mol Cell Biol 23:1304–1315

Chapter 35

Generation and Use of Antibody Fragments for Structural Studies of Proteins Refractory to Crystallization

Stephen J. Stahl, Norman R. Watts, and Paul T. Wingfield

Abstract

With the rapid technological advances in all aspects of macromolecular X-ray crystallography the preparation of diffraction quality crystals has become the rate-limiting step. Crystallization chaperones have proven effective for overcoming this barrier. Here we describe the usage of a Fab chaperone for the crystallization of HIV-1 Rev, a protein that has long resisted all attempts at elucidating its complete atomic structure.

Key words Fab, scFv, Phage-display, Crystallization, Chaperone, HIV, Rev

1 Introduction

The method of X-ray crystallography remains the gold standard for determining the structure of macromolecules. The number of protein atomic structures is increasing rapidly, but certain types of proteins remain resistant to this approach due to their inability to form diffraction quality crystals, notably membrane proteins, proteins with a strong propensity to polymerize, and intrinsically disordered proteins. Proteins resistant to the usual approach of simply screening many solution conditions can sometimes be crystallized by more informed means, including site-specific mutations, surface entropy reduction (SER), trimming of flexible regions, addition or insertion of chaperones (fusion with proteins with high crystallization potential), and use of non-covalent chaperones. Several such auxiliary chaperones have been developed, including monoclonal Fab, scFv, camelid single-chain antibodies, DARPINS, and others. Excellent reviews are provided in [1–4]. Here we describe methods which result in the crystallization of a protein that has both a strong tendency to polymerize and aggregate and has an intrinsically disordered domain.

Rev is an HIV-1 regulatory protein that mediates the nuclear export of unspliced and partially spliced viral mRNA, thereby

Vincent Ossipow and Nicolas Fischer (eds.), *Monoclonal Antibodies: Methods and Protocols*, Methods in Molecular Biology, vol. 1131, DOI 10.1007/978-1-62703-992-5_35,

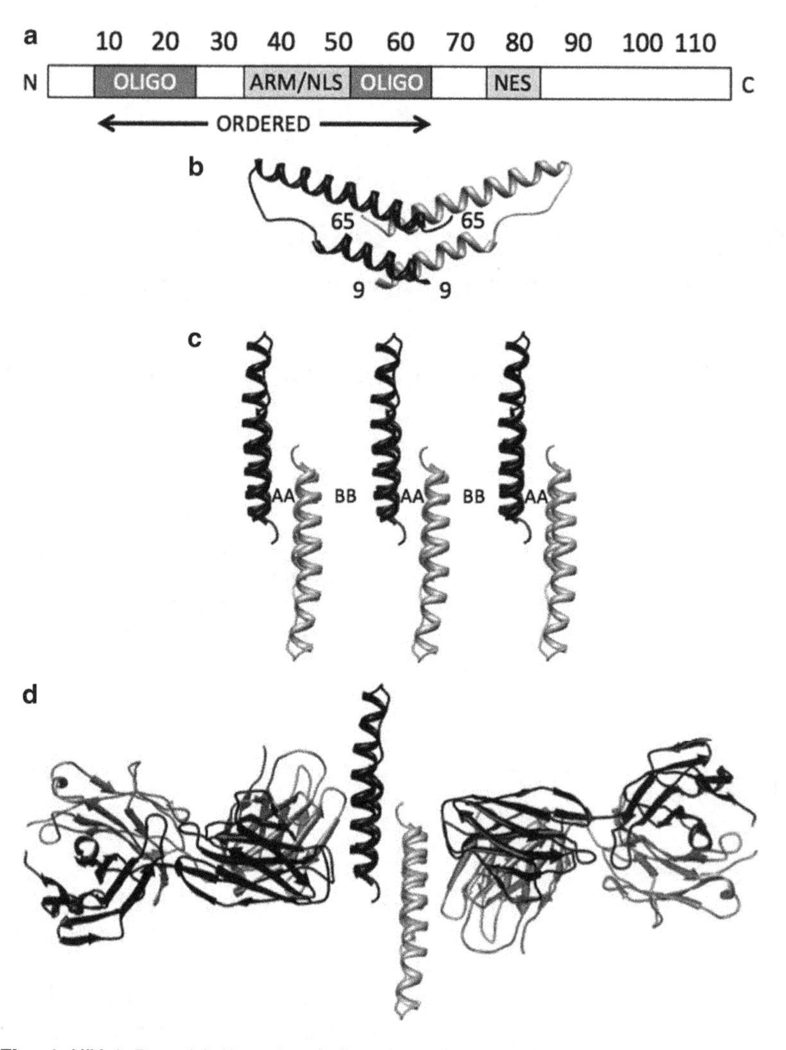

Fig. 1 HIV-1 Rev. (**a**) Functional domains of Rev include two oligomerization sequences, an arginine-rich motif (ARM) that partially coincides with a nuclear localization sequence (NLS), and a nuclear export sequence (NES). Residues 9 to ~65 are ordered, the C-terminal domain is predicted to be intrinsically disordered. (**b**) The helix-turn-helix structure of the N-terminal domain, shown as the dimer observed by crystallography. (**c**) Dimers associate via AA interactions and are proposed to oligomerize via BB interactions. (**d**) Fab–Rev complex

initiating the transition to the late phase of viral replication. The 13-kDa protein has two domains: an N-terminal domain, comprising half the protein, with a helix-turn-helix structure, and a C-terminal domain that is predicted to be intrinsically disordered (Fig. 1a, b). In vivo, Rev initially interacts with viral mRNAs by binding, via an arginine-rich motif (ARM) in its N-terminal domain, to a highly structured region in the RNA named the Rev response element (RRE) and then oligomerizing in a highly cooperative manner to form a 200–300 kDa ribonucleoprotein

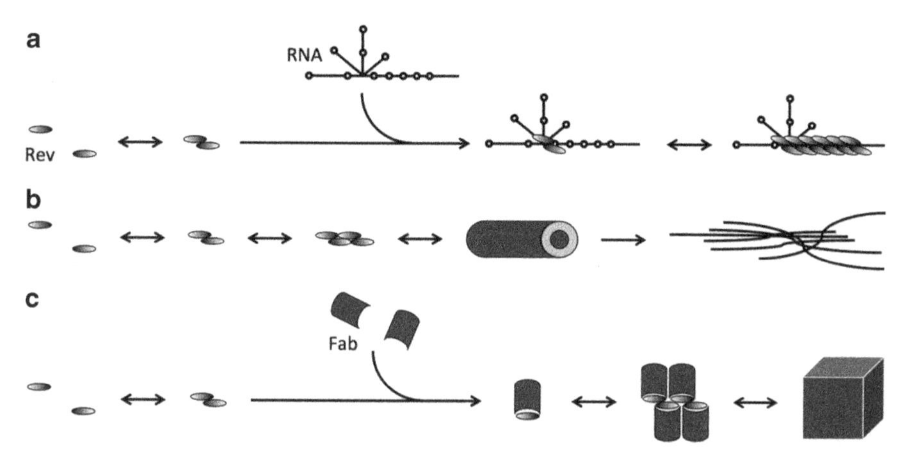

Fig. 2 Assemblies of Rev. (**a**) In vivo, Rev monomers associate with the viral mRNA, first at the site of the Rev response element (RRE) and then at secondary sites, usually up to 10–12 monomers. In vitro, in the presence of heterologous RNA, Rev can form irregular filaments with a length proportional to the mass of the RNA. (**b**) In vitro, in the absence of RNA, Rev forms dimers that polymerize into regular hollow filaments. Such filaments are of indeterminate length and tend to aggregate, preventing protein crystallization. (**c**) Rev monomers can associate in an antiparallel manner via the A-faces of their helix-turn-helix N-terminal domains and then further associate via their B-faces to form polymers. When Rev is mixed with a monoclonal Fab directed against the B-face of the N-terminal domain a 1:1 Rev–Fab complex forms and Rev can no longer polymerize. These Rev–Fab complexes can then associate via Fab:Fab interactions and thus be crystallized, as can the corresponding Rev–scFv complexes

complex containing 8–12 Rev monomers (Fig. 2a). Rev binds, via a nuclear export sequence (NES) in its C-terminal domain, to the host protein Crm1 (exportin-1) to direct the mRNA to the cytoplasm in a manner that avoids default splicing. After dissociating from the RNA, Rev binds to importin-β via the nuclear localization sequence (NLS) in its N-terminal domain and returns to the nucleus for further rounds of mRNA export [5–7]. More recently it has been recognized that Rev also performs several other functions, including the prevention of multiple viral integration events in the early phase of infection and enhancement of viral translation and genome encapsidation of genomic RNA into virions in the late phase [8, 9].

In vitro, in the absence of RNA, Rev monomers form dimers at a concentration of ~80 μg/ml and then polymerize into long hollow filaments. These filaments have a diameter of ~15 nm and are of indefinite length [10, 11]. Such filaments have a strong tendency to associate laterally and form aggregates (Fig. 2b). This association can be readily prevented by the presence of 50 mM citrate ion. However, the formation of the filaments effectively prevents the formation of crystals. Agents such as the divalent cation Mg^{+2}; poly-anions such as RNA (both cognate and heterologous), DNA, and oligo-glutamate ($n=8$ or 10); and detergents such as CHAPS cause disassembly to variously sized oligomers, their heterogeneity precluding crystallization. Rev filaments also react rapidly and

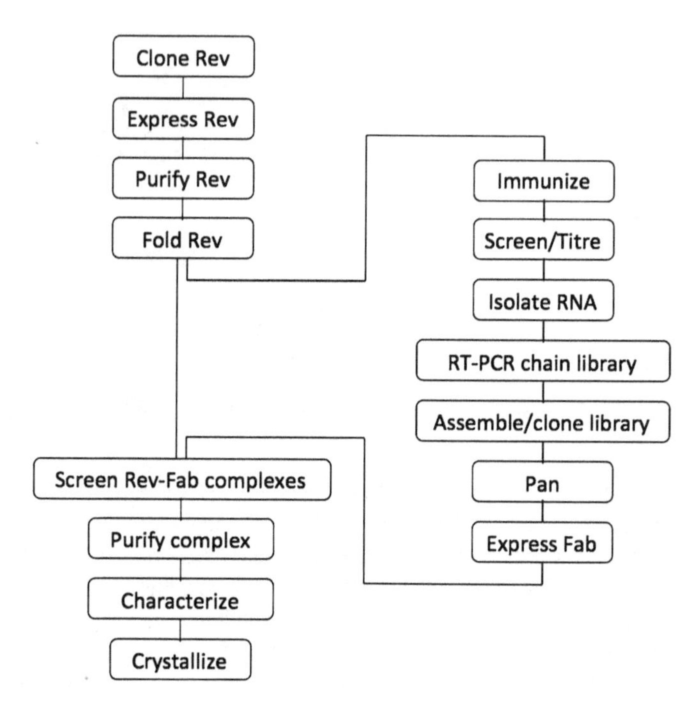

Fig. 3 Flow diagram for crystallizing Rev with a Fab chaperone

stoichiometrically with unpolymerized tubulin (or microtubules) to form heterogeneous ring complexes [12] and with tubulin preformed into monodisperse rings with the macrolide cryptophycin into homogeneous ring complexes. Both types of Rev–tubulin rings associate laterally in an antiparallel manner. The latter double rings have eightfold symmetry, are exceedingly homogeneous, and readily form square arrays, but they do not stack, and they do not form crystals (our unpublished observations). However, a mutated and truncated form of Rev has been crystallized and the structure solved [13].

Full-length Rev can be crystallized with the aid of a Fab chaperone (Figs. 2c and 3). The rationale and method [14] are summarized as follows. By immunizing rabbits with Rev at concentrations below that at which it polymerizes one can induce antibody production against epitopes otherwise buried in the polymer. Following affinity maturation in the animal one can then screen and further increase the affinity of the corresponding Fab by means of phage display. High-affinity Fab are then rescreened for the ability to form well-behaved soluble complexes with Rev, the complexes are purified and characterized, and then screened for crystallization. In retrospect, with the atomic structure in hand [15], the Fab functions as follows: Rev monomers associate in an antiparallel manner via the A-faces of their helix-turn-helix N-terminal domains and then further associate via their B-faces [13, 15]. The Fab binds to and blocks the B-face of the Rev dimer, preventing helical polymerization and allowing crystallization (Fig. 1c, d).

It should be noted that whilst the method described here produces material that can yield diffraction quality crystals from the full-length Rev protein, only the N-terminal domain is ordered—the C-terminal domain remains disordered. We are investigating other approaches to resolve this issue. These include rescreening the above libraries for Fab that can bind and potentially order the C-terminal domain, forming complexes with Rev's normal binding partner proteins, and further exploiting by NMR and electron microscopy the order present in the Rev polymers themselves.

2 Materials

2.1 Rev Folding

1. Folding buffer 1: 50 mM sodium phosphate, 150 mM sodium chloride, 1 mM DTT, 1 mM EDTA, 600 mM ammonium sulfate, pH 6.8.
2. Folding buffer 2: 50 mM sodium phosphate, 150 mM sodium chloride, 1 mM DTT, 1 mM EDTA, 10 mM ammonium sulfate, pH 6.8.
3. Millex-GS 0.2 μm filter units (Millipore).

2.2 Rabbit Immunization

1. Ribi Adjuvant System (RAS) is a stable oil-in-water emulsion that may be used as an alternative to the water-in-oil emulsions (Sigma Aldrich).
2. TRIzol Reagent (Invitrogen).

2.3 RNA Isolation

1. 1-Bromo-3-chloro-propane (BCP; Molecular Research Center).
2. RNA storage buffer: RNase-free 1 mM sodium citrate (pH 6.4) (Life Technologies/Ambion).

2.4 RT-PCR Antibody V_L and V_H Gene Library Preparation

1. SuperScriptII PCR Kit: RT enzyme mix. This includes SuperScriptIII RT and reaction mixture; oligo(dT)$_{20}$ random hexamers; dNTPs and *E. coli* RNase H (Invitrogen).
2. Qiagen MinElute Gel Extraction Kit (Qiagen).

2.5 Assembly and Cloning of the Fab Gene Library

1. Qiagen PCR Purification Kit (Qiagen).
2. XL1-Blue cells (Stratagene).
3. PEG 8000: Polyethylene glycol average molecular weight 8 kDa (Sigma-Aldrich).

2.6 Selection Anti-Rev Fab Clones by Phage Display

1. Streptavidin magnetic beads (Dynal).
2. TBS: Tris buffered saline (Quality Biologics Inc.).
3. PBS: Phosphate buffered saline (Life Technologies).

4. LB Broth (Quality Biologics Inc.).
5. IPTG: Isopropylthio-β-galactoside (Life Technologies).
6. Peroxidase-conjugated goat anti-human IgG (Jackson Laboratories).

2.7 Production of Anti-Rev Fab

1. Ni-Sepharose 6 Fast Flow (GE Healthcare).
2. Amicon stirred ultrafiltration cell (Millipore): Model 8200, 200 ml (63.5 mm membrane), or Model 8400, 400 ml (76 mm membrane). Ultrafiltration Disc (membranes), PM30, Amicon, 30 kDa NMWL (Millipore).

2.8 Crystallization and Structural Determination

Amicon Ultra-15 Centrifugal Filter Unit with Ultracel-30 membrane (Millipore).

3 Methods

3.1 Expression and Purification of HIV-1 Rev (See Note 1)

1. If Rev has been stored frozen, thaw the tube rapidly under warm water.
2. Add additional solid urea to a final concentration of 6 M and DTT to a final concentration of 5 mM. Adjust the protein concentration to 1 mg/ml.
3. Fold the protein by diluting with an equal volume of cold folding buffer 1, and then dialyzing overnight at 4 °C versus a large volume (1:1,000) of folding buffer 1, and then remove the ammonium sulfate by two sequential dialyses against folding buffer 2.
4. Confirm successful folding of Rev by monitoring α-helical content by far-UV circular dichroism [10] and the formation of long, regular filaments by negative-stain electron microscopy [11].
5. Sterile-filter with a 0.2 μm Millex-GS filter unit, dilute the Rev to below 80 μg/ml, and snap freeze 0.5 ml aliquots in liquid nitrogen (*see* **Note 2**).

3.2 Rabbit Immunization

1. Inoculate rabbits with 0.5 ml of 66 μg/ml purified Rev protein mixed with an equal volume of Ribi Adjuvant.
2. Give boosts at 3-week intervals with the same amount of antigen and adjuvant.
3. After four boosts and when serum titers have stabilized in the rabbits, a final boost is given.
4. Euthanize the rabbits 5 or 6 days later, and collect the bone marrow and spleens. Store in TRIzol at −80 °C.

3.3 RNA Isolation

1. Homogenize the spleens and bone marrow separately in 30 ml of TRIzol and then centrifuge at 2,500 × *g* at 4 °C.
2. Mix the supernatants with 3 ml of BCP and, after standing 15 min at room temperature, centrifuge at 17,500 × *g* for 15 min at 4 °C.
3. Transfer the upper aqueous phase to a fresh centrifuge tube and add 15 ml isopropanol.
4. After 10 min at room temperature, centrifuge at 17,500 × *g* for 10 min at 4 °C.
5. Discard the supernatant and add 30 ml of 70 % ethanol. Centrifuge at 17,500 × *g* for 10 min at 4 °C.
6. Discard the supernatant, and air-dry the pellets for ~5 min. Dissolve the pellets in 250 μl RNA storage buffer.
7. Determine the RNA concentration by measuring A_{260} nm (1 OD_{260} = 40 μg/ml).

3.4 RT-PCR Antibody V_L and V_H Gene Library Preparation

1. Dilute the RNA to 20 μg in a final volume of 32 μl.
2. Add 4 μl of a solution at 2.5 mM of each of the four dNTPs and 4 μl 50 μM Oligo $(dT)_{20}$.
3. Add 40 μl of SuperScriptIII 2× RT reaction. Mix (8 μl 10× reaction buffer, 16 μl 25 mM $MgCl_2$, 8 μl 0.1 M DTT, 4 μl 40 units/μl RNaseOUT, 4 μl 200 units/ml SuperScriptIII RT).
4. Incubate for 50 min at 50 °C and then 5 min at 85 °C.
5. Cool on ice, then add 4 μl of 2 units/μl RNase H, and incubate for 20 min at 37 °C.
6. The rabbit immunoglobulin V_L and V_H genes are PCR amplified from this SS DNA library using 11 primer pairs (*see* **Note 3**).
7. The PCR products are purified with 1 % agarose gel electrophoresis and eluted using the Qiagen MinElute Kit.

3.5 Assembly and Cloning of the Fab Gene Library

1. Combine the collection of V_L and V_H gene fragments with the C_K containing stuffer fragment. The primers used to amplify the 3′ end of the V_L gene fragments and the 5′ end of the V_H gene fragments in Subheading 3.4 are designed to overlap with the ends of the C_K coding region. PCR primers that contain an *Sfi*I restriction site and overlap with the 5′ end of the V_L gene fragments or the 3′ V_H gene fragments are used to generate large *Sfi*I restriction fragments comprising the V_L–C_K–V_H coding sequences (*see* **Note 4**).
2. Purify the V_L–C_K–V_H PCR products using the Qiagen PCR Purification Kit, then digest with *Sfi*I, and then separate as above with 1 % agarose gel electrophoresis.

3. Ligate these purified *Sfi*I DNA fragments into phage display vector pC3C. (This vector provides the coding sequence for C_H, resulting in a complete V_L–C_K–V_H–C_H coding sequence as well as the regions needed for expression and transport of the Fabs.) Transfect the ligation mixture into XL-1-Blue cells using electroporation.
4. Use these transfected cells to prepare 200 ml phagemid prep, using VCSM helper phage (*see* **Note 4**).
5. Harvest the phagemid from the culture supernatant with PEG 8000 precipitation (*see* **Note 4**).

3.6 Selection of Anti-Rev Fab Clones by Phage Display

1. Mix 1 ml of the phagemid prep with 20 μl of 0.25 mg/ml biotinylated Rev (*see* **Note 5**). Incubate for 1 h on a roller at 37 °C.
2. Wash streptavidin magnetic beads (100 μl/1 mg) 3× with 100 μl 1 % BSA in TBS.
3. Add the phagemid/Rev mixture to the beads and continue rolling at 37 °C for 1 h.
4. Isolate the beads with a magnet and wash 5–10× at 5 min per wash on the roller at 37 °C with TBS + 0.05 % Tween-20.
5. Isolate the phage bound to the beads by incubating with 50 μl of 10 mg/ml trypsin in PBS for 30 min at 37 °C.
6. Use this "panned" preparation of phagemid to transfect 2 ml of log-phase XL1-Blue cells in SB. Grow this culture, and continue with another four more rounds of panning as described (*see* **Note 4**).
7. Isolate individual colonies, inoculate 2 ml of LB broth, and shake at 37 °C until the culture is slightly turbid. Add 20 μl of 0.1 M IPTG, and continue shaking overnight. Freeze and thaw the cultures, and then remove the bacteria with centrifugation for 10 min at 4,000 × *g*.
8. To identify anti-Rev Fab-producing clones, assay the culture supernatants with ELISA. Microtiter plate wells are coated with Rev, washed with H_2O, and then incubated with blocking solution (5 % milk powder in PBS). Dilute the culture supernatants 1:1 with blocking solution and add to the Rev-coated wells. After 3 h at 37 °C, wash the wells with H_2O and add peroxidase-conjugated goat anti-human IgG (anti-human IgG is used as the constant regions in these Fabs are derived from human IgG). After another 1 h of incubation, wash the plates and develop with a peroxidase substrate.

3.7 Production of Anti-Rev Fab

1. Subclone the bi-cistronic coding sequence (OmpA leader-V_L-CL and pelB leader-V_H-C_H-polyHis) of the identified anti-Rev Fabs into *E. coli* expression vector pET11a [16].

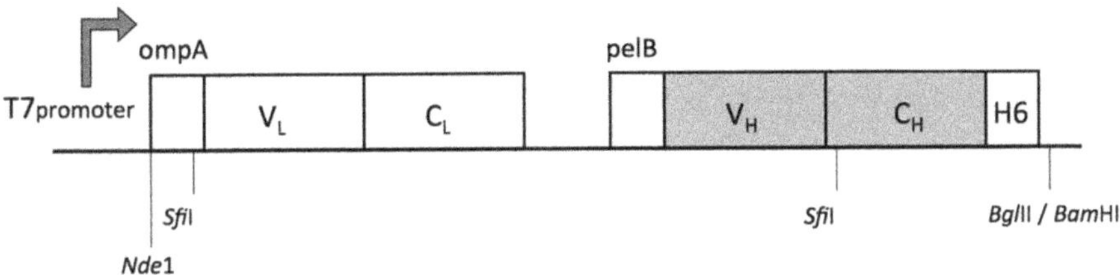

Fig. 4 A single T7 promoter drives synthesis of both the Fab light-chain V_L-C_L (*white*) and heavy-chain fragment V_H-C_H (*grey*). Transport of both chains to the periplasm is mediated by the leader peptides ompA and pelB. A His-tag (H6) on C-terminus of V_H-C_H is used for affinity purification

2. Grow cells transfected with the pET11-RevFab plasmid (*see* Fig. 4) in 1 L of LB broth, and induce Fab production with the addition of IPTG to 1 mM (*see* **Note 6**).
3. After 3–5 h of IPTG induction, freeze the cultures at −80 °C.
4. Thaw the culture, centrifuge at 11,000 × *g* for 20 min, and add the supernatant to ~75 ml Ni-Sepharose 6 Fast Flow equilibrated with PBS. Gently shake for 30 min at 4 °C.
5. Collect the resin with a sintered glass funnel and wash with 250 ml PBS and then with 300 ml 50 mM imidazole in PBS.
6. Elute the Fab with 250 ml of 0.5 M imidazole in PBS. Concentrate to about 25 ml using an Amicon stirred ultrafiltration cell and dialyze against PBS. The yield is typically about 10 mg (*see* **Notes 7** and **8**).

3.8 Rev:RevFab Complex Formation and Purification

1. Combine 10 mg of RevFab (in 25 ml PBS) with 6 mg Rev (in 10 ml 50 mM Tris, 20 mM ammonium sulfate, 5 mM DTT, 2 M urea, pH 7.5) (*see* **Note 9**).
2. Incubate for 30 min at 4 °C, and then dialyze against PBS at 4 °C.
3. Remove any precipitate (mostly excess Rev) by centrifugation at 18,000 × *g* for 15 min, and then load the supernatant on a Ni-Sepharose 6 Fast Flow column (1.6 × 0.8 cm). Wash with 25 ml of PBS and then with 50 ml of 50 mM imidazole in PBS.
4. Elute the Rev:RevFab complexes with 0.5 M imidazole in PBS, collecting 10 ml fractions.
5. Monitor the fractions with SDS-PAGE, and combine the fractions containing the Rev:RevFab complex (Fig. 5).
6. Dialyze the pooled fractions against 20 mM Hepes, pH 8.0.

3.9 Crystallization and Structural Determination

1. Characterize the protein complex using some of the biochemical and biophysical methods described in [14] (*see* **Note 10**).
2. Concentrate the Rev:RevFab complex to ~10 mg/ml using a Millipore Ultra-15 centrifugal concentrator (*see* **Note 11**).

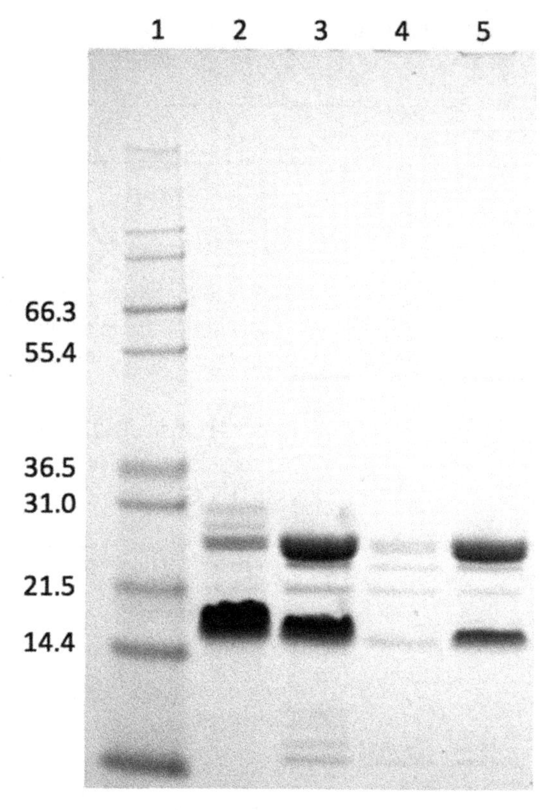

Fig. 5 SDS-PAGE of affinity purification of Rev–Fab complex. (Lane 1) molecular weight standards, (Lane 2) Rev, (Lane 3) Rev–Fab mixture applied to Ni-affinity column, (Lane 4) low-imidazole wash, (Lane 5) Rev–Fab complex eluted with high-imidazole buffer

3. Remove any precipitate by centrifugation at 12,000×*g* for 10 min.
4. Crystallize the protein complex as described by [15] (*see* **Notes 12–15**).

4 Notes

1. The expression in *E. coli* and purification of HIV-1 Rev are described in [10]. Rev has 115 residues and a mass of 12,905 Da.
2. Rev folds and polymerizes well at 0.5 mg/ml, forming long filaments. As the protein concentration is increased nucleation will increase and the filaments will become short, making visual assessment of successful folding by microscopy more difficult. The filaments are very prone to lateral aggregation and massive precipitation, but this can be prevented by the inclusion of 25–50 mM sodium citrate in all buffers and by keeping the

Table 1
Primer pairs for chimeric rabbit/human Fab libraries

	5′ Sense primer	3′ Sense primer	Product
1	RSCλ1	RHybL-B	Vλ
2	RSCVK1	RHybK1-B	Vκ
3	RSCVK1	RHybK2-B	Vκ
4	RSCVK1	RHybK3-B	Vκ
5	RSCVK2	RHybK1-B	Vκ
6	RSCVK2	RHybK2-B	Vκ
7	RSCVK2	RHybK3-B	Vκ
8	RHyVH1	RHyIgGCH-1B	C_H
9	RHyVH2	RHyIgGCH-1B	C_H
10	RHyVH3	RHyIgGCH-1B	C_H
11	RHyVH4	RHyIgGCH-1B	C_H

These are chimeric Fabs and have variable regions from immunized rabbits and light- and heavy-chain constant regions from cloned human Fab. *See* [17], Chapter 9, Protocol 9.6

protein cold at all times. The extinction coefficient is 0.7 at 280 nm for a 1 mg/ml solution.

3. *See* Table 1: described in detail [17]: *see* Chapter 9, Protocol 9.6.
4. Described in detail by [18].
5. Biotinylation of Rev: A 14-residue peptide biotin ligase substrate domain (Avitag: GGGLNDIFEAQKIEWHE) is appended to the C-terminus [19], expressed in *E. coli*, and purified by ion-exchange and gel filtration chromatographies in buffers supplemented with 2 M urea. Biotinylation with biotin ligase (Avidity, LLC) is done according to the manufacturer's protocol. Following the reaction, the protein is gel filtrated on Superdex S200 using 20 mM Tris–HCl, pH 7.4, containing 2 M urea. The proteins are characterized by mass spectrometry.
6. Cells are grown using a 1-L shake flask or as described [14] using a Biostat B 2-L bench top fermentor (Braun Biotech) using minimal media and a glycerol carbon source. Cells were grown in the fermentor at 37 °C and induced with IPTG with a typical cell yield of 50–60 g/L.
7. Further purification, if required, can be performed by gel filtration using Superdex S200. Urea (1 M) was included in the PBS column buffer to increase the solubility of the Fab [14].
8. The sequences for the Fab L- and H-chains have been deposited in GenBank with accession numbers GU223201 and GU223202, respectively.

9. This is a molar ratio of ~2 Rev:RevFab.
10. Rev binds to the Fab with a 1:1 molar stoichiometry and with a very high affinity ($K_d < 10^{-10}$ M).
11. The complex in this buffer can be concentrated to about 10 mg/ml, whereas in PBS the complex starts to precipitate above concentrations of ~1 mg/ml.
12. The methods used for protein crystallization and structure determination are beyond the scope of this chapter. Some of the research tools and methods used can be found at http://www.hamptonresearch.com.
13. The protein complex is crystallized without removal of the His-tag. The presence of the tag does not interfere with crystallization. For studies involving the binding of RNA to the complex the tag is removed as polyhistidine can bind nonspecifically to nucleic acid.
14. A single-chain (scFv) version of the anti-Rev Fab is constructed where the anti-Rev V_L and V_H chains are linked with an 18-residue linker. This construct is expressed and purified analogous to the Fab. The anti-Rev scFv also acts as an efficient crystallization chaperone with the advantage that the crystals diffract to higher resolution (DiMattia et al., unpublished observations).
15. The structure of the Rev–Fab complex has the PDB accession code 2X7L.

Acknowledgement

This work was supported by the Intramural Research Program of NIAMS, National Institutes of Health.

References

1. Derewenda ZS (2010) Application of protein engineering to enhance crystallizability and improve crystal properties. Acta Crystallogr D Biol Crystallogr 66(Pt 5):604–615. doi:10.1107/S090744491000644X
2. Koide S (2009) Engineering of recombinant crystallization chaperones. Curr Opin Struct Biol 19(4):449–457. doi:10.1016/j.sbi.2009.04.008
3. Kim J, Stroud RM, Craik CS (2011) Rapid identification of recombinant Fabs that bind to membrane proteins. Methods 55:303–309
4. Lieberman RL, Culver JA, Entzminger KC, Pai JC, Maynard JA (2011) Crystallization chaperone strategies for membrane proteins. Methods 55(4):293–302. doi:10.1016/j.ymeth.2011.08.004
5. Pollard VW, Malim MH (1998) The HIV-1 Rev protein. Annu Rev Microbiol 52:491–532
6. Cullen BR (2003) Nuclear mRNA export: insights from virology. Trends Biochem Sci 28(8):419–424. doi:10.1016/S0968-0004(03)00142-7
7. Strebel K (2003) Virus-host interactions: role of HIV proteins Vif, Tat, and Rev. AIDS 17 Suppl 4:S25–S34
8. Groom HC, Anderson EC, Lever AM (2009) Rev: beyond nuclear export. J Gen Virol 90(Pt 6):1303–1318. doi:10.1099/vir.0.011460-0

9. Grewe B, Uberla K (2010) The human immunodeficiency virus type 1 Rev protein: menage à trois during the early phase of the lentiviral replication cycle. J Gen Virol 91(Pt 8):1893–1897. doi:10.1099/vir.0.022509-0
10. Wingfield PT, Stahl SJ, Payton MA, Venkatesan S, Misra M, Steven AC (1991) HIV-1 Rev expressed in recombinant Escherichia coli: purification, polymerization, and conformational properties. Biochemistry 30:7527–7534
11. Watts NR, Misra M, Wingfield PT, Stahl SJ, Cheng N, Trus BL, Steven AC, Williams RW (1998) Three-dimensional structure of HIV-1 Rev protein filaments. J Struct Biol 121(1):41–52
12. Watts NR, Sackett DL, Ward RD, Miller MW, Wingfield PT, Stahl SS, Steven AC (2000) HIV-1 rev depolymerizes microtubules to form stable bilayered rings. J Cell Biol 150(2):349–360
13. Daugherty MD, Liu B, Frankel AD (2010) Structural basis for cooperative RNA binding and export complex assembly by HIV Rev. Nat Struct Mol Biol 17(11):1337–1342. doi:10.1038/nsmb.1902
14. Stahl SJ, Watts NR, Rader C, DiMattia MA, Mage RG, Palmer I, Kaufman JD, Grimes JM, Stuart DI, Steven AC, Wingfield PT (2010) Generation and characterization of a chimeric rabbit/human Fab for co-crystallization of HIV-1 Rev. J Mol Biol 397(3):697–708. doi:10.1016/j.jmb.2010.01.061
15. Mattia MA, Watts NR, Stahl SJ, Rader C, Wingfield PT, Stuart DI, Steven AC, Grimes JM (2010) Implications of the HIV-1 Rev dimer structure at 3.2Å resolution for multimeric binding to the Rev response element. Proc Natl Acad Sci U S A 107(13):5810–5814, doi:0914946107 [pii] 10.1073/pnas.0914946107
16. Studier FW, Moffatt BA (1986) Use of bacteriophage T7 RNA polymerase to direct selective high-level expression of cloned genes. J Mol Biol 189(1):113–130
17. Andris-Widhopf J, Steinberger P, Fuller R, Rader C, Barbas CF (2001) Generation of antibody libraries: PCR amplification and assembly of light- and heavy-chain coding sequences. In: Barbas CF, Burton DR, Scott JK (eds) Phage display: a laboratory manual. Cold Spring Harbor, New York
18. Rader C (2009) Generation and selection of rabbit antibody libraries by phage display. Methods in molecular biology 525:101–128, xiv. doi:10.1007/978-1-59745-554-1_5
19. Beckett D, Kovaleva E, Schatz PJ (1999) A minimal peptide substrate in biotin holoenzyme synthetase-catalyzed biotinylation. Protein Sci 8(4):921–929. doi:10.1110/ps.8.4.921

Chapter 36

Antibody Array Generation and Use

Carl A.K. Borrebaeck and Christer Wingren

Abstract

Affinity proteomics, represented by antibody arrays, is a multiplex technology for high-throughput protein expression profiling of crude proteomes in a highly specific, sensitive, and miniaturized manner. The antibodies are individually deposited in an ordered pattern, an array, onto a solid support. Next, the sample is added, and any specifically bound proteins are detected and quantified using mainly fluorescence as the mode of detection. The binding pattern is then converted into a relative protein expression map, or protein atlas, delineating the composition of the sample at the molecular level. The technology provides unique opportunities for various applications, such as protein expression profiling, biomarker discovery, disease diagnostics, prognostics, evidence-based therapy selection, and disease monitoring. Here, we describe the generation and use of planar antibody arrays for serum protein profiling.

Key words Antibody array, Protein expression profiling, Microarray, Nanoarray, Planar arrays, Serum profiling

1 Introduction

The antibody microarray technology can be used for high-throughput protein expression profiling of crude, non-fractionated biological samples, targeting high- and low-abundant (pg/ml range) analytes in a true multiplex fashion [1–3]. The concept of generating antibody arrays is based on dispensing minute amounts (≥fmole protein in 300 pL sized drops) of numerous (a few to several thousands) antibodies with the desired specificities one by one in discrete positions (100 nm to 200 μm sized spots) in an ordered pattern, an array, onto a solid support (Fig. 1). The arrayed antibodies will act as specific probes, or catcher molecules, for the target protein analytes. To date, a variety of platforms, based on monoclonal, polyclonal, and recombinant antibodies, have been developed and established, within both the academia and the industry [4]. The setups are mainly produced on planar arrays, although bead-based arrays have also been established [5, 6]. These miniaturized arrays (<1 cm^2 in size) are then incubated with microliter-scale crude, non-fractionated biological sample. Next, specifically bound

Vincent Ossipow and Nicolas Fischer (eds.), *Monoclonal Antibodies: Methods and Protocols*, Methods in Molecular Biology, vol. 1131, DOI 10.1007/978-1-62703-992-5_36, © Springer Science+Business Media New York 2014

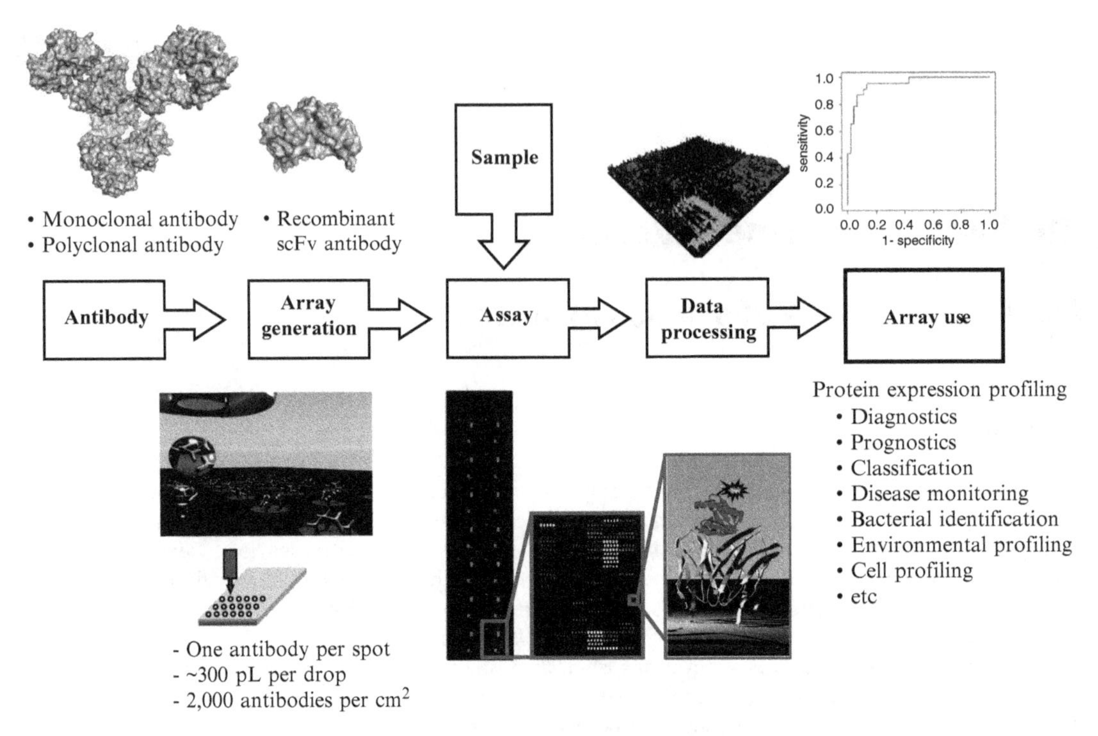

Fig. 1 Schematic illustration of the generation and use of antibody arrays

analytes are detected and quantified, mainly using fluorescence as the mode of detection [7]. The complete assay is run within less than 4 h, whereafter the microarray images are transformed into protein expression profiles, or protein maps, revealing the detailed composition of the sample. Depending on the application at hand, different bioinformatic strategies are then applied to further explore the wealth of data generated, e.g., pin-pointing differentially expressed protein analytes between patients suffering from a given indication and healthy controls [1].

Typical applications of antibody-based microarrays include (1) detection of disease-associated biomarker signatures for diagnosis, prognosis, classification, phenotyping, evidence-based therapy selection, disease monitoring, and prediction of the risk for, e.g., relapse and survival; (2) phenotyping of intact cells; (3) glycan profiling; (4) detection and identification of bacteria, bacterial proteins, and bacterial disease; (5) environmental profiling; and (6) delineation of signaling pathways [1–3, 8–10]. Any sample format can be analyzed as long as the protein content is exposed and available (e.g., cell surface membrane proteins) and/or can be extracted and solubilized, including plasma, serum, urine, cerebrospinal fluid, tissue extracts, intact cells, cell lysates, and cell supernatants [1]. In this protocol, we describe the generation and use of planar antibody arrays for serum protein expression profiling.

2 Materials

2.1 Preparation of Antibodies

1. Intact monoclonal and polyclonal antibodies as well as recombinant antibody fragments, predominantly single-chain Fv (scFv) antibodies, can be used as specific catcher antibodies on the arrays, but their on-chip performances must be thoroughly evaluated [4, 5] (*see* **Note 1**).
2. PBS stock solution: 140 mM NaCl, 2.7 mM KCl, and 10 mM sodium phosphate buffer, pH 7.4.

2.2 Clinical Sample Preparation

1. Collect the serum samples using a standardized protocol, and store them at −20 or −80 °C prior to use (*see* **Note 2**).
2. PBS stock solution.
3. Micro bicinchoninic acid (BCA™) Protein assay kit (Pierce, Rockford, IL, USA).

2.3 Labeling of Samples

1. EZ-link Sulfo-NHS-LC-Biotin (Pierce) [7] (*see* **Note 3**).
2. PBS stock solution.
3. 3.5 kDa molecular weight cut-off dialysis units (Thermo Scientific, Rockford, IL, USA).

2.4 Printing of Antibody Microarrays

1. Printing buffer: PBS stock solution.
2. Polymer Maxisorp microarray slides (NUNC, Roskilde, Denmark) (*see* **Note 4**).
3. Protein Printers BiochipArrayer1 (PerkinElmer Life Sciences) and SciFlexarrayer 11 (Scienion, Berlin, Germany) (*see* **Note 5**).
4. Positive control: Biotinylated bovine serum albumin (BSA) (Sigma) (*see* **Note 6**).
5. Negative control: PBS stock solution (*see* **Note 7**).
6. Hydrophobic pen (DakoCytomation Pen, DakoCytomation, Glostrup, Denmark).
7. Nexterion IC 16 incubation chamber with 16 well silicone superstructure (Schott-Nexterion, Mainz, Germany).

2.5 Antibody Microarray Assay

1. Blocking buffer: 1 % (w/v) fat-free milk powder and 1 % (v/v) Tween-20 in PBS (PBS-MT).
2. Washing buffer: 0.05 % (v/v) Tween-20 in PBS (PBS-T).
3. Sample buffer: PBS-MT.
4. Alexa647-conjugated streptavidin (Invitrogen, Carlsbad, CA, USA).
5. Incubation chamber (box with lid and moist wettex cloth).
6. Nitrogen gas.

2.6 Data Analysis

1. Confocal fluorescence slide scanner (ScanArray Express scanner, PerkinElmer Life Sciences) (*see* **Note 8**).
2. ScanArray Express software v4.0 (PerkinElmer Life & Analytical Sciences, Wellesley, MA, USA) (*see* **Note 9**).
3. Statistical computing environment R.

3 Methods

3.1 Preparation of Antibodies

1. Long-time store the antibody stock solutions at 4 °C or −20 °C based on the recommendations given for each antibody preparation (*see* **Note 10**).
2. Preferentially use only antibodies for which the on-chip array performance (specificity, functionality, and stability) has been validated prior to the array assay (*see* **Note 1**).
3. Dilute the antibodies to 0.1 mg/ml in PBS (prior to array generation) (*see* **Note 11**).

3.2 Clinical Sample Preparation

1. Collect the serum samples using a standardized protocol (*see* **Note 2**).
2. Long-time store the serum samples as aliquots at −20 or −80 °C (*see* **Note 2**).
3. Thaw the samples on ice and centrifuge at 16,000×*g* for 20 min at 4 °C.
4. If so required, determine the total protein concentration in the samples using a micro BCA protein assay kit.
5. Dilute the sample to 2 mg/ml in PBS (45× dilution assuming a concentration of 90 mg/ml).
6. Mix the serum with EZ-link Sulfo-NHS-LC-Biotin to a final concentration of 0. 6 mM and incubate on ice for 2 h (*see* **Note 12**).
7. Remove free biotin by extensive dialysis against PBS for 72 h at 4 °C, using 3.5 kDa molecular weight cutoff dialysis units.
8. Aliquot the samples, and store them at −20 °C prior to use.

3.3 Printing of Antibody Microarrays

1. Centrifuge the antibodies and controls to be arrayed in order to remove any precipitates, etc.
2. Dilute the antibodies (0.1 mg/ml) (*see* **Note 11**) and the positive control (10 µg/ml) in spotting buffer (i.e., PBS) (*see* **Note 13**).
3. Set the print layout (*see* **Note 14**).
4. Print the arrays using a protein printer (*see* **Note 5**).
5. Use the slides directly, or store them at RT overnight (*see* **Note 15**).

3.4 Antibody Microarray Assay

All incubation and washing steps are performed at RT with gentle shaking. A humidity chamber, generated by placing a moist wettex cloth in a box with a lid, can be used in order to ensure that the arrays do not dry out during the incubation steps.

1. Create individual sub-arrays either by drawing a hydrophobic barrier around each sub-array using a hydrophobic pen or by mounting the slide in a multi-well incubation chamber (e.g., 16 sub-arrays/slide). The steps below assume that a 16-well incubation chamber is used.
2. The individual sub-arrays are blocked with 150 μL blocking buffer for 1 h.
3. Wash the arrays with 4 × 150 μL washing buffer.
4. Add 100 μL labeled serum sample (diluted ten times in sample buffer, thus resulting in a final concentration of about 0.2 mg/ml) and incubate for 1 h (or 2 h) (*see* **Note 16**).
5. Wash the arrays with 4 × 150 μL washing buffer.
6. Add 100 μL 1 μg/ml Alexa647-conjugated streptavidin and incubate for 1 h.
7. Wash the arrays with 4 × 150 μL washing buffer.
8. Dismount the chip holder, and immerse the entire chip in distilled water, whereafter the chips are directly dried under a stream of nitrogen gas.
9. Store the slides in the dark at RT prior to scanning.

3.5 Data Analysis

1. Slides are scanned with the confocal ScanArray Express scanner or an equivalent (scan settings 5 or 10 μm resolution, laser power 90 (fixed), and a PMT gain of 50 or higher) (*see* **Note 17**).
2. Image analysis (quantification) can be done using the ScanArray Express software v4.0 using the fixed circle method. Use only non-saturated spots (*see* **Note 17**).
3. Subtract the local background, and to compensate for any possible local defects, the highest and lowest replicate values are automatically removed, and each data point represents the mean value of the remaining data points (*see* **Note 18**).
4. Perform chip-to-chip normalization using a semi-global normalization approach [11], conceptually similar to the normalization developed for DNA microarrays (*see* **Note 19**). Calculate and rank the coefficient of variation (CV) for each. Fifteen percent of the analytes that display the lowest CV values over all samples are identified and used to calculate a chip-to-chip normalization factor. The normalization factor N_i is calculated by the formula $N_i = S_i/\mu$, where S_i is the sum of the signal intensities for the selected analytes for each sample and μ is the sum of the signal intensities for the selected analytes

averaged over all samples. Each dataset generated from one sample is divided with the normalization factor N_i.

5. For the intensities, use log2 values.
6. The signal intensity values can be seen as a protein expression profile, or protein maps, indicating which proteins are present and at what relative level.
7. Comparison of protein expression profiles for different sample cohorts, e.g., healthy versus disease, can be done using statistical computing environment R (*see* **Note 20**). Differentially expressed analytes ($p < 0.05$) can be pin-pointed using Wilcoxon signed-rank test (assuming non-normally distributed data) or student *t*-test (assuming normally distributed data). The ability to classify two groups can be estimated using support vector machine (SVM) in R and be described by a receiver operating characteristic (ROC) curve.

4 Notes

1. Different antibody formats, including intact monoclonal and polyclonal antibodies as well as recombinant scFv antibody fragments, have been extensively used as probes for antibody-based microarrays. Apart from validating the antibody specificity, special attention should be placed upon evaluating their on-chip performances (e.g., stability and functionality) [4, 12]. The performance of arrayed antibodies has been shown to vary to a great extent, reflecting the fact that the antibodies are deposited onto a solid support and stored in a dried-out condition. This could potentially result in denaturing of the protein molecules and subsequently loss of functionality.
2. The serum samples should preferentially be collected using a standardized protocol, ensuring that the integrity of the sample (proteins) is conserved. Next, the sample should be aliquoted and rapidly frozen at −20 or −80 °C, and the thawing-freezing cycle of each vial should be kept at a minimum. The collection protocol depends on the organization responsible for collecting the sample.
3. The sample can be labeled using a variety of reagents, ranging from direct labeling with a fluorescent dye to indirect labeling with, e.g., biotin and subsequent visualization using fluorescently labeled streptavidin. We have based our approach on biotinylation of the serum samples [7, 13]. The labeling protocol depends on the particular labeling reagent.
4. A wide variety of surfaces can be used as planar, solid supports for antibody microarrays. We have used black polymer Maxisorp slides from Nunc [13]. The array assay protocol, ranging from

the amount of each antibody deposited to choice of buffers (e.g., blocking buffer), depends on the particular solid support.

5. Different printers can be used for the preparation of antibody microarrays. We have used two non-contact inkjet printers, BiochipArrayer1 (Perkin Elmer) and SciFlexarrayer 11 (Scienion, Berlin, Germany). The preparation protocol depends on the particular spotter.
6. Different reagents (proteins or dyes) can be used as positive control. We used biotinylated BSA as positive control.
7. Different reagents can be used as negative control, including an antibody directed against an analyte not present in the sample ("control antibody"), a completely different protein, or simply spotting buffer. We have used a control antibody and/or spotting buffer as negative control.
8. Different scanners can be used for scanning the slides. We have used the confocal ScanArray Express scanner. The scanner setting depends on the particular scanner.
9. Different software can be used for detecting and quantifying each individual spot. We used the ScanArray Express software v4.0. The settings depend on the particular software and spot intensity.
10. The storage condition depends on the particular antibody and antibody format. While some should be stored frozen in a certain buffer, others are preferentially stored at 4 °C. This should be optimized per antibody.
11. The spotting concentration depends on the particular antibody and solid support. This will have to be optimized for each antibody and surface. We have used an antibody concentration of 0.1 mg/ml as starting concentration. A too low concentration will result in partial spots, while too high concentration will result in blurry spots. Dilute the antibodies just prior to use, i.e., store them as concentrated as possible.
12. The samples can be labeled at different molar ratios of biotin:protein. This is a key feature, as a too high ratio may result in over-labeling and thus reduced immunoreactivity due to epitope masking, and a too low ratio will give very weak signals in particular for low-abundant analytes. We have used a molar ratio of biotin:protein of 15:1 assuming an average molecular weight of serum proteins of 50 kDa [13, 14].
13. The concentration of the positive control depends on the particular positive control.
14. The print layout depends on several factors, such as the particular chip size and application. The array should include the target antibodies, positive and negative controls, deposited with an adequate number of replica spots. We have commonly used five

to eight replicate spots of each individual antibody/control. The replicate spots should be preferentially spread across the entire array. In order to facilitate the subsequent spot finding and quantification process, the positive control should be included at regular intervals, e.g., every 20th row. If multi-well incubation chamber is used up to 16 sub-arrays/slide could be generated. If the sub-arrays are created using a hydrophobic pen, two to eight sub-arrays per slide could normally be generated.

15. The storage time for printed arrays depends on the on-chip stability of the deposited antibodies. We have normally used the slide the day after production in order to standardize and minimize any impact of storage time on the performances of the arrays. The storage conditions depend on the particular combination of antibodies and solid support. We have normally stored our slide in the dark at RT.
16. The concentration of serum sample depends on several factors, including concentration of the target analytes, choice of labeling reagent, solid support, and blocking buffer, and signal-to-noise ratios will have to be optimized for each setup. We have used a concentration of 0.2 mg/ml directly biotinylated (non-fractionated) serum.
17. The scanner settings depend on the observed signal intensity, reflecting several factors, such as concentration of the target analyte and how well the support can be blocked for nonspecific background binding. Non-saturated spots only should be used. We usually scan the slide at several predefined scanner settings, e.g., a laser power of 90 combined with several different PMT gain settings. The quenching needs to be evaluated for the particular setup in order to avoid biased data due to quenching effects.
18. The spot quality is vital for the data analysis. In order to facilitate the QC control, a system is frequently adopted where the replica spots with the highest and lowest signal intensities automatically are removed in order to compensate for any technical issues (e.g., dust particles), and a mean value of the remaining replicates is then reported. We have commonly generated eight replicate spots, automatically removed the two highest and two lowest signal intensities, and then reported the mean value of the four remaining spots.
19. A normalization is required in order to readily enable a comparison of data generated on different sub-arrays, slides, and days. Different normalization procedures can be used. We have normally used the semi-global CV normalization process described here.
20. Different statistical software can be used for (1) identifying any differentially expressed ($p<0.05$) analytes between two or

more sample cohorts and (2) determining how well two or more sample cohorts can be differentiated or classified. We have frequently used statistical computing environment R for delineating differentially expressed analytes and for determining how well two groups can be classified (using SVM, a supervised learning method in R). The software settings depend on the particular question and dataset.

Acknowledgements

This work was supported by grants from the Swedish National Science Council (VR-NT and VR-M), the SSF Strategic Center for Translational Cancer Research (CREATE Health), and Vinnova.

References

1. Borrebaeck CAK, Wingren C (2009) Design of high-density antibody microarrays for disease proteomics: key technological issues. J Proteomics 72:928–935
2. Haab BB (2006) Applications of antibody array platforms. Curr Opin Biotechnol 17(4):415–421
3. Sanchez-Carbayo M (2011) Antibody microarrays as tools for biomarker discovery. Methods Mol Biol 785:159–182
4. Borrebaeck CAK, Wingren C (2011) Recombinant antibodies for the generation of antibody arrays. Methods Mol Biol 785: 247–262
5. Borrebaeck CAK, Wingren C (2007) High-throughput proteomics using antibody microarrays: an update. Expert Rev Mol Diagn 7:673–686
6. Wingren C, Borrebaeck CAK (2007) Progress in miniaturization of protein arrays—a step closer to high-density nanoarrays. Drug Discov Today 12:813–819
7. Wingren C, Borrebaeck CAK (2008) Antibody microarray analysis of directly labelled complex proteomes. Curr Opin Biotechnol 19:55–61
8. Haab BB (2005) Antibody arrays in cancer research. Mol Cell Proteomics 4:377–383
9. Hanash S (2003) Disease proteomics. Nature 422:226–232
10. MacBeath G (2002) Protein microarrays and proteomics. Nat Genet 32 Suppl:526–532
11. Carlsson A, Wuttge DM, Ingvarsson J, Bengtsson AA, Sturfelt G, Borrebaeck CAK, Wingren C (2011) Serum protein profiling of systemic lupus erythematosus and systemic sclerosis using recombinant antibody microarrays. Mol Cell Proteomics 10:M110 005033
12. Haab BB, Dunham MJ, Brown PO (2001) Protein microarrays for highly parallel detection and quantitation of specific proteins and antibodies in complex solutions. Genome Biol 2:RESEARCH0004
13. Wingren C, Ingvarsson J, Dexlin L, Szul D, Borrebaeck CAK (2007) Design of recombinant antibody microarrays for complex proteome analysis: choice of sample labeling-tag and solid support. Proteomics 7:3055–3065
14. Ingvarsson J, Larsson A, Sjoholm AG, Truedsson L, Jansson B, Borrebaeck CAK, Wingren C (2007) Design of recombinant antibody microarrays for serum protein profiling: targeting of complement proteins. J Proteome Res 6:3527–3536

Index

A

ADCC. *See* Antibody-dependent cell-mediated cytotoxicity (ADCC)
Adjuvants....34–37, 47, 49, 50, 54, 55, 66, 67, 72, 74, 77, 192, 201, 553, 554
Adsorption....129, 241, 242, 248, 293, 390
Affinity chromatography....4, 227–228, 287–289, 297, 298, 308, 313, 384, 401, 534
Affinity maturation....114, 134, 141, 151–180, 407–419
Affinity measurements....384–389
Aggregates....56, 67, 260, 280, 290, 507–511, 534, 549, 551
Alkaline salt elution....242
Allergen....483, 484
Antibody array....563–571
Antibody-dependent cell-mediated cytotoxicity (ADCC)....226, 274, 315–317, 381, 525–527, 529, 532, 537–538
Antibody discovery....114
Antibody engineering....152, 525–539
Antibody fragment....114, 148, 152, 153, 174, 177, 273, 283, 285, 297–313, 315, 396, 401, 454, 549–571
Antibody humanization....114, 208, 337
Antibody isoforms....502, 504
Antibody labeling....489–490, 497–498
Antibody library....114, 115, 118–121, 126–127, 148, 174, 489–490, 497–498
Antigen....3, 22, 33, 47, 72, 81, 113, 133, 152, 183, 192, 207, 229, 273, 298, 315, 338, 383, 395, 407, 421, 428, 447, 483, 486, 536, 542, 554, 563
Antigenic determinant....34

B

B cell epitope....477
B cells....47, 49, 61, 68, 101, 113, 114, 179, 183–188, 191, 192, 197, 207, 230, 318, 477, 529, 531
Binding specificity....33, 34, 261

C

CaptureSelect™ affinity resins....300, 304–305
CDC. *See* Complement-dependent cytotoxicity (CDC)
CDR$_S$. *See* Complementary determining region (CDR$_S$)
CELLine bioreactor....12–13, 18
Cell line development....264
Cell surface antigens....81–103, 392
Cell surface display....485–500
Ceramic hydroxyapatite....241–251
Chaperone....549, 552, 560
CHO cells....264–267, 270, 531
Chromatography....4, 7, 13, 15, 241–251, 269, 274–278, 280, 282, 284, 286–289, 293, 297, 298, 300, 304, 308, 313, 320, 326, 384, 396, 401, 402, 411, 501–539
Complementary determining region (CDR$_S$)....114, 115, 118, 120, 121, 124, 141, 147, 208, 321, 329, 330, 401, 408, 409, 413
Complement-dependent cytotoxicity (CDC)....226, 315–318, 525–527, 529, 532, 537–538
Conformational epitope....82, 477
Crystallization....18, 395–399, 401–403, 505, 510, 549–560
Cytoskeleton....544–546

D

Desalting....7, 14–15, 17, 19, 128, 224, 255, 257, 260, 515, 518, 519
Desorption....241, 242, 297, 298
Diversity....114–116, 118, 121, 123, 126, 157, 159, 160, 166, 168, 175, 176, 179, 180, 192, 338, 351, 372

E

EBV. *See* Epstein Barr virus (EBV)
E. coli....6, 11, 49, 114, 117–121, 124–127, 129, 135, 139, 141, 143, 148, 153, 155, 169, 179, 225, 274–276, 278–281, 289, 304, 309, 310, 318, 320, 321, 324, 325, 330, 331, 430, 449, 450, 452–455, 458, 459, 463, 464, 468, 470–473, 477, 486–488, 491–493, 553, 556, 558, 559
Enzyme-linked immunosorbent assay (ELISA) screening....49, 52–53, 62–63, 85, 448, 450–451, 459–461, 464, 468, 475
Episomal vector....4, 8
Epitope....22, 23, 29, 48, 82, 146, 152, 230, 298, 311, 312, 338, 369, 371–372, 395, 421–445, 447–500, 569
  excision....427–445
  mapping....421–445, 447–500
Epitope-based vaccines....434
Epstein Barr virus (EBV)....183–188, 332
Equilibrium binding measurements....386

V. Ossipow and N. Fischer (eds.), *Monoclonal Antibodies: Methods and Protocols*, vol. 1131,
DOI 10.1007/978-1-62703-992-5, © Springer Science+Business Media, New York 2014

F

Fab....................116, 118, 120, 121, 137, 141, 148, 226, 274, 299, 301–304, 308–313, 316, 395–397, 399–404, 549–557, 559, 560
FACS. *See* Fluorescence-activated cell sorting (FACS)
Fc-engineering 525–539
Fc fusion....................274, 277–278, 287–289, 303
FLAG-tag 120, 275–277, 279, 280, 283, 285, 300, 309
Flow cytometry....................55, 68, 81–103, 151, 153, 156, 160, 164–171, 175, 178, 202, 485, 529, 535, 536
Fluorescence-activated cell sorting (FACS)....................7, 12, 13, 18, 79, 87, 92, 133, 264, 265, 267, 384, 390, 485, 486, 498, 500, 531, 535, 536
Fluorochromes....................87–91, 96–98, 102, 103
Fragmentation 22, 269, 486, 490–492, 499, 515
Fusion....................4, 24, 25, 33–39, 41–44, 46–69, 71–78, 81, 85, 87, 100, 102, 120, 141, 191, 207, 225, 230, 237, 303, 467
Fusion PCR....................230, 231, 233, 235, 237, 239
Fusion protein 273–294, 311, 318, 374, 526, 528, 531–535

G

Glycan
characterization 513–523
identification....................515
labeling 521
Glycopeptide 514–516, 520–521, 523
Glycoprotein....................48, 513–520
Gram-positive 486

H

HEK293....................192, 234, 238, 254, 259, 274–276, 280–283, 288, 291, 318, 320, 325–327, 332, 530, 532–533
High-throughput homogeneous assay....................81–103
High-throughput selection....................106
High-throughput sequencing....................338
HIV-1. *See* Human immunodeficiency virus type 1 (HIV-1)
Homologous recombination....................153, 156–160, 179
Human immunodeficiency virus type 1 (HIV-1)............395, 549, 550, 554, 558
Humanized antibody....................109, 111, 118
Hybridoma....................24, 28, 33–45, 47–69, 71–74, 76–79, 81–103, 109, 113, 114, 151, 191, 207–227, 229–239
Hybridoma screening....................81, 82, 85, 88, 90
Hydrogen/deuterium exchange....................427–445

I

IAC. *See* Immunoaffinity chromatography (IAC)
IMAC. *See* Immobilized metal ion affinity chromatography (IMAC)
IMGT Collier de Perles....................338–340, 342, 343, 352, 359, 362, 364, 367, 369–371, 373, 375
IMGT-ONTOLOGY....................338, 339, 341, 369
IMGT unique numbering....................338–343, 350–352, 354, 356, 363, 364, 369, 372
Immobilized metal ion affinity chromatography (IMAC)....................4, 274–277, 279, 280, 285–286, 289, 290, 293, 534
Immortalization....................34, 36–38, 183–188, 191, 192
Immune repertoire....................191–203
Immunization....................3, 24, 33–45, 47, 48, 50, 54–56, 66, 67, 72, 74, 113, 133, 151, 191–195, 201, 202, 207, 553, 554
Immunoaffinity chromatography (IAC)...........298–300, 311
Immunoblotting....................24–27, 68, 535
Immunocytochemistry....................22, 24, 26–28
Immunofluorescence 49, 72, 529, 531, 536, 539, 543–547
Immunogen....................48, 49, 54, 55, 66, 67
Immunogenetics....................337, 338
Immunoinformatics....................338
In situ regeneration....................242
Ion exchange-high performance liquid chromatography (IEX-HPLC)....................501–506
Isotypes..........41, 94, 101, 102, 202, 207, 208, 218, 222, 223, 226–227, 231, 299, 302, 311, 477, 479, 484, 525–527

K

KinExA....................384–388, 392
Kunkel mutagenesis....................124, 448

L

Label-free method....................383–392
LCL. *See* Lymphoblastoid cell lines (LCL)
LCMS....................513–515
Lymphoblastoid cell lines (LCL)....................183–188

M

Magnetic microspheres....................253–361
Mammalian cells....................3, 4, 208, 212, 218, 231, 235, 238, 239, 258, 260, 263–266, 275–279, 282–289, 327, 330, 408, 526, 532
Microarray....................49, 563–569
Monoclonal antibody production process....................263
Monomer....................316, 507–511, 551, 552
Myeloma....................33–36, 38, 39, 41–44, 47, 49, 51, 57–61, 66–68, 207, 230, 235, 237, 263

N

Natural killer (NK) cell......101, 274, 317, 526, 529, 532, 535
Nuclear antigens....................543–547

O

On-cell binding....................383
Overlapping peptides....................421–426

P

Patient serum....426
Peak integrations.... 505, 508, 510, 512
Peptide-panning techniques....427–445
Phage ELISA.... 459–461, 475, 483
Phage panning....134, 137–139, 141–142, 144
Phosphate elution....242
Phosphopeptide....22, 23, 25
Planar arrays....563
Plasma cells....34, 192, 193, 196, 197, 202
Primary antibodies....100, 156, 490, 545–547
Proliferation....183, 186
Protein engineering....526, 529, 536
Protein expression profiling....563, 564

R

Randomization....115, 121, 329, 449, 463, 464, 466, 475, 476
Random phage display library....477–484
Recombinant antibodies....201, 230, 237–239, 563
Recombinant protein purification....3
Resin in situ regeneration....242
Rev....549–554, 556–560

S

S. carnosus....488–490, 494–495, 497–499
scFv. *See* Single-chain variable fragments (scFv)
scFv-Fc format....315–333
scFv reformatting....315–333
Secondary antibodies....29, 156, 421, 544–547
Semi-solid media....106
Semi-stable cell line....12, 227
Serum profiling....564
Single cell cloning....267
Single-chain variable fragments (scFv)....137, 141, 152, 153, 157, 160, 167, 169–171, 173, 177, 179, 180, 273, 274, 278, 279, 289, 290, 299, 315–333, 395, 454, 534, 549, 560, 564, 565, 568
Single expression plasmid....230, 231
Single site biotinylation....4, 8
Site- and phosphorylation state-specific antibody....22
Site directed-mutagenesis....408–411, 413–416, 418, 419, 449–451, 456–459, 470, 532
Size exclusion high performance liquid chromatography (SEC-HPLC)....507–512
Small scale expression....207–227
Stable transfection....264, 533
Strep-tag....275, 277, 280, 286–287, 293
Structure....3, 97, 114, 122, 180, 225, 337–340, 342, 343, 364, 365, 371, 372, 375, 377, 395–404, 418, 438, 439, 442, 443, 445, 456, 461, 466, 484, 486, 513, 514, 528, 544, 549, 550, 552, 560, 565
Surface neutralization....242
Systems immunology....192

T

Transformation....34, 129, 134–135, 138–140, 148, 157–160, 169, 174, 179, 183, 221, 225, 233, 235, 239, 278, 289, 384, 415, 416, 419, 459, 463, 473, 475, 476, 496, 499

U

UPLC-ESI TOF MS....518, 520, 522

V

Variable gene cloning....63, 114, 115, 118, 191–203, 210–212, 225, 278, 298–300, 309, 311, 332, 344, 346, 347, 351, 352, 354, 374, 375, 408
Variable region CDR3....199, 200, 203, 329, 408, 409, 413, 415, 416, 419

X

X-ray....26, 29, 399, 403, 447, 461, 485, 549

Y

Yeast surface display....151–180

CPSIA information can be obtained
at www.ICGtesting.com
Printed in the USA
LVHW061017020619
619865LV00003B/11/P